Management and Control of Invertebrate Crop Pests

Management and Control of Invertebrate Crop Pests

Editor:

GORDON E. RUSSELL

Emeritus Professor of Agricultural Biology,
University of Newcastle upon Tyne

Intercept

Andover, Hampshire

ISBN 0 946707 25 1

Published in September 1989 by Intercept Limited, PO Box 716, Andover, Hampshire, SP10 1YG, UK

Typeset by Roger Booth Associates, Newcastle upon Tyne
Printed in Great Britain by
Athenaeum Press, Newcastle upon Tyne

Contributors

NELLO P. D. ANGERILLI, *Entomology/Plant-Pathology Section, Agriculture Canada Research Station, Summerland, BC, Canada V0H 1Z0*

R. L. BRANDENBURG, *Department of Entomology, Box 7613, North Carolina State University, Raleigh, North Carolina 27695-7613, USA*

J. B. CARTER, *Department of Biology, Liverpool Polytechnic, Byrom Street, Liverpool, L3 3AF, UK*

T. H. COAKER, *Department of Applied Biology, University of Cambridge, Pembroke Street, Cambridge, CB2 3DX, UK*

A. F. COCKBURN, *Department of Biology, University of California, San Diego, La Jolla, California 92093, USA*

C. P. DUFAULT, *Department of Applied Biology, University of Cambridge, Pembroke Street, Cambridge, CB2 3DX, UK*

A. J. HOWELLS, *Department of Biochemistry, The Faculties, Australian National University, Canberra, ACT 2601, Australia*

G. G. KENNEDY, *Department of Entomology, Box 7613, North Carolina State University, Raleigh, North Carolina 27695-7613, USA*

A. R. KHAN, *Department of Agricultural and Environmental Science, University of Newcastle upon Tyne, NE1 7RU, UK*

Z. R. KHAN, *International Centre of Insect Physiology and Ecology (ICIPE), PO Box 30772, Nairobi, Kenya*

E. P. LLOYD, *Boll Weevil Research Laboratory, USDA-ARS, Mid South Area, PO Box 5367, Mississippi State, MS 39762, USA*

C. M. PORT, *Agricultural Development and Advisory Service, Government Buildings, Kenton Bar, Newcastle upon Tyne, NE1 2YA, UK*

G. R. PORT, *Department of Agricultural Biology, The University, Newcastle upon Tyne, NE1 7RU, UK*

R. D. RIGGS, *Department of Plant Pathology, University of Arkansas, Fayetteville, AR 72701, USA*

BERNARD D. ROITBERG, *Behavioural Ecology Research Group and Centre for Pest Management, Department of Biological Sciences, Simon Fraser University, Burnaby, BC, Canada V5A 1S6*

R. C. SAXENA, *Department of Entomology, International Rice Research Institute (IRRI), PO Box 933, Manila, Philippines*

D. P. SCHMITT, *Department of Plant Pathology, Box 7631, North Carolina State University, Raleigh, NC 27695, USA*

B. J. SELMAN, *Department of Agricultural and Environmental Science, University of Newcastle upon Tyne, NE1 7RU, UK*

M. J. WHITTEN, *Division of Entomology, CSIRO, GPO Box 1700, Canberra, ACT 2601, Australia*

Contents

ix

1

The Boll Weevil: Recent Research Developments and Progress Towards Eradication in the USA

E. P. LLOYD

Boll Weevil Research Laboratory, Agricultural Research Service, United States Department of Agriculture, Mississippi State, Mississippi 39762, USA

Introduction
Disruption of overwintering mechanisms
Developments contributing to more efficient use of insecticides
Identification of the boll weevil pheromone and development of dispensing systems
Design and development of pheromone traps
Use of pheromone traps for detection and suppression
Sterilization of the boll weevil
Development of technologies for mass-rearing large numbers of boll weevils for sterilization
Development of insect growth regulators as foliar sprays for boll weevil population suppression
Large-scale field trials of boll weevil eradication technology
       Pilot Boll Weevil Eradication Experiment 1971–73—Boll Weevil Eradication Trial 1978–80
Conclusions
References

Introduction

The boll weevil, *Anthonomus grandis* Boheman (Coleoptera: Curculionidae), is an introduced pest of cotton in the United States. The first report of the occurrence of the boll weevil in the United States was received in the US Department of Agriculture in the fall of 1894. A careful survey by Townsend (1895) indicated that it had been causing damage near Brownsville, Texas, since 1892. The boll weevil crossed the Mississippi River in 1907 and by 1922 the insect had infested the cotton in the remainder of the south-eastern US (Hunter and Coad, 1923). The first field experiments which gave effective control of boll weevil were with calcium arsenate, as reported by Coad (1918). Coad and Cassidy (1920) recommended that treatments with calcium arsenate should begin when 15–20% of the cotton squares (flower buds)

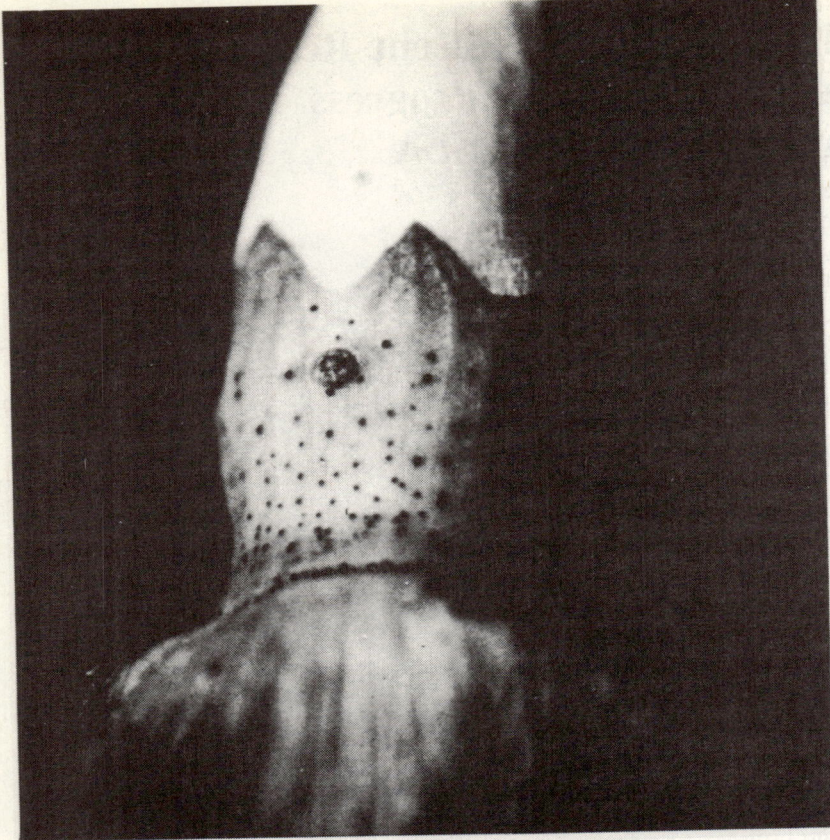

Figure 1.   Boll weevil oviposition puncture on cotton square (flower bud).

were punctured (*Figure 1*) and that these treatments should be continued at 4- or 5-day intervals until the crop was mature. Subsequently, Coad, Johnson and McNeil (1924) demonstrated the effectiveness of calcium arsenate when applied by aircraft. The use of calcium arsenate made it possible to grow cotton in the boll weevil-infested south and southeast. However, the calcium arsenate caused undesirable side-effects by destroying the natural enemies of such secondary pests as the bollworm and the cotton aphid (Ewing and Ivy, 1943).

Subsequently, synthetic organic insecticides were introduced for the control of the boll weevil and other cotton pests. Ivy (1944) tested DDT for the control of the cotton leafworm and boll weevil but this compound did not control either pest. Later, Ivy and Ewing (1946) reported excellent control of the boll weevil with benzene hexachloride (BHC) dust in laboratory, cage, and field tests. Ewing, Parencia and Ivy (1947) and Gaines and Dean (1947) confirmed the effectiveness of gamma isomer of BHC against the boll weevil. Gaines and Wipprecht (1948) reported that a mixture of gamma benzene hexachloride, DDT and sulphur was more effective than calcium arsenate in controlling late-season cotton pests including the boll weevil, bollworm and spider mites. As a result of these successful tests and the commercial intro- duction of synthetic organic insecticides, there was almost complete reliance upon insecticides for the control of the boll weevil and other pests of cotton. Ewing and Parencia (1950) proposed a community-wide programme for boll weevil suppression based on the early-season use of insecticides to kill emerging overwintered weevils. These workers recognized the value of pre- serving naturally occurring beneficial arthropods for the control of bollworms, rather than relying entirely on the use of broad-spectrum insec- ticides. The programme that they advocated gave early-season control of the boll weevil, then allowed time for parasites and predators to re-establish in the cotton fields and to provide natural control of the bollworm until late- season boll weevil infestations had to be treated.

In 1954, Roussel and Clower (1955) reported high levels of resistance to chlorinated insecticides in boll weevils in several areas in Louisiana. Similar resistance in the boll weevil to chlorinated insecticides was reported in Texas. The next year, however, several organophosphorous insecticides and calcium arsenate were reported to be effective against these populations (Walker *et al.*, 1956). The development of insecticide resistance and annual losses caused by the boll weevil resulted in Congress directing USDA to review the boll weevil problem and to identify new research approaches. A working group appointed to study the matter concluded that the future of conventional chemicals was threatened and recommended the establishment of a new interdisciplinary laboratory to initiate a broad-based research programme to find ways to reduce losses to a minimum or to eliminate the problem altogether (USDA, 1958). The threat posed by insecticide resistance to the cotton industry was intensified when resistance to DDT was reported in the tobacco budworm (Brazzel, 1963) and the bollworm (Graves, Roussel and Phillips, 1963) and resistance to methyl parathion in the tobacco budworm was reported by Nemec and Adkisson (1969).

In addition to insecticide resistance developing in major pests of cotton and the use of insecticides disrupting the natural control systems of secondary pests, concerns about the potential environmental and health effects of insecticides were increasing. The publication of *Silent Spring* greatly increased the expression of that concern (Carson, 1962). Congressional hearings that followed resulted in an increase in appropriations that were to be used to develop improved methods for controlling insects. Subsequently, the registration of DDT and its use on cotton was cancelled on 31 December 1972 (Dunlap, 1981): this effective and economical insecticide was, therefore, no longer available for use by growers in the south and south-eastern US where it still provided control of the bollworm and tobacco budworm.

Because of these problems a number of new insect strategies emerged that provided different approaches to insect control, such as management and/or eradication. Rabb (1972) described the evolution of insect-pest control actions which included three basic strategies for insect management: man- agement of localized populations, population management over a large area, and eradication of an insect pest over a wide area. Two basic strategies, IPM (integrated pest management) and TPM (total population management), were focused on by many, as distinctly different approaches to insect management. Ridgway and Lloyd (1983) proposed that these two strategies should be merged into a single step-by-step strategy beginning with management of localized populations, then proceeding into management of large-area populations and finally to eradication if technically feasible.

A major research effort was initiated with a view to eradication of the boll weevil, with the establishment of the USDA–ARS Boll Weevil Research Laboratory at Mississippi State, Mississippi, in 1962 and strengthening research at other federal and state laboratories. Research was focused on a number of areas: (1) overwintering mechanisms (diapause) and ways of disrupting the diapause phenomenon; (2) the development of more efficient use of insecticides (ultra-low-volume sprays); (3) the development of systemic insecticides (this was not, however, included in the final eradication technology); (4) the identification of the boll weevil pheromone and developing systems for dispensing it; (5) the design of pheromone traps; (6) the use of pheromone traps for detection and suppression; (7) progress in the development of chemosterilants for sexual sterilization of the boll weevil; (8) the development of technologies for mass-rearing large numbers of boll weevils for sterilization, and (9) the development of insect growth regulators for boll weevil population suppression.

Disruption of overwintering mechanisms

Brazzel and Newsom (1959) found that boll weevils must enter a physiological condition called diapause if they are to survive adverse environmental periods. (Diapause in the boll weevil is characterized by the cessation of gametogenesis and atrophy of gonads, increase in fat content, decrease in water content and decrease in respiratory rate.) Brazzel and Hightower (1960) reported that weevils with the diapause condition were first observed from late July through August in populations from selected fields in central Texas. Lloyd and Merkl (1961) reported that diapause weevils were first observed in Mississippi cotton fields as early as 11 August and in other fields as late as 8 September. Subsequent studies by Lloyd, Laster and Merkl (1964) showed that the initial entry of a segment of the boll weevil population was coincident with cessation of flowering by cotton plants in individual fields. In other laboratory studies Lloyd, Tingle and Gast (1967) associated the initiation of diapause with boll feeding by adult weevils, limited numbers of squares available to adult weevils, low night temperatures, exposure of immature stages to a short photophase and larval feeding on bolls. In the field, these workers observed two periods of entry of the boll weevil into diapause: the first occurred in August or September (under Mississippi conditions) when fields ceased flowering and bolls remained as the only food source; the second period occurred during October when night temperatures decreased and only squares were available as food (*Figure 2*).

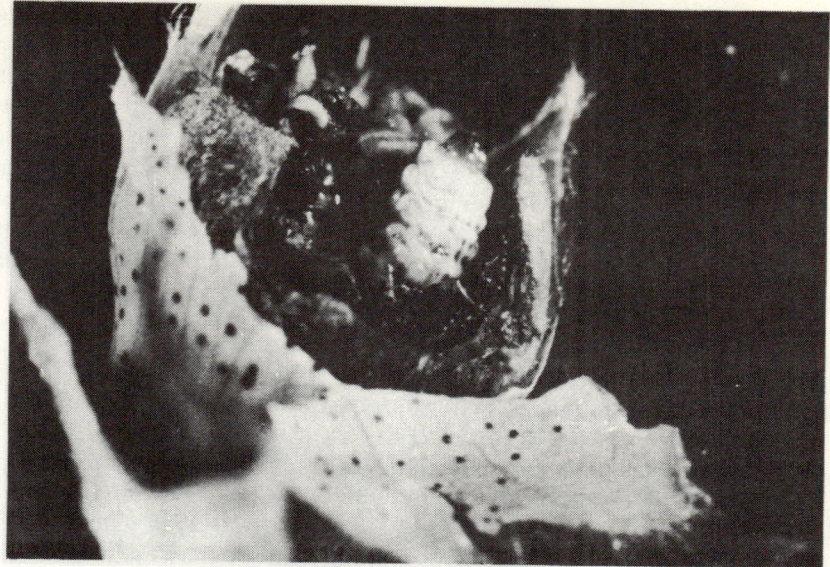

Figure 2. Boll weevil larvae feeding on cotton square (flower bud).

Several workers conducted field experiments using insecticides to disrupt entry into diapause. During 1959, a large-scale boll weevil diapause-control experiment (Brazzel, Davich and Harris, 1961) was conducted in the Big Bend area of Texas to determine whether weevils could be killed before they attained diapause. Four applications of methyl parathion were made at a rate of 0.5 pound per acre ($\approx$ 0.5 kg/ha) from 30 September to 8 November on a 525 acre ($\approx$ 2126 ha) isolated cotton planting. As a result of these late-season insecticide treatments, only one of the 10 treated fields required the insecticide treatment in the next crop season for boll weevil control before insecticide treatments were needed for bollworm–budworm control. However, when a similar test was conducted in Mississippi (Lloyd, Laster and Merkl, 1964), insecticide treatments spaced a week or more apart in September permitted sufficient deposition of boll weevil eggs between treatments for sizable boll weevil populations to develop in mid- to late-October when cooler temperatures reduced the effectiveness of insecticide treatments. Nevertheless, fall populations of boll weevils were significantly reduced by weekly treatments.

Knipling (1963) suggested improved timing of insecticide treatments against the last reproducing generation (treatments spaced close together) in order to limit egg deposition. Harris, Lloyd and Baker (1966) found that boll weevil eggs laid after about 25 September at Mississippi State, MS, would not develop into adult weevils before the first killing frost. Tingle and Lloyd (1969) observed that newly-emerged adult weevils had to feed for 2–4 weeks to attain firm diapause at temperatures simulating those of late October. Lloyd *et al.* (1966), in field tests conducted in 1963–64, applied six insecticide treatments at 5-day intervals against the last reproducing generations of boll weevils during September and one insecticide treatment in October against weevils that had survived or emerged following the September treatments. The results of this so-called 'reproduction–diapause' programme were to limit severely reproduction by the last reproducing generation of weevils in the late summer and to kill surviving and late-emerging adults from late September until frost if treatments were applied at 10- to 14-day intervals during this period. Adkisson *et al*. (1966) and Fye *et al*. (1968) subsequently demonstrated that boll weevil populations in Texas could be suppressed by 98% or more in the fall by applying about seven reproduction–diapause control treatments.

Developments contributing to more efficient use of insecticides

A major development in insecticide-application methodology was the application of ultra-low-volume (ULV) undiluted insecticides as foliar sprays for controlling insect pests. Messenger (1965) reported that when technical malathion was used in a control programme against the cereal leaf beetle, *Oulema melanopus* (L.), rangeland grasshoppers, *Melanopus* spp. and the boll weevil, *Anthonomus grandis* Boheman, the total cost per acre of applying the low-volume technical grade materials from aircraft was 50–60% less than that with conventional methods. Burt, Smith and Lloyd (1966) devised a rotary disc device for ground application of ULV sprays of undiluted insecticides. Using azinphosmethyl (0.25 lb/acre ≈ 0.1125 kg/0.4 ha) as a ULV spray, they found equivalent boll weevil control to that obtained with the water emulsion standard of methyl parathion (0.5 lb/acre ≈ 0.225 kg/0.4 ha). Since these findings, ULV sprays have found wide usage against many pests including the boll weevil.

Identification of the boll weevil pheromone and development of dispensing systems

Cross and Mitchell (1966) reported that male boll weevils feeding on cotton plants attracted female boll weevils at distances of 9.1 m or more. Hardee, Cross and Mitchell (1969) observed that, in the spring, traps baited with live males were more attractive to both sexes of overwintered weevils than were weevil-free cotton plants. Mitchell and Hardee (1974) then reported that when pheromone traps were placed in fruiting cotton fields, nearly all of the field-generation weevils captured during mid-season were unmated females. Later, Mitchell *et al*. (1976) found that before the squaring phase of cotton development, more males than females were caught in traps. However, during mid-season, many more females than males were caught in traps placed inside cotton fields; then, as the crop matured, substantial numbers of males were again captured in traps. Therefore, during the emergence of the overwintered population the boll weevil pheromone functions as an aggregation pheromone; during the fruiting period of the cotton plant it functions as a sex pheromone; in the fall it again appears to function as an aggregating pheromone.

Keller *et al.* (1964) found a substance that was attractive to female boll weevils in the laboratory, which could be collected by drawing air over males, passing the air through activated charcoal, and extracting the charcoal with chloroform. Hedin (1976) found that the extraction of males with dichloromethane produced a substance that was consistently attractive to females. The substance was more attractive when steam distilled and solvent extracted. Tumlinson *et al.* (1969) found that when the frass of males was steam distilled it was highly attractive to males. Tumlinson *et al.* (1971) isolated, identified and synthesized four terpenoid compounds: ((+)-cis-2-isopropenyl-1-methylcyclobutaneethanol; (*Z*)-3-3-dimethyl-$\Delta^{1,\beta}$-cyclohexane-acetaldehyde; (*Z*)-3,3-dimethyl-$\Delta^{1,\alpha}$-cyclohexaneacetaldehyde; and (*E*)-3,3-dimethyl-$\Delta^{1,\alpha}$-cyclohexaneacetaldehyde) as the components of the pheromone of live male boll weevil. The synthesis and structural assignments of the four compounds are also reported by Tumlinson *et al.* (1971).

McKibben *et al.* (1977) steam distilled frass of female boll weevils and obtained an extract that was more attractive to males than to females in the laboratory. These active components appeared to be alcohols and hydrocarbons; however, females produced only one-hundredth as much pheromone as the males.

For practical use under field conditions, the boll weevil pheromone must be released slowly to be an effective attractant. Initial field tests consisted of applying solutions of pheromone to chromatography supports such as firebrick: these release systems were attractive only for a few hours. McKibben *et al.* (1971) screened a number of polymers as controlled release additives and found polyethylene glycol the most effective in controlling the release of the boll weevil pheromone (grandlure). When grandlure was added to polyethylene glycol heated to approximately 40°C and then poured into gelatin capsules, the responsiveness of test weevils increased as the concentration increased. Hardee *et al.* (1972) then found that a formulation of grandlure containing glycerol, water and methanol dispensed from cotton dental rolls and later from cigarette filters was an effective dispensing system. A commercial gel formulation of grandlure was prepared by Zoecon Corporation in 1972 (Hardee *et al.*, 1974). McKibben (1976) also developed a cottonseed oil-based gel formulation. Bull *et al.* (1973) were able to slow the rate of release of grandlure from the cigarette filter dispenser by placing it in an open glass vial (filter-in-vial system) that functioned as a physical barrier. McKibben (1972, 1974) devised and improved a device that made it possible for one operator to inject grandlure into about 3000 cigarette filters in 1 hour. A laminated plastic dispensing system for grandlure was developed by the Health Chem Corporation as described by Hardee, McKibben and Huddleston (1975). McKibben *et al.* (1980) developed a polyester-wrapped cigarette-filter dispenser which replaced the filter in a vial system described by Bull *et al.* (1973).

Design and development of pheromone traps

Boll weevil pheromone traps are widely used for monitoring and suppressing boll weevils in management and/or eradication programmes. The first field tests with traps baited with male weevils were conducted by Cross *et al.* (1969). Of the 12 types of traps tested, the solid wing trap and an oblique-funnel trap appeared most efficient: wing traps were coated with Stikem® whereas, in the oblique-funnel traps, weevils were captured in a box on the trap. Initially, Stikem-coated plywood wing traps were painted dark green; however, Cross, Leggett and Hardee (1971) reported that white or bright-yellow traps were more attractive to boll weevils than darker-green traps. In 1968, Cross, Mitchell and Hardee (1976) found that traps painted a fluorescent yellow were the most attractive. Leggett and Cross (1971) developed the so-called Leggett trap, a non-sticky trap constructed from a floral liner (i.e. a papier maché vase) 29.21 cm high. It was painted daylight fluorescent yellow and capped with a screen cone (held just off the inverted liner with glass beads or other spacers) that had a small hole in the apex, which opened into a 5 cm³ plastic box. The completed trap was mounted on a stake about 1.22 m above the ground.

   Mitchell and Hardee (1974) designed an infield trap that was similar to the Leggett trap but with a smaller base suitable for use inside cotton fields. The construction was similar to the Leggett trap except that the base was a 1 litre soda cup painted fluorescent yellow, instead of the inverted floral liner. Dickerson *et al.* (1981) modified the Mitchell trap in order to reduce costs and facilitate mass production. Dickerson (1985) subsequently developed a manufactured trap with interchangeable components to facilitate storage and assembly under field conditions (*Figure 3*).

Use of pheromone traps for detection and suppression

The discovery that the male boll weevil emits a pheromone and the subsequent isolation, identification, and synthesis of the compound prompted many researchers to make efforts to use the pheromone traps to capture portions of the overwintered populations as they emerged in the spring. Hardee, Lindig and Davich (1971), in an area with very low populations of native weevils, compared the following six treatments: (1) one wing trap per 0.4 ha, around the field; (2) one wing trap per 0.4 ha, in the field; (3) one-half wing trap per 0.4 ha (i.e. one trap for every 2 acres) in the field and one-half wing trap per 0.4 ha, around the field; (4) one wing trap per 0.4 ha, around the field plus untreated trap plots of cotton; (5) one wing trap per 0.4 ha, around the cotton field, plus aldicarb-treated trap plots of cotton; and (6) one wing trap per 0.4 ha, around the field in three tiers. During the trapping period from May until 22 August, the traps captured an average of 0.64 weevils per trap from field populations estimated at 0.08–0.21 weevils per 0.4 ha. Although the male-baited traps suppressed the boll weevil popu-

Figure 3. The Dickerson pheromone trap.

lation by 63–100%, the small size of the populations made it difficult to distinguish between treatments.

Lloyd *et al.* (1972) in an area-wide trapping experiment in the spring of 1969, installed approximately one trap per 0.4 ha around approximately 1600 ha of cotton in Monroe County, Mississippi following a voluntary grower-sponsored reproduction–diapause control programme conducted in the fall of 1968. Stikem-coated wing traps painted yellow and baited with males were placed around cotton fields in April. Because this control programme was voluntary and implemented by individual growers, effectiveness varied. The larger populations of weevils reduced the effectiveness of the trapping pro-gramme, but these different population densities provided an opportunity to assess the efficiency of traps against populations of varying sizes. Clearly, as the population increased, trap efficiency decreased: traps suppressed low populations, but large populations were not significantly affected.

Pheromone-baited trap crops have been tested for suppression of emerging overwintered weevils. Mally (1901) had observed that early-planted cotton fields were the first in an area to be infested by the boll weevil: he suggested that a few rows of an early maturing variety should be planted so that emerged weevils would concentrate there where they could be trapped. When Cross and Mitchell (1966) found the responsiveness of female weevils to the male pheromone, interest was renewed in attracting and concentrating overwintered boll weevils into trap crops where they could be killed without treating the entire field. Ridgway *et al.* (1968) found that the systemic insecticide, aldicarb, was highly effective in suppressing populations of over-wintered boll weevils. However, soil application of aldicarb at rates effective against the boll weevil did produce outbreaks of bollworms and tobacco budworms. Scott *et al.* (1974) evaluated the use of trap crops, baited with grandlure and treated with aldicarb. They found that trap crops, with or without grandlure, proved to be much more attractive to overwintered weevils than the normal plantings prior to squaring. However, after squaring began in the normal plantings, trap crops with grandlure were more attractive than those without grandlure. The use of trap crops has not had wide acceptance in the Cotton Belt because the northern parts have cool weather early in the season and growers in warmer regions have heavy demands on their time when the trap crops should be planted. However, in many sections of the Cotton Belt, early plantings of an early fruiting variety and subsequent baiting of trap crops with grandlure will attract a very high percentage of emerged overwintered weevils until squaring begins in the normal plantings. Trap crops should therefore be considered as one method for improved management of boll weevil populations.

As indicated earlier, during the mid-season many more females than males are captured in pheromone traps, indicating that the male boll weevil pheromone functions as a sex pheromone at this time: detection of repro-ducing populations during an eradication programme can therefore be accomplished most efficiently during the reproduction period with infield traps. In 1978, in special detection experiments with infield traps, clumps of 20 infested squares were placed in weevil-free fields (Leggett, Lloyd and Witz, 1981). In fields with one trap per 0.4 ha, the trap catches allowed detection of four of seven of the reproducing clumps in the F_1 generation. In fields with four traps per 0.4 ha, the trap catches allowed detection of seven of seven clumps during the F_1 generation. Subsequently, in 1979, Lloyd *et al.* (1980) conducted a series of similar field experiments in Surry County, NC, a non-cotton area. In 1.28 ha cotton plots, clumps of 20 infested squares were randomly placed, and traps were spaced 64, 45, 36 and 32 m apart. An average of 2, 3, 5.5 and 6.7 F_1 female weevils were captured, respectively, at these spacings. These authors estimated that 80% or more of the F_1 female weevils were captured when traps were 36 m apart and probably more than 90% of the F_1 females were captured when traps were 32 m apart. These results indicate the potential value of pheromone-baited infield traps for both detection and suppression.

Sterilization of the boll weevil

Knipling (1955) suggested that insect control or eradication could be achieved by introducing sterile male insects into natural populations of these species. Davich and Lindquist (1962) reported that gamma irradiation was not a satisfactory method of sterilization because sterilization was not permanent and the dose required to produce temporary sterility was lethal. Haynes (1963) reported that an alkylating agent, apholate, reduced the hatch of eggs when male weevils were dipped in a 2% aqueous solution of apholate. Davich *et al.* (1967) using apholate-sterilized males in a field experiment in Baldwin County, Alabama, against very low populations of boll weevils, found that 93.6% of punctured squares were not infested where sterile weevils were released between 22 June and 31 August 1964; however, some males survived and regained fertility. Klassen, Norland and Borkovec (1968) and Klassen and Earle (1970), used a chemosterilant, busulfan, to sterilize male boll weevils without appreciably reducing vigour or mating ability of sterilized males. Haynes *et al.* (1972) showed that a combination of busulfan and hempa was superior to busulfan alone. Lloyd, McCoy and Haynes (1976) in the Pilot Boll Weevil Eradication Experiment (PBWEE) found that feeding busulfan or busulfan-hempa in the adult diet resulted in 93% sterility in males released in the PBWEE. Reduction in egg hatch from different zones in the PBWEE ranged from 70.6% to 98.5%; however, some of the sterilized males regained fertility 2–3 weeks after sterilization.

Following the PBWEE, three new procedures were developed and tested for the sexual sterilization of the boll weevil. Borkovec, Woods and Terry (1978) and McCoy and Wright (1979) used the chemosterilant, bisazir, administered as a fumigant to sterilize male boll weevils. Haynes *et al.* (1977) found that fractionated doses of irradiation to pupae or newly emerged adults resulted in 98–100% sterility. Earle *et al.* (1978) confirmed the unpublished results of R. A. Leopold and D. T. North that acute irradiation of adults plus diflubenzuron was an effective procedure for sterilizing adults. Wright *et al.* (1979) then examined the effects of ageing the adults before exposing them to sterilizing procedures and compared the mating ability, sterility, and survival of weevils treated with irradiation or fumigation. Wright *et al.* (1980a) selected acute irradiation of adults in a nitrogen atmosphere plus dipping in a 0.1% acetone solution of diflubenzuron as the best of three candidate sterilization methods. Earle, Nilakhe and Simmons (1979) reported that when newly emerged adults were fed or dipped in an acetone solution of diflubenzuron, mating was reduced by 40–50%. Wright *et al.* (1980b) therefore administered diflubenzuron in the adult diet (100 ppm) for five days prior to irradiation with 10 krad of gamma irradiation from a caesium-137 source in a nitrogen atmosphere. This was the procedure used to sterilize 11 million boll weevils released in the North Carolina Boll Weevil Eradication Trial in 1979.

Villavaso, Earle and Hollier (1977) developed procedures using 0.2 ha isolated cotton plots for measuring competitiveness of sterile boll weevils.

After plants had started squaring, equal numbers of sterilized and normal males were released on three rows per plot. Virgin females were released 3–5 rows from the male rows. On the seventh day after release of weevils on to the plots, cotton squares which appeared to have oviposition punctures (*see Figure 1*) were collected. After 4–5 days' incubation the squares were dissected and examined under a binocular microscope for eggs and larvae (live or dead). Competitiveness was computed from these data using the method of Fried (1971). Villavaso (1981), using the field-competitive procedures which he had developed (Villavaso, Nilakhe and McGovern, 1979), determined that the competitiveness value of sterile boll weevils released in the Boll Weevil Eradication Trial was about 6% that of a native strain, and that the sterile weevils should be released every 3–4 days. In field tests in 1980 and 1981, Villavaso (1982) found that diflubenzuron-fed irradiated weevils were about 24% as competitive as an untreated laboratory strain on days 0–4 and about 1.5% as competitive 4–7 days after release. Villavaso and Thompson (1984) determined that weevils sterilized with ecdysteroid and 10 krad gamma irradiation were 43.7% as competitive, for 7 days, as an untreated laboratory strain. However, Villavaso *et al.* (1986) observed that when weevils were sterilized with an ecdysteroid followed by 10 krad of gamma irradiation, using scaled-up procedures, the competitiveness value was only about 17% of that of untreated laboratory weevils. E. J. Villavaso and J. L. Roberson (personal communication) found that when 5-day-old adult boll weevils were sterilized by dipping in an aqueous solution of diflubenzuron (0.4%) and irradiated with 10 krad of gamma irradiation, the competitiveness values were 19% for days 1–4 and 17% for days 5–7 following sterilization. The LT_{50} for these sterile weevils held on squares of growing cotton plants was 7.7 days.

In the 1973 PBWEE and the Boll Weevil Eradication Trial, sterile weevils were released free-fall from an aircraft into cotton fields. Specially designed equipment metered chilled weevils from the aircraft. Wright (1985) found that holding adult boll weevils at 2.7°C for periods exceeding 24 hours significantly reduced mobility. In 1983 Villavaso and Roberson reported that mass-reared boll weevils released on loose sand became entrapped and that when sterile weevils were released under midday conditions, they were killed when they fell on the hot soil surfaces (Villavaso, 1984). Reinecke *et al.* (1986) reported that, in a 1985 field experiment, sterile boll weevils were released in shotgun-shell-size containers before soil-surface temperatures became elevated; as weevils were containerized before shipment, they were not chilled, but were shipped in ventilated boxes.

When costs were estimated for the use of sterile boll weevils versus insecticide treatments for a 4-week period following initiation of squaring, the cost of releasing sterile boll weevils was found to be 30% that of applying insecticides.

Development of technologies for mass-rearing large numbers of boll weevils for sterilization

Large numbers of sterile boll weevils will be required if they are to be used in eradication programmes and therefore technology has been developed for mass-rearing of boll weevils. The first of several developments that eventually made it possible to mass rear boll weevils in the laboratory was the development, by Vanderzant and Davich (1958), of a suitable diet, which was later modified by Earle, Gaines and Roussel (1959), Gast and Davich (1966), Lindig and Malone (1973), and Lindig, Roberson and Wright (1979).

In order to rear larger numbers of weevils, mechanical equipment was designed not only to increase numbers but also to reduce costs. These initial mechanization developments were made by Gast (1961), Gast (1965), and Gast and Vardell (1963). In order to house developing and rearing facilities, the Robert T. Gast Rearing Laboratory was constructed at Mississippi State, Mississippi, during 1971 and 1972. As the facility was completed, a number of changes were made that significantly improved rearing efficiency: (1) high-efficiency particular air (HEPA) filters were installed over outlets of the air-conditioning supply in the egg planting and larval development rooms; (2) adult weevils were moved into rooms that were further away from the egg planting and larval development rooms; (3) the larval development area was expanded from one room to two adjoining rooms so there could be more space and better air circulation around the carts that held the trays of diet and the eggs and larval stages; (4) a cabinet was installed for fumigating the plastic and tyvek (porous bedding material) used in constructing rearing traps; (5) antimicrobial agents were added to the sand used to cover the diet and eggs in the rearing trays; (6) storage and processing operations relating to the diet were moved to a separate building to reduce the possibility of microbial contamination (Griffin, Sikorowski and Lindig, 1983). Detailed descriptions of all rearing operations are provided by Griffin, Sikorowski and Lindig (1983).

In the summer of 1977, a 6-week production run was conducted with the goal of producing 5 million boll weevils per week: this goal was, in fact, exceeded during this trial operational run.

Development of insect growth regulators as foliar sprays for boll weevil population suppression

Insect growth regulators (IGRs) are highly selective chemical pesticides that interfere with the normal development of certain life stages of arthropods. One IGR, diflubenzuron, (1-(4-chlorophenyl)-3-(2,6-difluorobenzoyl)urea), has been approved for use on cotton against the boll weevil (Bull, Ables and Lloyd, 1983). It was introduced by Philips-Duphar BV, Amsterdam, Holland and developed in the United States by Thompson-Hayward Chemical Company.

The biological activity of diflubenzuron results from the disruption by the insect of the chemical synthesis of chitin, an essential structural component of insect cuticle (Mulder and Gijswijt, 1973). Taft and Hopkins (1975) provided the first evidence of the effects of diflubenzuron against field populations of the boll weevil. Mixtures of diflubenzuron with invert sugar and molasses cause significant reductions in the hatch of boll weevil eggs. Lloyd, Wood and Mitchell (1977) subsequently evaluated formulations of diflubenzuron in the greenhouse and determined that mixtures of diflubenzuron with certain oils, especially cottonseed oil, were as effective as the sugar–molasses insect bait and potentially more convenient for practical field application. Simple mixtures of water and diflubenzuron were ineffective: however, when sprays of a suspension of diflubenzuron (280 g a.i./ha, 16 applications) in raw cottonseed oil were applied in the field, boll weevil reproduction was suppressed so that no field generations were detected until 20 August, near the end of the crop season. Ganyard et al. (1977) conducted a field study in isolated cotton plots in a non-cotton area of North Carolina. They obtained 99% suppression of reproduction with repeated treatments of diflubenzuron applied in raw cottonseed oil.

Although it is effective against the boll weevil, diflubenzuron is generally ineffective against *Heliothis* spp. (Taft and Hopkins, 1975; House *et al.*, 1978; Nemec, 1978; Bull *et al.*, 1979). As diflubenzuron appears to have little effect on *Heliothis* spp. and other pests of cotton, its effects on the natural enemies of these pests were investigated.

Ables, Jones and Bee (1977), Wilkerson *et al.* (1978), and House *et al.* (1980) tested diflubenzuron in the laboratory against several species of entomophagous arthropods. Laboratory sprays applied against adult convergent lady beetles (*Hippodomia convergens* Guérin-Méneville) reduced egg hatch and prevented larval development, but these harmful effects gradually subsided after treatments were terminated (Ables, Jones and Bee, 1977). Keever, Bradley and Ganyard (1977) reported similar effects on the reproduction of *H. convergens* adults collected from diflubenzuron-treated cotton fields and held in the laboratory. Wilkerson *et al.* (1978) reported that topical applications of diflubenzuron had no apparent adverse effects on adult convergent lady beetles. Ables, Jones and Bee (1977) and Ables *et al.* (1980) in central Texas indicated that diflubenzuron had little or no impact on predacious arthropods when applied at 5-day intervals in crop oil and water: levels of predators were slightly higher in the untreated plots, but plots receiving the highest dose of diflubenzuron had a greater abundance of predators than any of the plots treated with lower dosages. Diflubenzuron thus appears to be much more selective in controlling the boll weevil than are most conventional insecticides; its use will probably conserve many entomophagous arthropods.

Large-scale field trials of boll weevil eradication technology

Two large-scale field trials were conducted to test and evaluate combinations of suppression and detection technology for eradication or elimination of the boll weevil from geographical areas of the Cotton Belt. The first of these was the Pilot Boll Weevil Eradication Experiment (PBWEE) conducted from 1971 to 1973 in southern Mississippi and adjoining areas in Louisiana and Alabama. The second was the Boll Weevil Eradication Trial (BWET) conducted in north-eastern North Carolina from 1978 to 1980.

THE PILOT BOLL WEEVIL ERADICATION EXPERIMENT, 1971–73

The Pilot Boll Weevil Eradication Experiment (PBWEE) was conducted in an area that covered all or part of 30 counties in southern Mississippi, five parishes in Louisiana, and two counties in Alabama (*Figure 4*) (Boyd, 1976). The PBWEE was conducted in an area where the boll weevil was well established; populations were consistently high and it was difficult to carry out suppression measures. Most fields were small, surrounded by high trees and difficult to treat, with an average field size of about 4 ha. The area was selected as being representative of the worst conditions likely to be encountered in the boll weevil belt. The experiment was organized geographically and with an eradication zone located in the centre of a series of suppression zones. The eradication zone consisted of an area within a radius of 40 km of Columbia, Mississippi, and concentric buffer zones of approximately 80 km in depth were established around the eradication zone; the

Figure 4. Location of the Pilot Boll Weevil Eradication Experiment in Southern Mississippi, Alabama, and Louisiana, 1971–73.

evaluation of the experiment was conducted in the eradication zone. Suppression techniques used in the PBWEE were (1) in-season control with insecticides, (2) reproduction–diapause control, (3) cultural control—defoliation, (4) cultural control—stalk destruction, (5) pheromone traps, (6) trap crops, (7) pinhead square treatment and (8) sterile-male releases.

In 1971, the in-season boll weevil control programme was voluntary and the Cooperative Extension Services promoted the need for good in-season control by the producers. Although considerable effort was put forth to encourage producers to carry out an adequate in-season control programme, approximately 50% of the acreage received no in-season insecticide treatments. In the late summer and fall a total of 13 insecticides were applied at 5- to 12-day intervals, beginning in early August and continuing until the plants were killed by frost on 1 November. Cotton was defoliated in the fall to reduce the food supply for weevils that survived the insecticide treatments. When harvest was complete, cotton stalks were shredded so that fields would not have to receive additional insecticide treatments.

In 1972, pheromone traps were placed around cotton fields in the eradication and buffer zone fields in the spring and summer to monitor populations of overwintered and field generations of boll weevils, and to locate those areas in which the boll weevil populations were concentrated. In addition to the pheromone-baited traps placed on field borders, trap crops were planted in all fields. These four-row strips of trap cotton extended the length of the cotton fields and were of the Qua Paw early fruiting variety. The trap crops were planted 2–3 weeks earlier than the normal planting and were baited with the boll weevil pheromone. At planting time, the trap crop received aldicarb at the rate of 1 lb/acre ($\approx$ 0.45 kg/0.4 ha) active ingredient of aldicarb in-furrow and 2 lb/acre ($\approx$ 0.9 kg/0.4 ha) sidedress 6–7 weeks later. Foliar insecticide treatments of malathion or azinphosmethyl were also applied to some of the trap crops to ensure adequate boll weevil control in these early planted strips.

A single insecticide treatment of azinphosmethyl (0.25 lb/acre ($\approx$ 0.225 kg/0.4 ha) was applied to all cotton fields in the eradication and first buffer zones when the cotton reached the pinhead square stage. Sterile male boll weevils (sterilized with busulfan) were released aerially in all fields in the eradication zone (Boyd, 1976).

In-season control in 1972 was implemented by the Animal and Plant Health Inspection Service, USDA (APHIS) operational team as the 1971 voluntary programme had not effectively suppressed the boll weevil population. Five applications of azinphosmethyl at the rate of 0.25 lb/acre ($\approx$ 0.1125 kg/0.4 ha) or toxaphene + DDT + methyl parathion (2.0 lb + 1.0 lb + 0.5 lb/acre ($\approx$ 0.9 kg + 0.45 kg + 0.225 kg/0.4 ha) were used for in-season control. Subsequently, 13 reproduction-diapause control treatments were applied between 7 August and the first killing frost in November. Cultural control was identical to that described for 1971.

In 1973, pheromone traps were used, as were trap crops as in 1972. In 1973, instead of planting the variety Qua Paw (which was adapted for Mississippi), Stoneville 213 was used. Pheromone-bait stations were installed at 31 m intervals the entire length of the trap crop. Pinhead insecticide treatments were applied to cotton fields where two or more weevils were captured per acre with traps prior to the appearance of pinhead squares (squares too small for oviposition by the boll weevil). Sterile boll weevils (sterilized with busulfan and busulfan-hempa) were released in the eradication zone and the adjoining 8 m in the first buffer zone from 4 June to 10 August. The experiment was completed on 10 August 1973.

At the completion of the PBWEE, Knipling (1976) concluded in a report of the Technical Guidance Committee for the Pilot Experiment:

> Based on the results and experiences gained in the Pilot Boll Weevil Eradication Experiment conducted in south Mississippi and adjacent areas in Alabama and Louisiana, and mindful that the experiment was conducted in an area representative of the most severe boll weevil conditions likely to be encountered in the boll weevil belt, the Technical Guidance Committee has reached the conclusion that it is technically and operationally feasible to eliminate the boll weevil as an economic pest in the United States by the use of techniques that are ecologically acceptable. The economic and environmental benefits of achieving this goal far exceed the costs that will be involved. For such programs to be successful, it must be carried out with thoroughness and precision. The participation of a number of agencies will be required. Complete cooperation and participation by all cotton growers in the boll weevil belt is essential.

Recommendations included improving mass-rearing procedures to ensure the capability of producing adequate numbers of high-quality boll weevils for sterilization and release, improved methods of sterilization of both sexes and continued development of practical uses of the boll weevil pheromone.

Concurrent with the report of the Technical Guidance Committee of the PBWEE, a report of a review committee of the Entomological Society of America (ESA) (Eden, 1976) concluded that the eradication zone of the PBWEE was smaller than desired and that eradication had not been achieved in the core area. The ESA committee concluded that the decision regarding an attempt at eradication should be sociopolitical: they recommended that a detailed summary of the programme should be published and that all concerned members of ESA should inform themselves about the long-term environmental and economic benefits that would result from a successful eradication programme, and should weigh those against the costs involved.

Because of the lack of agreement on the outcome of the PBWEE, a beltwide eradication programme was not initiated: instead, a Boll Weevil Eradication Trial (BWET) was conducted in North Carolina from 1978 to 1980.

BOLL WEEVIL ERADICATION TRIAL, 1978–80

The Boll Weevil Eradication Trial conducted in north-eastern North Carolina encompassed 29 counties and extended from Fayetteville, North Carolina, to southern Virginia (*Figure 5*) (USDA, 1981). During year 1 (1978), pheromone traps, one per 0.4 ha ($\approx$ 1 acre) of cotton, were installed on field borders and monitored from April through November. In-season insecticide

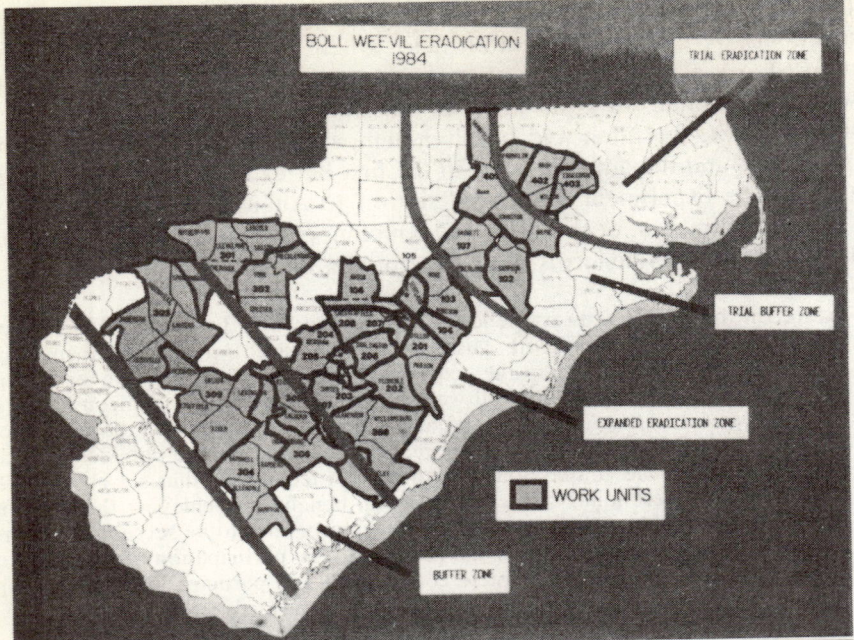

Figure 5. Location of Boll Weevil Eradication Trial (Trial Eradication Zone and Trial Buffer Zone) in Northeastern North Carolina and Expanded Southeastern Boll Weevil Eradication Program in Southern North Carolina and South Carolina, 1978–84.

treatments for control of the boll weevil and other pest species were applied as a part of the operational activities of the BWET. During late August, September and October, five applications of insecticide were made to all cotton plantings by the APHIS operational team. Those areas that were difficult to treat were treated with ground equipment (i.e. mist blowers mounted in 4-wheel drive vehicles were used to treat field borders where aerial spraying could not be applied effectively). In year 2 (1979), 11.2 million sterile boll weevils (139 sterile weevils per acre (0.4 ha), per week for four weeks) were released, beginning at the pinhead square stage. Four applications of the insect growth regulator diflubenzuron were applied to 11% of the acreage at weekly intervals during June and early July. A single application of an organophosphorous insecticide was applied after the fourth diflubenzuron treatment. In April, one pheromone trap per 0.4 ha was installed around all cotton fields from April until early July; then two pheromone traps per 0.4 ha were installed within the cotton fields from early July to late August, to detect the presence of reproducing populations; subsequently, one pheromone trap per 0.4 ha was installed on field borders and serviced until the first killing frost.

In year 3 (1980) pheromone traps (one per 0.4 ha) were installed on field borders in late April, moved inside the field from early July until late August, and then placed on field borders for the remainder of the crop season.

At the beginning of the BWET, the population of weevils was relatively low because of the intensive use of insecticides for control of bollworms and budworms and two successive unusually cold winters in years preceding the initiation of the trial. During the second year, after the fall applications of the first year, only seven native overwintered boll weevils were trapped in border traps placed around the cotton fields. As no boll weevils were captured in the infield traps, there was no evidence that reproduction occurred within the valuation area. In the fall of year 2, two boll weevils were trapped at widely separated locations on the southern edge of the evaluation area: one of these was adjoining a major interstate highway which is located in the evaluation area; because these were widely separated weevil captures, they were assumed to be migrants. In year 3, a single headless boll weevil was found during the first inspection of traps in May. This weevil was believed to have been carried over in the winter in a stored trap not properly cleaned before spring installation. Again in year 3, four weevils were captured between 18 August and 28 October in traps away from cotton fields just inside the evaluation area; these appeared to be migrants from outside the evaluation area. Subsequently, starting on 11 September, one pupa and nine adult boll weevils were detected in a clump within a single field near the northern limits of the evaluation area; these weevils appeared to be the progeny of one introduced female. This infestation was eliminated by intensive trapping and cultural practices, and traps were placed there at a very high intensity and operated until 15 November; no additional weevils were captured. McKibben and Cross (1984), using the expected rate of increase of a boll weevil population and the sensitivity of pheromone traps, indicated a probability of at least 0.9983 that the occurrence of a reproductive population would have been detected during the evaluation period.

In the BWET, the average number of insecticide applications for control of all cotton pests within the evaluation area decreased by 88% whereas the comparison area outside the Trial area decreased by 25% compared with 1974–77 pre-trial averages: the BWET was, therefore, judged to be a technical as well as a biological success (USDA, 1981).

In an independent review of the BWET, a specially appointed committee of the National Research Council of the National Academy of Sciences recommended an indefinite postponement of a BWE programme and encouraged the private sector, the academic community and government agencies to assist the development and adoption of private integrated pest management so that its potential could be more fully realized (NRC/NAS, 1981).

Although the National Research Council did not recommend proceeding with boll weevil eradication, cotton producers in southern North Carolina and in South Carolina voted in 1983 to support a boll weevil eradication programme in their area by providing 70% of the cost. The Federal Government appropriated funds to pay 30% of the cost and the USDA Animal and Plant Health Inspection Service was charged with the overall management of the programme through a co-operative agreement with the Southeastern Boll Weevil Eradication Foundation (Ridgway *et al.*, 1985). Dickerson *et al.* (1986) reported that, at the end of the 1985 crop season, there was no evidence of reproduction in 97% of the fields in the eradication areas of southern North Carolina and South Carolina. Growers in Georgia, Florida Panhandle, and south-eastern Alabama are considering expansion of boll weevil eradication into their areas.

Conclusions

The following conclusions may be drawn with regard to the research strategies, developments and eradication programmes intended to combat the boll weevil in the USA (the role of resistant cotton varieties has not been considered here):

1. Research to develop technology for boll weevil eradication has resulted in integration of technical components which are designed to prevent reproduction by native boll weevil populations in selected geographical areas of the Cotton Belt of the United States.
2. Technical components currently in use in the Southeastern Boll Weevil Eradication Program include suppression of late season (diapausing) boll weevil populations with well-timed reproduction–diapause control treatments (insecticides), surveillance and suppression of emerging over-wintered populations the next spring and early summer with pheromone traps, and suppression of incipient overwintered field populations of boll weevils with insecticides or an insect growth regulator, diflubenzuron. Where reproduction occurs, reproduction–diapause treatments are applied during the second crop season of the eradication programme to reduce further the size of the population. During the third crop season, surveillance and suppression of surviving overwintered weevils is accomplished with pheromone traps, insecticides, and insect growth regulators.
3. Future expansions of boll weevil eradication will be determined by the willingness of cotton producers to pay 70% of the cost of eradication. When the boll weevil is a persistent and costly economic problem, growers are likely to elect to pay the cost of the eradication programme because economic benefits in areas where eradication has been achieved have been substantial. However, in the northern boll weevil-infested area of the Cotton Belt where the problem is sporadic, growers will probably be much less supportive than those in areas where the problem is persistent.
4. In these areas that are more lightly infested, additional cost-effective technology is needed to ensure grower acceptance and to eliminate these areas as sources of reinfestation for eradicated areas.
5. Effective containment barriers must be established and maintained for all eradicated areas.

References

Ables, J. R., Jones, S. L. and Bee, M. J. (1977). Effect of diflubenzuron on beneficial arthropods associated with cotton. *Southwestern Entomologist* 2, 66–72.

ABLES, J. R., HOUSE, V. S., JONES, S. L. AND BULL, D. L. (1980). Effectiveness of diflubenzuron on boll weevils in central Texas River Bottoms area. *Southwestern Entomologist* **5** (Supplement Number 1), 15–21.

ADKISSON, P. L., RUMMEL, D. R., STERLING, W. L. AND OWEN, W. L., JR (1966). *Diapause Boll Weevil Control: A Comparison Of Two Methods. Texas Agricultural Experiment Station B-1054*, p. 11.

BORKOVEC, A. B., WOODS, C. W. AND TERRY, P. H. (1978). Boll weevil: chemosterilization by fumigation and dipping. *Journal of Economic Entomology* **71**, 862–866.

BOYD, F. J. (1976). Operational plan and execution of the pilot boll weevil eradication experiment. In *Proceedings of Conference on Boll Weevil Suppression, Management, and Elimination Technology, 13–15 February 1974. Memphis, TN. ARS-S-71*, pp. 62–69. USDA, Washington, DC.

BRAZZEL, J. R. (1963). Resistance to DDT in *Heliothis virescens. Journal of Economic Entomology* **56**, 571–574.

BRAZZEL, J. R. AND HIGHTOWER, B. G. (1960). A seasonal study of diapause, reproductive activity, and seasonal tolerance to insecticides in the boll weevil. *Journal of Economic Entomology* **53**, 41–46.

BRAZZEL, J. R. AND NEWSOM, L. D. (1959). Diapause in *Anthonomus grandis* Boheman. *Journal of Economic Entomology* **52**, 603–611.

BRAZZEL, J. R., DAVICH, T. B. AND HARRIS, L. D. (1961). A new approach to boll weevil control. *Journal of Economic Entomology* **54**, 723–730.

BULL, D. L., ABLES, J. R. AND LLOYD, E. P. (1983). Insect growth regulators with emphasis on the use of benzophenyl ureas. In *USDA Handbook No. 589. Cotton Insect Management with Special Reference to the Boll Weevil* (R. L. Ridgway, E. P. Lloyd and W. H. Cross, Eds), pp. 207–235. USDA, Washington, DC.

BULL, D. L., COPPEDGE, J. R., RIDGWAY, R. L., HARDEE, D. D. AND GRAVES, T. M. (1973). Formulations for controlling the release of synthetic pheromone (grandlure) of the boll weevil. 1. Analytical studies. *Environmental Entomology* **2**, 829–835.

BULL, D. L., HOUSE, V. S., ABLES, J. R. AND MORRISON, R. K. (1979). Selective methods for managing insect pests of cotton. *Journal of Economic Entomology* **72**, 841–846.

BURT, E. C., SMITH, D. B. AND LLOYD, E. P. (1966). A rotary disc device for applying ultra-low-volume (undiluted) pesticides with ground equipment. *Journal of Economic Entomology* **59**, 1487–1489.

CARSON, RACHEL (1962) *Silent Spring*. Fawcett Books Group, Brooklyn, NY, 304 pp.

COAD, B. R. (1918). *Recent Experimental Work On Poisoning Cotton Boll Weevils. US Department of Agriculture Bulletin 731*, 15 pp. USDA, Washington, DC.

COAD, B. R. AND CASSIDY, T. P. (1920). *Cotton Boll Weevil Control By The Use Of Poison. US Department of Agriculture Bulletin 875*, 31 pp. USDA, Washington, DC.

COAD, B. R., JOHNSON, E. AND MCNEIL, G. L. (1924). *Dusting Cotton From Aeroplanes. US Department of Agriculture Bulletin 1204*, 40 pp. USDA, Washington, DC.

CROSS, W. H. AND MITCHELL, H. C. (1966). Mating behavior of the female boll weevil. *Journal of Economic Entomology* **59**, 1503–1507.

CROSS, W. H., LEGGETT, J. E. AND HARDEE, D. D. (1971). *Improved Traps For Capturing Boll Weevils. US Department of Agriculture, Cooperative Economic Insect Report 21*, pp. 367–368. USDA, Washington, DC.

CROSS, W. H., MITCHELL, H. C. AND HARDEE, D. D. (1976). Boll weevils: response to light sources and colors on traps. *Environmental Entomology* **5**, 565–571.

CROSS, W. H., HARDEE, D. D., NICHOLS, F., MITCHELL, H. C., MITCHELL, E. B., HUDDLESTON, P. M. AND TUMLINSON, J. H. (1969). Attraction of female boll weevil to traps baited with males or extracts of males. *Journal of Economic Entomology* **62**, 154–161.

DAVICH, T. B. AND LINDQUIST, D. A. (1962). Exploratory studies on gamma radiation for the sterilization of the boll weevil. *Journal of Economic Entomology* **55**, 164–167.

DAVICH, T. B., MERKL, M. E., MITCHELL, E. B., HARDEE, D. D., GAST, R. T., McKIBBEN, G. H. AND HUDDLESTON, P. M. (1967). Field experiments with sterile males for eradication of the boll weevil. *Journal of Economic Entomology* **60**, 1533–1538.

DICKERSON, W. A. (1985). A manufactured componentized boll weevil pheromone trap. Patent pending.

DICKERSON, W. A., McKIBBEN, G. H., LLOYD, E. P., KEARNEY, J. F., LAM, J. J. JR AND CROSS, W. H. (1981). Field evaluation of a modified infield boll weevil trap. *Journal of Economic Entomology* **74**, 280–282.

DICKERSON, W. A., RIDGWAY, R. L., PLANER, F. R., BRAZZEL, J. R. AND BRADWAY, T. J. (1986). Pheromone trap captures in southeastern boll weevil eradication program and insecticide use in the Boll Weevil Eradication Trial zone. In *Proceedings of Beltwide Cotton Production Research Conferences, 5–8 January, 1986, Las Vegas, NV* in press.

DUNLAP, T. R. (1981). *DDT: Scientists, Citizens and Public Policy*. Princeton University Press, Princeton, NJ. 318 pp.

EARLE, N. W., GAINES, R. C. AND ROUSSEL, J. S. (1959). A larval diet for boll weevils containing an acetone powder of cotton squares. *Journal of Economic Entomology* **52**, 710–712.

EARLE, N. W., NILAKHE, S. S. AND SIMMONS, L. A. (1979). Mating ability of irradiated male boll weevils treated with diflubenzuron or penfluron. *Journal of Economic Entomology* **72**, 334–336.

EARLE, N. W., SIMMONS, L. A., NIKAKLE, S. S., VILLAVASO, E. J., McKIBBEN, G. H. AND SIKOROWSKI, P. P. (1978). Pheromone production and sterility in boll weevils: effect of acute and fractionated gamma irradiation. *Journal of Economic Entomology* **71**, 591–595.

EDEN, W. G. (1976). Report of the Entomological Society of America Review Committee on the Pilot Boll Weevil Eradication Experiment. In *Proceedings of a Conference on Boll Weevil Suppression, Management, and Elimination Technology, Memphis, TN, 13–15 February 1974. ARS-S-71*, p. 126. USDA, Washington, DC.

EWING, K. P. AND IVY, E. E. (1943). Some factors influencing bollworm populations and damage. *Journal of Economic Entomology* **36**, 602–606.

EWING, K. P. AND PARENCIA, C. R., JR. (1950). *Early-Season Applications of Insecticides on a Community-wide Basis for Cotton-insect Control in 1950*. US Department of Agriculture, Bureau of Entomology and Plant Quarantine E-810, Washington, DC. 8 pp.

EWING, K. P., PARENCIA, C. R., JR. AND IVY, E. E. (1947). Cotton insect control with benzene hexachloride, alone or in mixture with DDT. *Journal of Economic Entomology* **40**, 374–381.

FRIED, M. (1971). Determination of sterile insect competitiveness. *Journal of Economic Entomology* **64**, 869–872.

FYE, R. E., COLE, C. L., TINGLE, F. C., STONER, A., MARTIN, D. F. AND CURL, L. F. (1968). A reproductive–diapause control program for boll weevil in the Presidio, Texas-Ojinaga, Chihuiahua Area, 1965–67. *Journal of Economic Entomology* **61**, 1660–1666.

GAINES, J. C. AND DEAN, H. A. (1947). New insecticides for boll weevil, bollworm and aphid control. *Journal of Economic Entomology* **40**, 363–370.

GAINES, J. C. AND WIPPRECHT, R. (1948). Effect of dusting schedules on yield of cotton during 1947. *Journal of Economic Entomology* **41**, 410–412.

GANYARD, M. C., BRADLEY, J. R., BOYD, J. F. AND BRAZZEL, J. R. (1977). Field evaluation of diflubenzuron (Dimilin) for control of boll weevil reproduction. *Journal of Economic Entomology* **70**, 347–350.

GAST, R. T. (1961). Some shortcuts in laboratory rearing of boll weevils. *Journal of Economic Entomology* **54**, 395–396.

GAST, R. T. (1965). Methods for mass production of diet pellets for adult boll weevils. *Journal of Economic Entomology* **58**, 1024–1025.

GAST, R. T. AND DAVICH, T. B. (1966). Boll weevils. In *Insect Colonization and Mass Production*, (C. N. Smith, Ed.), pp. 405–419. Academic Press, New York.

GAST, R. T. AND VARDELL, H. (1963). *Mechanical Devices to Expedite Boll Weevil Rearing in the Laboratory*. US Department of Agriculture, ARS 33-89. 10 pp.

GRAVES, J. B., ROUSSEL, J. S. AND PHILLIPS, J. R. (1963). Resistance to some chlorinated hydrocarbon insecticides in the bollworm, *Heliothis zea*. *Journal of Economic Entomology* **56**, 442–444.

GRIFFIN, J. G., SIKOROWSKI, P. P. AND LINDIG, O. H. (1983). Mass rearing boll weevils. In *USDA Handbook No. 589. Cotton Insect Management with Special Reference to the Boll Weevil* (R. L. Ridgway, E. P. Lloyd and W. H. Cross, Eds), pp. 265–301. USDA, Washington, DC.

HARDEE, D. D., CROSS, W. H. AND MITCHELL, E. B. (1969). Male boll weevils are more attractive than cotton plants to boll weevil. *Journal of Economic Entomology* **62**, 165–169.

HARDEE, D. D., LINDIG, O. H. AND DAVICH, T. B. (1971). Suppression of populations of boll weevils over a large area in west Texas with pheromone traps in 1969. *Journal of Economic Entomology* **64**, 928–933.

HARDEE, D. D., McKIBBEN, G. H. AND HUDDLESTON, P. M. (1975). Grandlure for boll weevils: Control release with a laminated plastic dispenser. *Journal of Economic Entomology* **68**, 447–479.

HARDEE, D. D., McKIBBEN, G. H., GUELDNER, R. C., MITCHELL, E. B., TUMLINSON, J. H. AND CROSS, W. H. (1972). Boll weevils in nature respond to grandlure, a synthetic pheromone. *Journal of Economic Entomology* **65**, 97–100.

HARDEE, D. D., GRAVES, T. M., McKIBBEN, G. H., JOHNSON, W. L., GUELDNER, R. C AND OLSEN, C. M. (1974). A slow-release formulation of grandlure, the synthetic pheromone of the boll weevil. *Journal of Economic Entomology* **67**, 44–46.

HARRIS, F. A., LLOYD, E. P. AND BAKER, D. N. (1966). Effects of the fall environment on the boll weevil in northeast Mississippi. *Journal of Economic Entomology* **59**,1327–1330.

HAYNES, J. W. (1963). *Chemical Sterility Agents As They Affect The Boll Weevil, Anthonomus grandis Boheman*. MS thesis, Mississippi State University, State College, MS, 38 pp.

HAYNES, J. W., MITLIN, N., SLOAN, C. E. AND DAWSON, J. R. (1972). Busulfan: development of improved methods of sterilizing boll weevils. *Journal of Economic Entomology* **66**, 619–622.

HAYNES, J. W., MITLIN, N., DAVICH, T. B., DAWSON, J. R., McGOVERN, W. L. AND McKIBBEN, G. H. (1977). Sterilization of boll weevil pupae with fractionated doses of gamma irradiation. *Entomologia experimentalis et applicata* **21**, 57–62.

HEDIN, P. A. (1976). Grandlure development. In *Proceedings, Conference on Boll Weevil Suppression, Management, and Elimination Technology, Memphis, TN, 13–15 February 1974. ARS-S-71*. pp. 31–34. US Department of Agriculture, Washington, DC.

HOUSE, V. S., ABLES, J. R., JONES, S. L. AND BULL, D. L. (1978). Diflubenzuron for control of the boll weevil in unisolated cotton fields. *Journal of Economic Entomology* **71**, 797–800.

HOUSE, V. S., ABLES, J. R., MORRISON, R. K. AND BULL, D. L. (1980). Effect of diflubenzuron formulations on the egg parasite *Trichogramma pretiosum*. *Southwestern Entomologist* **5**, 133–138.

HUNTER, W. D. AND COAD, B. R. (1923). The boll weevil problem. *US Department of Agriculture, Farmers' Bulletin* pp. 1329–1330.

IVY, E. E. (1944). Tests with DDT on the more important cotton insects. *Journal of*

Economic Entomology **37**, 142.

IVY, E. E. AND EWING, K. P. (1946). Benzene hexachloride to control cotton insects. *Journal of Economic Entomology* **39**, 38–41.

KEEVER, D. W., BRADLEY, J. R. JR AND GANYARD, M. C. (1977). Effects of diflubenzuron (Dimilin) on selected beneficial arthropods in cotton fields. *Environmental Entomology* **6**, 732–736.

KELLER, J. C., MITCHELL, E. B., McKIBBEN, G. AND DAVICH, T. B. (1964). A sex attractant for female boll weevils from males. *Journal of Economic Entomology* **57**, 609–610.

KLASSEN, W. AND EARLE, N. W. (1970). Permanent sterility produced in boll weevils with busulfan without reducing production of pheromone. *Journal of Economic Entomology* **63**, 1195–1198.

KLASSEN, W., NORLAND, J. F. AND BORKOVEC, A. B. (1968). Potential chemosterilants for boll weevils. *Journal of Economic Entomology* **61**, 401–407.

KNIPLING, E. F. (1955). Possibilities of insect control of eradication through the use of sexually sterile males. *Journal of Economic Entomology* **48**, 459–462.

KNIPLING, E. F. (1963). *An Appraisal of the Relative Merits of Insecticidal Control Directed at Reproducing and Diapausing Boll Weevils in Efforts to Develop Eradication Procedures.* Entomology Research Division, ARS, USDA. (Mimeographed, 22 pp).

KNIPLING, E. F. (1976). Report of the Technical Guidance Committee for the Pilot Boll Weevil Eradication Experiment. In *Proceedings of a Conference on Boll Weevil Suppression, Management, and Elimination Technology, Memphis, TN, 13–15 February 1974, ARS-S-71*, pp. 122–125. USDA, Washington, DC.

LEGGETT, J. E. AND CROSS, W. H. (1971). A new trap for capturing boll weevils. In *US Department of Agriculture, Cooperative Economic Insect Report 21*, pp. 773–774. USDA, Washington, DC.

LEGGETT, J. E., LLOYD, E. P. AND WITZ, J. A. (1981). Efficiency of infield traps in detecting and suppressing low population levels of boll weevils. *Environmental Entomology* **10**, 125–130.

LINDIG, O. H. AND MALONE, O. L. (1973). Oviposition of boll weevils fed diets containing germinated puree or cottonseed meats puree. *Journal of Economic Entomology* **66**, 566–567.

LINDIG, O. H., ROBERSON, J. AND WRIGHT, J. E. (1979). Evaluation of three larval and adult boll weevil diets. *Journal of Economic Entomology* **72**, 450–452.

LLOYD, E. P. AND MERKL, M. E. (1961). Seasonal occurrence of diapause in the boll weevil in Mississippi. *Journal of Economic Entomology* **54**, 1214–1218.

LLOYD, E. P., LASTER, M. L. AND MERKL, M. E. (1964). A field study of diapause, diapause control, and population dynamics of the boll weevil. *Journal of Economic Entomology* **57**, 433–438.

LLOYD, E. P., McCOY, J. R. AND HAYNES, J. W. (1976). Release of sterile boll weevils in the pilot boll weevil eradication experiment in 1972–73. In *Proceedings of a Conference on Boll Weevil Suppression, Management, and Elimination Technology, Memphis, TN, 13–15 February 1974. ARS-S-71*, pp. 95–102. USDA, Washington, DC.

LLOYD, E. P., TINGLE, F. C. AND GAST, R. T. (1967). Environmental stimuli inducing diapause in the boll weevil. *Journal of Economic Entomology* **60**, 99–102.

LLOYD, E. P., WOOD, R. H. AND MITCHELL, E. B. (1977). Boll weevil: suppression with TH-6040 applied in cottonseed oil as a foliar spray. *Journal of Economic Entomology* **70**, 442–444.

LLOYD, E. P., TINGLE, F. C., McCOY, J. R. AND DAVICH, T. B. (1966). The reproduction–diapause approach to population control of the boll weevil. *Journal of Economic Entomology* **59**, 813–816.

LLOYD, E. P., MERKL, M. E., TINGLE, F. C., SCOTT, W. P., HARDEE, D. D. AND DAVICH, T. B. (1972). Evaluation of male-baited traps for control of boll weevils following a reproduction–diapause program in Monroe County, Mississippi.

Journal of Economic Entomology **65**, 552–555.

LLOYD, E. P., MCKIBBEN, G. H., KNIPLING, E. F., WITZ, J. A., HARTSTACK, A. W., LEGGETT, J. E. AND LOCKWOOD, D. F. (1980). Mass trapping for detection, suppression, and integration with other suppression measures against the boll weevil. In *Proceedings, International Colloquium on the Management of Insect Pests with Semio-chemicals, 24–28 March, 1980, Gainesville, Florida* (E. R. Mitchell, Ed.), pp. 191–203. Plenum Press, New York and London.

MCCOY, J. R. AND WRIGHT, J. E. (1979). Evaluation of bisazir and penfluron as sterilants for the boll weevil. *Southwestern Entomologist* **4**, 209–215.

MCKIBBEN, G. H. (1972). A device for injecting grandlure into cigarette filters. *Journal of Economic Entomology* **65**, 1509–1510.

MCKIBBEN, G. H. (1974). An improved device for dispensing pheromone solutions. *Journal of Economic Entomology* **67**, 558.

MCKIBBEN, G. H. (1976). *Composition for Attracting the Cotton Boll Weevil.* US Patent 3,803,303.

MCKIBBEN, G. H. AND CROSS, W. H. (1984). Use of pheromone traps to estimate probability of zero populations of boll weevils. *Southwestern Entomologist* **9**, 371–374.

MCKIBBEN, G. H., HARDEE, D. D., DAVICH, T. B., GUELDNER, R. C. AND HEDIN, P. A. (1971). Slow-release formulations of grandlure, the synthetic pheromone of the boll weevil. *Journal of Economic Entomology* **64**, 317–319.

MCKIBBEN, G. H., HEDIN, P. A., MCGOVERN, W. L., WILSON, N. M. AND MITCHELL, E. B. (1977). A sex pheromone for male boll weevils from females. *Journal of Chemical Ecology* **3**, 331–335.

MCKIBBEN, G. H., JOHNSON, W. L., EDWARDS, R., KOTTER, E., KEARNEY, J. F., DAVICH, T. B., LLOYD, E. P. AND GANYARD, M. C. (1980). A polyester-wrapped cigarette filter for dispensing grandlure. *Journal of Economic Entomology* **73**, 250–251.

MALLY, F. W. (1901). The Mexican cotton-boll weevil. *US Department of Agriculture, Farmers' Bulletin No. 130.* pp. 11–12.

MESSENGER, L. K. (1965). Liquid concentrates for insect control. In *Proceedings Beltwide Cotton Production Mechanization Conference, 1965.* pp. 27–28. National Cotton Council of America, Memphis, TN.

MITCHELL, E. B. AND HARDEE, D. D. (1974). Infield traps: a new concept in survey and suppression of low populations of boll weevils. *Journal of Economic Entomology* **67**, 506–508.

MITCHELL, E. B., LLOYD, E. P., HARDEE, D. D., CROSS, W. H. AND DAVICH, T. B. (1976). Infield traps and insecticides for suppression and elimination of populations of boll weevils. *Journal of Economic Entomology* **69**, 83–88.

MULDER, R. AND GIJSWIJT, M. J. (1973). The laboratory evaluation of two promising new insecticides which interfere with cuticle deposition. *Pesticide Science* **4**, 737–745.

NEMEC, S. (1978). How a consultant looks at Dimilin. In *Dimilin: Breakthrough in Pest Control*, pp. 19–20. Agri-Fieldman and Consultant, Willoughby, OH.

NEMEC, S. AND ADKISSON, P. L. (1969). *Laboratory Tests of Insecticides for Boll Worm. Tobacco Budworm and Boll Weevil Control. Texas Agricultural Experiment Station Progress Report 2674*, 4 pp.

NRC/NAS (1981). *Cotton Boll Weevil: An Evaluation of USDA Programs, National Research Council/National Academy of Sciences*, pp. 1–130. National Academy Press, Washington, DC.

RABB, R. L. (1972). Principles and concepts of pest management. In *Proceedings of the National Insect Pest Management Workshop, Purdue University, West Lafayette, IN*, pp. 6–29. Indiana Cooperative Extension Service, West Lafayette, IN.

REINECKE, J. P., ROBERSON, J. L., VILLAVASO, E. J. AND LLOYD, E. P. (1986). Rearing, sterilizing, containerizing, and quality assessing of sterile boll weevils released in a large-scale field experiment in South Carolina. In *Proceedings of*

Cotton Insect Research and Control Conference. Las Vegas, NV. 8 January 1986, pp. 238–241. National Cotton Council of America, Memphis, TN.

RIDGWAY, R. L. AND LLOYD, E. P. (1983). Evaluation of cotton insect management in the United States. In *USDA Handbook No. 589: Cotton Insect Management with Special Reference to the Boll Weevil* (R. L. Ridgway, E. P. Lloyd and W. H. Cross, Eds), pp. 3–27. USDA, Washington, DC.

RIDGWAY, R. L., JONES, S. L., COPPEDGE, J. R. AND LINDQUIST, D. A. (1968). Systemic activity of 2-methyl-2-(methylthio)propionaldehyde *O*-(methyl-carbamoyl)oxime (UC-21149) in the cotton plant with special reference to the boll weevil. *Journal of Economic Entomology* **61**, 1705–1712.

RIDGWAY, R. L., DICKERSON, W. A., BRAZZEL, J. R., LEGGETT, J. E., LLOYD, E. P. AND PLANER, F. R. (1985). Boll weevil pheromone trap captures for treatment thresholds and population assessments. In *Proceedings of Beltwide Cotton Production Research Conferences, 6–11 January 1985, New Orleans, LA*, pp. 138–142. National Cotton Council of America, Memphis, TN.

ROUSSEL, J. S. AND CLOWER, D. F. (1955). *Resistance to the Chlorinated Hydrocarbon Insecticides in the Boll Weevil. Louisiana Agricultural Experiment Station, Circular 41*, 9 pp.

SCOTT, W. P., LLOYD, E. P., BRYSON, J. O. AND DAVICH, T. B. (1974). Trap plots for suppression of low density overwintered populations of boll weevils. *Journal of Economic Entomology* **67**, 281–283.

TAFT, H. M. AND HOPKINS, A. R. (1975). Boll weevils: field populations controlled by sterilizing emerging overwintered females with a TH-6040 sprayable bait. *Journal of Economic Entomology* **68**, 551–554.

TINGLE, F. C. AND LLOYD, E. P. (1969). Influence of temperature and diet on the attainment of firm diapause in the boll weevil. *Journal of Economic Entomology* **62**, 596–599.

TOWNSEND, C. H. T. (1895). Report on the Mexican cotton boll weevil in Texas (*Anthonomus grandis* Boh.). *Insect Life* **7**, 295–309.

TUMLINSON, J. H., HARDEE, D. D., GUELDNER, R. C., THOMPSON, A. C., HEDIN, P. A. AND MINYARD, J. P. (1969). Sex pheromone produced by the male boll weevil: isolation, identification, and synthesis. *Science* **166**, 1010–1012.

TUMLINSON, J. H., GUELDNER, R. C., HARDEE, D. D., THOMPSON, A. C., HEDIN, P. A. AND MINYARD, J. P. (1971). Identification and synthesis of the four compounds comprising the boll weevil sex attractant. *Journal of Organic Chemistry* **36**, 2616–2621.

USDA (1958). *The Boll Weevil Problem and Research and Facility Needs to Meet the Problem.* A report prepared at the request of the US Congress. 30 December. US Department of Agriculture, Washington, DC. 50 pp.

USDA (1981). *Biological Evaluation of Beltwide Boll Weevil/Cotton Insect Management Programs.* SEA-AR Staff Report. Attachment c. pp. 117–142.

VANDERZANT, E. S. AND DAVICH, T. B. (1958). Laboratory rearing of boll weevils: a satisfactory larval diet and oviposition studies. *Journal of Economic Entomology* **51**, 288–291.

VILLAVASO, E. J. (1981). Field competitiveness of sterile male boll weevils released in the Boll Weevil Eradication Trial, 1979. *Journal of Economic Entomology* **71**, 591–595.

VILLAVASO, E. J. (1982). Boll weevil (Coleoptera: Curculionidae): field competitiveness of diflubenzuron-fed, irradiated males – 1980, 1981. *Journal of Economic Entomology* **75**, 662–664.

VILLAVASO, E. J. (1984). New methods for sterilizing and releasing boll weevils. *Proceedings Mississippi Entomological Society* **3**, 28.

VILLAVASO, E. J. AND THOMPSON, M. J. (1984). Field competitiveness of boll weevils (Coleoptera: Curculionidae) sterilized by the feeding of chemosterilants followed by irradiation or fumigation. *Journal of Economic Entomology* **77**, 583–587.

VILLAVASO, E. J., EARLE, N. W. AND HOLLIER, D. D. (1977). Boll weevils: field and laboratory assessment of mating ability and sperm content after irradiation with or without diflubenzuron treatments. *Journal of Economic Entomology* **70**, 562–564.

VILLAVASO, E. J., NILAKHE, S. S. AND McGOVERN, W. L. (1979). Field competitiveness of sterile male boll weevils. *Journal of Georgia Entomological Society* **14**, 113–120.

VILLAVASO, E. J., ROBERSON, J. L., SEWARD, R. W. AND THOMPSON, M. J. (1986). Effectiveness of sterile boll weevils against naturally occurring populations in commercially grown cotton. *Journal of Economic Entomology* **79**, in press.

WALKER, J. K., JR., HIGHTOWER, B. G., HANNA, R. L. AND MARTIN, D. F. (1956). *Control of Boll Weevils Resistant to Chlorinated Hydrocarbons. Texas Agricultural Experiment Station, Progress Report 1902.*

WILKERSON, J. D., BIEVER, K. D., IGNOFFO, C. M., PONS, W. J., MORRISON, R. K. AND SEAY, R. S. (1978). Evauation of diflubenzuron formulations on selected parasitoids and predators. *Journal of the Georgia Entomological Society* **13**, 227–236.

WRIGHT, J. E. (1985). Mobility of boll weevils as influenced by holding temperature and length of storage. *Journal of Agricultural Entomology* **2**, 155–160.

WRIGHT, J. E., HAYNES, J. W., McCOY, J. R. AND DAWSON, J. R. (1979). Boll weevil: mating ability, sterility, and survival of irradiated and fumigated adults of different ages. *Southwestern Entomologist* **4**, 53–58.

WRIGHT, J. E., McCOY, J. R., DAWSON, J. R., ROBERSON, J. AND SIKOROWSKI, P. P. (1980a). Boll weevil sterility: effects of different combinations of antibiotics, fumigation and irradiation. *Southwestern Entomologist* **5**, 84–89.

WRIGHT, J. E., MOORE, R., McCOY, J., WIYGUL, G. AND HAYNES, J. (1980b). Comparison of three sterilization procedures on the quality of the male boll weevil. *Journal of Economic Entomology* **73**, 493–496.

2
Management of Temperate-Zone Deciduous Fruit Pests: Applied Behavioural Ecology

BERNARD D. ROITBERG* AND NELLO P. D. ANGERILLI**

*Behavioural Ecology Research Group and Centre for Pest Management, Department of Biological Sciences, Simon Fraser University, Burnaby, BC, Canada V5A 1S6 and **Entomology/Plant-Pathology Section, Agriculture Canada Research Station, Summerland, BC, Canada V0H 1Z0

Introduction

Behavioural ecology has been defined as 'the study of the survival value of behaviour' (Krebs and Davies, 1981) and as such draws upon several different disciplines including: (1) ecology, (2) ethology, (3) economics and (4) evolutionary theory. Pest management, on the other hand, has been defined as 'a program in which all available techniques are evaluated into a unified program to manage pest populations so that economic damage is avoided and adverse side effects on the environment are minimized' (FAO, 1967). Pest management is also multidisciplinary and derives from four major disciplines: (1) ecology, (2) economics, (3) sociology and (4) systems science.

 At first glance, the goals of the behavioural ecologist and pest manager may appear disparate even though they often employ similar analytical techniques (e.g. dynamic optimization models: Mangel and Clark, 1986;

Shoemaker, 1979). Indeed, both disciplines appear to be developing independently of one another. We argue, however, that such differences are primarily a function of narrow vision by those individuals engaged in their studies and, in fact, behavioural-ecological principles offer great utility when employed within pest-management programmes. Although we illustrate our points with examples from temperate deciduous fruit-crop systems, much of our argument is applicable also to other resource-management systems.

Behavioural ecology is unique in that it combines both 'how' and 'why' questions, or what Tinbergen (1963) called proximate and ultimate approaches to animal behaviour. This 'dual approach' (Charnov and Skinner, 1985) has proved to be a more powerful and a better predictive tool than the ethological method, which is more mechanistic in nature and lacks the evolutionary framework to explain why a particular behaviour exists. For example, while an ethogram of pheromone-mediated flight behaviour might describe mate search by a single species it may not provide a general explanation for the origin or maintenance of the phenomenon (e.g. Lorenz, 1981). By contrast, behavioural ecology begins with the premise that males will attempt to maximize their reproductive fitness through efficient mate finding, and then asks how such solutions might be achieved, given the constraints (primarily physiological and environmental: Crews and Moore, 1986) within which the animal operates (Thornhill and Alcock, 1983; Krebs and McCleery, 1984). In other words, we think of mate finding by males bound within particular physiological and environmental limits as a design problem (Alexander, 1982). Using this approach, we can predict animal behaviour and test those predictions through experimentation.

With regard to the behavioural-ecological approach, three further important points should be noted. First, behavioural ecologists neither expect nor test whether an animal is perfect (at one or all times) but rather, we use both the fitness maximization premise and our knowledge of constraints to make (preferably testable) predictions (cf. Janetos and Cole, 1981; Maynard Smith, 1978). Further, biological insight is required to formulate the appropriate questions relative to the animal and its environment. For example, we would be unlikely to ask why nocturnal mate-seeking moths do not (a) improve their vision or (b) become diurnal, and then seek mates through sight; rather, we ask how they might search, given the constraints under which they operate. Whereas a purely mechanistic approach to mate finding would require empirical confirmation of the behaviour in each and every species tested. a behavioural-ecological approach would provide a general, testable explanation for mate search for all species facing similar constraints (Pyke, 1984) and thus would provide an opportunity for behaviour exploitation for pest-management purposes.

Second, any biological design problem should also take into account how other individuals in the population behave (Maynard Smith, 1982). Under some conditions, the best solution to a problem may be very dependent upon the frequency with which it is employed within a population (e.g. sex ratio allocation: Charnov, 1982). Thus, populations may be composed of subgroups with radically different life histories. Such subgroups will be predicted to vary markedly in how they respond to particular stimuli (e.g. alternative mating strategies: Thornhill and Alcock, 1983). By contrast, analyses which are not based on evolution will probably not predict such variance, even though it is known to occur (Thornhill, 1981). Although expression of such variation can and does present problems for employment of some behaviour-based pest-management programmes, it may be anticipated usefully through an evolutionary approach (Brady, 1985).

Third, because behavioural ecology is explicitly derived from natural selection theory, it requires knowledge of population genetics and population dynamics (Wilson, 1975; Sibly and Smith, 1985). Thus, the study of behavioural ecology problems can lead to increased understanding of pest population structure and thereby improve pest-management decisions (Carey, 1983).

Finally, a behavioural-ecological approach can stimulate new ways of thinking about the problems that beset pest insects. The process of attempting to specify the relevant design features can result in reformulation of the original questions (Waddington, 1977), thus greatly enhancing the power of management decisions.

In this chapter we describe the behavioural ecology approach for pest management in temperate fruit orchards. Rather than attempting an exhaustive review of orchard-pest management (e.g. Hoyt and Burts, 1974; Croft and Hoyt, 1983), our aim is to establish a framework for discussion of such an approach.

The deciduous fruit system

In recent years it has become increasingly evident that pest-control tactics must be analysed in terms of the entire resource system, from both an economic and a biological perspective and as such, simply becomes one component of resource management (e.g. Norton and Holling, 1979). Thus, before we can examine current pest-management programmes for deciduous fruit, we must define the system itself. In the following brief description of the major components and their interactions we concentrate on apple crops because they have been extensively studied, although many of the principles apply to most temperate-zone fruit-production systems.

A temperature fruit orchard typically consists of regularly spaced rows of trees and usually includes a cover crop of grass between but not within the rows. Modern orchards utilize growth-controlling rootstocks to limit the ultimate size of the tree in order to maximize the density of trees per unit area and to eliminate the need for ladders or mechanical devices for pruning or picking.

Management intensity is often much greater in an orchard than in an annual crop. For example, the trees originate from seedlings grown in greenhouses where the desired fruit variety is grafted to the appropriate rootstalk. The rootstalk must be compatible with the soil type, irrigation regime, planting density and tree-training system to be utilized. Planting is a labour-, time- and machine-intensive operation which can take up to 2 days per hectare to complete.

Several other features of orchards distinguish them from annual crops, including:

1. The value of the crop per unit area is much higher (*Table 1*);
2. Initial installation costs are very high compared with those of annual crops;
3. The time to an economic return is much longer, e.g. for apples it begins 3–5 years after planting;
4. The consequences of management practices may be long term. For example, the effects of pruning errors may not be realized for several years after they are committed. In addition, selection of a particular variety/rootstalk combination normally must be endured for the life of the orchard even though it may not be economically optimal;
5. The economic consequences of management errors are especially large in high-density orchards where the fruit-to-shoot ratio is high.

The principal orchard fruit crops grown in temperate areas include apples, pears, peaches, nectarines, apricots, plums, and cherries. Gross returns per hectare for British Columbia, Canada are indicated in *Table 1*. Apples in North America represent the largest area planted for such orchard crops.

Table 1.   Relative values of various deciduous fruit crops and animal feed in the Okanagan Valley of British Columbia

Crop	$/Hectare
Apples	5388
Peaches	5585
Apricots	4930
Cherries (sweet)	4368
Cherries (sour)	4265
Hay	83

ORCHARD PESTS

The persistence of orchards in space and time allows for the accumulation of large pest complexes. *Table 2* lists those arthropod pests that can normally be found in orchards in various parts of North America, and specifies the crop attacked, our subjective designation of the pest as either key, secondary or sporadic and the damaging stage of the organism. Key pests are defined here as those pests around which control programmes are designed and which must be controlled in order to avoid economic damage; secondary pests are defined as those organisms that cause damage as a result of management practices aimed at key pests; sporadic pests are those organisms that occasionally become abundant, independent of current management practices, but which require control in order to prevent the occurrence of economic damage.

Several points emerge from the Table. First, the designation of key pest is area-dependent. For example, in western North America, codling moth is the key orchard pest whereas lygus bugs are rare. By contrast, in Quebec apple orchards (Boivin and Stewart, 1983) codling moth is sporadic (Anonymous, 1986) but lygus bugs are a key pest. Second, there is no apparent relationship between the type of damage caused (i.e. direct or indirect) and the designation of the organism as key, secondary or sporadic (*Table 3*). Third, of the 50 organisms listed in the Table, only two (one-spotted stink bug and the western flower thrips) are exclusively damaging in the adult stage, whereas 23 other species are exclusively damaging in the immature stage; however, in 14 of those 23 species, the damage potential and population levels are assessed by monitoring adults. Such assessment schemes render imprecise the prediction of potential damage and can influence a grower's propensity to apply controls (Hall, 1983).

Economics of orchard systems

As pointed out by Headley (1982), the goal of the orchardist should be to maximize returns on investment, whether that investment be directed toward direct yield enhancement (e.g. fertilizer treatment) or reduction of pest-related damage. An important difference between yield-enhancement investment and damage-reduction investment is that, in the former, maximum crop yield increases with investment whereas, in the latter, maximum yield is fixed and increased investment leads to yields which approach that fixed maximum, often in a decelerating fashion; in fact, in some instances, pesticide side-effects on crop plants may cause greater yield reductions than would the pest population being controlled (Toscano *et al.*, 1982a, b). In other words, in pest-damage control one attempts to minimize losses as long as it is economically feasible to do so. This concept is particularly important in orchard crops where individual units of fruit may be high in value when blemish-free and low or nil when blemished (Pimentel *et al.*, 1977). Thus, damage by direct fruit pests such as apple maggot flies, as a function of pest density, may best be described by a step function where each item of fruit may be valued by a small, finite set of grades, e.g. top grade, low grade, cull. Under such conditions, damage caused by small numbers of direct-fruit pests may be considerable and orchardists will tend to behave in a risk-adverse manner, i.e. they are likely to be more sensitive to the variance in pest densities than to average yearly pest levels (Webster, 1977; Headley, 1985). In fact, damage due to orchard pests may be substantial: Canadian entomologists estimated that, in the absence of controls, losses in apple-crop yields to pests would vary between 50% and 100% per year (Stemeroff and George, 1983). Given the kinds and sources of information available to growers regarding potential pest threat (Farnsworth and Moffitt, 1984), those orchardists are likely to invest heavily in insurance measures (i.e. application of chemical biocides based upon calendar dates) that reduce the chance of

Table 2. Some characteristics of arthropod pests associated with temperate deciduous fruit orchards in North America

Pest	Genus and species	Crop (in order of apparent preference)	Importance†	Damage*
1. American plum borer	Euzophera semifuneralis (Walker)	apples, apricots, peaches, pears, plums, cherries	2	indirect
2. Apple aphid	Aphis pomi DeGeer	apples, pears	2	indirect
3. Apple curculio	Tachypterellus quadrigibbus (Say)	apples	3	direct
4. Apple grain aphid	Rhopalosiphum fitchii (Sanderson)	apples, pears, plums	3	indirect
5. Apple maggot	Rhagoletis pomonella (Walsh)	apples, pears, cherries, plums, apricots	1	direct
6. Apple mealybug	Phenacoccus aceris (Signoret)	apples, pears, cherries, peaches, apricots, prunes	1	direct, indirect
7. Apple rust mite	Aculus schlechtendali (Nalepa)	apple	3	indirect
8. Black cherry aphid	Myzus cerasi (Fabricius)	cherries	3	indirect
9. Bruce spanworm	Operophtera bruceata (Hulst.)	apples, pears, cherries, plums, peaches, apricots	2	direct
10. California pear sawfly	Pristophora abbreviata (Hartig)	pear	3	indirect
11. Cherry fruit flies	Rhagoletis cingulata (Loew)	cherries, pears, plums	1	direct
12. Cherry fruitworm	Grapholita packardi Zeller	apples, plums, cherries, peaches	3	direct, indirect
13. Climbing cutworms		apples, peaches, pears, plums	3	direct, indirect
14. Codling moth	Laspeyresia pomonella (Linnaeus)	apple, pear, cherry, apricot, plum, peach	1	direct
15. Dock sawfly	Ametastegia glabrata (Fallon)	apple	3	direct
16. European apple sawfly	Hoplocampa testudinea Klug.	apple	1	direct
17. European fruit scale	Quadraspidiotus ostraeformis (Curtis)	apples	2	direct
18. European leafroller	Archips rosanus (Linnaeus)	apple, pear, cherry	2	direct
19. European red mite	Panonychus ulmi (Koch)	apples, pears, peaches	1	indirect
20. Eye-spotted budmoth	Spilonota ocellana (Denis & Schiffermuller)	apple	3	direct
21. Fall webworm	Hyphantria cunea (Drury)	apples, pears, cherries, peaches, apricots, prunes	3	indirect
22. Fruit-tree leafroller	Archips argyrospilus (Walker)	apple, pear, cherry	1	direct
23. Fruitworms	Lithophane georgii	apples, pears	3	direct
24. Green peach aphid	Myzus persicae (Sulzer)	peaches, apricots, plums, cherries	3	direct, indirect

No.	Common name	Scientific name	Hosts	Key†	Damage*
25.	Lesser appleworm	*Grapholita prunivora* (Walsh)	apples, plums, cherries	3	direct
26.	Lesser peach-tree borer	*Synyanthedon pictipes* (Grote & Robinson)	peaches, cherries, plums	1	indirect
27.	McDaniel spider mite	*Tetranychus mcdanieli* McGregor	apple, pear	2	indirect
28.	Mullein bug	*Campylomma verbasci* (Meyer)	apple	2	direct
29.	Oblique-banded leafroller	*Choristoneura rosaceana* (Harris)	apples, pears, cherries	1	direct
30.	One-spotted stink bug	*Euschistus variolarius* (Palisot de Beauvois)	pears	3	direct
31.	Oriental fruit moth	*Grapholitha molesta* (Busck)	peaches, apples, apricots, plums, cherries, pears	1	direct, indirect
32.	Oystershell scale	*Lepidosaphes ulmi* (Linnaeus)	apple, pears, peaches, apricots, cherries	3	direct
33.	Peach-tree borer	*Synanthedon exitiosa* (Say)	peaches, apricots, cherries, nectarines, plum	1	indirect
34.	Pear psylla	*Psylla pyricola* Foerster	pears	1	indirect, direct
35.	Pear rust mite	*Epitrimerus pyri* (Nalepa)	pears	1	direct
36.	Pear sawfly	*Caliroa cerasi* (Linnaeus)	pears, cherries	2	indirect
37.	Pear-leaf blister mite	*Phytoptus pyri* Pagenstecher	apples, pears	3	direct, indirect
38.	Plum curculio	*Conotrachelus nenuphar* (Herbst)	plums, cherries, peaches, apricots, apple	1	direct
39.	Red-banded leafroller	*Argyrotaenia velutinana* (Walker)	apples, plums, prunes, peaches, cherries	2	direct
40.	Rosy apple aphid	*Dysaphis plantaginea* (Passerini)	apples	3	indirect, direct
41.	San José scale	*Quadraspidiotus perniciosus* (Comstock)	apples, pears, plums, apricots, peaches, cherries	1	direct, indirect
42.	Spotted tentiform leafminer	*Phyllonorycter blancardella* (Fabricius)	apples, plums	2	indirect
43.	Tarnished plant bug	*Lygus lineolaris* (Palisot de Beauvois)	apple	3	direct
44.	Tentiform leafminer	*Phyllonorycter elmaella* Doganlar	apple	3	indirect
45.	Three-lined leafroller	*Pandemis* spp.	apples, pears, cherries	2	direct
46.	Tufted apple budmoth	*Platynota idaeusalis*	apple	3	direct
47.	Two-spotted spider mite	*Tetranychus urticae* (Koch)	apples, pears, peaches, apricots	2	indirect
48.	Western flower thrips	*Frankliniella occidentalis* (Pergande)	apple	3	direct
49.	White apple leafhopper	*Typhlocyba pomaria* McAtee	apples, pears, prunes, peaches, cherries	2	indirect
50.	Woolly apple aphid	*Eriosoma lanigerum* (Hausmann)	apples, pears	3	indirect

* Direct damage refers to direct fruit damage.

† 1: key; 2: secondary; 3: sporadic

Table 2. *Cont'd*

Pest*	Damaging stage	Damage to	Control options	Susceptible stages	Insect life stage, sex and method of monitoring for management
1.	larvae	trunk cambium	chemical	larva	larva/visual inspection
2.	nymph, adult	foliage, fruit (honeydew)	biological, chemical	nymph, adult	all/visual
3.	adult, larva	fruit	chemical	adult	adult/visual/limb tap
4.	nymph, adult	foliage, fruit (honeydew)	biological, chemical	egg, nymph, adult	all/visual
5.	larva	fruit	chemical	larva	adult female/sticky trap
6.	nymph, adult	fruit (honeydew – apples), tree	chemical	nymph	none
7.	nymph, adult	foliage	biological, chemical	nymph, adult	all/leaf brushing
8.	nymph, adult	foliage, growth distortion	biological, chemical	egg, nymph, adult	all/visual
9.	larva	fruit, foliage	chemical	larva	larva/limb tap
10.	larva	foliage	chemical	larva	none
11.	larva	fruit	chemical	adult, larva	adult/sticky trap
12.	larva	fruit, foliage	chemical	larva	none
13.	larva	fruit buds	chemical	larva	none
14.	larva	fruit	chemical, SIR, pheromone disruption	larva	adult/male/pheromone trap
15.	larva	fruit	cultural	larva	none
16.	larva	fruit	chemical, cultural	adult	adult/both/visual trap
17.	nymphs, adult female	fruit, cambium (minor)	chemical, physical	overwintering nymph, crawler	all/visual
18.	larva	fruit, foliage	chemical	larva	adult/male/pheromone trap or fruit damage at harvest
19.	nymph, adult	foliage	biological, physical, chemical	egg, nymph, adult	all/leaf brushing
20.	larva	fruit, foliage	chemical	larva	larva/visual
21.	larva	foliage	chemical, cultural	larva	larva/visual
22.	larva	fruit, foliage	chemical	larva	larva/limb tap, adult/male/ pheromone, harvest damage
23.	larva	fruit, foliage	chemical	larva	larva/limb tap
24.	nymph, adult	foliage, fruit (honeydew)	chemical	egg, nymph, adult	all/visual
25.	larva	fruit	chemical	larva	adult/male/pheromone, harvest damage

No.					
26.	larva	trunk, scaffold limbs, branches	chemical	larva	adult/male/pheromone, larva/visual
27.	nymph, adult	foliage	biological, chemical	egg, nymph, adult	all/leaf brush
28.	nymph	fruit	chemical, cultural	nymph	nymph/limb tap
29.	larva	fruit, foliage	chemical	larva	adult/male/pheromone, harvest damage
30.	adult	fruit	cultural	nymph	adult/limb tap
31.	larva	fruit, twigs	chemical	larva	adult/male/pheromone, larva/visual
32.	nymph, adult	fruit	chemical	crawler, egg	all/visual
33.	larva	trunk cambium	chemical	larva	adult/male/pheromone
34.	nymph, adult	foliage, fruit (honeydew)	chemical, behavioural	adult, nymph	all/limb tap/leaf brush
35.	nymph, adult	fruit	chemical	nymph, adult	all/visual
36.	larva	foliage	chemical	larva	none
37.	nymph, adult	fruit, foliage	chemical	nymph, adult	none
38.	larva, adult	fruit	chemical	adult	adult/visual/limb tap
39.	larva	fruit, foliage	chemical	larva	egg/visual, larva/visual, adult/male/pheromone
40.	nymph, adult	foliage, fruit (distortion)	chemical	egg, nymph, adult	none
41.	nymph, adult	fruit, cambium	physical, chemical	nymph, male	all/harvest sample, adult/male/pheromone
42.	larva	foliage	chemical	larva	larva/visual, adult/male/pheromone
43.	nymphs, adult	buds, flowers, terminal growth	chemical	adult	adult/visual, adult/sticky trap
44.	larva	foliage	chemical	larva	larva/visual, adult/male/pheromone
45.	larva	fruit, foliage	chemical	larva	adult/male/pheromone, harvest damage
46.	larva	fruit, foliage	chemical	larva	larva/visual, adult/male/pheromone
47.	nymph, adult	foliage	biological, chemical	egg, nymph, adult	all/leaf brush
48.	adult (egg)	fruit	chemical	nymph	harvest damage
49.	nymphs, adults	foliage, fruit (honeydew)	chemical	nymph	nymph/leaf count
50.	nymph, adult	twigs, branches, fruit (honeydew)	biological, chemical	nymph, adult	none

*For names of pests see corresponding numbers in left-hand column of first part of Table 2 (pages 142–143).

Table 3. Chi-square comparison of damage and importance categories from Table 2.

	Damage type	
Importance	Direct	Indirect
Key	10	7
Expected	9.44	7.56
Secondary	6	7
Expected	7.22	5.78
Sporadic	14	10
Expected	13.33	10.67

Chi-square = 0.6139 NS: d.f. = 2

heavy losses due to pests (Hall, 1983). There also appears to be a relationship between the size of the investment a grower has made in his crop and his tendency to use 'insurance sprays' as opposed to a management programme that implements control actions based on pest numbers (N.P.D. Angerilli, personal observation). Such behaviour has led to extremely heavy use of chemical biocides in orchard systems, with apple orchards receiving the highest rate of insecticide use per hectare of any commercial crop (Pimentel et al., 1977; Croft, 1983): for example, the average cost/ha for chemical insecticides on a calendar basis (not including application costs) in apple orchards in the Okanagan Valley in British Columbia, Canada during 1983 was $353.40 (Anonymous, 1983).

Whether employment of chemical control is cost effective is not clear, as there is controversy over (1) how production costs should be evaluated (Low and Kemp, 1977; Hoffman and Gustafson, 1983) and (2) how and which indirect costs should be evaluated (Pimentel et al., 1980). Nevertheless, direct dollar returns from application of chemical pesticides are generally greater than direct dollar costs of such treatments (Pimentel et al., 1980). Whether, however, application of simple calendar-based chemical control programmes is an optimal management strategy, is a quite different question. In fact, management programmes that rely on alternative methods of control and application of chemical biocides at lower than calendar rates, appear to reduce pest-control costs greatly, with no apparent change in effectiveness. For example, in British Columbia apple orchards, control of apple rust mite outbreaks through applications of one-half of the label rate of acaricide will control the outbreak without reducing the number of predator mites and thereby will prevent subsequent outbreaks of other phytophagous mites (N.P.D. Angerilli, unpublished data.)

As maximum yields are unlikely to increase with different types or intensities of control efforts, in order to assess the relative value of different control measures we could compare the costs required to attain a particular yield (notwithstanding the positive yield effects of aldicarb: Marshall, 1985). Unfortunately, however, as pointed out by Whalon and Croft (1983), most economic analyses of pest-managment programmes have been incomplete in that their various components have been evaluated in isolation from the rest of the system. Furthermore, Norton (1985) has pointed out that very little information is available for comparison of the economics of different management tactics. Finally, in their enthusiasm for pest-management research, few pest managers have considered the relationship between production costs, yields, market demands (including the timing of crop delivery and its storage potential) and market value. It may frequently be the case that pest management will lead to consumer savings and reduced environmental damage but without increasing the income of the fruit producer, although increased benefits to the latter group is generally the rationale under which these programmes are sold (Taylor, 1980). Frequently, only when some but not all individuals or aggregates employ cost-reducing methodologies/ programmes will those individuals achieve increased net benefits (Frisbie and Adkisson, 1985). Thus, decisions to implement particular pest-management programmes should be based upon clearly defined budget priorities and target groups (Taylor *et al.*, 1983).

Both direct and indirect costs for pest-damage reduction, in temperate orchards, can be substantial (i.e. several hundreds of dollars/hectare even when pest-management schemes are employed: e.g. Thompson, How and White, 1980; Whalon and Croft, 1983). This means that alternative methods of pest-damage reduction that might be deemed too expensive in other resource systems (e.g. rangelands) might be cost effective in orchard systems. Finally, because orchards tend to accumulate complexes of pests, alternative damage-reduction tactics (for a given pest) that do not destabilize management schemes for other pests, may prove cost effective when the system is analysed as a whole (e.g. Headley, 1982). In short, methodologies such as behaviour manipulation (discussed below) that may be expensive to initiate may prove to be cost effective in resource systems where damage thresholds are low and potential losses due to pests are high.

ALTERNATIVES AND ADJUNCTS TO CHEMICAL CONTROLS

Given the potentially high environmental, sociological and economic costs of applying chemical biocides for control of orchard pests, alternative methodologies have been sought; these include biological, physical and cultural controls. A fourth potentially useful category involves behaviour exploitation, which we define as 'any methodology which utilizes some aspect(s) of pest behaviour to reduce pest damage'. Some of these categories necessarily overlap: for example reflective mulches (physical) interfere with aphid host selection (behavioural) (Adlerz and Everett, 1968).

Behavioural exploitation, like any other control methodology, has costs and benefits associated with it, particularly in orchard-pest complexes, where actions against one pest may influence damage levels of others (Gruys, 1982; Pimentel *et al.*, 1984; Easterbrook *et al.*, 1985). On the one hand, such methodologies are highly selective, environmentally sound and in some instances have been more effective at reducing crop damage than chemical control (Proverbs, Newton and Campbell, 1982); further, such approaches usually eliminate the problems of pest resistance and the development of secondary pests. On the other hand, employment of behaviour-exploitation programmes requires great understanding of pest biology and behaviour and may be technically and logistically difficult to implement. In addition, unlike the effects of chemical biocides, the efficacy of behaviour exploitation may be difficult to demonstrate, thus leading to increased costs and other impediments to registration. On balance, however, behaviour exploitation is a desirable alternative to chemical control and is likely to be of increasing value in pest-management programmes.

Behaviour exploitation has been employed, or suggested for employment, in several areas of pest management including: (1) population monitoring for population estimation and optimization of timing of control application; (2) mating disruption; (3) oviposition disruption; (4) biological control. Within these categories are several approaches: (1) ethological; (2) physiological, and (3) behavioural-ecological. As pointed out in the Introduction, the first two approaches are useful but the behavioural-ecological is the most powerful and offers the greatest potential for fully understanding insect behaviour. The utility of employing the behavioural-ecological approach to behaviour exploitation in pest management is examined in the remainder of this chapter.

The behavioural-ecological approach to behaviour exploitation in pest management

EXPLOITATION OF MATING BEHAVIOUR

An understanding of mating systems, from an evolutionary perspective, is essential if we are to exploit them fully for pest-management purposes. Evolutionary forces are likely to act in different ways upon the two sexes: thus, potentially mating males and females are likely to have very different interests (Thornhill and Alcock, 1983). This arises from the differential investment required for gamete production: sperm is cheap to produce, compared with eggs. Thus, the reproductive success of males is less likely to be gamete-limited than that of females, despite Bateman's principle (Bateman, 1948; Nakatsuru and Kramer, 1982). Males can therefore increase their reproductive potential by increasing the number of females that they inseminate (i.e. by intrasexual selection). The intensity of male competition for female insemination is exemplified by the destruction of conspecific sperm by male *Calopteryx* (damsel fly) (Waage, 1979a). Conversely, females, when egg-limited, can increase their reproductive potential only by increasing the quality of offspring (i.e. by choice of mate).

For many species of insects, sex pheromones facilitate the mating process. Such compounds are thought to represent a mechanism by which search time and recognition time (within species and between sexes) can be greatly reduced. It is important to note that pheromones do not cause behaviour, but rather, insects use these 'pieces of information' to modify their behaviour in a manner appropriate to the context in which they are received. Moreover, as noted above, males and females may exploit these compounds in different ways. Thus it follows that exploitation of mating behaviour *per se*, or through use of pheromone technology, for the purpose of fruit-pest management, must take these differences into account.

Exploitation of the mating behaviour of orchard fruit pests has been explored in four principal ways:

1. Population monitoring using sex pheromone traps;
2. Male removal by mass trapping;
3. Mating disruption;
4. Autocidal control.

These are discussed below in an examination of the exploitation of mating behaviour of the codling moth, a key orchard pest throughout most of the world.

Population monitoring with sex pheromone traps

A sex pheromone ((E,E)-8,10-dodecadien-1-ol (EEOH)) produced by the female codling moth was identified by Roelofs *et al.* (1971) and other components have subsequently been reported (e.g. McDonough *et al.*, 1972). Since then, synthesized EEOH has been utilized in various traps to detect and monitor the presence of males.

The use of traps baited with synthetic pheromones to monitor codling moth presence, absence and potential for damage has been accomplished by 'calibrating' trapping results with end-of-season crop damage (Riedl and Croft, 1974; Vakenti and Madsen, 1976). Here, it is assumed that there is some constant numerical relationship between the number of males searching for females and the number of females in the local population. The trap type, trap placement, number of traps per unit area and male-moth-capture treatment threshold normally need to be determined for each crop-growing area. Not surprisingly, the method sometimes fails and extensive economic damage can occur, even when the treatment thresholds are strictly adhered to (N.P.D. Angerilli, personal observation). Several different explanations have been proposed, but most include the lack of understanding of the basic behaviour of the animal: for example, it is not known why male codling moths sometimes disperse over long distances nor why they are variably responsive to EEOH over time (Howell and Clift, 1974; Proverbs, Logan and Newton, 1975). Codling moth females are generally believed to be short-range dispersers (Rothschild, 1982; White, Hutt and Butt, 1973) but this is very difficult to quantify and it is, therefore, possible that on some occasions the number of males captured in traps is not at all related to the number of females in the area, because of differential male and female immigration and emigration.

Several other uncertainties are associated with sex-pheromone traps. First, we do not know how many males approach traps but do not land. Second, we know little of the importance of habitat and/or mate-associated visual cues for either males or females, beyond the fact that different trap designs and their placement capture different numbers of males (e.g. Madsen and Vakenti, 1973; Howell and Clift, 1974). Thus, although the numbers of males captured in EEOH-baited traps are similar to those in traps baited with live virgin females (Batiste, Olson and Berlowitz, 1973), such results do not explain codling moth pheromone responses beyond the 'arousal, oriented flight and landing' phases (Rothschild, 1982). In other words, pheromone-baited traps work because males blunder into them in their search for females.

A behavioural-ecological approach could reduce a number of areas of uncertainty with regard to monitoring systems based on sex pheromones. First, an understanding of female sex-pheromone calling behaviour, from the perspective of a female's future reproductive success, needs to be explored. For example, how do timing, temperature and photoperiod contribute to male arrival rates and how do such rates affect offspring production? In fact, high arrival rates of males may lead to lower female survival through courtship-enhanced mortality (Hathaway, Butt and Lydin, 1970), while very low arrival rates may lead to low insemination rates and thus low fecundity. Thus, for females there will be an optimal male arrival rate which is likely to be different from that for males; furthermore, such optimal rates should vary as a function of female age and mating status. When designing traps for monitoring as opposed to mass trapping, for example, it is perhaps inadvisable to attempt to mimic females, as their goals with regard to male arrivals probably differ from those of pest managers; however, these considerations are rarely borne in mind when traps are designed.

Second, it has been noted that male codling moths are variably responsive to females as a function of female mating status: i.e. virgin vs. recently mated vs sperm-depleted females (Proverbs, Logan and Carty, 1973). It is difficult to determine whether such responses are a function of differential male response, differential female pheromone display, or both: however, bearing in mind the mate value mentioned previously, the latter is most likely (e.g. virgin females are reproductively more valuable to males than recently mated ones.)

Third, because males and females require different resources to achieve reproductive success (i.e. males require females whereas females require sperm and oviposition sites) their foraging patterns might be very different. This leads to difficulties in interpreting male catches as indices of female population size.

Finally, male mating success will be a function of tactics, both olfactory and and non-olfactory, that they employ to locate females. These, in turn depend upon habitat structure, resource (females) distribution and intensity of conspecific competition which can be quantitatively analysed using the foraging models already developed for host-seeking parasitoids (*see* Biological control, pages 155–158) and mate-seeking males of other species (Thornhill and Alcock, 1983). Through such analyses it should be possible to predict male search tactics (Willis and Baker, 1984) and therefore trap efficiency.

Mass trapping of males

Some of the reasons for the variable success of male removal through mass trapping have already been outlined. Proverbs, Logan and Newton (1975), using a synthethic pheromone (EEOH) at 34 traps/ha in a lightly infested orchard did not prevent the codling moth population from increasing over a 3-year period. They attributed failure to the polygamous nature of codling moth males and the low efficiency of the traps. With a better understanding of male mate-foraging behaviour, trap efficiency could be improved through the deployment of a design based upon an evolutionary perspective.

Mating disruption

Numerous experiments have been conducted to disrupt codling moth mating by permeating the air with synthetic pheromone (*see* review by Rothschild, 1982) or with antipheromone (e.g. Hathaway, Moffitt and George, 1985). Although successful mating disruption (as measured by the number of males captured in traps baited with virgin females or EEOH) is regularly achieved, crop protection is less frequent (e.g. Moffitt, Westigard and Hathaway, 1979, Rothschild, 1982; Hathaway, Moffitt and George, 1985). Explanations for this inconsistency vary from immigration of mated females into treated sites, to mechanical and meteorological effects (Rothschild, 1982). Again, knowledge of the evolutionary forces underlying codling moth mating systems is necessary to confirm these explanations and to develop a consistently effective method.

Mass release of sterile males

In contrast to mating disruption, sterile insect release (SIR) has been very successful at eliminating crop damage by codling moth (Proverbs, Newton and Campbell, 1982). By inundating wild populations with radiation-sterilized males (at a ratio of 40:1, sterile:wild) it is possible to reduce populations to near-extinction. The mechanisms through which the technique operates are varied but explicable from a behavioural-ecological perspective. As discussed earlier, males have a strong propensity to mate with as many females as possible and various mechanisms have evolved to enhance this process; in addition, female behavioural mechanisms exist that lead to reproductive maximization.

First, irradiated males produce sterile sperm that elicit normal oviposition of sterile eggs, which do not hatch. However, radiation dose is important in determining the competitiveness of sperm, and high radiation doses (40 krad ^{60}Co or more) lead to the development of sperm that are sufficiently abnormal not to elicit normal female behaviour. Proverbs, Logan and Carty (1973) found that wild females mated with such males regained much of their original virginal attractiveness whereas wild females mated with males irradiated with 25 krad, or with unirradiated males, were never as attractive as virgins. The same authors also report that sperm from males irradiated with 40 krad fails to induce normal oviposition and it seems likely that this lack of 'ovipositional satiation' is one explanation for the resurgence of virginal attractiveness by mated females. Both of these responses are predicted by an evolutionary analysis, in that females would be expected to judge male quality on both a pre- and post-insemination basis.

Second, females appear to use sperm from the last male with which they mated (sperm precedence: Parker, 1970). Proverbs, Logan and Carty (1973) showed that in various orders of sequential matings of single females and males irradiated with 25 krad or unirradiated, the last mating determined the number of viable eggs laid. Such behaviour is predicted by evolutionary theory, in that it allows already-mated females to increase the quality of their offspring by utilizing only sperm from males of higher quality than previous mates. The implication is that in populations with a high sterile: fertile male ratio, it is likely that only sterile eggs (contribution from last mating) will be produced.

Third, males will court and attempt to copulate with all calling females. It has been determined that by releasing only sterile males, as opposed to sterile males and females, the rate of male dispersal is much higher than in mixed releases in which released males mate with nearby sterile females rather than searching for wild females (Proverbs, Logan and Carty, 1973).

Fourth, it is probable that limited female dispersal leads to the development of clumped populations that serve as 'foci' for the activities of males and lead to even further exaggeration of the operational sex ratio. This clumping enhances the effect of the mechanisms described above.

Finally, SIR requires the production of large numbers of individuals for sterilization and release. Laboratory rearing can impart strong selection to aspects of mating (e.g. mate selection under enhanced crowding) which may differ greatly from the natural pattern (*see* Biological control, pages 155–158) and can lead to 'unexpected' performance in steriles (Jackson and Lee, 1985).

MANIPULATION OF OVIPOSITION BEHAVIOUR

Because it is the larvae of several key orchard pests (e.g. apple maggot, codling moth, European apple sawfly) that are the fruit-damaging stages, any methodology that reduces rates of deposition of eggs/larvae in or on fruit should be of interest to orchard-pest managers. One of the most promising of such methodologies employs marking or oviposition-deterring pheromones to reduce the propensity of females to oviposit in such marked fruit (Prokopy, 1981a). Marking pheromones are deposited by females at or after oviposition and are known to be produced by at least one key orchard pest, the apple maggot (Prokopy, 1972) and possibly several others (e.g. the apple sawfly: Roitberg and Prokopy, 1984a), although not necessarily by all orchard pests (e.g. not the codling moth: Roitberg and Prokopy, 1983a). Such compounds are thought to be relatively non-toxic and highly selective, so their incorporation into current management programmes would probably be non-disruptive.

To incorporate marking pheromones into a management programme, first the pheromonal components would be identified, then synthesized and finally applied to orchard fruit at different concentrations and in different patterns until a particular level of performance is achieved. While such an approach may be successful (e.g. Katsoyannos and Boller, 1976, 1980; Klijnstra, 1985), it is also likely to be fraught with difficulties for several reasons, including (1) insect 'habituation' to marking pheromone, (2) insect change in response to pheromone as a function of oviposition deprivation (Roitberg and Prokopy, 1983b) and (3) abiotic effects (Prokopy, 1981b). These difficulties, if analysed from the non-evolutionary or phenomenological perspective, would be considered to be physiological or ethological constraints to be tolerated or overcome in a management programme (Prokopy, 1981b). Using this approach, one would attempt to overcome such problems by solving empirically how pest responses to marking pheromone change in relation to the factors described above. However, with this phenomenological approach, there is no *a priori* reason for predicting how or why pests respond to marking pheromones under different ecological circumstances, and such responses will probably be determined only through careful observation.

Such an approach can be contrasted with that which would be employed from a behavioural-ecological perspective. Starting with a premise of fitness maximization, which assumes that a host-foraging insect will desposit its eggs in such a fashion as to maximize the number of offspring produced (Charnov and Skinner, 1985), one asks whether there are any situations in which the reproductive fitness of a female is likely to increase if she were to deposit a second egg in an already-infested (i.e. pheromone-marked) host, where offspring survival is some inverse funcion of larval density (Iwasa, Suzuki and Matsuda, 1984; Parker and Courtney, 1984; Smith and Lessells, 1985). Analyses based upon life history parameters, pest–host encounter rates and larval competition coefficients lead one to the conclusion that, under some circumstances (in this case, low availability of unmarked hosts), females would oviposit in such marked fruit (Roitberg and Prokopy, 1986; Mangel, 1986). In other words, the behavioural-ecological approach can predict those ecological situations in which deterrence based on marking pheromones will fail. For example, M. Mangel (unpublished work) used the reproductive maximization premise to predict the shape and magnitude of an empirically derived marked-host acceptance (i.e. oviposition) vs. host-deprivation-time curve (Roitberg and Prokopy, 1983b). Using such theoretical models we may soon design management programmes that deal directly with problems of pest 'non-response' to marking pheromone. It is true that deriving the rigorous predictions from the behavioural-ecological models requires extensive ecological, physiological and behavioural data whereas the phenomonological approach would, initially, be far less labour intensive. However, it is likely that the ability to understand rather than to demonstrate a phenomenon will ultimately produce far greater benefits (Pyke, 1984). The following section describes the use of synthetic marking pheromone in an orchard-pest management programme.

MODELLING APPLE-MAGGOT MANAGEMENT WITH MARKING PHEROMONE

Will an understanding of the important elements in pest response to marking pheromone and the inevitability that such compounds will soon be identified and synthesized (Boller and Hurter, 1985) make it feasible to incorporate synthetic marking pheromone into management programmes? Roitberg (1985a) recently attempted to answer this question for apple maggot in commercial orchards in Northeastern United States. As noted previously, the response by females to pheromone-marked fruit is likely to depend upon their age, history of oviposition events, and finally the availability of fruit resources that they 'discern' to be present. Roitberg (1985b) suggested that the latter can be elucidated by evaluating the following components of fly search:

1. Host (both marked and unmarked) distribution in space and time;
2. Fly movement patterns;
3. Fly perceptive ability;
4. Fly decision rules for search persistence within trees.

These four components can be employed to predict the frequency and sequence of encounters between flies and hosts, unprotected fruits and those fruits that pest managers might wish to treat with marking pheromone. In an evaluation of these search components, Roitberg and colleagues have observed the following: (1) flies adjust the amount of time that they spend searching within individual trees as a function of rate and sequence of encounter with marked and unmarked fruit (Roitberg and Prokopy, 1984b) and distance to nearest neighbour trees (Roitberg and Prokopy, 1982); (2) fly visual acuity varies as a function of distance and angle to host fruit (Roitberg, 1985b); (3) fly movement patterns can be readily predicted (Roitberg, Elkinton and Prokopy, 1982), and (4) fly acceptance of marked hosts as a function of host deprivation can be described by a convex curve (Roitberg and Prokopy, 1983b). Together, these factors suggest that a fly searches for hosts within trees, and that the more time a fly goes without finding a suitable host, then the more likely it is to abandon that tree. However, continued host deprivation increases the fly's willingness to accept a pheromone-marked host for oviposition. From this we can predict that application of synthetic marking pheromone to apple trees would lead to extensive movement of flies among trees and that occasional ovipositions in treated fruit would occur. Behaviour-rich computer simulations of flies in orchards harbouring semi-dwarf trees treated with marking pheromone showed this to be the case (Roitberg, 1985a). Further analysis of those data suggested that synthetic marking pheromone could be more effectively employed if applied in conjunction with sticky-coated fruit-mimic traps. The object of such a strategy would be to capture flies before they emigrate from trees and before they begin to oviposit in pheromone-marked fruit. Preliminary results from further simulations (and partially validated against data from Reissig *et al.*, 1984), suggest that a single trap per tree, in conjunction with an application of marking pheromone, is likely to provide protection from maggot damage equal to that provided by chemical biocide (Roitberg, 1985a). Thus through the behavioural-ecological approach, knowledge of pest behaviour can be used to derive effective management tactics.

BIOLOGICAL CONTROL

One technique thought to be important to orchard-pest managers is the incorporation of natural enemies of pests into management programmes. The degree of impact by natural enemies can be increased by increasing densities of natural enemies or increasing the efficiency of individuals (*see*, e.g., Lewis and Nordlund, 1985); in this chapter we concentrate on the latter.

Parasitoids have been imported and (according to the following citations) successfully attack several important apple pests. These parasitoids include *Allotropa utilis* on apple mealybug (LeRoux, 1971); *Blastothrix sericea* on Lecanium scale (*Lecanium corni* Bouché) (MacPhee *et al.*, 1976); *Epilampsis laricinellae* on pistol casebearer (*Coleophora malivorella* Riley) (LeRoux, 1971); *Macrocentrus ancylivorus* on oriental fruit moth (MacPhee *et al.*, 1976); *Prospaltella perniciosus* on San José scale (De Bach, 1964); and *Trichogramma* sp. on codling moth (MacLellan, 1962; Yu, Laing and Hagley, 1984). Endemic parasitoids include a large native complex on tentiform leafminers (Maier, 1982; van Driesche and Taub, 1983); *Aliotum curculionum* on plum curculio (Croft and Bode, 1983); *Aphelinus mali* on woolly apple aphid (LeRoux, 1971); *Aphytis mytilapsidis* on oystershell scale (MacPhee *et al.*, 1976); *Ascogaster quadridentatis* on codling moth (MacLellan, 1962); *Lysiphlebus* sp. and *Praon* sp. on apple aphid (Carroll and Hoyt, 1984); *Opiine* sp. on *Rhagoletis* sp. (AliNiazee, 1985); and *Typhlocyba pomari* on leafhoppers (Seyedoleslami and Croft, 1979). DeBach (1964), LeRoux (1971), MacPhee *et al.* (1976), and Croft and Bode (1983) discuss these organisms and their import in much greater detail.

Establishment of imported biological control agents is likely to depend upon several criteria, including (1) the number of individuals released (Beirne, 1985), (2) quality problems from mass rearing (Messenger, Wilson and Whitten, 1976) (*see below*), (3) release procedures (Beirne, 1985), and (4) habitat management (Rabb, Stinner and van den Bosch, 1976). Establishment of parasites does not, however, guarantee biological control, such cases being termed 'ecologic' rather than 'economic' successes (Hall and Ehler, 1979). The actual impact of parasitoids, in terms of the proportion of pests parasitized, varies both among species and within species among different orchards. Probable explanations for this variation include variation in frequency and intensity of chemical biocide use (Maier, 1982), habitat differences (Doutt and Nakata, 1973) and genetically based differences in pests (Diehl and Bush, 1984) and/or their parasites (Caltagirone, 1985).

The conclusion to be drawn from the studies cited above is that, under some conditions, parasitoids may act as very effective control agents for some orchard pests (e.g. leafminers: Van Driesche and Taub, 1983). At present, however, it is very difficult to predict those conditions under which biological control through natural enemies is likely to occur, thus adding uncertainty to pest-management decisions (*see* Economics of orchard systems, pages 141, 146–148). Tauber, Hoy and Herzog (1985) have blamed a lack of understanding qf natural enemy ecology and behaviour for control unpredictability and van Lenteren (1980) recently lamented the lack of evolutionary theory employed in biological control. In addition, parasite–host-dynamics theory is currently an area of controversy (cf. Hassell and May, 1974; Murdoch *et al.*, 1984; Reeve and Murdoch, 1985).

Messenger, Wilson and Whitten (1976) have pointed out that many biocontrol researchers tend to confuse fitness, in its biological sense (i.e. relative contribution to gene pool) with effectiveness (i.e. ability to reduce pest damage). Such confusion can lead to difficulties when attempting to apply biological control, as outlined next, with the evolutionary perspective in mind. Rather than dwell upon why some parasitoids are more effective than others, we explore some ways in which a behavioural-ecological perspective is likely to enhance our understanding of parasitoid–host systems and thus their employment in management programmes.

As with the frugivorous insects (which may be termed 'fruit parasites'; Price, 1977) discussed in the previous section, parasitoids must make several 'decisions' upon encountering a potential host. These decisions include not only whether or not to attempt to oviposit in or on that host, but also how many eggs to lay (Skinner, 1985). In addition, hymenopterous parasitoids must also decide which sex each of those offspring will be (Waage and Godfray, 1985). These 'decision' processes have important implications for (1) mass rearing of parasitoids and (2) manipulation of parasitoid behaviour, in orchard systems.

Consider a mass-rearing programme designed to produce large numbers of parasitoids, imported for release against a larval pest of apple. To provide sufficient numbers of parasitoids for release, a rearing system is developed, wherein pest larvae are systematically dispersed over, and presented on, trays where parasitoids oviposit in or on them. This system may proceed for several or many generations before releases are made.

Such a rearing system often begins with relatively small numbers of individuals, which means that genetic problems (e.g. genetic drift: Unruh *et al.*, 1983; founder effects: Mackauer, 1976) are likely to occur. In addition, and probably critical to the success or failure of biocontrol efforts, is the strong artificial selection that mass-rearing programmes impart to physiology (e.g. diapause: House, 1967) and behaviour (Boller, 1972; Chambers *et al.*, 1983). Such effects can occur in several ways.

First, as previously noted, behavioural-ecological analysis predicts that decisions by parasitoids regarding acceptance or rejection of hosts for oviposition is influenced by relative availability of parasitized and unparasitized hosts (Iwasa, Suzuki and Matsuda, 1984). Behavioural ecologists do not expect parasitoids to employ mathematical tools to allow them to make oviposition decisions, but rather expect that parasitoids employ 'simple' rules of thumb (e.g. if the previous three hosts encountered were already parasitized, then oviposit in the next host visited) which have been shaped by natural selection. If the availability of parasitized and unparasitized hosts differs greatly between rearing conditions and natural habitats, there will probably be selection for changes in parasitoid host-acceptance patterns over time. How such changes would affect biological control programmes will differ, depending upon the programme structure and goals. If parasitoids are released on an intermittent basis (i.e. analogous to chemical biocides) and the goal is to reduce pest damage in defined areas (e.g. individual orchards), one may wish to produce parasitoids that more readily super-parasitize hosts, as long as such behaviour leads to parasites remaining longer within—and eventually parasitizing more hosts in—that treatment area than might otherwise be the case (i.e. reduced efficiency is acceptable if, overall, more hosts are parasitized). By contrast, if the goal is to establish, over a large area, a parasitoid population that would maintain itself, behaviour modified in this way will lead to a greater (local) average number of para-sitoid larvae per parasitized host and generally lower larval survival and fecundity, which may preclude the production of sufficient offspring for the parasite to establish and/or maintain itself. Thus, our understanding of the effects, on fitness, of parasitoid behaviour and pest-management goals are crucial to the success of biocontrol programmes.

Second, arguments similar to those above can be made for sex ratio allocation decisions by parasitoids which may be based upon host quality, availability and density of conspecifics. Waage and Godfray (1985) discuss an example in which biological control was poor because of a male-biased allocation in the treatment site where hosts were smaller than those found in the native habitat..

Third, parasite search behaviour is likely to be influenced by the distribution of its hosts in nature (Iwasa, Higashi and Yamamura, 1981; Strand and Vinson, 1982). Rearing systems that provide different spatial distributions of hosts to those that parasites normally experience in nature may select for modification to the 'rules' employed by parasites that direct search paths and search-time allocation (Waage, 1979b; Morrison and Lewis, 1981; Roitberg and Prokopy, 1984b).

While behavioural ecology can have a major impact on the rearing of biological control agents, it is likely to be equally important during release and post-release phases. Recent attempts to manipulate parasite behaviour through semiochemicals are outlined below.

During the past few years, host-emitted chemical compounds (kairomones) have been shown to elicit and maintain parasite host-seeking behaviour (Weseloh, 1981). The demonstrated existence of these compounds has given rise to the idea that parasitoid efficiency might be manipulated for man-agement purposes through application of kairomone in the field (Lewis *et al.*, 1975; Greenblatt and Lewis, 1983; Gross *et al.*, 1984). Such applications have met with mixed success but have been most successful when applied at high host densities (Lewis *et al.*, 1975; Gross *et al.*, 1984). Experiments have demonstrated that patterns of kairomone application (i.e. continuous vs. intermittent; Beevers *et al.*, 1981), host density (Lewis *et al.*, 1975), and parasitoid learning (Dmoch *et al.*, 1985) may all be important in explaining parasitism rates in the field.

Most of the studies cited above have attempted to explain parasitoid response to semiochemicals on a mechanistic or phenomenological basis. As pointed out by Gardner and van Lenteren (1986), such an approach will probably lead only to speculation on how kairomones might be used effectively within management programmes. By contrast, an evolutionary-based approach will attempt to understand why and how parasitoids search for hosts, how they use kairomonal cues (within that context) and finally how pest managers might take advantage of such processes. Surprisingly, as noted by Waage (1983) and Morrison (1986), few laboratory or field studies have actually attempted to solve how parasitoids allocate search effort within host-containing patches, thus making it difficult to explain patterns of parasitism and how one might best attempt to manipulate parasitoid behaviour.

Finally, pest insects are under pressure to escape their natural enemies; their spatial and temporal distribution patterns and the tactics that parasitoids employ to locate them are therefore likely to reflect this (Stewart-Oaten, 1982). Pest distribution patterns may, however, be altered significantly when present in agricultural vs. non-agricultural situations. Understanding the effects of such alterations on parasitoid–pest interactions may provide an insight on how best to manage agroecosystems to provide optimal control (Waage and Hassell, 1982).

Conclusions

We have attempted, through examples, to demonstrate how an evolutionary and ecological perspective can aid in management of orchard insect pests. Thus, we advocate that pest managers first develop a clear understanding of pest behaviour and ecology before attempting to apply management tactics. This is not a new argument: during the late nineteenth century, Professor Stephen A. Forbes emphasized the need for an ecological approach to controlling pest insects in agricultural crops (Smith, Apple and Bottrell, 1976).

Acknowledgements

This paper is dedicated to the memory of Mr Ronnie Smith. We thank J. Cossentine and B. Lalonde for reviewing the manuscript. We also thank G. Geldart of the British Columbia Ministry of Agriculture and Food for the information contained in Table 1. Roitberg is supported by an NSERC operating grant.

References

Adlerz, W. and Everett, P. (1968). Aluminum foil and white polyethylene mulches to repel aphids and and control watermelon mosaic virus. *Journal of Economic Entomology* **61**, 1276–1279.

ALEXANDER, R. (1982). *Optima For Animals*. Edward Arnold, London.

ALINIAZEE M.T., (1985). Opiine (Hymenoptera: Braconidae) parasitoids of *Rhagoletis pomonella* and *R. zephyria* (Diptera: Tephritidae) in the Willamette Valley, Oregon. *Canadian Entomologist* **117**, 163–166.

ANONYMOUS (1983). *Tree Fruit Production Costs*. British Columbia Ministry of Agriculture and Food, Victoria.

ANONYMOUS (1986). *Tree Fruit Production Guide*. British Columbia Ministry of Agriculture and Food, Victoria.

BATEMAN, A.J. (1948). Intra-sexual selection in *Drosophila*. *Heredity* **2**, 349–368.

BATISTE, W.C., OLSON, W.H. AND BERLOWITZ, A. (1973). Codling moth: diel periodicity of catch in synthetic sex attractant vs. female-baited traps. *Environmental Entomology* **2**, 673–676.

BEEVERS, M., LEWIS, W.J:, GROSS, H.R. AND NORDLUND, D.A. (1981). Kairomones and their use for management of entomophagous insects. X. Laboratory studies on manipulation of host-finding behaviour of *Trichogramma pretiosum* Riley with a kairomone extracted from *Heliothis zea* (Boddie) moth scales. *Journal of Chemical Ecology* **7**, 635–648.

BEIRNE, B.P., (1985). Avoidable obstacles to colonization in classical biological control of insects. *Canadian Journal of Zoology* **63**, 743–747.

BOIVIN, G. AND STEWART, R.K. (1983). Seasonal development and interplant movements of phytophagous mirids (Hemiptera: Miridae) on alternate host plants in and around an apple orchard. *Annals of the Entomological Society of America* **76**, 776–780.

BOLLER, E.F. (1972). Behavioural aspects of mass rearing of insects. *Entomophaga* **17**, 9–26.

BOLLER, E.F. AND HURTER, J. (1985). Oviposition-deterring pheromone in *Rhagoletis cerasi*: Behavioral laboratory test to measure pheromone activity. *Entomologia experimentalis et applicata* **39**, 163–169.

BRADY, R.M. (1985). Optimization strategies gleaned from biological evolution. *Nature* **317**, 804–806.

CALTAGIRONE L.E. (1985). Identifying and discriminating among biotypes of parasites and predators. In *Biological Control in Agricultural IPM Systems* (M.A. Hoy and D.C. Herzog, Eds), pp. 189–200. Academic Press. Orlando.

CAREY, J.R. (1983). A life table examination of growth rate and age structure trade-offs in Mediterranean fruit fly populations. In *Fruit Flies of Economic Importance* (R. Cavalloro, Ed.), pp. 315–320. A.A. Balkema, Rotterdam.

CARROLL, D.P. AND HOYT, S.C. (1984). Natural enemies and their effects on apple aphid, *Aphis pomi* DeGeer (Homoptera: Aphididae), colonies on young apple trees in central Washington. *Environmental Entomology* **13**, 469–481.

CHAMBERS, D.L., CALKINS, C.O., BOLLER, E.F., ITO, Y. AND CUNNINGHAM, R.T. (1983). Measuring, monitoring and improving the quality of mass-reared Mediterranean fruit flies, *Ceratitis capitata* Wied. 2. Field tests for confirming and extending laboratory results. *Zeitschrift für angewandte Entomologie* **95**, 285–303.

CHARNOV, E.L. (1982). *The Theory of Sex Allocation*. Princeton University Press, Princeton.

CHARNOV, E.L. AND SKINNER, S.W. (1985). Complementary approaches to the understanding of parasitoid oviposition decisions. *Environmental Entomology* **14**, 383–391.

CREWS, D. AND MOORE, M.C. (1986). Evolution of mechanisms controlling mating behavior. *Science* **231**, 121–125.

CROFT, B.A. (1983). Introduction. In *Integrated Management of Insect Pests of Pome and Stone Fruits* (B.A. Croft and S.C. Hoyt, Eds), pp. 1–18. John Wiley & Sons, New York.

CROFT, B.A. AND BODE, W.M. (1983). Tactics for deciduous fruit IPM. In *Integrated Management of Insect Pests of Pome and Stone Fruits* (B.A. Croft and S.C. Hoyt, Eds), pp. 219–270. John Wiley & Sons, New York.

CROFT, B.A. AND HOYT, S.C. (EDS) (1983). *Integrated Management of Insect Pests of Pome and Stone Fruits.* John Wiley & Sons, New York.

DEBACH, P. (ED.) (1964). *Biological Control of Insect Pests and Weeds.* Reinhold, New York.

DIEHL, S.R. AND BUSH, G.L. (1984). An evolutionary and applied perspective of insect biotypes. *Annual Review of Entomology* **29**, 471–504.

DMOCH, J., LEWIS, W.J., MARTIN, P.B. AND NORDLUND, D.A. (1985). Role of host-produced stimuli and learning in host selection behavior of *Cotesia marginventris* Cresson. *Journal of Chemical Ecology* **11**, 453–463.

DOUTT, R.L. AND NAKATA, J. (1973). The Rubus leafhopper and its egg parasitoid: An endemic biotic system useful in pest management. *Environmental Entomology* **2**, 381–386.

EASTERBROOK, M.A., SOLOMON, M.G., CRANHAM, J.E. AND SOUTER, E.F. (1985). Trials of an integrated pest management programme based on selective pesticides in English apple orchards. *Crop Protection* **4**, 215–230.

FAO (1967). *Report of the First Session of the FAO Panel of Experts on Integrated Pest Control.* Food and Agriculture Organization, Rome.

FARNSWORTH, R.L. AND MOFFITT, L. J. (1984). Farmer's perceptions and information sources: A quantitative analysis. *Agricultural Economics Research* **36**, 8–11.

FRISBIE, R.E. AND ADKISSON, P.L. (1985). IPM: Definitions and current status in U.S. agriculture. In *Biological Control in Agricultural IPM Systems* (M.A. Hoy and D.C. Herzog, Eds), pp. 41–50. Academic Press, Orlando.

GARDNER, S.M. AND VAN LENTEREN, J.C. (1986). Characterisation of the arrestment responses on *Trichogramma evanescens. Oecologia* **68**, 265–270.

GREENBLATT, J.A. and LEWIS, W.J. (1983). Chemical environment manipulation for pest insects control. *Environmental Management* **7**, 35–41.

GROSS, H.R., LEWIS, W.J., BEEVERS, M. AND NORDLUND, D.A. (1984). *Trichogramma pretiosum* (Hymenoptera: Trichogrammitidae): Effects of augmented densities and distribution of *Heliothis zea* (Lepidoptera: Noctuidae) host eggs and kairomones on field performance. *Environmental Entomology* **13**, 981–985.

GRUYS, P. (1982). Hits and misses. The ecological approach to pest control in orchards. *Entomologia experimentalis et applicata* **31**, 70–87.

HALL, F.R. (1983). Pesticide usage patterns for Ohio apple orchards. *Journal of Economic Entomology* **76**, 584–589.

HALL, R.W. AND EHLER, L.E. (1979). Rate of establishment of natural enemies in classical biological control. *Bulletin of the Entomological Society of America* **26**, 111–113.

HASSELL, M.P. AND MAY, R.M. (1974). Aggregation of predators and insect parasites and its affect on stability. *Journal of Animal Ecology* **43**, 567–594.

HATHAWAY, D.O., BUTT, B.A. AND LYDIN, L.V. (1970). Reduction of sexual aggressiveness of male codling moths treated with tepa or gamma irradiation. *Journal of Economic Entomology* **63**, 1881–1883.

HATHAWAY, D.O., MOFFITT, H.R. AND GEORGE, D.A. (1985). Codling moth (Lepid.: Tortricidae): Disruption of sexual communication with an antipheromone ((E,E)-8,10-dodecadien-1-ol acetate). *Journal of the Entomological Society of British Columbia* **82**, 18–22.

HEADLEY, J.C. (1982). The economics of pest management. In *Introduction to Pest Management* (R.L. Metcalf and W.H. Luckman, Eds), pp. 69–91. Wiley Interscience, New York.

HEADLEY, J.C. (1985). Cost–benefits analysis: defining research needs. In *Biological Control in Agricultural IPM Systems* (M.A. Hoy and D.C. Herzog, Eds), pp. 53–62. Academic Press, Orlando.

HOFFMAN, G. AND GUSTAFSON, C. (1983). A new approach to estimating agricultural costs of production. *Agricultural Economics Research* **35**, 9–14.

HOUSE, H.L. (1967). The decreasing occurrence of diapause in the fly *Pseudosarcophaga affinis* through laboratory-reared generations. *Canadian Journal of Zoology* **45**, 149–153.

HOWELL, J.F. AND CLIFT, A.E. (1974). The dispersal of sterilized codling moths released in the Wenas Valley, Washington. *Environmental Entomology* 3, 75–81.

HOYT, S.C. AND BURTS, E.C. (1974). Integrated control of fruit pests. *Annual Review of Entomology* 19, 231–252.

IWASA, Y., HIGASHI, M. AND YAMAMURA, N. (1981). Prey distribution as a factor determining the choice of optimal foraging strategy. *American Naturalist* 117, 710–723.

IWASA, Y., SUZUKI, Y. AND MATSUDA, H. (1984). Theory of oviposition strategy of parasitoids. I. Effect of mortality and limited egg number. *Theoretical Population Biology* 26, 205–227.

JACKSON, D.S. AND LEE. B.G. (1985). Medfly in California 1980–1982. *Bulletin of the Entomological Society of America* 31, 29–37.

JANETOS, A.C. AND COLE, B.J. (1981). Imperfectly optimal animals. *Behavioural Ecology and Sociobiology* 9, 203–209.

KATSOYANNOS, B.I. AND BOLLER, E.F. (1976). First field application of oviposition-deterring marking pheromone of European cherry fruit fly. *Environmental Entomology* 5, 151–152.

KATSOYANNOS, B.I. AND BOLLER, E.F. (1980). Second field application of oviposition-deterring pheromone of the European cherry fruit fly. *Zeitschrift für angewandte Entomologie* 89, 278–281.

KLIJNSTRA, J.W. (1985). *Oviposition behaviour as influenced by the Oviposition-deterring Pheromone in the Large White Butterfly*, Pieris brassicae. PhD thesis, Agricultural University, Wageningen. 135 pp.

KREBS, J.R. AND DAVIES, N. (EDS) (1981). *An Introduction to Behavioural Ecology*. Blackwell Scientific Publications, Oxford.

KREBS, J.R. AND MCCLEERY, R.H. (1984). Optimization in behavioural ecology. In *Behavioural Ecology: An Evolutionary Approach* (J.R. Krebs and N.B. Davies, Eds), pp. 91–122. Sinauer, Sunderland.

LEROUX, E.J. (1971). Biological control attempts on pome fruit (apple and pear) in North America, 1860–1970. *Canadian Entomologist* 103, 963–974.

LEWIS, W.J. AND NORDLUND, D.A. (1985). Behavior–modifying chemicals. In *Biological Control in Agricultural IPM Systems* (M.A. Hoy and D.C. Herzog, Eds), pp. 89–101. Academic Press, Orlando.

LEWIS, W.J., JONES, R.L., NORDLUND, D. A. AND GROSS, H.R. (1975). Kairomones and their use for management of entomophagous insects. II. Mechanisms causing increased rates of parasitism in the field. *Journal of Chemical Ecology* 1, 343–347.

LORENZ, K.Z. (1981). *The Foundations of Ethology*. Simon and Schuster, New York.

LOW, A.R.C. AND KEMP, R.L. (1977). A management-oriented approach to the estimation of farm production costs and returns in rural development projects. *Journal of Agricultural Economics* 28, 129–138.

MCDONOUGH, L.M., GEORGE, D.A., BUTT, B.A., RUTH, J.M., AND HILL, K.R. (1972). Sex pheromone of the codling moth: structure and synthesis. *Science* 177, 177–178.

MACKAUER, M. (1976). Genetic problems in the production of biological control agents. *Annual Review of Entomology* 21, 369–385.

MACLELLAN, C.R. (1962). Mortality of codling moth eggs and young larvae in an integrated control orchard. *Canadian Entomologist* 94, 655–666.

MACPHEE, A.W., CALTAGIRONE, L.E., VAN DE VRIE, M. AND COLLYER, E. (1976). Biological control of pests of temperate fruit and nuts. In *Theory and Practice of Biological Control* (C.B. Huffaker and P.S. Messenger, Eds), pp. 337–358. Academic Press, New York.

MADSEN, H.F. AND VAKENTI, J.M. (1973). The influence of trap design on the response of codling moth (Lepidoptera: Olethreutidae) and fruittree leafroller (Lepidoptera: Tortricidae) to synthetic sex attractants. *Journal of the Entomological Society of British Columbia* 70, 5–8.

MAIER, C.T. (1982). Parasitism of the apple blotch leafminer, *Phyllonorycter crataegella* on sprayed and unsprayed apple trees in Connecticut. *Environmental Entomology* **11**, 603–610.

MANGEL, M. AND CLARK, C.W. (1986). Unified foraging theory. *Ecology*, in press.

MARSHALL, E. (1985). The rise and fall of Temik. *Science* **229**, 1369–1371.

MAYNARD SMITH, J. (1978). Optimization theory in evolution. *Annual Review of Ecology and Systematics* **9**, 31–56.

MAYNARD SMITH, J. (1982). *Evolution and the Theory of Games*. Cambridge University Press, Cambridge.

MESSENGER, P.S., WILSON, F. AND WHITTEN, M.J. (1976). Variation, fitness and adaptability of natural enemies. In *Theory and Practice of Biological Control* (C.B. Huffaker and P.S. Messenger, Eds), pp. 209–231. Academic Press, New York.

MOFFITT, R.R., WESTIGARD, P.H. AND HATHAWAY, D.O. (1979). Pheromonal control of the codling moth and biological control of the pear psylla. *Proceedings of the Oregon Horticultural Society* **70**, 95–96.

MORRISON, G. (1986). 'Searching time aggregation' and density-dependent parasitism in a laboratory host–parasitoid interaction. *Oecologia* **68**, 298–303.

MORRISON, G. AND LEWIS, W.J. (1981). The allocation of searching time by *Trichogramma pretiosum* in host-containing patches. *Entomologia experimentalis et applicata* **30**, 31–39.

MURDOCH, W.W., REEVE, J.D., HUFFAKER, C.B. AND KENNETT, C.E. (1984). Biological control of scale insects and ecological theory. *American Naturalist* **123**, 371–392.

NAKATSURU, K. AND KRAMER, D.L. (1982). Is sperm cheap? Limited male fertility and female choice in the lemon tetra (Pisces, Characidae). *Science* **216**, 753–755.

NORTON, G. (1985). Review of *Integrated Management of Insect Pests of Pome and Stone Fruits* (B.A. Croft and S.C. Hoyt, Eds). *Crop Protection* **4**, 403–404.

NORTON, G. AND HOLLING, C.S. (EDS) (1979). *Pest Management: Proceedings of an International Conference*. Pergamon, Oxford.

PARKER, G.A. (1970). Sperm competition and its evolutionary consequences in the insects. *Biological Reviews of the Cambridge Philosophical Society* **45**, 525–568.

PARKER, G.A. AND COURTNEY, S.P. (1984). Models of clutch size in insect oviposition. *Theoretical Population Biology* **26**, 27–48.

PIMENTEL, D., TERHUNE, E.C., DRITSCHILO, W., GALLAHAN, D., KINNER, N., NAFUS, D., PETERSON, R., ZAREH, N., MISITI, J. AND HABER-SCHAIM, O. (1977). Pesticides, insects in foods and cosmetic standards. *BioScience* **27**, 178–185.

PIMENTEL, D., ANDOW, D., DYSON-HUDSON, R., GALLAHAN, D., JACOBSON, S., IRISH, M., MOSS, A., SHREINER, I., SHEPARD, M., THOMPSON, T. AND VINZANT, B. (1980). Environmental and social costs of pesticides: a preliminary assessment. *Oikos* **34**, 125–140.

PIMENTEL, D., GLENISTER, C., FAST, S. AND GALLAHAN, D. (1984). Environmental risks of biological pest controls. *Oikos* **42**, 283–290.

PRICE, P.W. (1977). General concepts on the evolutionary biology of parasites. *Evolution* **31**, 405–420.

PROKOPY, R.J. (1972). Evidence for a marking pheromone deterring repeated oviposition in apple flies. *Environmental Entomology* **1**, 326–332.

PROKOPY, R.J. (1981a). Epideictic pheromones that influence spacing patterns of phytophagous insects. In *Semiochemicals: Their Role in Pest Control* (D.A. Nordlund, R.L. Jones and W.J. Lewis, Eds), pp. 181–213. John Wiley & Sons, New York.

PROKOPY, R.J. (1981b). Oviposition-deterring pheromone system of apple maggot flies. In *Management of Insect Pests with Semiochemicals* (E.R. Mitchell, Ed.), pp. 477–497. Plenum Press, New York.

PROVERBS, M.D., LOGAN, D.M. AND CARTY, B.E. (1973). Some biological observations related to codling moth control by the sterility principle. In *Computer*

Models and Application of the Sterile Male Technique, pp. 149–163. International Atomic Energy Agency, Vienna.

PROVERBS, M.D., LOGAN, D.M. AND NEWTON, J.R. (1975). A study to suppress codling moth (Lepidoptera: Olethreutidae) with sex pheromone traps. *Canadian Entomologist* 107, 1265–1269.

PROVERBS, M.D., NEWTON, J.R. AND CAMPBELL, C.J. (1982). Codling moth: A pilot program of control by sterile insect release in British Columbia. *Canadian Entomologist* 114, 363–376.

PYKE, G.H. (1984). Optimal foraging: A critical review. *Annual Review of Ecology and Systematics* 15, 523–576.

RABB, R.L., STINNER, R.E. AND VAN DEN BOSCH, R. (1976). Conservation and augmentation of natural enemies. In *Theory and Practice of Biological Control* (C.B. Huffaker and P.S. Messenger, Eds), pp. 233–254. Academic Press, New York.

REEVE, J.D. AND MURDOCH, W.W. (1985). Aggregation by parasitoids in the successful control of California red scale: a test of theory. *Journal of Animal Ecology* 54, 797–816.

REIDL, H. AND CROFT, B.A. (1974). A study of pheromone trap catches in relation to codling moth (Lepidoptera: Olethreutidae) damage. *Canadian Entomologist* 106, 525–537.

REISSIG, W.H., WEIRES, R.W., FORSHEY, C.G., ROELOFS, W.L., LAMB, R.C., ALDWINCKLE, H.S. AND ALM, S.R. (1984). Management of the apple maggot, *Rhagoletis pomonella* (Walsh) (Diptera: Tephritidae), in disease-resistant dwarf and semi-dwarf apple trees. *Environmental Entomology* 13, 684–690.

ROELOFS, W., COMEAU, A., HILL, A., AND MILICEVIC, G. (1971). Sex attractant of the codling moth: characterization with electroantennagram technique. *Science* 174, 297–299.

ROITBERG, B.D. (1985a). Search dynamics in fruit-parasitic insects. *Journal of Insect Physiology* 31, 865–872.

ROITBERG, B.D. (1985b). Ecological theory and management of tephritid flies. Paper presented at the symposium on *Total Management of Tephritids* at the Annual Meeting of the Pacific Branch of the Entomological Society of America.

ROITBERG, B.D. AND PROKOPY, R.J. (1982). Influence of intertree distance on the foraging behaviour of *Rhagoletis pomonella* in the field. *Ecological Entomology* 7, 437–443.

ROITBERG, B.D. AND PROKOPY, R.J. (1983a). Resource assessment by adult and larval codling moths. *Journal of the New York Entomological Society* 90, 258–265.

ROITBERG, B.D. AND PROKOPY, R.J. (1983b). Host deprivation influence on response of *Rhagoletis pomonella* to its oviposition-deterring pheromone. *Physiological Entomology* 8, 69–72.

ROITBERG, B.D. AND PROKOPY, R.J. (1984a). Host visitation sequence as a determinant of search persistence in fruit parasitic tephritid flies. *Oecologia* 62, 7–12.

ROITBERG, B.D. AND PROKOPY, R.J. (1984b). Host discrimination by adult and larval European apple sawflies. *Environmental Entomology* 13, 1000–1003.

ROITBERG, B.D. AND PROKOPY, R.J. (1986). Behavioural ecology of marking pheromones in insect herbivores. *BioScience*, in press.

ROITBERG, B.D., ELKINTON, J.S. AND PROKOPY, R.J. (1982). Foraging behaviour of *Rhagoletis pomonella*: a model for fruit parasites. In *Insect–Plant Relationships* (A.K. Minks and J.H. Visser, Eds), p. 184. Pudoc, Wageningen.

ROTHSCHILD, G.H.L. (1982). Suppression of mating in codling moths with synthetic sex pheromone and other compounds. In *Insect Suppression with Controlled Release Pheromone Systems. Vol. 2* (A.F. Kydonieus and M. Beroza, Eds), pp. 117–134. CRC Press, Boca Raton.

SEYEDOLESLAMI, H. AND CROFT, B.A. (1979). Spatial distribution of overwintering eggs of the white apple leafhopper, *Typhlocyba pomaria*, and parasitism by

Anagrus epos. Environmental Entomology **9**, 624–628.

SHOEMAKER, C.A. (1979). Optimal management of an alfalfa ecosystem. In *Pest Management: Proceedings of an International Conference* (G.A. Norton and C.S. Holling, Eds), pp. 301–315. Pergamon, Oxford.

SIBLY, R.M. AND SMITH, R.H. (EDS) (1985). *Behavioural Ecology: Ecological Consequences of Adaptive Behaviour.* Blackwell Scientific Publications, Oxford.

SKINNER, S.W. (1985). Clutch size as an optimal foraging problem for insects. *Behavioural Ecology and Sociobiology* **17**, 231–238.

SMITH, R.F., APPLE, J.L. AND BOTTRELL, D.G. (1976). The origins of integrated pest management concepts for agricultural crops. In *Integrated Pest Management* (J.L. Apple and R.F. Smith, Eds), pp. 1–14. Plenum Press, New York.

SMITH, R.H. AND LESSELLS, C.M. (1985). Oviposition, ovicide and larval competition in granivorous insects. In *Behavioural Ecology: Ecological Consequences of Adaptive Behaviour* (R.M. Sibly and R.H. Smith, Eds), pp. 423–448. Blackwell Scientific Publications, Oxford.

STEMEROFF, M. AND GEORGE, J.A. (1983). The benefits and costs of controlling destructive insects on onions, apples and potatoes in Canada, 1960–1980: Summary. *Bulletin of the Entomological Society of Canada* **15**, 91–97.

STEWART-OATEN, A. (1982). Minimax strategies for a predator–prey game. *Theoretical Population Biology* **22**, 410–424.

STRAND, M.R. AND VINSON, S.B. (1982). Behavioural response of the parasitoid *Cardiochiles nigriceps* to a kairomone. *Entomologia experimentalis et applicata* **31**, 308–315.

TAUBER, M.J., HOY, M.A. AND HERZOG, D.C. (1985). Biological control in agricultural IPM systems: a brief overview of the current status and future prospects. In *Biological Control in Agricultural IPM Systems* (M.A. Hoy and D.C. Herzog, Eds), pp. 3–9. Academic Press, Orlando.

TAYLOR, C.R. (1980). The nature of benefits and costs of use of pest control methods. *American Journal of Agricultural Economics* **32**, 1007–1011.

TAYLOR, C.R., CARLSON, G.A., COOKE, F.T., REICHELDERFER, K.H. AND STARBIRD, I.R. (1983). Aggregate economic effects of alternative boll weevil management strategies. *Agricultural Economics Research* **35**, 19–22.

THOMPSON, P., HOW, R.B. AND WHITE, G.B. (1980). An economic evaluation of grower savings in a pest management program. *HortScience* **15**, 639–640.

THORNHILL, R. (1981). Panorpa (Mecoptera: Panorpidae) scorpionflies: systems for understanding resource-defense polygyny and alternative male reproductive effects. *Annual Review of Ecology and Systematics* **12**, 355–386.

THORNHILL, R. AND ALCOCK, J. (1983). *The Evolution of Insect Mating Systems.* Harvard University Press, Cambridge.

TINBERGEN, N. (1963). On the aims and methods of ethology. *Zeitschrift für Tierpsychologie* **20**, 410–433.

TOSCANO, N.C., SANCES, F.V., JOHNSON, M.W. AND LAPRE, L.F. (1982a). Effect of various pesticides on lettuce physiology and yield. *Journal of Economic Entomology* **75**, 738–741.

TOSCANO, N.C., VAN STEENWYK, R.A., KIDO, K., MCCALLEY, N.F., BARNETT, W.W. AND JOHNSON, M.W. (1982b). Yield responses in lettuce plants at various density treatment levels of lepidopterous larvae. *Journal of Economic Entomology* **75**, 916–920.

UNRUH, T.R., WHITE, W., GORDH, G. AND LUCK, R.F. (1983). Heterozygosity and effective size in laboratory populations of *Aphidius ervi* (Hym.: Aphidiidae). *Entomophaga* **28**, 245–258.

VAKENTI, J.M. AND MADSEN, H.F. (1976). Codling moth (Lepidoptera: Olethreutidae); Monitoring populations in apple orchards with sex pheromone traps. *Canadian Entomologist* **108**, 433–438.

VAN DRIESCHE, R.G. AND TAUB, G. (1983). Impact of parasitoids on *Phyllonorycter* leafminers infesting apple in Massachusetts USA. *Protection Ecology* **5**, 303–317.

VAN LENTEREN, J.C. (1980). Evaluation of control capabilities of natural enemies: does art have to become science? *Netherlands Journal of Zoology* **30**, 369–381.

WAAGE, J.K. (1979a). Dual function of the damsel fly penis: Sperm removal and transfer. *Science* **203**, 916–918.

WAAGE, J.K. (1979b). Foraging for patchily distributed hosts by the parasitoid, *Nemeritis canescens*. *Journal of Animal Ecology* **48**, 353–371.

WAAGE, J.K. (1983). Aggregation in field parasitoid populations: foraging time allocation of *Diadegma* (Hymenoptera: Ichneumonidae). *Ecological Entomology* **8**, 447–453.

WAAGE, J.K. AND GODFRAY, H.C.J. (1985). Reproductive strategies and population ecology of insect parasitoids. In *Behavioural Ecology: Ecological Consequences of Adaptive Behaviour* (R.M. Sibly and R.H. Smith, Eds), pp. 449–470. Blackwell Scientific Publications, Oxford.

WAAGE, J.K. AND HASSELL, M.P. (1982). Parasitoids as biological control agents. *Parasitology* **84**, 241–268.

WADDINGTON, C.H. (1977). *Tools for Thought*. Jonathan Cape, London.

WEBSTER, J.P.G. (1977). The analysis of risky farm management decisions: advising farmers about the use of pesticides. *Journal of Agricultural Economics* **28**, 243–259.

WESELOH, R.M. (1981). Host location by parasitoids. In *Semiochemicals: Their Role in Pest Control* (D.A. Nordlund, R.L. Jones and W.J. Lewis, Eds), pp. 79–95. John Wiley & Sons, New York.

WHALON, M.E. AND CROFT, B.A. (1983). Implementation of apple IPM. In *Integrated Management of Insect Pests of Pome and Stone Fruits* (B.A. Croft and S.C. Hoyt, Eds), pp. 411–448.

WHITE, L.D., HUTT, R.B. AND BUTT, B.A. (1973). Field dispersal of laboratory-reared fertile female codling moths and population suppression by release of sterile males. *Environmental Entomology* **2**, 66–69.

WILLIS, M.A. AND BAKER, T.C. (1984). Effects of intermittent and continuous pheromone stimulation on the flight behaviour of the oriental fruit moth, *Grapholita molesta*. *Physiological Entomology* **9**, 341–358.

WILSON, E.O. (1975). *Sociobiology*. Harvard University Press, Cambridge.

YU, D.S.K., LAING, J.E. AND HAGLEY, E.A.C. (1984). Dispersal of *Trichogramma* spp. in an apple orchard after inundative releases. *Environmental Entomology* **13**, 371–374.

3
Factors Affecting Resistance of Rice Varieties to Planthopper and Leafhopper Pests

R. C. SAXENA* AND Z. R. KHAN†

*Department of Entomology, International Rice Research Institute (IRRI), PO Box 933, Manila, Philippines and †International Centre of Insect Physiology and Ecology (ICIPE), PO Box 30772, Nairobi, Kenya; based at IRRI

Introduction

Rice, *Oryza sativa* L., is a primary staple food for more than two billion people in Asia and several hundred million people in Africa and Latin America. Most of the world's rice production comes from irrigated and rain-fed lowland rice fields where insect pests are severe constraints. Also, more conducive for the build-up of insect pests are heavily fertilized, high-tillering plants and the practice of growing rice throughout the year. The rice crop is attacked by more than 100 insect species that cause varying degrees of crop damage in various rice-growing environments (Pathak and Saxena, 1980). The loss in rice yield due to insect pests is estimated to be 32% in Asia and 21% in North and Central America (Cramer, 1967). The intensity of insect pest problem in rice was demonstrated at the International Rice Research Institute (IRRI), Philippines, where in 114 separate experimental trials conducted from 1964 to 1979, yield loss from insects averaged 40% (Pathak and Dhaliwal, 1981).

Table 1.   Major planthopper and leafhopper pests of rice (modified after Pathak, 1968)

Name	Common name	Distribution	Vector of
Delphacidae (planthoppers)			
Laodelphax striatellus (Fallén)	Small brown planthopper	Japan, China, Korea, Palearctic regions	Rice stripe, rice black-streaked dwarf
Nilaparvata lugens (Stål)	Brown planthopper	South and Southeast Asia, China, Japan	Grassy stunt, ragged stunt
Sogatella furcifera (Horváth)	Whitebacked planthopper	South and Southeast Asia, Japan, Korea, China, South Pacific Islands, northern Australia	—
Sogatodes oryzicola (Muir)	Rice delphacid	Southern United States, Caribbean islands, South America	'Hoja blanca'
Cicadellidae (leafhoppers)			
Cofana spectra (Distant)	White leafhopper	Africa, Australia, China, South and Southeast Asia	—
Nephotettix cincticeps (Uhler)	Rice green leafhopper	Japan, Taiwan, Korea, China	Rice dwarf, yellow dwarf
Nephotettix virescens (Distant)	Rice green leafhopper	South and Southeast Asia	Yellow dwarf, tungro, 'penyakit merah', yellow-orange leaf
Nephotettix nigropictus (Stål)	Rice green leafhopper	China, South and Southeast Asia	Rice dwarf, yellow dwarf, transitory yellowing, tungro, yellow-orange leaf, rice gall dwarf
Recilia dorsalis (Motschulsky)	Zigzag leafhopper	South and Southeast Asia, Japan, Taiwan	Rice dwarf, yellow-orange leaf

Among various insect pests that attack rice, planthoppers and leafhoppers are of major importance. Several species are serious pests of rice and are of world-wide distribution (*Table 1*). In many areas they frequently occur in numbers large enough to cause complete drying of the crop or 'hopperburn'. The more damaging species are the brown planthopper *Nilaparvata lugens* (Stål), the small brown planthopper *Laodelphax striatellus* (Fallén), the whitebacked planthopper *Sogatella furcifera* (Horváth), the rice delphacid *Sogatodes oryzicola* (Muir), the green leafhopper *Nephotettix* spp., and the zigzag leafhopper *Recilia dorsalis* (Motschulsky). These species occur in Asia except *S. oryzicola* which is found in the southern United States and in north Central South America. In addition to direct feeding damage, planthoppers and leafhoppers are also vectors of most presently known rice virus diseases.

The use of insect-resistant rice varieties is an ideal method of controlling these pests. Aside from the undesirable effects of pesticides, many farmers in South and Southeast Asia, where most rice is grown, have limited access to capital, pesticides and application equipment (Pathak and Saxena, 1980). Various studies have demonstrated the existence of natural resistance to planthoppers and leafhoppers in several rice varieties and wild rices (Pathak, Cheng and Fortuno, 1969; Saxena and Pathak, 1979; Heinrichs, Medrano and Rapusas, 1985; Razzaque and Heinrichs, 1985c; Heinrichs, 1986; Velusamy *et al.*, 1986; Saxena, 1989). Such resistance can be transferred to the high-yielding rice varieties by conventional methods or by innovative breeding techniques (Khush and Jena, 1986).

Breeding for planthopper and leafhopper resistance has now become a major research objective in most of the rice-growing countries of Asia, and Central and South America. Current breeding programmes include developing resistance to *N.lugens*, *S. furcifera*, *L. striatellus* and *N. virescens* in Asia, and to *S. oryzicola* in Central and South America (Jennings and Pineda, 1983; Heinrichs, 1986). The germplasm bank of IRRI consisting of more than 82 000 accessions is being evaluated for *N. lugens*, *S. furcifera*, *N. virescens* and *R. dorsalis* resistance (*Table 2*).

Dramatic progress has been made in the development of rice varieties with resistance to planthoppers and leafhoppers, and rice varieties resistant to these pests are now being grown over millions of hectares in South and Southeast Asia, and in Central and South America. However, the development of new biotypes of these insects capable of surviving on resistant varieties has often limited their full potential in insect control. The threat of selection and spread of such virulent biotypes of planthoppers and leafhoppers cannot be ruled out as most of the presently used pest-resistant varieties have a rather narrow genetic base. Also, the factors that confer insect resistance in rice are, for the most part, not well understood.

A sound basis for developing resistant varieties should, therefore, be oriented toward identifying the resistance-imparting plant characters and using them as cues in breeding programmes. However, determination of these factors often becomes difficult if complete information on behavioural and physiological responses of the pest to susceptible and resistant host

Table 2. Status of screening and breeding for varietal resistance to major planthopper and leafhopper pests of rice (after Pathak and Saxena, 1980; Heinrichs, 1986; Saxena, 1989)

Insect	Status of resistance					
	Screening methods developed	Resistance sources identified	Resistant breeding lines available	Resistant varieties released	Genes for resistance identified	Biotypes encountered
Nilaparvata lugens (Stål)	+	+	+	+	+	+
Sogatella furcifera (Horváth)	+	+	+	+	+	?
Laodelphax striatellus (Fallén)	+	+	+	+	−	−
Sogatodes oryzicola (Muir)	+	+	+	+	−	?
Nephotettix virescens (Distant)	+	+	+	+	+	+
Nephotettix cincticeps (Uhler)	+	+	+	+	−	−
Nephotettix nigropictus (Stål)	+	+	−	−	−	−
Recilia dorsalis (Motschulsky)	+	+	+	+	+	−
Cofana spectra (Distant)	+	+	−	−	−	−

+ = yes; − = no; ? = biotypes suspected.

plants is lacking. The continual evolution of biotypes of certain insect pest species capable of eroding specific host-plant resistance also underscores the need for a better understanding of the basis of host-plant resistance (Saxena, 1986).

Susceptibility or resistance of plants is the result of a series of interactions between plants and insects which influence the ultimate degree of establishment of insect populations on plants (Saxena, 1969; Saxena, Gandhi and Saxena, 1974; Saxena and Pathak, 1979). The factors which determine insect establishment on plants can be categorized into two groups—insect responses to plants, and plant characters influencing insect responses. The insect responses include orientation, feeding, metabolic utilization of ingested food, growth of larvae or nymphs to adult stage, adult longevity, egg production, oviposition and hatching of eggs. Unfavourable biophysical or biochemical plant characters may interrupt one or more of these insect responses, inhibit the establishment of an insect population on a plant and render it resistant to infestation and injury.

The present paper discusses the status of host-plant resistance in rice against major planthopper and leafhopper pests.

Rice planthoppers

THE BROWN PLANTHOPPER

Nilaparvata lugens (Stål) has become a particularly damaging pest of rice in many Asian countries following the introduction of modern heavy-tillering varieties, which are usually treated with high rates of fertilizers. Light infestations reduce plant tillering, plant height, number of productive tillers and crop vigour, and induce the production of unfilled grain; heavy infestations cause 'hopperburn' of the crop (Bae and Pathak, 1970). The insect is also a vector of grassy stunt and ragged stunt viral diseases (Rivera, Ou and Iida, 1966; Ling, 1977).

The growing severity of *N. lugens* outbreaks in tropical Asia prompted rice scientists to develop resistant varieties as the most practical solution to the pest problem in the developing countries. Studies on varietal resistance to *N. lugens* were started at IRRI in 1965, where 1400 varieties selected from an evaluation of 10 000 different rice varieties and collections for stem borer resistance were field tested for resistance to *N. lugens* and *N. virescens* (IRRI, 1967). The selected varieties were tested more intensively for consistency and the nature of their resistance. Since then, more than 50 000 accessions, and an even greater number of breeding lines, have been evaluated against *N. lugens*. Several national rice improvement programmes in Bangladesh, China, India, Indonesia and Thailand also have screened a large volume of rice varieties and germplasm for resistance to *N. lugens*, and have made significant progress in identifying and utilizing natural resistance to the pest. The insect thrives on high-yielding susceptible varieties but fails to feed, grow, survive and reproduce adequately on resistant varieties (Cheng and Pathak, 1972; Saxena and Pathak, 1979).

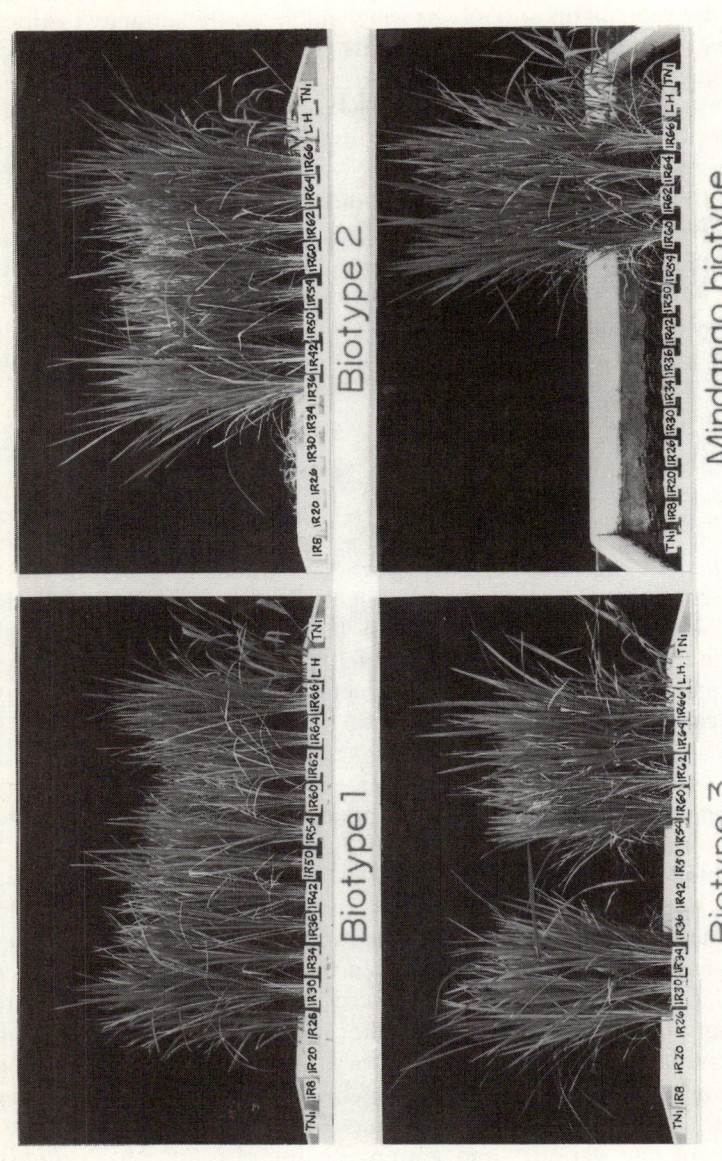

Figure 1. Resistance of rice to different biotypes of *Nilaparvata lugens*. Biotype 1 damages varieties with no genes for resistance. while Biotype 2 and Biotype 3 also damage those varieties with *Bph1* and *bph2* genes, respectively. A population of *N. lugens* collected from Mindanao, equally damages rice varieties with *Bph1* and *bph2* genes. None of these biotypes damages varieties with *Bph3*, *bph4*, *bph5*, *Bph6* or *bhp7* gene.

However, soon after the release of the first high-yielding, *N. lugens*-resistant rice variety 'IR26', it became apparent and alarming that new biotypes of *N. lugens* could overcome this plant resistance (IRRI, 1975). In 1975, based on differential varietal reactions, the existence of three distinct biotypes of *N. lugens* was demonstrated in the Philippines (*Figure 1*) (IRRI, 1976). Moreover, the allopatric brown planthopper populations in Australia, Bangladesh, China, India, Japan, Korea, Malaysia, Indonesia, Sri Lanka and Thailand belong to different biotypes (IRRI, 1986; Saxena, 1989). A total of seven genes that impart resistance to brown planthopper has so far been identified (Athwal *et al.*, 1971; Athwal and Pathak, 1972; Lakshminarayana and Khush, 1977; Khush, Rezaul Karim and Angeles, 1986; Kabir and Khush, 1988). The brown planthopper Biotype 1 can survive on and damage only those varieties that do not carry genes for resistance. Biotype 2 can thrive on resistant varieties carrying the *Bph*1 resistance gene and those susceptible to Biotype 1. Biotype 3 can infest varieties carrying the *bph*2 gene and those susceptible to Biotype 1. Recently, a population of *N. lugens* was collected from Mindanao, southern Philippines, which equally damages rice varieties with *Bph*1 and *bph*2 resistance genes. However, none of the Philippine biotypes survives on varieties with genes *Bph*3, *bph*4, *bph*5, *Bph*6 and *bph*7.

The possibility of occurrence or evolution of more biotypes of *N. lugens* cannot be excluded if resistant varieties with new genes for resistance are planted intensively. In such an eventuality, the sequential release of resistant varieties with varying genetic backgrounds will have to be relied upon. Also, two or more major genes can now be combined in the same improved variety (Saxena, 1989). *Bph*1 and *bph*2 genes could not be combined because of close linkage, but *Bph*3 and *bph*4 segregate independently of *Bph*1 and *bph*2. Crosses have now been made to combine *Bph*1 and *Bph*3, *bph*2 and *Bph*3, *Bph*1 and *bph*4, and *bph*2 and *bph*4 genes. Such gene combinations will be effective for a longer time and will help retard biotype selection. Fortunately, several wild rice accessions have high levels of resistance to all three biotypes of *N. lugens* (IRRI, 1983a). Recently, such resistance has been successfully transferred from a wild species (*Oryza officinalis* Wall.) into cultivated rice (Khush and Jena, 1986).

To monitor biotypes and to identify resistant materials, International Rice Brown Planthopper Nurseries have been established in many countries through the International Rice Testing Program (IRTP). Each nursery consists of a uniform set of varieties. If particular rice varieties are resistant in one area but susceptible in another, the insect populations at the two locations are suspected to be of different biotypes.

Various behavioural and physiological responses involved in the establishment of *N. lugens* on selected susceptible and resistant rice varieties were examined by Saxena and Pathak (1977). Results showed that the factors that determined feeding and metabolic utilization of ingested food and hatching of eggs were important in varietal resistance of rice. Reduced quantities of food ingested from resistant varieties, followed by their insufficient utilization, led to poor growth of nymphs and reduced longevity

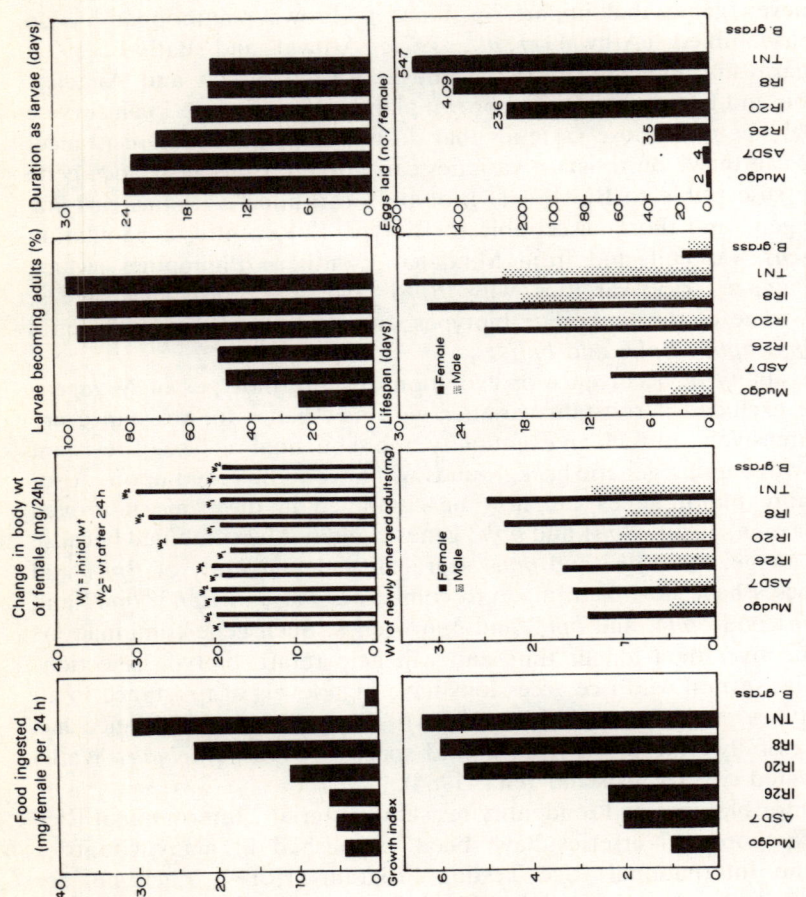

Figure 2. Intake, utilization of ingested food, growth, longevity and fecundity of *Nilaparvata lugens* (Biotype 1) on resistant and susceptible rice varieties, and on barnyard grass (Saxena and Pathak, 1979).

and egg production in adults (*Figure 2*). On susceptible varieties, on the other hand, greater ingestion of food and its better utilization satisfied all metabolic requirements and promoted better larval growth, adult survival and egg production. Indiscriminate egg-laying behaviour of *N. lugens* rendered all the test varieties equally suitable for oviposition, but fewer eggs hatched on the resistant rice varieties than on susceptible varieties.

Biotype 2 or Biotype 3 individuals of *N. lugens* ingested and utilized almost equal amounts of food from resistant rice varieties with *Bph*1 or *bph*2 genes, respectively, and on Biotype 1-susceptible rice varieties (*Figure 3*) (Saxena and Pathak, 1979). Such differential feeding response of the

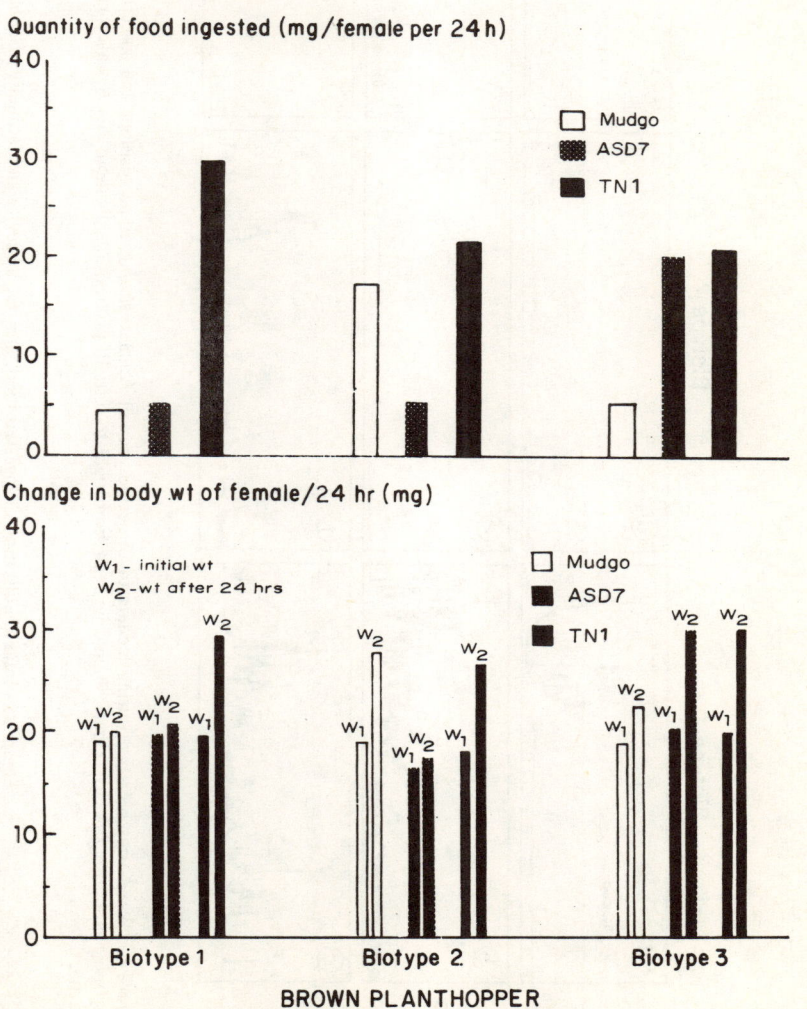

Figure 3. Quantity of food ingested and change in body weight of three biotypes of *Nilaparvata lugens* allowed to feed on 'Mudgo', 'ASD7' and 'TN1' rice varieties (Saxena and Pathak, 1979).

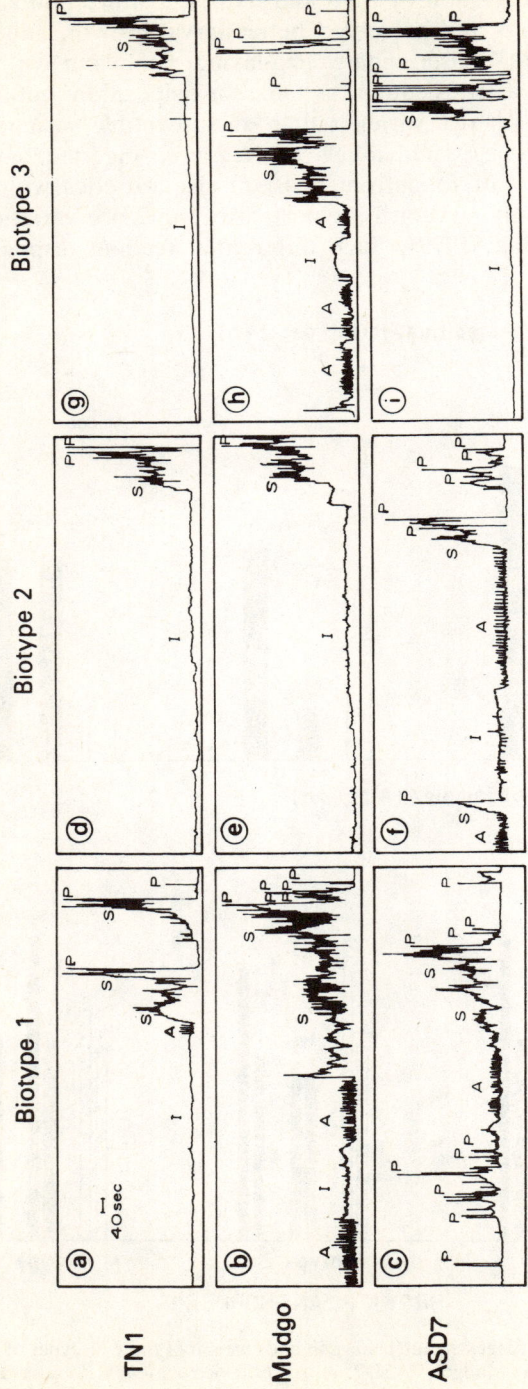

Figure 4. Waveforms recorded during feeding of three biotypes of *Nilaparvata lugens* on resistant and susceptible rice varieties using an electronic monitoring device. Charts should be read from right to left: P, probes; S, salivation; A, 'A waveform'; I, ingestion. (Khan and Saxena, 1988.)

three biotypes was also demonstrated using an electronic monitoring device (*Figure 4*) (Khan and Saxena, 1988). A homogeneous population of Biotype 1 was produced from single pair of males and females selected on the basis of their low food intake on rice varieties 'Mudgo' and 'ASD7' with *Bph*1 and *bph*2 genes, respectively. (Khan and Saxena, unpublished data). Nonetheless, resistance could not be traced to any morphological or anatomical peculiarities in the rice plant. Sogawa and Pathak (1970) suggested that reduced feeding by *N. lugens* on resistant rice varieties could be attributed to either a lack of phagostimulants or to the presence of antifeedants. The resistance of 'Mudgo' rice variety (*Bph*1 gene) to Biotype 1 was due to a somewhat low concentration of amino acids, particularly asparagine, which stimulates *N. lugens* feeding (Sogawa and Pathak, 1970). *trans*-Aconitic acid was identified as an antifeedant for *N. lugens* in the non-host barnyard grass *Echinochloa crus-galli* L., a common weed in rice fields (Kim *et al.*, 1976). On the other hand, soluble silicic acid and oxalic acid in the rice plant were reported to inhibit *N. lugens* feeding (Yoshihara *et al.*, 1979, 1980). But soon it was found that silicic acid was a general sucking inhibitor that occurs both in susceptible and resistant varieties. Likewise, oxalic acid was found to occur both in resistant and susceptible varieties, although its concentration was slightly higher in some resistant varieties (Yoshihara and Sogawa, 1979). Although both these acids are water soluble, their occurrence in the phloem sap has not yet been demonstrated. Silica, which is the elemental form of silicic acid, occurs in the soil and is more likely to be transported through the xylem vessels. On the other hand, oxalic acid, which is a product of the plant's cellular metabolism, is highly toxic even to the plant tissue and, therefore, less likely to occur in the phloem. Plants rich in oxalic acid (e.g. *Oxalis*) sequester this metabolite as crystals in the tonoplasm of vacuoles of the cells and not as free inclusions in the cell sap (Buvat, 1969).

Shigematsu *et al.* (1982) identified asparagine as a sucking stimulator, and β-sitosterol as a sucking inhibitor of *N. lugens*. However, none of these studies demonstrated that antifeedants extracted from the whole plants occurred principally in the phloem. Pure phloem sap from rice plants was collected by Kawabe, Fukumorita and Chino (1980) by severing, with a laser beam, the stylets of planthoppers and leafhoppers feeding on rice plants. But the analysis of the phloem sap indicated that, among carbohydrates, sucrose was the main translocate in the phloem of rice plants.

Systematic studies on the biochemical bases of resistance of rice varieties to *N. lugens* were conducted by Saxena and Pathak (1979), Saxena and Puma (1979) and Saxena and Okech (1985). The steam-distillate extracts of resistant rice varieties and of the non-host barnyard grass were found to repel *N. lugens*, and when applied topically, caused high mortality even at low doses. In contrast, extracts of susceptible varieties possessed moderate to high attractance and were relatively non-toxic to the insect. Obata *et al.* (1983) isolated and identified constituents of *N. lugens* attractant in the steam distillate of the Japanese rice cultivar 'Nihonbare'. Saxena and Okech (1985) demonstrated that *N. lugens* settling response on tillers of the susceptible 'TN1' rice variety sprayed with the steam-distillate extracts of resistant varieties showed the same pattern of response as when actual resistant plants were used. This indicated that treatment of the susceptible variety with steam-distillate extracts of resistant varieties conferred resistance, at least temporarily. The low amount of honeydew excreted by *N. lugens* females on the tillers of 'TN1' plants treated with the extract of resistant 'ARC 6650' or 'Ptb 33' varieties confirmed that the insect was unable to settle down for sustained feeding. Nymphs caged on similarly treated 'TN1' plants were unable to settle on them and suffered high mortality. Thus, the restlessness of *N. lugens* nymphs and adults on resistant plants could be attributed to exposure to the plant volatiles which have a repellent or toxic effect on the insect.

N. lugens Biotypes 1, 2 and 3 maintained on 'TN1,' 'Mudgo' and 'ASD7' rice varieties, respectively, showed distinct differences in relative vulnerability to the steam-distillate extracts of their respective resistant varieties and the barnyard grass (*Figure 5*) (IRRI, 1979; Saxena, 1986).

The exact identity of these allelochemicals in *N. lugens*-resistant varieties is not yet known. However, a large group of low-molecular-weight compounds, such as essential oils, particularly terpenoids, alcohols, aldehydes, fatty acids, esters, waxes, etc. would be obtained by steam distillation (Gunther, 1952; Robinson, 1983). Obata *et al.* (1983) identified a mixture of 14 esters, seven carbonyl compounds, five alcohols, and one isocynurate in the volatile attractant fraction of a *N. lugens*-susceptible Japanese rice cultivar 'Nihonbare'.

Cook *et al.* (1987) observed that *N. lugens* probed less frequently after surface exploration on resistant rice varieties 'IR46' and 'IR62' than on

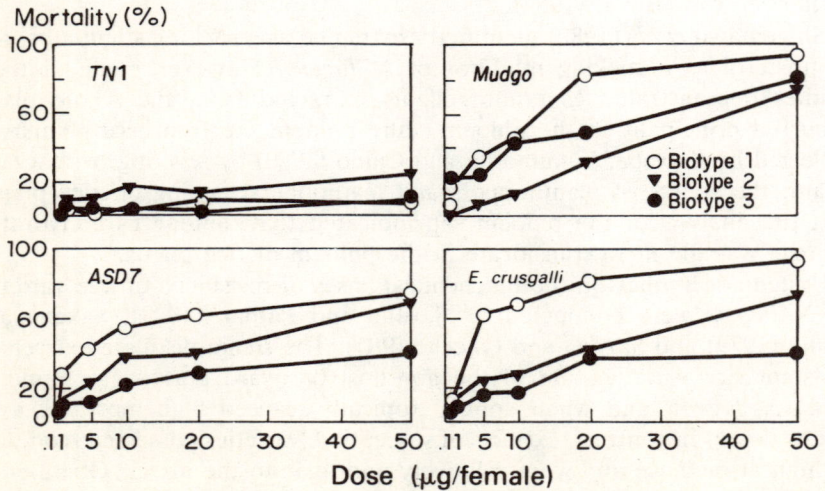

Figure 5. Mortality of brachypterous females on *Nilaparvata lugens* biotypes 24 h after topical application of steam-distillate extracts of different rice varieties and barnyard grass (Saxena, 1986).

susceptible 'IR22'. Reduced settling and probing on the resistant varieties was attributed to chemical cues, mainly the hydrocarbon- and carbonyl-containing fractions of the surface wax (Woodhead and Padgham, 1988).

While no major differences in total sugars and starch in *N. lugens*-susceptible and -resistant varieties have been reported, a comparison of the free amino acids occurring in the leafsheath tissues of 'TN1', 'Mudgo' and 'ASD7' varieties showed some quantitative differences (Saxena, 1986). Bioassays of seven of the major amino acids indicated differences in their relative degree of phagostimulation to the three *N. lugens* biotypes (*Figure 6*). Thus, certain nutrients may affect the suitability of the rice plant for *N. lugens* feeding. Allelochemicals and the nutritive balance of rice varieties are therefore important in eliciting optimal or suboptimal responses, and in affecting the ability of *N. lugens* to establish on rice plants.

Panda and Heinrichs (1983) evaluated selected varieties with moderate levels of resistance to *N. lugens* to determine the mechanisms of resistance. Variety 'Utri Rajapan' had a high level of tolerance and no antibiosis. On the other hand, resistance in 'IR46,' 'Kencana' and 'Triveni' varieties was due to a combination of antibiosis and tolerance. The causal mechanism for such tolerance in 'Utri Rajapan' is not clearly understood.

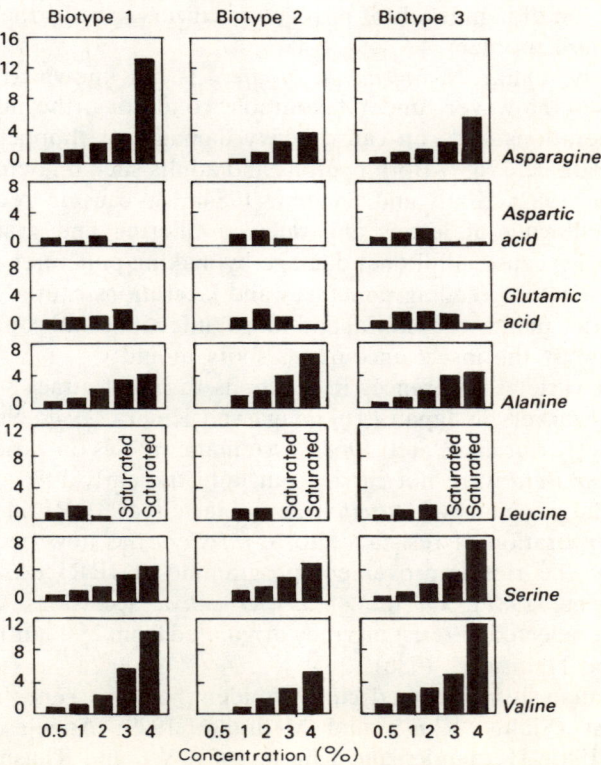

Figure 6. Relative intake of seven amino-acid solutions of different concentrations by three biotypes of *Nilaparvata lugens* (Saxena, 1986).

Using an electronic monitoring device, Velusamy and Heinrichs (1986) recorded feeding by *N. lugens* Biotype 3 females on resistant 'IR56,' field resistant 'IR36' and 'IR42', tolerant 'Utri Rajapan' and susceptible 'TN1' rice plants. Phloem ingestion from 'Utri Rajapan' was significantly less than from susceptible 'TN1', but significantly more than from resistant 'IR56' and field resistant 'IR36' and 'IR42'. Likewise, although fewer *N. lugens* Biotype 2 females settled on and fed significantly less from tolerant 'Utri Rajapan' than susceptible 'TN1', the insect's growth and development, longevity and rate of population increase were as high on 'Utri Rajapan' as on 'TN1' (Saxena, 1989). The allelochemical and nutritional factors that influence *N. lugens* behaviour on 'Utri Rajapan' are being investigated.

THE WHITEBACKED PLANTHOPPER

Sogatella furcifera (Horváth) is emerging as a serious pest of rice in many Asian countries, particularly in areas where varieties resistant to *N. lugens* have been grown successfully (Heinrichs and Rapusas, 1983). Generally, *N. lugens* maintains a numerical superiority over *S. furcifera*, but on varieties resistant to *N. lugens*, *S. furcifera* tends to multiply faster. Continuous cropping; reduced genetic variability of short-statured, high-yielding varieties; and application of high levels of nitrogen fertilizers have further accentuated the *S. furcifera* problem.

Fortunately, unlike *N. lugens, S. furcifera* is not known to transmit rice virus diseases. However, under favourable conditions, the insect produces several generations and can cause heavy damage, or 'hopperburn', in the rice crop (Pathak, 1968). Both nymphs and adults suck phloem sap (Auclair and Baldos, 1982; Khan and Saxena, 1984a,b), causing reduced vigour, stunting, yellowing of leaves and delayed tillering and grain formation. Gravid females cause additional damage by making punctures for egg-laying in the leaf sheaths. Feeding punctures and lacerations caused by ovipositor predispose rice plants to bacterial and fungal infections and copious excretion of honeydew by the insect encourages sooty mould.

Although varietal differences in reactions to insect attack were noted by some early workers in Japan (Tokunaga and Kidera, 1948; Suenaga, 1950) and in India (Kittur and Patel, 1968), systematic studies on varietal resistance against *S. furcifera* were not carried out until the early 1970s. Screening of rice germplasm against *S. furcifera* was started at IRRI in 1971 (IRRI, 1972). Incorporation of resistance to *S. furcifera* has now become a major objective of the rice improvement programme at IRRI (Angeles, Khush and Heinrichs, 1981). Of the 48 554 *O. sativa* accessions screened, 401 (0.8%) were selected as resistant; they originated from 21 countries (Romena, Rapusas and Heinrichs, 1986).

Genetic analysis of selected cultivars identified five genes for resistance to this pest (Sidhu, Khush and Medrano, 1979; Angeles, Khush and Heinrichs, 1981; Hernandez and Khush, 1981; Wu and Khush, 1985). Four of these are dominant and designated as *Wbph*1, *Wbph*2, *Wbph*3 and *Wbph*5; one is recessive and designated as *wbph*4. In addition to major genes, minor genes also contribute to resistance against *S. furcifera* in some rice varieties (IRRI, 1983b) and probably retard the selection of biotypes. Because no modern variety is highly resistant to *S. furcifera*, attempts are being made to incorporate resistance genes into improved breeding lines. By wide crossing of cultivated rice and wild rice *Oryza officinalis*, and through subsequent backcrossing, breeding lines have been obtained which are highly resistant to the pest.

The insect thrives on high-yielding, susceptible rice cultivars, but fails to feed, grow, survive and reproduce adequately on resistant cultivars (Choi *et al.*, 1973b; Lee and Park, 1976; Heinrichs and Rapusas, 1983; Khan and Saxena, 1985c) (*Figure 7*). No mechanical barriers to *S. furcifera* feeding have been identified in resistant plants. Reduced feeding on them was attributed to lack of phagostimulates or to the presence of deterrents in the phloem tissue (Pablo, 1977).

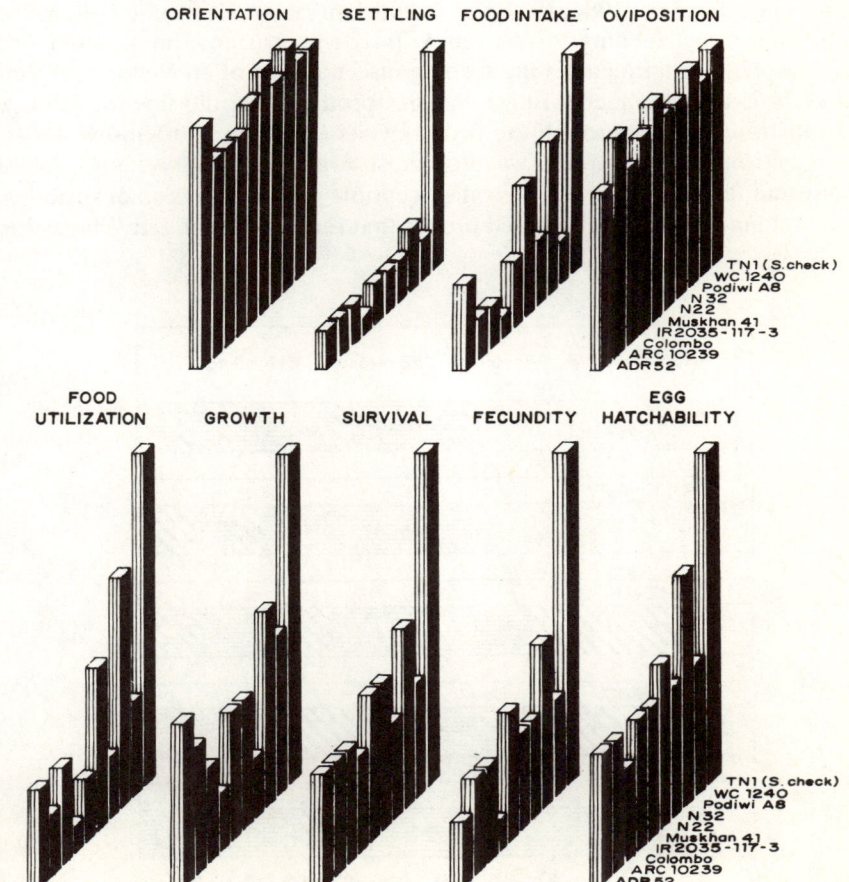

Figure 7. Relative intensity of various behavioural and physiological responses of *Sogatella furcifera* towards resistant and susceptible rice varieties (modified from Khan and Saxena, 1985c).

In resistant rice varieties, the mechanisms that operate against the pest are non-preference for settling and feeding, antibiosis, or a combination of these factors. A greater preference for settling on susceptible varieties than on resistant ones has been reported for *S. furcifera* adults (Rodriguez-Rivera, 1972; Pablo, 1977; Choi *et al.*, 1982; Khan and Saxena, 1985c) and nymphs (Choi *et al.*, 1982; Saxena and Khan, 1984). The insect made more salivary marks in leaf sheaths and leaf blades of resistant plants than in susceptible plants (Rodriguez-Rivera, 1972; Pablo, 1977). Newly emerged females exhibited a distinct non-preference for feeding on resistant varieties (IRRI, 1977). On susceptible 'TN1' plants, they settled down quickly and fed continuously. Feeding was not sustained on resistant plants; often, the insect did not feed for the first 20–40 minutes and, thereafter, feeding periods were short and interrupted (*Figure 8*) (IRRI, 1977). The assimilation of the ingested food was greatly reduced on resistant varieties (Khan and Saxena, 1985c).

Khan and Saxena (1984b) used a DC variant of an electronic system for monitoring insect feeding to confirm *S. furcifera* resistance in selected rice varieties. The technique uses the feeding insect as part of an electrical circuit and records the amplified voltage changes produced by the flow of salivary and substrate liquids through the insect stylets (McLean and Kinsey, 1964). The electronically recorded waveforms showed that *S. furcifera* probed readily and fed for longer periods on susceptible plants; on resistant varieties, the insect made brief and repeated probes that reduced the effective ingestion

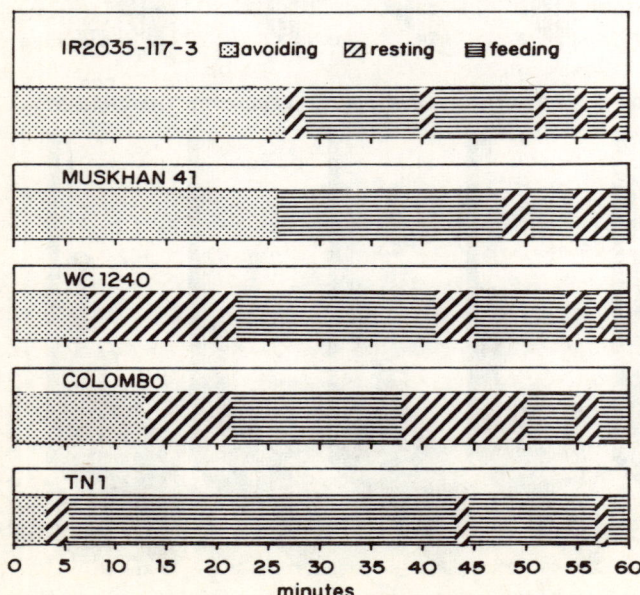

Figure 8. Arrival and duration of feeding of *Sogatella furcifera* on susceptible and resistant rice varieties (IRRI, 1977).

period (*Figure 9*). Thus, in addition to other established screening procedures, electronic recording of the feeding activity of *S. furcifera* could also be used as a powerful tool for evaluating and confirming resistance in seleted rice germplasm.

The resistant varieties adversely affected *S. furcifera* nymphal growth and adult longevity and fecundity (Choi *et al.*, 1973b, 1982; Lee and Park, 1976; Heinrichs and Rapusas, 1983; Khan and Saxena, 1985c). The insect's ovipositional preference differed from variety to variety (Choi *et al.*, 1982), but there was no definite trend. Pablo (1977) also reported that in an ovipositional preference test, 20 of 25 resistant and susceptible varieties received almost equal numbers of eggs. Khan and Saxena (1985c) found that the insect's ovipositional response was identical in susceptible and resistant varieties. Despite indiscriminate egg-laying by females on resistant and susceptible varieties, egg hatchability was markedly reduced on resistant varieties (Rodriguez-Rivera, 1972; Khan and Saxena, 1985c). In contrast, Heinrichs and Rapusas (1983) indicated that hatchability of *S. furcifera* eggs was not affected by varietal resistance. Nevertheless, all reports confirmed that the rate of insect's population increase was higher on susceptible than on resistant varieties (Choi *et al.*, 1982; Heinrichs and Rapusas, 1983; Khan and Saxena, 1985c).

Pablo (1977) correlated *S. furcifera* resistance with the nutritive value of the host plant. He indicated that plants containing a high ratio of sugar to amino nitrogen were susceptible. The resistant varieties were deficient in certain amino acids, such as alanine, asparagine and aspartic acid.

Kim and Heinrichs (1982) investigated the role of silica levels in resistance to *S. furcifera*. They observed that seedlings of variety 'N22' grown in a culture solution of silica (SiO_2) adversely affected *S. furcifera* development and survival. However, the investigation would have been more meaningful if a susceptible variety instead of the resistant 'N22' had been used, as the insect mortality is high on 'N22' plants anyway.

Khan and Saxena (1986) investigated the role of volatile allelochemicals from resistant rice varieties in imparting resistance to *S. furcifera*. Their findings indicated that odoriferous and volatile chemicals had a profound effect on the behaviour and biology of *S. furcifera*. Odours of extracts of susceptible rice plants attracted *S. furcifera* females, whereas the odours of resistant varieties repelled them. Extracts of resistant rice varieties were more toxic to first instars and newly emerged females than extracts of susceptible plants. Ingestion and assimilation of food by *S. furcifera* were reduced on susceptible plants treated with resistant extract, whereas application of susceptible-plant extract on resistant plants did not increase feeding or assimilation. Mature virgin females treated topically with resistant-plant extract stopped emitting mating signals and failed to mate with normal responsive males. Khan and Saxena (1986) also demonstrated that even external application of resistant plant extract to susceptible plants conferred resistance to the insect, indicating that susceptibility or resistance of rice cultivars to *S. furcifera* is determined by allelochemicals present in rice plant volatiles. Using a resistant wild rice, *Oryza officinalis*, and two of its

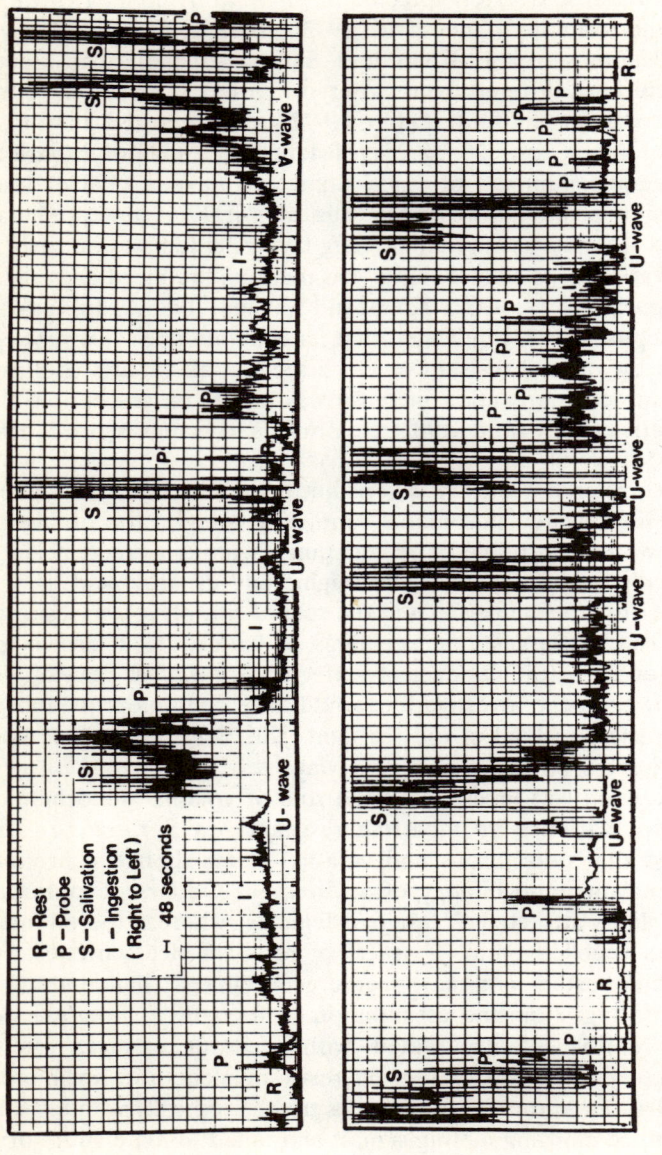

Figure 9. Electronically recorded waveforms associated with the feeding behaviour of *Sogatella furcifera* females on susceptible 'TN1' (top) and resistant 'IR2035-117-3' (bottom) rice varieties. The insect probed readily and fed longer on the susceptible variety. In contrast, the insect made brief and repeated probes on the resistant variety, consequently reducing the effective ingestion period (Khan and Saxena, 1984b).

elite lines, 'IR54742-23-1-29-18' and 'IR54751-2-41-10-5', Ye (1988) also demonstrated that food intake and assimilation, growth, longevity, fecundity and population increase of *S. furcifera* were adversely affected.

THE SMALL BROWN PLANTHOPPER

Laodelphax striatellus (Fallén) is widely distributed in temperate areas of China, Japan and Korea. The insect causes considerable direct damage to the rice crop in those countries. It is the vector of rice stripe, the most serious disease of rice in East Asian countries, and also transmits the black-streaked dwarf virus.

None of the *japonica* varieties has resistance to *L. striatellus*. Several resistant 'Tong-il' types derived from 'IR667' have been released (Choi and Song, 1974). Another rice variety, 'Milyang 30', has multiple resistance to *L. striatellus*, *N. lugens* and *N. cincticeps* (Kim, Lee and Park, 1983). 'Milyang 30' and other 'Tong-il' type varieties are widely grown in the southern part of Korea where *L. striatellus* vectors black-streaked dwarf virus from winter barley or wheat crops to rice in spring. In addition, several varieties such as 'ASD7', 'Vellailangalayan', 'Manavari', 'CO22', 'Mudgo', 'Ptb18', 'IR8', 'Suweon 213', 'Suweon 214' and 'Suweon 215' are highly resistant to the insect in Korea (Choi and Song, 1974; Choi *et al.*, 1974; Khush, 1977). 'Murunga 137' and 'Pannetti' are highly resistant in Taiwan (Khush, 1977), whereas 'Konaso', 'Tadukan' and 'Hu-nan-tsao' varieties are resistant in Japan (Okamoto and Inoue, 1967).

The causes of resistance to *L. striatellus* have not been intensively investigated. Insect resistance was suggested to be highly related with non-preference for feeding and antibiosis (Choi *et al.*, 1974). The planthopper nymphs caged on resistant varieties showed poor growth, high mortality and longer growth periods (Okamoto and Inoue, 1967; Choi *et al.*, 1974). Adults reared on the resistant varieties were small in body size, had a shorter lifespan, and females laid fewer eggs as compared with those reared on susceptible varieties. However, the ovipositional response of the insect was identical on susceptible and resistant rice varieties. Resistance of the 'Tong-il' variety to *L. striatellus* was attributed to the insect's reduced feeding and to greater tolerance of 'Tong-il' plants to damage than was found in 'Jinheung' and other susceptible rice varieties (Choi, 1974; Choi and Song, 1974).

Kurata and Sogawa (1976) concluded from a laboratory bioassay that aromatic amines deterred continuous feeding of *L. striatellus* on resistant rice plants. The aromatic amines—phenethylamine hydrochloride, tyramine hydrochloride and hordenine sulphate—at 1000 p.p.m. in a 5% sucrose solution inhibited sucking by the insect. On the other hand, Kim, Koh and Fukami (1985) isolated from rice plants eight *C*-glycosylflavones that stimulated the planthopper's probing activity.

No biotypic variation has been reported in the insect population, but genetic variation in the ability of the planthopper to acquire the rice stripe virus has been demonstrated (Kisimoto, 1967). Inheritance of resistance has not been studied, but Korea has an active programme of breeding for *L. striatellus* resistance (Choi *et al.*, 1976).

THE RICE DELPHACID OR SOGATA

Sogatodes oryzicola (Muir) is an important pest of rice in Central America, northern South America and the Caribbean islands. It is the only significant vector of the 'hoja blanca' virus and also causes appreciable loss by direct feeding (Galvez, 1969).

All tropical American land races and all *japonica* rice varieties are highly susceptible to sogata. Resistance gene sources from Asia, which is beyond the distribution areas of *S. oryzicola*, have been utilized in the breeding of *S. oryzicola*-resistant varieties (Jennings and Pineda, 1970a). Varieties resistant to *S. oryzicola* were adopted in 1972, after which insect populations in rice fields declined dramatically (Jennings and Pineda, 1983). Thereafter, no serious feeding damage has been reported from Latin America. Three resistant rice varieties, 'CICA4,' 'CICA6' and 'CICA8', were released jointly by Centro Internacional de Agricultura Tropical (CIAT) and Instituto Colombiano Agropecuario. Presently, CIAT and Cuba have excellent breeding programmes against sogata. Scientists in Peru and Ecuador are also evaluating promising materials for *S. oryzicola* resistance. CIAT also has a breeding programme for resistance to 'hoja blanca' virus which reappeared in 1981 in epidemic form. Resistance to *S. oryzicola* and 'hoja blanca' virus are inherited independently.

Resistance to *S. oryzicola* is apparently not associated with any morphological trait in the rice plant, including height, pubescence or any other plant character (Jennings and Pineda, 1983). Deleterious effects of resistant rice plants on the insect revealed that antibiosis is a major component in *S. oryzicola* resistance (Jennings and Pineda, 1970b). All phases of the insect life cycle were adversely affected on resistant varieties. Resistance reduced oviposition and number of eggs hatching, decreased nymphal survival, prolonged the growth period and reduced adult longevity. Strong nonpreference for oviposition on some resistant rice varieties was also evident (Jennings and Pineda, 1983).

All varieties released by CIAT are still highly resistant, with no indication of biotype development (Jennings and Pineda, 1983). However, it is suspected that *S. oryzicola* populations in Cuba and Colombia are different biotypes (Orellana, Jennings and Perez, 1982).

Inheritance of resistance to *S. oryzicola* has not been intensively studied. Jennings and Pineda (1970a) noted that the resistance trait was highly heritable and easily combined with other desirable agronomic traits. As the few genetic studies attempted have given confusing and inconsistent results, the genetics of resistance to the pest is not well understood (Jennings and Pineda, 1983).

Rice leafhoppers

Like planthoppers, leafhoppers are also important pests of rice. They remain confined to the upper parts of the plants but cause considerable damage by direct feeding, resulting in reduced vigour of the crop and a decreased number of filled grains. Although only a few cases of 'hopperburn' caused by leafhoppers have been reported, occasionally, under heavy population pressure, the crop turns yellowish.

Leafhoppers are vectors of several serious virus diseases of rice in Asia. Being active, they readily spread virus diseases, particularly those caused by stylet-borne viruses. Research on the control of leafhoppers through host-plant resistance has been increasingly emphasized in recent years, especially due to the outbreak of virus diseases. Discussion of resistance to the major leafhopper species which attack rice follows.

THE RICE GREEN LEAFHOPPER [*NEPHOTETTIX VIRESCENS* (DISTANT)]

This is the most important pest of rice in tropical Asia. This pest has caused heavy crop losses throughout South and Southeast Asia as a vector of tungro viruses (Alam and Islam, 1959; Britton, Dam and Tao, 1962; Ling, 1969; Banerjee, 1971; Mishra and Kaushik, 1976; Bergonia, 1978; Heinrichs, 1979; Hibino, 1987).

A major strategy in the management of *N. virescens* populations and tungro disease is the development of *N. virescens*-resistant rice varieties. Resistance to this pest in rice varieties is now well documented. Systematic screening of rice varieties against *N. virescens* began in 1965 and two rice varieties—'Pankhari 203', a tall Indica, and 'IR8', a dwarf, high-yielding IRRI rice variety—were identified as resistant to the insect (IRRI, 1967, 1968; Pathak, Cheng and Fortuno, 1969). Later on, several resistant sources were identified after screening several thousand rice varieties (Pathak, 1969, 1972; Cheng and Pathak, 1972; Rezaul Karim, 1978). In the genetic evaluation of 50 000 rices from the International Rice Germplasm Bank (IRGC), using the standard seedbox method for screening 7-day-old seedlings, 1200 accessions have been selected for *N. virescens* resistance (Heinrichs, 1986). Several resistant sources in India were identified by Shastry *et al.* (1971), Seshu *et al.* (1974) and Chelliah and Hanifa (1981). Resistant germplasm also comes from several other countries of South and Southeast Asia. So far, seven genes for resistance to green leafhopper have been identified. Of these seven genes, six are dominant and are designated as *Glh*1, *Glh*2, *Glh*3, *Glh*5, *Glh*6 and *Glh*7, while the one recessive gene was designated as *glh*4 (Athwal *et al.*, 1971; Siwi and Khush, 1977; Rezaul Karim and Pathak, 1982). The first five genes convey resistance to Philippine populations of green leafhopper, while *Glh*6 and *Glh*7 provide resistance to Bangladesh populations of the insect (Rezaul Karim and Pathak, 1982).

Although there are no confirmed reports of sympatric biotypes of *N. virescens* in the Philippines, virulent populations were selected at **IRRI** on resistant rice varieties 'Pankhari 203' (*Glh1*), 'IR8' (*Glh3*), 'Ptb8' (*glh4*), 'TAPL 796' (*Glh6*), and 'Moddai Karuppan' (*Glh7*) (IRRI, 1983c; Rapusas and Heinrichs, 1985). However, no such population could be selected on 'ASD7' (*Glh2*) and 'ASD8' (*Glh5*) rice varieties (Heinrichs and Rapusas, 1984). *N. virescens* populations in different countries show consistent differences in their virulence to resistant rice varieties. For example, *N. virescens* populations in Bangladesh and the Philippines were different (Rezaul Karim and Pathak, 1979; Saxena, Barrion and Soriano, 1985). The varieties 'Pankhari 203', 'ASD7', 'IR8' and 'ASD8' resistant to green leafhopper in the Philippines were susceptible in Bangladesh. Recently, high levels of resistance to the green leafhopper have been reported in many wild rices (Razzaque and Heinrichs, 1985c; Velusamy *et al.*, 1986). However, there is no information regarding the resistance of these wild rices to rice tungro virus.

The mechanism of resistance to *N. virescens* was studied by Pathak, Cheng and Fortuno (1969), Cheng and Pathak (1972) and Rezaul Karim (1978) in several rice varieties. The resistant varieties always suffered distinctly less (or no) damage as compared to the susceptible varieties (*Figure 10*). Both adults and nymphs exhibited preference for susceptible varieties and caused greater damage to them than to the non-preferred resistant varieties. Nymphs suffered higher mortality, and their developmental period was more prolonged on resistant varieties than on susceptible ones. Thus, only 0–3% of first-instar *N. virescens* nymphs reached the adult stage on resistant varieties, while 76–90% of nymphs become adult on susceptible varieties (*Figure 10*) (Cheng and Pathak, 1972).

Auclair, Baldos and Heinrichs (1982) reported that *N. virescens* damages resistant rice varieties less because the leafhopper's intake of phloem sap from such varieties is low and its drinking is restricted to the xylem tissue. Using a lignin-specific dye as well as an electronic monitoring device, Khan and Saxena (1984a, 1985b) confirmed that *N. virescens* is primarily a phloem feeder on susceptible varieties, but on resistant varieties it drinks mainly from the xylem tissue. Auclair, Baldos and Heinrichs (1982) speculated that these differences in feeding sites were due to the presence of a feeding deterrent in tissues adjacent to or within the sieve elements. If a leafhopper feeds less in the phloem, its chances of acquiring or transmitting the virus are considerably reduced.

Khan and Saxena (1985a) monitored the leafhopper's feeding behaviour on susceptible 'TN1' plants sprayed with steam distillate extracts of resistant 'ASD7' plants, using the electronic monitoring device and the lignin-specific dye. Application of 'ASD7' extract to susceptible 'TN1' plants disrupted the normal feeding behaviour of the insect (*Figure 11*). Phloem feeding by the insect was significantly less on 'TN1' plants sprayed with 'ASD7' extract at 500, 1000, 2000 and 4000 p.p.m. than on control 'TN1' plants sprayed with acetone/water mixture. The reduced phloem-feeding on the extract-treated plants was associated with a significant increase in probing frequency and an increase in duration of salivation and xylem feeding.

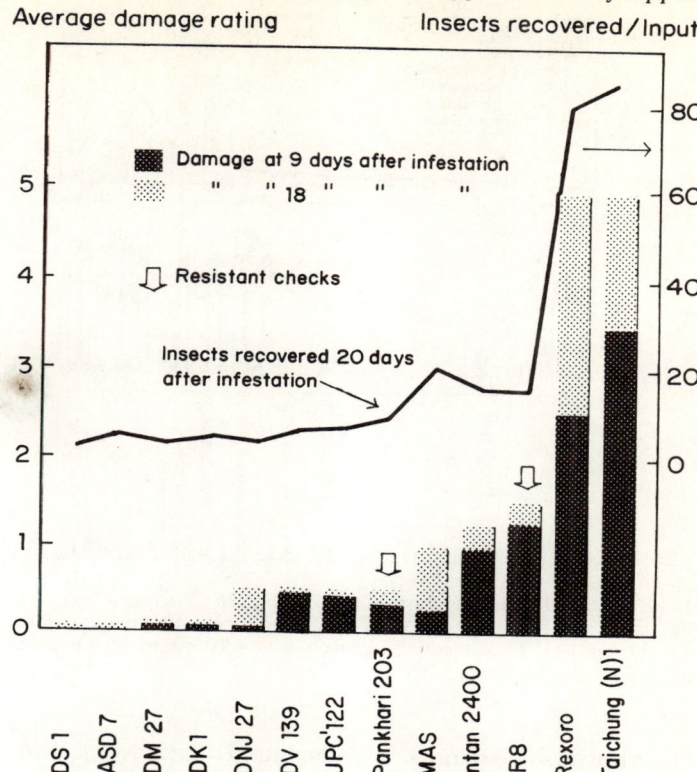

Figure 10. Survival and damage by 100 *Nephotettix virescens* second-instar nymphs on 20-day-old seedlings of different varieties of rice (Cheng and Pathak, 1972).

These results stressed the role of rice plant volatiles in determining insect behaviour. Most plant volatiles are sparingly soluble in water and are not likely to be translocated in the vascular bundles (McKey, 1979). However, their odoriferous and volatile nature makes them exert a strong influence on the total chemical environment of the rice plant, and, hence, is of ecological significance in determining the susceptibility or resistance of rice plants to insect pests.

THE RICE GREEN LEAFHOPPER [*NEPHOTETTIX CINCTICEPS* (UHLER)]

This leafhopper occurs in Taiwan, Korea, Japan and China and causes considerable damage to the rice crop in these countries. It is reported that the pest destroyed 240 000 tons of rice in Japan in 1940 (Yoshimeki, 1967). The insect is also a vector of the rice dwarf and yellow dwarf virus diseases.

Resistance to *N. cincticeps* is now well documented. Systematic screening of rice varieties against this pest started in 1960 in Japan. Through field

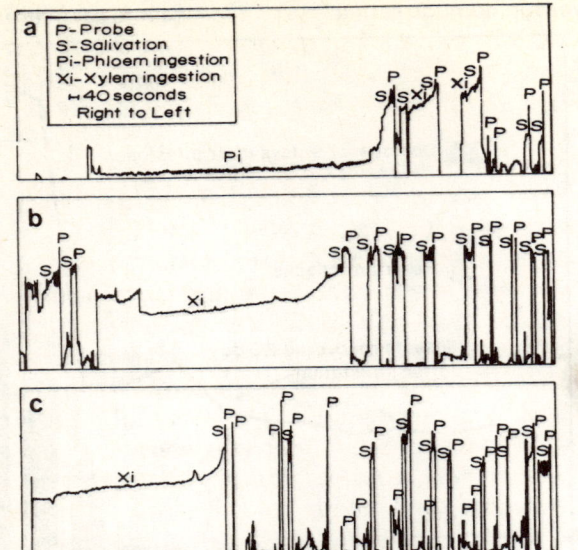

Figure 11. Electronically recorded waveforms during *Nephotettix virescens* feeding on (a) susceptible 'TN1' and (b) resistant 'ASD7' rice seedlings, both sprayed with acetone/water mixture (4 : 1); and (c) 'TN1' seedlings sprayed with 4000 p.p.m. steam-distillate extract of 'ASD7' plants (Khan and Saxena, 1985a).

tests and artificial inoculations, it was found that several *indica* varieties showed considerable degrees of resistance to this pest, while most of the *japonicas* were susceptible (Yasuo, Yamaguchi and Ishni, 1960; Ishii, Yasuo and Yamaguchi, 1969; Kimura *et al.*, 1969; Sakurai, 1969; Kobayashi, 1983). Breeding for resistance to *N. cincticeps* and dwarf virus disease has now become an important aspect for their control (Kobayashi, 1983). Several rice varieties such as 'Murunga 137', 'Pannetti', 'H105', 'IR4-93', 'Bir-tsan-3', 'MTU-15', 'DK-1', 'DV-139', 'ASD7' and 'Ptb18' have been identified as resistant in Japan, Korea and Taiwan (Ishii, Yasuo and Yamaguchi, 1969; Choi *et al.*, 1973a; Khush, 1977).

Several studies on the mechanism of resistance to *N. cincticeps* showed that antibiosis and non-preference are important resistant factors (Choi *et al.*, 1973a; Choi, 1975; Kishino and Ando, 1979; Ōya and Sato, 1980; Sekizawa and Ogawa, 1980). Nymphs caged on resistant rice varieties suffered higher mortality, had a lower rate of adult emergence and had a slower rate of growth than on susceptible varieties. The longevity of adults and fecundity of females were also distinctly reduced on resistant varieties. The resistant varieties at seedling stage were also less preferred for settling by the leafhopper, but ovipositional preference was not different between susceptible and resistant rice varieties (Choi, 1975; Ōya and Sato, 1980).

Using an electronic monitoring device for recording insect feeding, Kawabe (1985) demonstrated that on susceptible rice varieties, the insect ingested mainly from the phloem, whereas on resistant varieties, it fed from xylem vessels. Reduced intake from the phloem of resistant varieties was attributed to the possible presence of a feeding deterrent in the sieve elements of resistant plants.

Studies on the inheritance of resistance to *N. cincticeps* and dwarf virus disease were conducted in Japan by Kobayashi (1983). Resistance to *N. cincticeps* was found to be controlled by a single dominant gene in 'Kanto PL3' and '77-2059' rice varieties and by two dominant genes in 'Pe-bi-hun', 'Te-tep', 'Hachijuushisen', 'Chiem-chank', 'Doujinkyou' and 'Tadukan'. The resistance gene in 'IR24' was found to be allelic to one of the two resistance genes in 'Pe-bi-hun'. Linkage relationship studies of genes controlling insect resistance revealed that in varieties 'Kanto PL3' and '77-2059', non-preference and antibiosis were controlled by the same resistance gene, whereas in 'IR24', genes controlling non-preference and antibiosis were different.

Sato and Sogawa (1981) reported biotypic variation of the insect in Japan. Two distinct populations showing significant intraspecific variations with respect to their responses to rice varieties were collected from Joetsu and Chikugo (Sogawa and Sato, 1981). Morphological and physiological characteristics of the two biotypes confirmed such intraspecific variations (Sogawa and Sato, 1983).

THE RICE GREEN LEAFHOPPER [*NEPHOTETTIX NIGROPICTUS* (STÅL)]

N. nigropictus is widely distributed in rice-growing areas of tropical Asia (Sajjan, 1972; Dhawan and Sajjan, 1976; Razzaque, 1984). The insect directly damages the rice plant by sucking the sap and plugging the vascular bundles with stylet sheaths. Although *N. nigropictus* populations seldom reach the economic injury level, severe damage may be caused as the pest can transmit tungro viruses, yellow dwarf, yellow-orange leaf virus and viruses causing rice dwarf, transitory yellowing and rice gall dwarf diseases (Pathak, 1968; Ishihara, 1969; Surin, Kerdchokchai and Surin, 1974; Morinaka *et al.*, 1982; Siwi *et al.*, 1987). Although a less efficient vector than *N. virescens*, *N. nigropictus* has definitely been incriminated in the transmission of tungro viruses in the Philippines and Indonesia (Ling, 1970; Siwi *et al.*, 1987). *N. nigropictus* is reported to be more abundant in weedy rice fields than in weeded rice fields (Cheng and Pathak, 1972).

Although more than 1000 cultivars in the rice germplasm collection at IRRI have been identified as resistant to *N. virescens* (Heinrichs, 1986), no detailed study has been undertaken to investigate varietal resistance to *N. nigropictus*. Only a few cultivars and wild rices have been identified as resistant to *N. nigropictus* in limited tests conducted at IRRI and in India (Sajjan, 1972; Viswanathan and Kalode, 1981; Razzaque, 1984; Razzaque and Heinrichs, 1985a,b,c). Although IR varieties have not been bred for *N. nigropictus* resistance, most are resistant or moderately resistant to this pest.

Non-preference and antibiosis mechanisms contribute to resistance against *N. nigropictus* (Mukhopadhyay and Chattopadhyay, 1975; Razzaque, 1984; Viswanathan and Kalode, 1984) but the exact causes of resistance are not known.

Biotype variation in the insect has not been reported. Studies on inheritance of resistance and breeding for resistance have not been undertaken yet.

THE ZIGZAG LEAFHOPPER

Recilia dorsalis (Motschulsky) is a pest of rice in South and Southeast Asia, Japan and Taiwan (Pathak, 1968). It is one of the most abundant species in the northern areas of Thailand (Pongprasert, 1974). The insect is a vector of rice gall dwarf virus in Japan (Nasu, 1967) and Thailand (Morinaka *et al.*, 1982), and of the orange leaf virus in the Philippines (Rivera, Ou and Pathak, 1963; Wathanakul, 1964) and Thailand (Chaimongkol, Wathanakul and Lamey, 1971).

Screening for resistance to *R. dorsalis* began at IRRI in 1973. About 2500 rice accessions and 500 wild rices have been screened against this pest. A few national programmes are also involved in screening rice germplasm against the pest. Several rice varieties such as 'Ptb21', 'Ptb27', 'Ptb33', 'Balamawee' and 'Rathu Heenati' were selected as highly resistant at IRRI (Pongprasert, 1974), whereas 'Su-Yai 20', 'Muthumanikam', 'Ptb18' and 'Vellanlangayan' were found to be highly resistant in Korea (Choi, Song and Park, 1973). However, only few IR varieties such as 'IR56', 'IR58' and 'IR62' show some degree of resistance to the pest (Heinrichs, 1986).

The mechanism of resistance to *R. dorsalis* in rice varieties was identified mainly as non-preference for feeding and antibiosis (Choi, Song and Park, 1973; Pongprasert, 1974). Varieties preferred by the adults were also generally reported to be preferred by nymphs. However, preferences for feeding and oviposition were not related to each other. The survival rate of nymphs caged on resistant varieties was significantly less than that of caged nymphs on susceptible varieties. Also, adult longevity and female fecundity were adversely affected on resistant varieties. The cumulative effect of the resistant plants was reflected on the insect population—30 days after caging, the leafhopper population was several times greater on the susceptible than on resistant varieties. Leafhoppers caged on resistant plants made more probing punctures but excreted less honeydew than those feeding on a susceptible variety. The causes of insect resistance in rice varieties were not investigated.

The mode of inheritance of resistance to *R. dorsalis* in three rice varieties, 'Rathu Heenati', 'Ptb21' and 'Ptb33', was studied by Angeles, Khush and Heinrichs (1986). Resistance to the leafhopper in 'Rathu Heenati', 'Ptb21' and 'Ptb33' is conditioned by single dominant genes $Zlh1$, $Zlh2$ and $Zlh3$, respectively. Tests for allelism showed that these genes are non-allelic and segregate independently of each other.

THE WHITE LEAFHOPPER

Cofana spectra (Distant) is widely distributed throughout South and Southeast Asia (Heinrichs, Medrano and Rapusas, 1985). Nymphs and adults remove plant sap and cause leaf yellowing and typical 'hopperburn' of the crop. Unlike other leafhoppers, the white leafhopper does not transmit rice virus diseases.

Rearing and greenhouse screening methods have been developed by Sekar and Chelliah (1983). Field evaluation methods were developed by Velusamy *et al.* (1975). Seven sources of *C. spectra* resistance have been reported from India (Velusamy *et al.*, 1975; Sekar and Chelliah, 1983): 'ADT14', 'ASD3', 'ASD5', 'ASD8' 'CH2', 'CO2' and 'CO29'.

Non-preference and antibiosis components in the selected resistant rice varieties reportedly contributed to resistance (Sekar and Chelliah, 1983). Resistant varieties were less preferred by the adult leafhoppers than the susceptible ones. Nymphs caged on the resistant varieties had a longer nymphal period and reduced survival rates. Adults feeding on resistant plants had shorter longevity and reduced fecundity. Honeydew excreted by insects feeding on the resistant varieties contained more total sugars and reducing sugars than the honeydew excreted on susceptible varieties, which had a greater total free-amino-acid content.

Future considerations and conclusions

The use of resistant rice varieties will remain the most logical and economical way of reducing planthopper and leafhopper damage in rice. Therefore, practical implications of a thorough understanding of the causes of insect resistance in plants are tremendous. Identification of the factors that confer resistance or susceptibility and the study of their inheritance in rice plants would greatly improve breeding strategies for resistant varieties. A proper understanding of the mechanisms of host-plant resistance will also lead to breeding for long-term resistance. Increased understanding of resistance factors will pave the way for manipulation of insect behaviour for use in pest-management programmes.

Wild relatives of cultivated rice, which are rich sources of several useful genes for insect and disease resistance, should be identified and utilized in the breeding programmes. The resistance genes can be transferred to cultivated rice varieties using innovative breeding techniques. Such genes from distant relatives can be useful tools for fortifying the primary gene pool of *O. sativa*.

Although major genes will continue to be useful for years to come, efforts are needed in identifying and utilizing minor genes for planthopper and leafhopper resistance. Sources of resistance to all biotypes need to be identified and genetically analysed.

With the increasing importance of several planthoppers and leafhoppers in recent years, there is a great need to develop varieties with multiple resistance. Development of improved plant type varieties with multiple resistance to important insect pests and diseases has been the major objective of the IRRI varietal improvement programme for several years. However, the success of such a programme largely depends upon the availability of sources of resistance, a well-planned breeding strategy and a liberal exchange of breeding materials among rice-growing countries. International co-operation is also essential in collecting and evaluating germplasm, studying the insect biotypes and identifying the diverse genes for resistance.

The future of insect resistance in rice through host-plant resistance looks very promising as it will be supported by a full interdisciplinary programme of formal working groups of entomologists, plant breeders and chemists.

References

Alam, M. Z. and Islam, A. (1959). Biology of the rice leafhopper, *Nephotettix bipunctatus* Fab. in East Pakistan. *Pakistan Journal of Scientific Research* 2, 20–28.

Angeles, E. R., Khush, G. S. and Heinrichs, E. A. (1981). New genes for resistance to whitebacked planthopper in rice. *Crop Science* 21, 47–50.

Angeles, E. R., Khush, G. S. and Heinrichs, E. A. (1986). Inheritance of resistance to planthoppers and leafhoppers in rice. In *Rice Genetics*, pp. 537–549. International Rice Research Institute, Los Baños, Laguna, Philippines.

Athwal, D. S. and Pathak, M. D. (1972). Genetics of resistance to rice insects. In *Rice Breeding*, pp. 375–386. International Rice Research Institute, Los Baños, Laguna, Philippines.

Athwal, D. S., Pathak, M. D., Bacalangco, E. N. and Pura, C. D. (1971). Genetics of resistance to brown planthoppers and green leafhoppers in *Oryza sativa* L. *Crop Science* 11, 747–750.

Auclair, J. L. and Baldos, E. (1982). Feeding by whitebacked planthopper, *Sogatella furcifera* within susceptible and resistant rice varieties. *Entomologia Experimentalis et Applicata* 32, 200–203.

Auclair, J. L., Baldos, E. and Heinrichs, E. A. (1982). Biochemical evidence for the feeding sites of the leafhopper, *Nephotettix virescens* within susceptible and resistant rice plants. *Insect Science and its Application* 3, 29–34.

Bae, S. H. and Pathak, M. D. (1970). Life history of *Nilaparvata lugens* (Homoptera: Delphacidae) and susceptibility of rice varieties to its attack. *Annals of the Entomological Society of America* 63, 149–153.

Banerjee, S. N. (1971). Recent progress in rice insect research in India. *Tropical Agricultural Research Series* 5, 83–97.

Bergonia, H. T. (1978). Control measure to prevent tungro virus outbreak. *Plant Protection News* 7, 4–16.

Britton, G. L., Dam, N. V. and Tao, C. H. (1962). Preliminary observation on ecology of rice leafhopper (*Nephotettix bipunctatus cincticeps* Uhler) and whiteback planthopper (*Sogata furcifera* Horváth) and their control measures in Vietnam. *Plant Protection Bulletin* 4, 1–12.

Buvat, R. (1969). *An Introduction to Plant Protoplasm*. McGraw-Hill, New York.

Chaimongkol, B., Wathanakul, L. and Lamey, H. A. (1971). Orange leaf virus diseases. In *Rice Diseases and Pests of Thailand*. Rice Protection Research Centre, Rice Department, Ministry of Agriculture, Thailand, in co-operation with the United Nations Development Programme and the Food and Agriculture Organization.

Chelliah, S. and Hanifa, A. M. (1981). Resistance of rice varieties to the green leafhopper in Southern India. *International Rice Research Newsletter* 6 (6), 8–9.

Cheng, C. H. and Pathak, M. D. (1972). Resistance to *Nephotettix virescens* in rice varieties. *Journal of Economic Entomology* 65, 1148–1153.

CHOI, S. Y. (1974). Resistance of a new variety 'Tong-il' to the small brown planthopper, *Laodelphax striatellus*. *The Rice Entomology Newsletter* 1, 17–18.

CHOI, S. Y. (1975). Varietal resistance of rice to the green leafhopper, *Nephotettix cincticeps* Uhler. *The Korean Journal of Plant Protection* 14, 13–21.

CHOI, S. Y. AND SONG, Y. H. (1974). Resistance of 'Tong-il' variety to the smaller brown planthopper, *Laodelphax striatellus* Fallén. *The Korean Journal of Plant Protection* 13, 77–82.

CHOI, S. Y., SONG, Y. H. AND PARK, J. S. (1973). Studies on the varietal resistance of rice to zigzag-striped leafhopper, *Recilia (Inazuma) dorsalis* (Motschulsky). *The Korean Journal of Plant Protection* 12, 83–87.

CHOI, S. Y., SONG, Y. H., PARK, J. S. AND SON, B. I. (1973a). Studies on the varietal resistance of rice to the green rice leafhopper, *Nephotettix cincticeps* Uhler. *The Korean Journal of Plant Protection* 12, 47–53.

CHOI, S. Y., SONG, Y. H., LEE, J. O. AND PARK, J. S. (1973b) Studies on the varietal resistance to the whitebacked planthopper, *Sogatella furcifera* (Horváth). *The Korean Journal of Plant Protection* 12, 139–142.

CHOI, S. Y., SONG, Y. H., PARK, J. S. AND CHOI, K. Y. (1974). Studies on the varietal resistance of rice to the smaller brown planthopper, *Laodelphax striatellus* Fallén. *The Korean Journal of Plant Protection* 13, 11–12.

CHOI, S. Y., LEE, J. O., LEE, H. R. AND PARK, J. S. (1976). Resistance of new varieties of Milyang No. 21 and No. 23 to plant and leafhoppers. *Plant Protection* 15, 147–151.

CHOI, S. Y., LEE, S. W., CHUNG, B. K. AND KIM, J. W. (1982). Varietal resistance of Korean new rice cultivars to the whitebacked planthopper, *Sogatella furcifera* (Horváth). *Seoul National University, College of Agriculture Bulletin* 7, 125–128.

COOK, A. G., WOODHEAD, S., MAGALIT, V. F. AND HEINRICHS, E. A. (1987). Variation in the feeding behavior of *Nilaparvata lugens* on resistant and susceptible rice varieties. *Entomologia Experimentalis et Applicata* 43, 227–235.

CRAMER, H. H. (1967). Plant Protection and World Crop Protection. *Pflanzenschutz-Nachrichten* 20, 1–524.

DHAWAN, A. K. AND SAJJAN, S. S. (1976). Some observations on behavior of the rice green leafhopper, *Nephotettix nigropictus* (Stål) (Cicadellidae: Hemiptera). *Haryana Agricultural University Journal of Research* 6, 235–236.

GALVEZ, G. E. (1969). Transmission of hoja blanca virus of rice. In *The Virus Diseases of the Rice Plant*, pp. 155–163. John Hopkins Press, Baltimore, Maryland, USA.

GUNTHER, E. (1952). *The Essential Oils*, 2nd edn, volume V. Van Nostrand, New York.

HEINRICHS, E. A. (1979). Control of leafhopper and planthopper vectors of rice viruses. In *Leafhopper Vectors and Plant Disease Agents* (K. Maramorosch and K. F. Harris, Eds), pp. 529–560. Academic Press, New York.

HEINRICHS, E. A. (1986). Perspectives and directions for the continued development of insect-resistant rice varieties. *Agriculture Ecosystems and Environment* 18, 9–36.

HEINRICHS, E. A. AND RAPUSAS, H. (1983). Levels of resistance to the whitebacked planthopper, *Sogatella furcifera* (Homoptera: Delphacidae) in rice varieties with different resistance genes. *Environmental Entomology* 12, 1793–1797.

HEINRICHS, E. A. AND RAPUSAS, H. (1984). Feeding, development, and tungro virus transmission by the green leafhopper, *Nephotettix virescens* (Distant) (Homoptera: Cicadellidae) after selection on resistant rice cultivars. *Environmental Entomology* 13, 1074–1078.

HEINRICHS, E. A., MEDRANO, F. G. AND RAPUSAS, H. R. (1985). *Genetic Evaluation for Insect Resistance in Rice*. International Rice Research Institute, Los Baños, Laguna, Philippines.

HERNANDEZ, J. E. AND KHUSH, G. S. (1981). Genetics of resistance to whitebacked planthopper in some rice varieties. *Oryza* 18, 44–50.

HIBINO, H. (1987). Tungro status in the Philippines and Japan. In *Proceedings of the Workshop on Rice Tungro Virus*, pp. 55–57. Ministry of Agriculture, AARD-Maros Research Institute for Food Crops, Indonesia.

IRRI (INTERNATIONAL RICE RESEARCH INSTITUTE) (1967). Varietal resistance to leafhoppers and planthoppers. *Annual Report for 1966*, pp. 189–192. International Rice Research Institute, Los Baños, Laguna, Philippines.

IRRI (1968). Resistance of variety Pankhari-203 to the rice green leafhopper. *Annual Report for 1967*, pp. 199–200. International Rice Research Institute, Los Baños, Laguna, Philippines.

IRRI (1972). Whitebacked planthopper. *Annual Report for 1971*, pp. 119–120. International Rice Research Institute, Los Baños, Laguna, Philippines.

IRRI (1975). Insect biotypes capable of surviving on resistant plants. *Annual Report for 1974*, pp. 89–90. International Rice Research Institute, Los Baños, Laguna, Philippines.

IRRI (1976). Biotypes. *Annual Report for 1975*, pp. 107–108. International Rice Research Institute, Los Baños, Laguna, Philippines.

IRRI (1977). Sources of resistance to whitebacked planthopper. *Annual Report for 1976*, pp. 54–57. International Rice Research Institute, Los Baños, Laguna, Philippines.

IRRI (1979). Biochemical bases of resistance. Steam distillate extracts. *Annual Report for 1978*, pp. 68–71. International Rice Research Institute, Los Baños, Laguna, Philippines.

IRRI (1983a). Brown planthopper resistance in wild rice. *Annual Report for 1982*, pp. 58–61. International Rice Research Institute, Los Baños, Laguna, Philippines.

IRRI (1983b). Minor genes for whitebacked planthopper resistance. *Annual Report for 1982*, pp. 61–62. International Rice Research Institute, Los Baños, Laguna, Philippines.

IRRI (1983c). Tungro virus transmission by green leafhopper biotypes. *Annual Report for 1982*, pp. 62–63. International Rice Research Institute, Los Baños, Laguna, Philippines.

IRRI (1986). Morphometric variation in brown planthopper. *Annual Report for 1985*, p. 61. International Rice Research Institute, Los Baños, Laguna, Philippines.

ISHIHARA, T. (1969). Families and genera of leafhopper vectors. In *Viruses, Vectors and Vegetation* (K. Maramorosch, Ed.), pp. 235–254. John Wiley and Sons, New York.

ISHII, M., YASUO, S. AND YAMAGUCHI, T. (1969). Testing methods and analysis of the varietal resistance to rice dwarf disease. *Journal of the Central Agricultural Experiment Station (Japan)* **13**, 23–44.

JENNINGS, P. R. AND PINEDA, A. T. (1970a). Screening rice for resistance to the planthopper, *Sogatodes oryzicola* (Muir). *Crop Science* **10**, 687–689.

JENNINGS, P. R. AND PINEDA, A. T. (1970b). Effect of resistant rice plants on multiplication of the planthopper, *Sogatodes oryzicola* (Muir). *Crop Science* **10**, 689–690.

JENNINGS, P. R. AND PINEDA, A. T. (1983). *Management of the rice planthopper, Sogatodes oryzicola in Latin America*. Paper presented at the *Working Group Meeting on Stable Plant Resistance in IPC programmes*, 28 Nov.–2 Dec., 1983, Denpasar, Bali, Indonesia. Sponsored by the Intercountry IPC Rice Programme, FAO, Bangkok, Thailand.

KABIR, M. A. AND KHUSH, G. S. (1988). Genetic analysis of resistance to brown planthopper in rice (*Oryza sativa* L.). *Plant breeding* **100**, 54–58.

KAWABE, S. (1985). Mechanism of varietal resistance to the rice green leafhopper (*Nephotettix cincticeps* Uhler). *Japan Agricultural Research Quarterly* **19**, 115–124.

KAWABE, S., FUKUMORITA, T. AND CHINO, M. (1980). Collection of rice phloem sap from stylets of homopterous insects severed by YAG laser. *Plant and Cell Physiology* **21**, 1319–1327.

KHAN, Z. R. AND SAXENA, R. C. (1984a). A technique for demonstrating phloem or xylem feeding by leafhoppers (Homoptera: Cicadellidae) and planthoppers (Homoptera: Delphacidae) in rice plant. *Journal of Economic Entomology* **77**, 550–552.

KHAN, Z. R. AND SAXENA, R. C. (1984b). Electronically recorded waveforms associated with feeding behavior of *Sogatella furcifera* (Homoptera: Delphacidae) on susceptible and resistant rice varieties. *Journal of Economic Entomology* **77**, 1479–1482.

KHAN, Z. R. AND SAXENA, R. C. (1985a). Effect of steam distillate extract of a resistant rice variety on feeding behaviour of *Nephotettix virescens* (Homoptera Cicadellidae). *Journal of Economic Entomology* **78**, 562–566.

KHAN, Z. R. AND SAXENA, R. C. (1985b). Mode of feeding and growth of *Nephotettix virescens* (Homoptera: Cicadellidae) on selected resistant and susceptible rice varieties. *Journal of Economic Entomology* **78**, 583–587.

KHAN, Z. R. AND SAXENA, R. C. (1985c). Behavioral and physiological responses of *Sogatella furcifera* (Homoptera: Delphacidae) to selected resistant and susceptible rice cultivars. *Journal of Economic Entomology* **78**, 1280–1286.

KHAN, Z.R. AND SAXENA, R. C. (1986). Effect of steam distillate extracts of resistant and susceptible rice cultivars on behavior and biology of *Sogatella furcifera* (Homoptera: Delphacidae). *Journal of Economic Entomology* **79**, 928–935.

KHAN, Z. R. AND SAXENA, R. C. (1988). Probing behavior of three biotypes of *Nilaparvata lugens* (Homoptera: Delphacidae) on different resistant and susceptible rice varieties. *Journal of Economic Entomology* **81**, 1338–1345.

KHUSH, G. S. (1977). Disease and insect resistance in rice. *Advances in Agronomy* **29**, 265–341.

KHUSH, G. S. AND JENA, K. K. (1986). *Capabilities and limitations of conventional plant breeding and the role of distant hybridization*. Paper presented at the workshop on *Biotechnology for Crop Improvement: Potential and Limitations*, 13–17 October 1986. International Rice Research Institute, Los Baños, Laguna, Philippines.

KHUSH, G. S., REZAUL KARIM, A. N. M. AND ANGELES, E. R. (1986). Genetics of resistance of rice cultivar ARC 10550 to Bangladesh brown planthopper. *Journal of Genetics* **64**, 121–125.

KIM, H. S. AND HEINRICHS, E. A. (1982). Effect of silica level on whitebacked planthopper resistance. *International Rice Research Newsletter* **7** (4), 17.

KIM, M., KOH, M. S. AND FUKAMI, H. (1985). Isolation of *C*-glycosyflavones as probing stimulants of planthoppers in rice plant. *Journal of Chemical Ecology* **11**, 441–452.

KIM, Y. H., LEE, J. O. AND PARK, J. S. (1983). Resistance of recommended rice varieties to planthopper and leafhopper in Korea. *Research Department Office of Rural Development* **25**, 79–84.

KIM, M., KOH, M., OBATA, T., FUKAMI, H. AND ISHII, S. (1976). Isolation and identification of *trans*-aconitic acid as the antifeedant in barnyard grass against the brown planthopper, *Nilaparvata lugens* (Stål). *Applied Entomology and Zoology* **11**, 53–57.

KIMURA, T., KOGA, S., NISHIZAWA, T. AND NISHI, Y. (1969). A method for testing varietal resistance of rice plant to dwarf disease. *Proceedings of the Association for Plant Protection of Kyushu* **15**, 34–37.

KISHINO, K. AND ANDO, Y. (1979). Resistance of rice plant to the green rice leafhopper, *Nephotettix cincticeps* Uhler. 2. Fluctuation of antibiosis with the growing stages of the resistant rice varieties. *Japanese Journal of Applied Entomology and Zoology* **23**, 129–133.

KISIMOTO, R. (1967). Genetic variability of a planthopper vector; *Laodelphax striatellus* (Fallén) to acquire the rice stripe virus. *Virology* **32**, 144–152.

KITTUR, S. V. AND PATEL, R. K. (1968). Varietal screening for paddy pests. *Annual Report 1967*. Central Rice Research Station, Raipur, Madhya Pradesh, India.

KOBAYASHI, A. (1983). Inheritance of resistance to green rice leafhopper (*Nephotettix cincticeps* Uhler) and dwarf virus disease. In *Proceedings of the 4th International Sabrao Congress. Crop Improvement Research*, pp. 157–165.

KURATA, S. AND SOGAWA, K. (1976). Sucking inhibitory action of aromatic amines for the rice plant- and leafhoppers (Homoptera: Delphacidae, Deltocephalidae). *Applied Entomology and Zoology* **11**, 89–93.

LAKSHMINARAYANA, A. AND KHUSH, G. S. (1977). New genes for resistance to the brown planthopper in rice. *Crop Science* **17**, 96–100.

LEE, J. O. AND PARK, J. S. (1976). Studies on varietal resistance of rice to plant and leafhoppers. *Korean Office of Rural Development Project Report* **18**, 67–72.

LING, K. C. (1969). Testing rice varieties for resistance to tungro disease. In *Virus Diseases of Rice Plant*, pp. 255–277. John Hopkins Press, Baltimore, Maryland, USA.

LING, K. C. (1970). Ability of *Nephotettix apicalis* (Motschulsky) to transmit the rice tungro virus. *Journal of Economic Entomology* **63**, 582–586.

LING, K. C. (1977). Rice ragged stunt disease. *International Rice Research Newsletter* **2** (5), 6–7.

McKEY, D. (1979). The distribution of secondary compounds within plants. In *Herbivores – Their Interactions with Secondary Plant Metabolites* (G. A. Rosenthal and D. H. Janzen, Eds), pp. 56–133. Academic Press, New York.

McLEAN, D. L. AND KINSEY, M. G. (1964). A technique for electronically recording aphid feeding and salivation. *Nature* (London) **202**, 1358–1359.

MISHRA, U. S. AND KAUSHIK, U. K. (1976). Phosphomidon as an aerial spray against *Nephotettix virescens* in Madhya Pradesh, India. *Rice Entomology Newsletter* **4**, 22–23.

MORINAKA, T., PUTTA, M., CHETTANACHIT, D., PAREJAREARN, A., DISTHAPORN, S., OMURA, T. AND INOUE, H. (1982). Transmission of rice gall dwarf virus by cicadellid leafhoppers *Recilia dorsalis* and *Nephotettix nigropictus* in Thailand. *Plant Disease* **66**, 703–704.

MUKHOPADHYAY, S. AND CHATTOPADHYAY, K. (1975). Preferential feeding of green leafhopper. *International Rice Commission Newsletter* **24** (2), 76–80.

NASU, S. (1967). Rice leafhoppers. In *Major Insect Pests of Rice Plant*, pp. 493–523. John Hopkins Press, Baltimore, Maryland, USA.

OBATA, T., KOH, M., KIM, M. AND FUKAMI, H. (1983). Constituents of planthopper attractant in rice plant. *Applied Entomology and Zoology* **18**, 161–169.

OKAMOTO, D. AND INOUE, H. (1967). Studies on the smaller brown planthopper, *Laodelphax striatellus* Fallén, as a vector of rice stripe virus. 2. Varietal resistance of rice to the smaller brown planthopper. *Bulletin of the Chugoko Agricultural Experimental Station, Ministry of Agriculture and Forestry* **42**, 135–136.

ORELLANA, P., JENNINGS, P. R. AND PEREZ, Y. L. (1982). Preliminary results of resistance studies in rice varieties (*Oryza sativa*) with colonies of *Sogatodes oryzicola* of different colonies. *Ciencia y Tecnica en la Agricultura Arroz* **5**, 63–82.

ŌYA, S. AND SATO, A. (1980). Antibiosis and non-preference in resistant rice varieties to the green rice leafhopper, *Nephotettix cincticeps* Uhler. *Proceedings of the Association for Plant Protection of Hokuriku* **28**, 23–29.

PABLO, S. J. (1977). *Resistance to Whitebacked Planthopper*, Sogatella furcifera (*Horváth*) *in Rice Varieties*. Ph.D. thesis, Post-graduate School, Indian Agricultural Research Institute, New Delhi, India.

PANDA, N. AND HEINRICHS, E. A. (1983). Levels of tolerance and antibiosis in rice varieties having moderate resistance to the brown planthopper, *Nilaparavata lugens* (Stål) (Hemiptera: Delphacidae). *Environmental Entomology* **12**, 1204–1214.

PATHAK, M. D. (1968). Ecology of common insect pests of rice. *Annual Review of Entomology* **13**, 257–294.

PATHAK, M. D. (1969). Stem borer and leafhopper–planthopper resistance in rice varieties. *Entomologia Experimentalis et Applicata* **12**, 789–800.

PATHAK, M. D. (1972). Resistance to insect pests in rice varieties. In *Rice Breeding*, pp. 325–341. International Rice Research Institute, Los Baños, Laguna, Philippines.

PATHAK, M. D. AND DHALIWAL, G. S. (1981). *Trends and Strategies for Rice Insect Problems in Tropical Asia*. IRRI Research Paper Series No. 64. International Rice Research Institute, Los Baños, Laguna, Philippines.

PATHAK, M. D. AND SAXENA, R. C. (1980). Breeding approaches in rice. In *Breeding Plants Resistant to Insects* (F. G. Maxwell and P. R. Jennings, Eds), pp. 421–455. John Wiley & Sons, New York.

PATHAK, M. D., CHENG, C. H. AND FORTUNO, M. E. (1969). Resistance to *Nephotettix impicticeps* and *Nilaparvata lugens* in rice varieties. *Nature* (London) **223**, 502–504.

PONGPRASERT, S. (1974). *Resistance of the Zigzag Leafhopper*, Recilia dorsalis (*Motschulsky*) *in Rice Varieties*. M.S. thesis, University of the Philippines at Los Baños, Laguna, Philippines.

RAPUSAS, H. R. AND HEINRICHS, E. A. (1985). Virulence of *Nephotettix virescens* colonies on resistant rices. *International Rice Research Newsletter* **10** (5), 10–11.

RAZZAQUE, Q. M. A. (1984). *Resistance of* Oryza *species to rice green leafhoppers*, Nephotettix nigropictus (*Stål*) *and* Nephotettix virescens (*Distant*). M.S. thesis, University of the Philippines at Los Baños, Laguna, Philippines.

RAZZAQUE, Q. M. A. AND HEINRICHS, E. A. (1985a). Evaluation of germplasm accessions for green leafhopper (GLH) resistance. *International Rice Research Newsletter* **10** (3), 9–10.

RAZZAQUE, Q. M. A. AND HEINRICHS, E. A. (1985b). Screening for resistance of IR varieties to green leafhopper (GLH). *International Rice Research Newsletter* **10** (3), 10–11.

RAZZAQUE, Q. M. A. AND HEINRICHS, E. A. (1985c). Screening wild rices for resistance to green leafhopper (GLH). *International Rice Research Newsletter* **10** (3), 11–12.

REZAUL KARIM, A. N. M. (1978). *Varietal resistance of rice to green leafhopper*, Nephotettix virescens (*Distant*): *Sources, Mechanisms, and Genetics of Resistance*. Ph. D. thesis, University of the Philippines at Los Baños, Laguna, Philippines.

REZAUL KARIM, A. N. M. AND PATHAK, M. D. (1979). An alternate biotype of green leafhopper in Bangladesh. *International Rice Research Newsletter* **4** (6), 7–8.

REZAUL KARIM, A. N. M. AND PATHAK, M. D. (1982). New genes for resistance to green leafhopper, *Nephotettix virescens* (Distant) in rice, *Oryza sativa* L. *Crop Protection* **1**, 483–490.

RIVERA, C. T., OU, S. H. AND IIDA, T. T. (1966). Grassy stunt disease of rice and its transmission by the planthopper, *Nilaparvata lugens* (Stål). *Plant Disease Reporter* **50**, 453–456.

RIVERA, C. T., OU, S. H. AND PATHAK, M. D. (1963). Transmission studies of the orange leaf disease of rice. *Plant Disease Reporter* **42**, 1045–1048.

ROBINSON, T. (1983). *The Organic Constituents of Higher Plants: Their Chemistry and Interrelationships*, 5th edn. Cordus, North Amherst.

RODRIGUEZ-RIVERA, R. (1972). *Resistance to Whitebacked Planthopper*, Sogatella furcifera (*Horváth*) *in Rice Varieties*. M.S. thesis, University of the Philippines at Los Baños, Laguna, Philippines.

ROMENA, A. M., RAPUSAS, H. R. AND HEINRICHS, E. A. (1986). Evaluation of rice accessions for resistance to the whitebacked planthopper, *Sogatella furcifera* (Horváth) (Homoptera: Delphacidae). *Crop Protection* **5**, 334–340.

SAJJAN, S. S. (1972). *Varietal resistance of rice to* Nephotettix nigropictus (*Stål*). Report of the work done during the fellowship period, 18 November 1971 to 23 May 1972 (mimeographed). Department of Entomology, International Rice Research Institute, Los Baños, Laguna, Philippines.

SAKURAI, Y. (1969). Varietal resistance to stripe, dwarf and black-streaked dwarf.

In *Virus Diseases of the Rice Plant*, pp. 237–275. John Hopkins Press, Baltimore, Maryland, USA.

SATO, A. AND SOGAWA, K. (1981). Biotypic variations in the green rice leafhopper, *Nephotettix cincticeps* Uhler (Homoptera: Deltocephalidae), in relation to rice varieties. *Applied Entomology and Zoology* **16**, 55–57.

SAXENA, K. N. (1969). Patterns of insect plant relationships determining susceptibility or resistance of different plants to an insect. *Entomologia Experimentalis et Applicata* **12**, 751–766.

SAXENA, K. N., GANDHI, J. R. AND SAXENA, R. C. (1974). Patterns of relationships between certain leafhoppers and plants. I. Responses to plants. *Entomologia Experimentalis et Applicata* **17**, 303–313.

SAXENA, R. C. (1986). Biochemical bases of insect resistance in rice varieties. In *Natural Resistance of Plants to Pests: Role of Allelochemicals* (M. B. Green and P. A. Hedin, Eds), pp. 142–159. American Chemical Society, Washington, DC.

SAXENA, R. C. (1989). Durable resistance to insect pests of irrigated rice. *International Rice Research Conference 21–25 September, 1987*. International Rice Research Institute, Chinese Academy of Agricultural Sciences and China National Rice Research Institute, in press.

SAXENA, R. C. AND KHAN, Z. R. (1984). A comparison between free-choice and no-choice seedling bulk tests for evaluating resistance of rice cultivars to the whitebacked planthopper. *Crop Science* **24**, 1204–1206.

SAXENA, R. C. AND OKECH, S. H. (1985). Role of plant volatiles in resistance of selected rice varieties to brown planthopper, *Nilaparvata lugens* (Stål) (Homoptera: Delphacidae). *Journal of Chemical Ecology* **11**, 1601–1616.

SAXENA, R. C. AND PATHAK, M. D. (1977). Factors affecting resistance of rice varieties to brown planthopper, *Nilaparvata lugens* (Stål). In *Proceedings 8th Annual Conference, Pest Control Council of the Philippines, 18–22 May 1977, Bacolod City, Philippines*.

SAXENA, R. C. AND PATHAK, M. D. (1979). Factors governing susceptibility and resistance of certain rice varieties to the brown planthopper. In *Brown Planthopper: Threat to Rice Production in Asia*, pp. 303–317. International Rice Research Institute, Los Baños, Laguna, Philippines.

SAXENA, R. C. AND PUMA, B. C. (1979). Effect of *trans*-aconitic acid, a barnyard grass allelochemic, on hatching of eggs of brown planthopper and green leafhopper. In *Proceedings of 10th Annual Research Conference of Pest Control Council of the Philippines, Manila, 2–5 May, 1979*.

SAXENA, R. C., BARRION, A. A. AND SORIANO, M. V. (1985). Comparative morphometrics of male and female genital and abdominal characters of *Nephotettix virescens* (Distant) populations from Bangladesh and Philippines. *International Rice Research Newsletter* **10** (3), 27–28.

SEKAR, P. AND CHELLIAH, S. (1983). Varietal resistance to the white leafhopper. *International Rice Research Newsletter* **8** (2), 7.

SEKIZAWA, K. AND OGAWA, T. (1980). Studies on the breeding of rice varieties resistant to green rice leafhopper. 1. Varietal differences of resistance to green rice leafhopper. *Bulletin of the Chugoku National Agricultural Experiment Station Series A* **27**, 37–48.

SESHU, D. V., PRAKASARAO, P. S., KALODE, M. B., SHASTRY, S. V. S., SASTRY, M. V. S., SRINIVASAN, T. E., RAMAKRISHNA RAO, J., MISHRA, B. C., PRASAD, K., PRASAD RAO, U., KULSHRESHTA, J. P. AND ROY, J. K. (1974). Breeding for resistance to rice gall midge, leafhoppers and planthoppers in India. *International Rice Research Conference, 22–25 April, 1974*. International Rice Research Institute, Los Baños, Laguna, Philippines.

SHASTRY, S. V. S., SHARMA, S. D., JOHN, V. T. AND KRESHAIAR, K. (1971). New sources of resistance to pests and diseases in the Assam rice collection. *International Rice Commission Newsletter* **10** (3), 1–16.

SHIGEMATSU, V., MUROFUSHI, N., ITO, K., KANEDA, C., KAWABE, S. AND TAKAHASHI,

N. (1982). Sterols and asparagine in the rice plant, endogenous factors related to resistance against the brown planthopper (*Nilaparvata lugens*). *Agricultural and Biological Chemistry* **46**, 2877–2896.

SIDHU, G. S., KHUSH, G. S. AND MEDRANO, F. G. (1979). A dominant gene in rice for resistance to whitebacked planthopper and its relationship to other plant characteristics. *Euphytica* **28**, 233–237.

SIWI, B. H. AND KHUSH, G. S. (1977). New genes for resistance to the green leafhopper in rice. *Crop Science* **17**, 17–20.

SIWI, S. S., KARTOHARDJONO, A., HARNOTO, S. AND DIRATMAJA, A. (1987). The green leafhopper, genus *Nephotettix* Matsumura. In *Proceedings of the Workshop on Rice Tungro Virus*, pp. 35–50. Ministry of Agriculture, AARD-Matos Research Institute for Food Crops, Indonesia.

SOGAWA, K. AND PATHAK, M. D. (1970). Mechanisms of brown planthopper resistance in Mudgo variety of rice. *Applied Entomology and Zoology* **5**, 145–158.

SOGAWA, K. AND SATO, A. (1981). The green rice leafhopper, *Nephotettix cincticeps* Uhler (Hemiptera: Deltocephalidae) populations with differential reactions to rice varieties. *Japanese Journal of Applied Entomology and Zoology* **25**, 280–285.

SOGAWA, K. AND SATO, A. (1983). Differences in morphological and physiological characters between the Joetsu and Chikugo green rice leafhopper populations with differential reactions to rice varieties. *Japanese Journal of Applied Entomology and Zoology* **27**, 22–27.

SUENAGA, H. (1950). Effect of biochemical components of food plant on the abundance of rice leafhoppers. *Kyushu Nogyo Konkyu* **7**, 61–62.

SURIN, P., KERDCHOKCHAI, D. AND SURIN, A. (1974). Studies on other vectors of yellow orange leaf virus other than *Nephotettix* sp., *Recilia dorsalis* and mealy bug. *Annual Research Report 1971*, pp. 630–634. Thailand Rice Department, Thailand.

TOKUNAGA, M. AND KIDERA, Y. (1948). The relation between the differences of rice varieties and the outbreak of rice leafhopper, *Sogata furcifera* (Horváth). *Oyo-Kontyu* **4**, 210–217.

VELUSAMY, R. AND HEINRICHS, E. A. (1986). Electronic monitoring of feeding behavior of *Nilaparvata lugens* (Homoptera: Delphacidae) on susceptible and resistant rice cultivars. *Environmental Entomology* **15**, 678–682.

VELUSAMY, R., JANAKI, I. P., SWAMINATHAN, R. AND SUBRAMANIAM, T. R. (1975). Varietal resistance in rice to white leafhopper, *Cicadella spectra* (Distant). *Madras Agricultural Journal* **62**, 305–307.

VELUSAMY, R., NATARAJAMOORTHY, K., SUNDARA BABU, P. C. AND RANGASWAMI, S. R. S. (1986). Green leafhopper (GLH) resistance in wild rices. *International Rice Research Newsletter* **11** (6), 11.

VISWANATHAN, P. R. K. AND KALODE, M. B. (1981). Studies on varietal resistance and host specificity of rice green leafhoppers. *International Rice Research Newsletter* **6** (3), 7.

VISWANATHAN, K. AND KALODE, M. B. (1984). Comparative study on varietal resistance to rice green leafhoppers *Nephotettix virescens* (Distant) and *N. nigropictus* (Stål). *Proceedings of Indian Academy of Science* (Animal Sciences) **93**, 55–63.

WATHANAKUL, L. (1964). *A Study on the Host Range of Tungro and Orange Leaf Viruses of Rice*. M.S. thesis, University of the Philippines at Los Baños, Laguna, Philippines.

WOODHEAD, S. AND PADGHAM, E. E. (1988). The effect of plant surface characteristics on resistance of rice to brown planthopper, *Nilaparvata lugens*. *Entomologia Experimentalis et Applicata* **47**, 15–22.

WU, C. F. AND KHUSH, G. S. (1985). A new dominant gene for resistance to whitebacked planthopper (*Sogatella furcifera*) in rice (*Oryza sativa*). *Crop Science* **25**, 505–509.

YASUO, S., YAMAGUCHI, T. AND ISHNI, M. (1960). *Studies on the stripe and dwarf viruses of rice plant*. Experimental results in plant diseases (mimeographed). Japan Central Agricultural Experimental Station.

YE, Z. (1988). *Resistance to the Whitebacked Planthopper*, Sogatella furcifera (*Horváth*) *in Elite Line of* Oryza sativa *and* Oryza officinalis *Crosses*. M.S. thesis, University of the Philippines at Los Baños, Laguna, Philippines.

YOSHIHARA, T. AND SOGAWA, K. (1979). Soluble silicic acid and insoluble silica contents in leaf sheaths of rice varieties carrying BPH-resistance genes. *International Rice Research Newsletter* **4** (5), 12–13.

YOSHIHARA, T., SOGAWA, K., PATHAK, M. D., JULIANO, B. O. AND ISAKAMURA, S. (1979). Soluble silicic acid as a sucking inhibitory substance in rice against the rice brown planthopper (Delphacidae: Homoptera). *Entomologia Experimentalis et Applicata* **26**, 314.

YOSHIHARA, T., SOGAWA, K., PATHAK, M. D., JULIANO, B. O. AND SUKAMURA, S. (1980). Oxalic acid as a sucking inhibitor of the brown planthopper in rice (Delphacidae, Homoptera). *Entomologia Experimentalis et Applicata* **27**, 149–155.

YOSHIMEKI, M. (1967). A summary of the forecasting program for rice stem borer control in Japan. In *The Major Insect Pests of the Rice Plant*, pp. 181–194. The John Hopkins Press, Baltimore, Maryland, USA.

4
Biology and Control of the Carrot Fly, *Psila rosae* (F.)

C. P. DUFAULT AND T. H. COAKER

Department of Applied Biology, University of Cambridge, Pembroke Street, Cambridge, CB2 3DX, UK

Introduction
Nomenclature and distribution
Host-plant range and preference
Crop losses and damage assessment
Sampling methods
      Eggs—Larvae and pupae—Adults
Biology
      Generations and life cycle—Eggs—Larvae—Pupae—Adults
Control
      Natural—Applied
Future prospects for carrot fly control
References

Introduction

The carrot fly (*Psila rosae* (F.) (Diptera: Psilidae)) is a pest of temperate agriculture attacking umbelliferous crops. In North America and New Zealand it is known as the carrot rust fly, probably because of the characteristic rust coloration of the larval mines on the swollen tap root (Beirne, 1971), or the reddened foliage of attacked plants. It is the feeding activity of the larva which makes this insect a pest.

The considerable economic losses caused annually by the carrot fly have stimulated the increasingly detailed biological studies conducted in recent years and, in this context, aspects of the ecology, biology and control of the carrot fly are summarized below. A bibliography of the carrot fly covering publications up to 1985 has been published (Hardman, Ellis and Stanley, 1985).

Nomenclature and distribution

The carrot fly was first described by Fabricius (1794). Williams (1954) claimed that the phrase 'Habitat in Kiliae floribus' meant the carrot fly was first found on flowers in Kiel, Germany, but Smith (1921) interpreted it as being found on roses, and Hardman and Ellis (1982) in Kiliya, Bessarabia in the USSR. Initially called *Musca rosae* by Fabricius (1794), he later renamed it *Tephritis rosae* (Fabricius, 1805). Fallén (1820) redescribed it as *Scatophaga rosae*, followed by Meigen (1826) who called it *Psila rosae* Fabr. Hendel (1917) founded a genus *Chamaepsila* for *P. rosae*; however, this action was not in accordance with Zoological Commission opinion (Collin, 1944) and the name was little used. Several authors reported insect damage to carrots early in the nineteenth century, which was probably caused by the carrot fly. Kirby and Spence (1815) suspected a species of *Musca*; however, it was not until Bouché (1833) that larvae found in carrots were identified as that of *P. rosae*.

The carrot fly is now endemic in most European countries from 68 degrees 50 minutes N in Norway (Ausland, 1957) to 40 degrees N in Italy (Anonymous, 1957; Sabatino, 1976) and 36 degrees N in Spain (Encobet, 1911). In Britain it has been known since the early nineteenth century (Curtis, 1860), occurring in the main carrot-growing areas, especially the Isle of Ely, Norfolk and Huntingdonshire where approximately 70% of the national acreage of carrots is grown (Hinton, 1971; Coppock, 1974). Since its introduction into North America around 1885 (Fletcher, 1886), carrot fly has been recorded from most provinces and states lying between 40 and 50 degrees N (Whitcomb, 1929; Anonymous, 1957; Scott, 1958). In the southern hemisphere it has been reported from New Zealand (Muggeridge, 1933). The carrot fly appears to have ceased expanding its range in North America and Europe. It is found on these continents in virtually all areas that lie between the 25°C July isotherm and the −10°C January isotherm; 25°C is above the optimum temperature for egg development, and pupal development is halted at this temperature (Burn, 1980). As parts of other continents lie within these isotherms, the potential exists for expansion of its range. If introduced, *P. rosae* would probably establish successfully in Japan and North-East Asia, and the southern and western thirds of South America. It may also be possible for it to establish in south Australia, the south and south-west coasts of Africa, and tropical highlands.

Host-plant range and preference

Many species in the family Umbelliferae are attacked by *P. rosae* larvae (*Table 1*). Whitcomb (1929) recorded that the degree of infestation on umbelliferous vegetable crops by second-generation larvae decreased in the following order: carrots (*Daucus carota* ssp. *sativus* (Hoffm.) Arcangeli); parsnip (*Pastinaca sativa* ssp. *sativa* L.); parsley (*Petroselinum crispum* (Miller) A.W. Hill), and celery (*Apium graveolens* var. *dulce* (Miller) DC.). Umbelliferous herbs also attacked include dill (*Anethum graveolens* L.), caraway (*Carum carvi* L.), fennel (*Foeniculum vulgare* Miller), coriander (*Coriandrum sativum* L.) and lovage (*Levisticum officinale* Koch) (Tullgren, 1917; Goedewaagen, 1926; Whitcomb, 1938; van't Sant, 1961; Hardman and Ellis, 1982). Umbelliferous weeds may provide reservoirs of flies; in East Anglia, England, hemlock (*Conium maculatum* L.) is an important wild host (Petherbridge, Wright and Davies, 1942; Petherbridge and Wright, 1943). Wainhouse and Coaker (1981) estimated that up to 10% of a local population of first-generation flies could have arisen from it. Hardman and Ellis (1982) recovered more than two carrot flies per root from 30-plant samples of rough chervil (*Chaerophyllum temulentum* L.) and upright hedge parsley (*Torilis japonica* (Houtt.) DC.). Bohlen (1967) tested the oviposi-tional preferences of females to 20 umbelliferous species, and found that all but two species stimulated oviposition (*Table 1*). These species, *A. graveolens* and *F. vulgare*, have, however, been shown to support larval development (Tullgren, 1917; Whitcomb, 1938). Cow parsley (*Anthriscus sylvestris* (L.) Hoffm.), ground elder (*Aegopodium podagraria* L.), giant fennel (*Ferula communis* L.) and alexanders (*Smyrnium olusatrum* L.), although possibly important as food sources for adults, are not attacked by larvae (van't Sant, 1961; Hardman and Ellis, 1982).

Early records of larvae attacking non-umbelliferous plants including turnips (*Brassica rapa* L.), rape (*Brassica napus* L.), beets (*Beta vulgaris* L.), potatoes (*Solanum tuberosum* L.) and cabbage (*Brassica oleracea* var. *capi-tata* DC) (Zetterstedt, 1846; MacLeod, 1929; Pettit, 1931; Hennig, 1941) have not been confirmed. More recent observations, however, suggest that non-umbelliferous plants may be attacked when planted in soil containing *P. rosae* larvae. Examples include lettuce (*Lactuca sativa* L.), endive (*Cichorium endivia* L.) and chicory (*Cichorium intybus* L.) (Miles, 1956; van't Sant, 1961). Beirne (1971) observed larval movement from carrots to adjacent maize (*Zea mays* L.) where they completed their development. Bohlen (1967) tested 12 non-umbelliferous species (including all except *B. rapa* and *Z. mays* listed above) and found them not to stimulate oviposition, thus strengthening the argument that non-umbelliferous plants are not normally hosts to the carrot fly. Development from egg to adult on *C. intybus* was reported after adults had been caged on plants (van't Sant, 1961).

Crop losses and damage assessment

Although carrots suffer the greatest economic losses, celery, parsnips, parsley and most umbelliferous herbs are also at risk (Smith and Gardner, 1922; Whitcomb, 1929; Ausland, 1957; van't Sant, 1961; Brunel, 1971a; Anony-mous, 1983a). The second generation usually causes problems on celery (Roebuck, 1945); the outer stalks are tunnelled, the base destroyed and as new stalks appear they are also damaged. Carrots and parsnips suffer damage by larval mines on the swollen tap root (Whitcomb, 1929) which may be followed by fungal and bacterial attack (Whitcomb, 1929; Collingwood and Croxall, 1954; Stone, 1954; van't Sant, 1961).

Table 1.   Historical list of first records of carrot fly hosts in the Umbelliferae. Modified from Hardman and Ellis (1982) and Bohlen (1967)

Host-plant species	Years	Country	Reference
Aethusa cynapium L.. fool's parsley	1980	England	Hardman and Ellis (1982)
Anethum graveolens L.÷, dill	1916	Sweden	Tullgren (1917)
Angelica sylvestris L.*, wild angelica	1967	Germany	Bohlen (1967)
Anthriscus caucalis Bieb.. bur chervil	1979	England	Hardman and Ellis (1982)
Anthriscus cerefolium (L.) Hoffm.*, garden chervil	1961	Holland	van't Sant (1961)
Apium graveolens var. *dulce* (Miller) DC.*, celery	1829	Great Britain	Major (1829)
Apium graveolens var. *rapaceum* (Miller) DC.. celeriac	1904	Canada	Fletcher (1904)
Apium inundatum (L.) Reichenb.. lesser marshwort	1979	England	Hardman and Ellis (1982)
Apium nodiflorum (L.) Lag.. fool's watercress	1980	England	Hardman and Ellis (1982)
Bupleurum tenuissimum L.. slender hare's ear	1980	England	Hardman and Ellis (1982)
Carum carvi L.*, caraway	1926	Holland	Goedewaagen (1926)
Chaerophyllum temulentum L.. rough chervil	1980	England	Hardman and Ellis (1982)
Cicuta virosa L.. cowbane	1980	England	Hardman and Ellis (1982)
Conium maculatum L.*, hemlock	1941	England	Petherbridge, Wright and Davies (1942)
Coriandrum sativum L.*, coriander	1928	USA	Whitcomb (1938)
Daucus capillifolius Gilli	1980	England	Hardman and Ellis (1982)
Daucus carota ssp. *carota* (L.) Thell.. wild carrot	1918	USA	Crosby and Leonard (1918)
Daucus carota ssp. *gadecaei* (Rouy and Camus) Heywood	1979	England	Hardman and Ellis (1982)
Daucus carota ssp. *gummifer* Hooker fil.	1979	England	Hardman and Ellis (1982)

(cont'd)

The rust-coloured larval mines on carrots were described by Curtis (1860) and Ormerod (1881). Larvae of the first generation mine from the root tip to the hypocotyl region (Petherbridge and Wright, 1943) causing reddening and wilting of the carrot foliage (Anonymous, 1983a), sometimes resulting in stunted and fanged carrots and the death of the plant (Whitcomb, 1929). However, a light attack may produce superficial mining which heals over by the second generation (Whitcomb, 1929). Larvae of the second generation mine extensively in the outer secondary phloem of the root.

Consumer acceptance of damaged carrots is low because the edible part of the plant is affected; losses are mostly due to a reduction in quality rather than yield. When reduction in gross yield occurs, it is usually associated with such a high degree of quality impairment that the whole crop may be

Table 1 (*cont'd*)

Host-plant species	Years	Country	Reference
Daucus carota ssp. *sativus* (Hoffm.) Arcangeli*, cultivated carrot	1807	Scotland	Henderson (1814)
Daucus glochidiatus (Labill.) Fisch. Mey and Ave-Lall., Australian carrot	1980	England	Hardman and Ellis (1982)
Ferula communis L.*, giant fennel	1967	Germany	Bohlen (1967)‡
Foeniculum vulgare Miller†, fennel	1928	USA	Whitcomb (1938)
Heracleum mantegazzianum Sommier and Levier, giant hogweed	1980	England	Hardman and Ellis (1982)
Heracleum sphondylium L., hogweed	1961	Holland	van't Sant (1961)
Levisticum officinale Koch*, lovage	1961	Holland	van't Sant (1961)
Ligusticum scoticum L., Scots lovage	1980	England	Hardman and Ellis (1982)
Oenanthe crocata L., hemlock water dropwort	1980	England	Hardman and Ellis (1982)
Oenanthe pimpinelloides L.*, corky-fruited water dropwort	1967	Germany	Bohlen (1967)
Orlaya grandiflora (L.). Hoffm.*	1967	Germany	Bohlen (1967)
Pastinaca sativa ssp. *sativa* L.*, parsnip	1812	England	Kirby and Spence (1815)
Pastinaca sativa ssp. *sylvestris* (Miller) Rouy and Camus, wild parsnip	1961	Holland	van't Sant (1961)
Petroselinum crispum (Miller) A. W. Hill*, parsley	1916	Sweden	Tullgren (1917)
Pimpinella anisum L.*, anise	1967	Germany	Bohlen (1967)
Scandix pecten-veneris L.*, shepherd's needle	1967	Germany	Bohlen (1967)
Selinum carvifolia (L.) L.*, Cambridge milk parsley	1967	Germany	Bohlen (1967)
Sison amomum L.*, stone parsley	1979	England	Hardman and Ellis (1982)
Sium sisarum L.*, skirret	1967	Germany	Bohlen (1967)
Torilis japonica (Houtt.) DC.*, upright hedge parsley	1967	Germany	Bohlen (1967)

* Found to stimulate oviposition runs (Bohlen, 1967). Other host species not designated * or † not tested.
† Found not to stimulate oviposition runs (Bohlen, 1967); found to be a host by Hardman and Ellis (1982).
‡ *F. communis* found not to be a host by Hardman and Ellis (1982).

unmarketable (Coaker and Wheatley, 1970). In England, carrots intended for canning must be undamaged; for the commercial soup market, slight damage but no larvae are tolerated; for direct sale 5–10% of slightly attacked roots may be acceptable (Coaker and Wheatley, 1970). Examination of market samples of carrots in 1970 showed that 0–21% contained carrot fly mines, yet these were described as acceptable to the consumer (Wheatley, 1971). Grading standards are, however, becoming increasingly stringent and contrast with those of the 1940s when carrots were considered unsaleable only when one-third or more of the carrot had to be cut away to remove all damaged tissue (Wright and Ashby, 1946a). Coppock (1974) reported that the severity of mining on over-wintering carrots occasionally made the crop totally unsaleable. Coppock, Maskell and Gair (1975) recorded that up to

60% of untreated carrots in England may be damaged if not harvested by early January. Toms (1972) estimated that an average attack resulted in 30% unsaleable carrots and assuming a 90% reduction in this damage with insecticides, he calculated a cost/benefit ratio for their use of 1/5.4.

Mining activity by first-generation larvae is greatest at the root apices, regardless of root length; in the second generation it is more evenly distributed over the carrot, whatever its length (Whitcomb, 1929; Jones, 1979). Thus, when assessing larval damage, the whole length of the carrot must be examined. The cause of visible injury can be determined, if necessary, by cutting open roots and collecting larvae for identification (Coaker and Wheatley, 1970). Mining by carrot fly larvae may be confused with that of other species. In Europe, larvae of *Napomyza carotae* Spencer (Diptera: Agromyzidae) (Spencer, 1966) enter the root by mining downwards through the leaf petiole and then tunnel beneath the epidermis of the upper half of the carrot root. Subsequent growth of roots causes some tunnels to open up to resemble carrot fly mines (van't Sant, 1961; Gersdorf and Kraft, 1965; Hassan, 1971; Jones, 1975; Robert, Fischer and Freuler, 1985). In North America, larvae of the carrot weevil (*Listronotus oregonensis* (Le Conte) (Coleoptera: Curculionidae)) produce tunnels generally in the upper one-third of the root, with a conspicuous, darkened partly-open tunnel left in the crown after the larva has matured and left the root (Stevenson, 1981a). Other insects capable of causing damage include larval *Agriotes* sp. (Coleoptera: Elateridae) and Noctuidae (Lepidoptera) (Robert, Fischer and Freuler, 1985). Damage initiated by phytophagous species, mechanical injury or disease, may be followed by the establishment of secondary or tertiary species (Brooks, 1949).

Carrot fly attack is distributed unevenly over the crop, being greatest near the edges of the field and least in mid-field (Petherbridge, Wright and Davies, 1942; Wright and Ashby, 1946a). When estimating attack in carrot fields, allowance should be made for differences in attack between the headland and mid-field. Fifty random samples should be taken at 5 metres and at 25 metres parallel to the edge of the crop. Carrots should be taken in twos or threes at each sampling point (Wright and Ashby, 1946a; Coaker and Wheatley, 1970). This method is not suitable for bed-rows as there may be differences in attack between carrots on the edge or in the middle of the rows. Standard lengths of bed-rows must be taken at each sampling point and all carrots examined. The numbers of damaged and undamaged roots can be expressed as percentages (Wright and Ashby, 1946a). Such estimates are insensitive when infestations are low or very high (Wheatley and Freeman, 1982). Ellis, Wheatley and Hardman (1978) developed a more detailed method of damage assessment. By slicing carrots longitudinally and placing the two slices cut-surface down, it was possible to estimate the percentage of root area damaged to the nearest 5%. Six types of mines were also distinguished and described. To overcome the difficulty of comparing experiments with differing infestations, Wheatley and Freeman (1982) developed a method whereby the proportion of undamaged roots can be related to a measure of larval infestation.

Sampling methods

EGGS

The distribution of eggs in a carrot crop is determined by the distribution of ovipositing females (Wright and Ashby, 1946a). Baker *et al.* (1942) caught most mature females and found most eggs close to hedges with high populations of flies. Petherbridge, Wright and Davies (1942) also reported that eggs were common around the edges of a carrot crop with fewest in mid-field. A study of the distribution of larval attack on parsnips showed that larvae attack mainly along the lengths of rows and infestations on adjacent rows are apparently independent of one another (Coaker and Wheatley, 1962). This may reflect egg-laying behaviour and suggests that experimental plots of carrots should be orientated to have their long axes across the rows so that samples can be taken from short lengths of a large number of rows. Eggs are usually laid singly or in groups of two to three within 0.5 cm of the host plant but may be up to 5.0 or occasionally more than 10.0 cm away (Whitcomb, 1938; Petherbridge, Wright and Davies, 1942; Overbeck, 1978). They are normally laid in cracks or under small lumps of soil, 0.3–0.5 cm below the surface, only occasionally being found on the soil surface or stuck to the base of the plant. Baker *et al.* (1942) reported that flies may disappear into soil crevices in order to oviposit. Eggs may be sampled from single-row carrots by taking carrot crowns and the surrounding soil 10.0 × 10.0 × 2.5 cm deep using a sampling tool similar to that used for frit fly (*Oscinella frit* (L.) (Diptera: Chloropidae)) eggs by Webley (1957) (Coppock, 1974; Overbeck, 1978). For carrots sown in beds, which may be up to 15 cm or more wide, this method is unsuitable and whole sections of row should be removed using a sharp trowel. Most eggs and some egg-shells, if not saturated with water, may be separated from the soil and roots by flotation in brine using a Fenwick apparatus (Webley, 1957). The eggs are collected on a mesh sieve and separated on to a moist black filter paper for identification and counting.

LARVAE AND PUPAE

Larvae are found in both carrots and the surrounding soil, moving up to 60 cm both along and between carrot rows. Most larvae pupate in the soil (Whitcomb, 1929; Körting, 1940) although they are found occasionally in old larval mines (Balachowsky and Mesnil, 1936; Jørgensen and Thygesen, 1968). Distribution of pupae in the soil is also affected by soil conditions, pupation occurring nearer the surface in moist conditions (van't Sant, 1961). Estimates of depth of pupation vary from 67% in the top 15 cm (van't Sant, 1961) to 93% in the top 5 cm (Petherbridge and Wright, 1943) but pupae may be found to a depth of 30 cm in 'dry' soils (van't Sant, 1961; Burn, 1984). Petherbridge and Wright (1943) estimated that 92% of pupae occurred within 7.5 cm either side of a single row of carrots. Sample units for larvae and pupae 30 cm long, 20–30 cm deep and 10–15 cm either side of the row are suitable. A modified Salt and Hollick (1944) soil-washing device can be used for extraction of larvae and pupae (Cockbill *et al.*, 1945) which finally may be floated off on brine in a Ladell funnel (Ladell, 1936). A 1 mm mesh diameter sieve will retain all third-instar larvae (Burn, 1980); larvae can also be dissected from carrots. After removal of large larvae and pupae from samples, Wright and Ashby (1946b) suggested boiling the samples to render the smaller larvae white and opaque and conspicuous against a black background. Burn (1980) rendered first-instar larvae more visible by boiling for 5 minutes in a 5% solution of crystal violet.

ADULTS

Adult emergence in the field may be estimated by enclosing lengths of carrot row in emergence cages. The design in Southwood (1978) is suitable but carrot foliage should first be removed. Emergence from commercial carrot crops treated with insecticides may be low and large numbers of cages are needed for reliable results. Wainhouse and Coaker (1981) estimated adult emergence from the number of larval mines on carrots and a non-crop host, *C. maculatum*. As larvae may cause more than one mine in carrot roots (Jones and Coaker, 1980) such estimates can give only an approximate index of adult emergence; they are most useful when the infestation is low.

Early workers monitored populations of adults in the carrot crop and field surrounds with insect sweep-nets (Baker *et al.*, 1942; Petherbridge, Wright and Davies, 1942) or wooden blocks coated with a sticky material (Baker *et al.*, 1942). When coloured traps came into use, factors influencing their effectiveness were studied. Traps in which the spectrum of reflectance is rich in yellow (560 nm) and deficient in blue (460 nm) were the most attractive to the adult carrot fly (Bohlen, 1967; Brunel and Langouet, 1970). The quality of incident light was also thought to influence trap efficiency (Brunel and Langouet, 1970). Yellow water traps placed in hedgerows adjacent to carrot crops caught large numbers of flies (Missonier and Boulle, 1964; Brunel and Rabasse, 1975; Wainhouse and Coaker, 1981). Round traps caught more adults than square ones of equal surface area (Brunel and Rabasse, 1975). There appeared to be a change in responsiveness of the sexes to the yellow water traps between the first and second generations: approximately three times more males than females were trapped at field boundaries in the first generation, whereas nearly equal numbers were trapped in the second generation (Wainhouse and Coaker, 1981), this despite parity of the sexes in both generations. When trapping takes place in a carrot crop, the vegetative stage can affect the efficiency of the trap, maximum catches occurring when the carrot rows are distinguishable against the soil (Brunel, 1971a). Trap catches decline as carrots grow taller, and more flies move beneath the canopy than above it (Brunel and Blot, 1975).

Yellow sticky traps have also been used and require less servicing than water traps (Shaw, Allan and Inkson, 1961; Wakerley, 1963; Stevenson, 1977; Getzin, 1982). Stevenson (1977) developed a disposable two-sided yellow adhesive Bristol board trap and Judd, Vernon and Borden (1985a) a vertical, four-sided yellow sticky trap which was equally attractive to adults. Disposable yellow sticky traps and re-usable yellow acrylic sticky plates were found to be more efficient than yellow water traps, and were more readily accepted by vegetable growers (Freuler, Fischer and Bertuchoz, 1982a). Trapping at different heights at the edge of host-plant fields and in adjacent meadows (Städler, 1972) obtained over 30% of captures at 20 cm and 50% at 80 cm above the ground. Uvah and Coaker (1984), trapping 1 m outside carrot plots, caught 60–70% of adults at 37.5 cm and the remainder at 75 and 150 cm above the ground and Judd, Vernon and Borden (1985a) caught more adults on traps placed 5–10 cm above the carrot canopy than at other heights. Vertical sticky traps also caught significantly more flies than horizontal ones, and the numbers caught increased linearly with trap size. Philipsen (1986) obtained greater catches on yellow sticky traps by placing artificial 'tall weeds' made of plastic netting next to the sticky traps. More adults were attracted to traps baited with oil of coriander (Hills, 1965) and a mixture of leaf aldehydes and propylbenzenes (Guerin, Städler and Buser, 1983). A Dietrick-type suction sampler (Southwood, 1978) is also suitable for trapping adults that are concentrated in the herbaceous layer of field surrounds and caught approximately 40% more males than females in both the first and second generations (Wainhouse and Coaker, 1981).

Biology

GENERATIONS AND LIFE CYCLE

The number of generations of *P. rosae* per year depends on the geographical location and prevailing climatic conditions. 'Generation' in this paper describes the cycle starting at the adult and ending at the pupa (Smith, 1921). In northern USSR, there is one complete generation with a partial second (Savzdarg, 1927). Ausland (1957) reported two generations as far north as 63 degrees 28 minutes and there are three generations at the southern extent of its range (Glasgow and Gaines, 1929; Missonier *et al.*, 1964; Biernaux, 1972). In eastern England there are always two generations per year, although in most years there is a partial third (Wright and Ashby, 1946b; Coppock, 1974). First-generation adults in England emerge around mid-May and reach a peak between late May and early June. Second-generation flies generally emerge at the end of July, being most abundant in early August, and thereafter are found in varying numbers until December. Peak appearance of third-generation flies, when present, occurs in November (Coppock, 1974).

First eggs have been found in Canada 4–6 days after first emergence of flies (Scott, 1952) and in England after 6–14 days (Coppock, 1974). Hatching in 7–10 days (van't Sant, 1961), larvae move to the plant roots to feed. In England, the time from egg to adult takes at least 2 months (Petherbridge and Wright, 1943; Coppock, 1974).

EGGS

Description

Eggs are white and elongate, 0.15 mm in diameter by 0.6–0.7 mm in length (Ashby and Wright, 1946; Jørgensen and Thygesen, 1968); they may be as small as 0.46 mm (van't Sant, 1961). The chorion is sculptured in an irregular reticulate pattern superimposed upon longitudinal striations extending meridionally around the length of the egg. Its structure consists of three layers similar to the chorion of the cabbage root fly (*Delia radicum*, L. (Diptera: Anthomyiidae)) (Hinton and Cole, 1965) and the wheat bulb fly (*Delia coarctata* (Fall.) (Diptera: Anthomyiidae)) (Hinton, 1962). The micropylar cap has a small circular plug with eight sockets around its periphery, and on hatching, larvae emerge by rupturing the chorion just behind the micropylar cap along eight lines of weakness (Ashby and Wright, 1946).

Conditions for hatching

The incubation period, which depends on temperature, was 6–12 days in Russia (Savzdarg, 1927) and Canada (Dustan, 1930; Gorham, 1934), 6.2 days on average in the USA (Whitcomb, 1929) and 5–8 days in Germany (Körting, 1940). Under laboratory conditions eggs took 11.4 days to develop at 15°C compared with 5.5 days at 28°C; when kept at 10° or 30.5°C eggs did not hatch within 21 days but did so when the temperature was restored to 25°C (van't Sant, 1961). Threshold temperatures and day-degrees (D°) required to complete development of eggs in England were 6°C and 94.5 D° (Burn, 1980), and in Canada were 4.1°C and 102 D° (Stevenson, 1981b) and 4.5° C and 94 D° (McLeod, Whistlecraft and Harris, 1985). Exposure to 40°C for 1 h was lethal to newly laid carrot fly eggs, whereas at 35°C, a longer period of exposure was tolerated (Burn, 1980). Temperature sensitivity declined progressively as eggs aged. Although 100% relative humidity (r.h.) was optimum for egg survival, there was no significant reduction in hatching success at 80% r.h. and some eggs completed development and hatched at 65% r.h. (Burn, 1980). Values for relative humidity converted to saturation deficit and plotted on a log-probit scale gave a saturation deficit LD_{50} of about 7.5 mb (0.75 kPa).

LARVAE

Description

P. rosae has three larval instars each with a thin rounded tapering body; they are acephalous with a typical cyclorrhaphan cephalopharyngeal skeleton (Whitcomb, 1938; Ashby and Wright, 1946; Osborne, 1961; Jørgensen and Thygesen, 1968). Each instar has 13 body segments: a preoral (head) segment bearing the sensory papillae, three thoracic and nine abdominal segments (Ashby and Wright, 1946; Ryan and Behan, 1973; Jones, 1977).

The larval instars can be distinguished by several means. Body length is a somewhat unreliable parameter because of the overlap in size between instars (Jones, 1977); however, differences in other parameters can be distinguished under low magnification or by eye. The mean lengths of the cephalopharyngeal sclerites, from the anterior end of a mandibular sclerite to the posterior end of the ventral process of a thecal sclerite, for the three larval instars are 0.20, 0.38 and 0.64 mm respectively, while those for posterior spiracular diameters are 0.03, 0.06 and 0.10 mm (Jones, 1977). Posterior spiracles, of which there are two in all instars, have one spiracular opening in first-, and three in second- and third-instar larvae. In addition, two anterior spiracles are found in second and third instars only (Ashby and Wright, 1946; Jørgensen and Thygesen, 1968). Cuticular denticles, which assist in locomotion, occur on the anterior borders of the ventral surface of abdominal segments 1–7 and also on the dorsal surface of the first three abdominal segments. These denticles are found in all instars but are conspicuous only in the third instar where they occur on a raised reddish-pigmented area (Ashby and Wright, 1946). On hatching, the first larval instar is colourless and translucent (Jørgensen and Thygesen, 1968) changing to a creamy opaque colour in the second instar and to a yellowish colour in the third instar, due to the deposition of lipid material within the body.

Third-instar larvae can be distinguished from those of the closely related species, *Psila nigricornis* Mg. (Diptera: Psilidae), by several external morphological features (Osborne, 1961).

Development

First-instar larvae feed on the fibrous side roots of the carrot while older larvae feed on the main root (Pettit, 1931; Gorham, 1934; Chamberlain, Skillman and Stewart, 1937; Körting, 1940; van't Sant, 1961; Jørgensen and Thygesen, 1968). In laboratory and field conditions, Jones (1977) found that about 20% of second instars fed on the main root and the remainder on the side roots, while all third instars fed on the main root. During dry conditions, however, larvae fed entirely on lateral roots and pupated at depths of 20–30 cm, which was below the main carrot root (Jones, 1979).

Moulting occurs in the soil only, as cast skins were not found in damaged roots (Ashby and Wright, 1946). The threshold temperatures and D° above the thresholds necessary to complete larval development in Canada were 2.0°C and 642 D° (Stevenson, 1981b), and 2.02°C and 625 D° (McLeod, Whistlecraft and Harris, 1985). Development of the first instar in England is completed in a few days, the second instar in 2–3 weeks and the third instar in from 3 weeks to 4 months, depending on the temperature.

Distribution in the soil, movement and dispersal

In the laboratory, first-instar larvae enter the soil in response to temperature, high humidity and the absence of tactile stimulation; gravitational stress and light do not influence soil penetration (Städler, 1971a). Larvae avoid low moistures, this response becoming stronger as they approach pupation (Jones, 1979). First-instar larvae have a preferred temperature of about 18°C (Städler, 1971a) while that of the third is about 3°C lower (Jones, 1979). Temperatures between 30° and 40°C adversely affect movement and those over 40°C are lethal to third-instar larvae (Jones, 1979). Although first-instar larvae are not influenced by light, third instars are photonegative (Städler, 1971a; Jones, 1979).

Larvae move from root to root, both in the field (Petherbridge, Wright and Davies, 1942; Petherbridge and Wright, 1943; Wright and Ashby, 1946b, van't Sant, 1961; Jørgensen and Thygesen, 1968; Jones and Coaker, 1980) and in clamped carrots (Petherbridge, Wright and Davies, 1942; Petherbridge and Wright, 1943; van't Sant, 1961). During development, larvae can move at least 60 cm both along and between carrot rows, this movement being stimulated and enhanced by increased soil moisture. Moisture also appears to have the greatest influence on the vertical distribution of larvae in the soil and on carrot roots (Jones, 1979). Jones and Coaker (1980) showed that, after irrigation of a carrot plot, larvae moved from carrots into the soil; subsequently the number of mines, which prior to irrigation was equal to the number of larvae, increased by 30%. Although dispersal of larvae is greatest during the first generation (Jones, 1977), dispersal of second-generation larvae may be responsible for some of the increased damage to over-wintering carrots in England, previously attributed to a partial third generation (Wright and Ashby, 1946b; van't Sant, 1961; Coppock, 1974; Anonymous, 1983a). In general, peak numbers of second-generation third-instar larvae in roots occur from December to February in southern England (Petherbridge, Wright and Davies, 1942; Coppock, 1974). The proportions of larvae and pupae in soil and carrots over the winter depend on the time of onset of the second-generation attack as well as climatic conditions.

Responses to host-plant odour

Larvae are attracted to host-plant material and to a lesser degree to non-hosts (Städler, 1971a; Jones and Coaker, 1977), suggesting that specific attractants as well as a non-specific attractant (CO_2) are involved in host-plant finding. Jones and Coaker (1977) obtained orientated responses by all larval instars to CO_2, to a steam-volatile fraction of carrot roots and also to methyl-eugenol, a component of that fraction. Methyl-eugenol and CO_2 together at sub-optimal concentrations have an additive if not synergistic effect on larval responses (Jones, 1977). Several other components of the steam-volatile fraction of carrot roots are also attractive to larvae (Ryan and Guerin, 1982). Bornyl acetate, 2,4-dimethyl styrene, α-ionine, β-ionine and biphenyl elicited consistently positive responses (klinotaxis and klinokinesis). *Trans*-2-nonenal was the most consistently avoided compound; it has also demonstrated insecticidal activity (Guerin and Ryan, 1980). Larvae also respond to certain agar-soluble exudates by contact chemostimulation (Jones, 1977).

The sensory organs mediating these orientated responses have been little studied. Ashby and Wright (1946), using the light microscope, described the receptors found on the preoral segment of third-instar larvae, and Ryan and Behan (1973), using a scanning electron microscope, studied the external structure of these receptors, but could only postulate their involvement in host-plant finding from similar structures found in other insects. Ryan, Guerin and Behan (1978), from ultrastructural studies, suggested that ampullaceous sensilla could be olfactory, as the dome of at least one is perforated by approximately 50 000 pores adjacent to much-branched dendrites derived from groups of neurones. Recordings from electrolyte-filled capillaries applied to an ampullaceous sensillum presented with test compounds in an air stream, confirmed this (Ryan and Guerin, 1982). Jones (1977) also found trichoid sensilla in close association with the three spiracular apertures of the posterior spiracles of third-instar larvae. When CO_2, but not O_2 or N_2, was puffed over the spiracles, the larva responded by raising its posterior end and these receptors may therefore monitor the CO_2 level around the spiracular openings.

PUPAE

Description

Visible metamorphosis begins with a brief prepupal stage when the cuticle contracts and hardens to form a puparium about 5 mm in length (Whitcomb, 1929; Biernaux, 1972). The puparium changes from yellow to a golden-brown as it hardens, and is characterized by a sharply cut-off anterior end. This 'lid' from which the adult emerges is formed from the dorsal part of the meso- and metathoracic and first abdominal segments (Ashby and Wright, 1946) and is typical of Psilidae (Jørgensen and Thygesen, 1968). Puparia formed in autumn are smooth and pale yellow whereas spring forms are darker and larger, with a clearly marked segmentation (Wright and Ashby, 1946b). Wright and Ashby (1946b) designated the metamorphic activity within the puparium as occurring in three stages. Biernaux (1972), however, proposed four stages: the prepupa; fully formed pupa with no pigmentation; pupa with pigmented eyes, and pupa with both eyes and wing primordia pigmented.

Conditions for development

The threshold temperatures and thermal requirements for complete development of pupae in Canada were 3°C and 107 D° (Stevenson, 1981b), and 1.47°C and 374 D° (McLeod, Whistlecraft and Harris, 1985). The length of the pupal stage is also dependent on the prevailing environmental conditions. At temperatures of 19–20°C the pupal stage lasts about 25 days, but lower and higher temperatures lead to diapause or aestivation, respectively (van't Sant, 1961).

Diapause and aestivation

Some first-generation pupae (Körting, 1940; Ausland, 1957; Biernaux and Seutin, 1969) and larvae and pupae from the second and third generations overwinter (Körting, 1940; Wright and Ashby, 1946b). In the more northern part of its range at Landvik, Norway (58 degrees 20 minutes N), 70–80% of first-generation pupae overwinter (Ausland, 1957). Photoperiod does not influence the induction of diapause (Städler, 1970), being induced in the prepupal stage (Burn and Coaker, 1981) in response to cool temperatures. In Switzerland, 14% of pupae entered diapause at 15°C, 89% at 12°C and 96% at 10°C (Städler, 1970). In Canada, 40% entered diapause at 16°C, and 100% at 13° and 10°C (McLeod, Whistlecraft and Harris, 1985). As in some other locations, the carrot fly overwinters in Britain in both the larval and pupal stages, pupae comprising an increasingly large proportion of the population as winter progresses (Wright and Ashby, 1946b). Overwintering larvae continue to feed until early spring, when they pupate, but do not enter diapause despite soil temperatures of 2–8°C (Burn and Coaker, 1981). Thus the sensitivity to diapause-inducing conditions declines over winter. Brunel (1968) found the reverse situation to occur in summer, i.e. sensitivity to diapause-inducing conditions and the threshold temperature for diapause–induction increased from late summer to autumn. Spring-formed pupae tend to emerge earlier in spring than autumn-formed pupae (Biernaux, 1968; Burn and Coaker, 1981; McLeod, Whistlecraft and Harris, 1985). McLeod, Whistlecraft and Harris (1985) found that diapause terminated in 95% or more of pupae following exposure to 1°C for 20 weeks.

Aestivation and delayed emergence occurred at temperatures ranging from 21° to 25°C (van't Sant, 1961; McClanahan and Niemczyk, 1963; Missionnier and Boulle, 1964; Brunel, 1968; Städler, 1970).

ADULTS

Description

The adult is 6–8 mm long, has a shining black thorax and abdomen, a reddish brown head, yellowish legs and iridescent wings (Stevenson, 1981a; Anonymous, 1983a). The female has a pointed abdomen formed by a telescopic ovipositor, while the male has a more rounded abdomen (Jørgensen and Thygesen, 1968). The third antennal segment of P. rosae is usually partly yellow whereas in P. nigricornis, a species with which it can be confused, it is usually entirely black. These two species can be distinguished with certainty only by the size and structure of the male genitalia (Collin, 1944).

Emergence

Coppock (1974) calculated that peak emergence of flies of the first (overwintering) generation caught in sweep-nets and/or sticky traps in England was associated with 234 ± 53 D° of ambient air temperature above a threshold of 5.6°C from 1 April; however, warm conditions in February and March lead to an earlier emergence. This compares with first emergence trapped on yellow sticky traps in Ontario at 258 D° above 5°C after 1 March (Stevenson, 1983) and British Columbia at 326 ± 14 D° above 3°C after 1 February (Judd and Vernon, 1985). Adults of the second generation were first caught at 1148 D° above 5°C after 1 March (Stevenson, 1983) and 1125 ± 41 D° above 3°C after 1 February (Judd and Vernon, 1985). Adult emergence occurs mainly at 15°C and above (Jørgensen and Thygesen, 1968). Adults can emerge from depths of 30 cm (van't Sant, 1961; Burn, 1984) and possibly from up to 70 cm (Savzdarg, 1927).

Feeding requirements

Petherbridge, Wright and Davies (1942) caged newly emerged flies without a food-source over moist soil and carrot seedlings, and fertile eggs were laid despite no apparent adult feeding. Brunel (1979) also found that the first batch of eggs could develop without feeding, but that protein and carbohydrate were essential for the development of the second and third batches. Flies provided with water only, survived about 10 days, whereas when provided with protein, carbohydrate and water, they survived over 20 days. While laboratory-rearing carrot fly, Städler (1971b) obtained 75 eggs per female with a diet of 4 parts cane sugar to 1 part enzymatic yeast hydrolysate; McLeod, Whistlecraft and Harris (1985) obtained 109 eggs per female when provided with a 4:1 mixture of honey and yeast hydrolysate. Adults were observed feeding in the field on the blossoms of *A. sylvestris* and *C. maculatum* (Petherbridge, Wright and Davies, 1942) and on the honeydew of aphids on carrot plants (Watkins and Miner, 1943). In the laboratory, Wainhouse (1977) observed that feeding was much more intense in the early morning than at other times of the day.

Mating

In the laboratory, mating occurs mostly in the afternoon, pairs remaining coupled for at least 40 minutes (Wainhouse, 1977). Copulating pairs attract males which climb on to them without dislodging them. In the field, mating is thought to occur in the shelter of hedgerows. There is no evidence that pheromones are involved; however, high-pitched sound produced by males may have some role (Brunel, 1977; Städler, 1977). Males also vibrate their bodies with their legs very rapidly for about one second; these vibrations are repeated after short runs, regardless of the substrate. The conformity of the responses of males and females to volatile plant constituents suggests that they may provide aggregation cues for both sexes (Guerin and Visser, 1980).

Movement and dispersal

The orientation of newly emerged flies to the soil surface is influenced by gravity and, as they approach the soil surface, by light (Städler, 1972). After emergence, both sexes move to the surrounding trees and bushes for shelter, responding primarily to the silhouettes of these plants (Städler, 1972; Brunel, 1977). Städler (1972) suggested that they fly towards the highest silhouette; Wainhouse and Coaker (1981) found that most flies were not trapped at the tallest windbreaks; and D. R. Hartley (unpublished data) found the effect of trees to be of overriding importance to the distribution of flies. Carrot fly adults sheltered on a variety of plants in field surrounds, the vegetation providing food and shelter and sometimes wild host plants (Petherbridge, Wright and Davies, 1942; Wainhouse, 1975). When hedgerows were present at boundaries, flies sheltered on the leeward side (Baker *et al.*, 1942). Wakerley (1964) suggested that wind probably created a micro-environment of high humidity, and therefore had an indirect effect on the distribution of flies which preferred 85–100% r.h. In strong winds, flies left the upper part of the hedge for the undergrowth on the sheltered side; calm conditions caused an equal distribution of flies throughout the hedge. Most authors found that sides of hedges in full sun were avoided (Petherbridge and Wright, 1943; Watkins and Miner, 1943; Jørgensen and Thygesen, 1968). Wakerley (1963), however, found no evidence that the carrot fly sought shade during the sunny part of the day. Under laboratory conditions, flies aggregated in a temperature gradient of 18–24°C (Wakerley, 1964).

Flies actively discriminate between different types of field surrounds and do not accumulate at them passively. Their abundance is not proportional to the number of nearby carrot fields from which they emerged (Wainhouse and Coaker, 1981). In a study of the distribution of flies in a commercial carrot-growing area in East Anglia, Wainhouse and Coaker (1981) described the relative abundance of flies at field surrounds from which flies were emerging, with a 'boundary index'. This index was derived from the relative abundance of trees and bushes in windbreaks, common nettle (*Urtica dioica* L.), *C. maculatum*, and other flowering plants within the boundaries. Although the density of windbreaks was of some significance, the single most important component of the index was *U. dioica*; large populations of flies occurred at field surrounds without windbreaks but with abundant *U. dioica* which formed favourable resting sites (Petherbridge, Wright and Davies, 1942; Coppock, 1974). *C. maculatum* also contributed significantly to local concentrations of flies. Laboratory actographic records showed that 5-day-old females with mature eggs were 7–8 times more active than females on the day of emergence, suggesting that there is no innate pre-reproductive dispersal (Wainhouse, 1977).

Some flies probably disperse along hedgerows (Roebuck, 1945; Wainhouse, 1975) and females flying to fields to oviposit may return to different shelter sites (Städler, 1972), thus gradually dispersing away from emergence sites by trivial movements. Städler (1972) recaptured marked flies up to 130 m from the release point, estimating that only 1–2% had flown further than 80 m. Wainhouse (1975) observed that numbers of flies in hedgerows declined rapidly over the first 100 m from the source-field; however between-crop movement could still occur over distances of at least 400 m. Evidence for dispersal of perhaps at least 4 km comes from van't Sant (1961) who observed attack on carrots planted on islands from which the natural population had been removed by flooding.

Responses to host-plant odour

Carrot plants at different stages of growth in the field varied in attractiveness to carrot flies (Brunel and Blot, 1975), suggesting the presence of olfactory attractants. In laboratory choice tests, carrot flies made more landings and deposited more eggs in the vapour plume over host plants than over non-hosts (Guerin and Städler, 1980). In the field, the majority of flies approached carrot plots upwind, possibly in response to carrot odour (Uvah and Coaker, 1984). Using an inflatable polythene wind tunnel, Nottingham (1987) also observed an upwind anemotactic response to carrot-plant odour. Female flies demonstrated a higher electroantennogram (EAG) response to headspace vapour from host plants than from non-host plants (Guerin and Städler, 1980) and a greater response to carrot-plant volatiles than *D. radicum* or the onion fly (*Delia antiqua* (Mg.), Diptera: Anthomyiidae) (Guerin and Städler, 1982). Guerin and Visser (1980) recorded the best EAG responses to the aldehyde component of the 'general green leaf volatiles' and from *trans*-methyl-iso-eugenol, β-caryophyllene, linalool and *trans*-2-nonenal, and Guerin and Städler (1980) to *trans*-asarone, which elicited a response almost ten times greater than that to the green leaf volatiles or *trans*-methyl-iso-eugenol. The strongest responses, using a gas chromatography linked electroantennogram detector (GC-EAD), were to the propylbenzenes, *trans*-methyl-iso-eugenol, *trans*-asarone and to a lesser extent, to the leaf aldehydes, hexanal, (*E*)-2-hexenal and heptanal and to the terpenes, linalool and caryophyllene (Guerin, Städler and Buser, 1983).

Bohlen (1967) suggested that host-plant recognition was probably chemical, resulting from contact by the tarsi and proboscis. Methyl-iso-eugenol was identified as an oviposition stimulant in carrots and anise aldehyde and estragol in fennel and celery (Städler, 1972). More recently, the propylbenzenes, *trans*-methyl-iso-eugenol and *trans*-asarone; the furanocoumarins, bergapten and xanthotoxin; the substituted coumarin, osthol; and the polyacetylene, falcarindiol identified in the surface wax of carrot leaves have been shown to stimulate oviposition (Städler and Buser, 1984).

Guerin, Städler and Buser (1983) suggested that, through its selective sensitivity to foliar volatiles, carrot fly orientates to the host plant and makes contact with the less volatile leaf surface components which then result in selection of an oviposition site. Insects are likely to depend on the 'chemical signature' on the leaf surface for successful host-plant recognition; with the carrot fly, this appears to be a complex mixture rather than a key compound (Städler and Buser, 1984).

Responses to host-plant appearance

No effect on host-plant selection by the carrot fly could be demonstrated as being attributable to leaf shape (pinnate vs. non-pinnate) or leaf surface (hairy vs. smooth) (Bohlen, 1967). Städler (1972), using pinnate celery leaf dummies and non-pinnate round dummies, was also unable to demonstrate the effect of the pinnate leaf shape on oviposition. Although flies could distinguish between the shades of green of carrot leaves and two non-umbelliferous species, tomato (*Lycopersicon esculentum* Miller) and bean (*Vicia faba* L.), leaf colour did not appear to provide a specific stimulus for landing (Bohlen, 1967). On the other hand, catches of flies in four umbelliferous crops correlated with the intensity of reflectance of these crops at 560 nm (Brunel, Freuler and Duchesne, 1981). Catches were also found to be related to the intensity of reflectance from different ages of carrot foliage, and inversely related to the height of carrot foliage.

Oviposition

Eggs may first be laid in the field 1–4 days after emergence under favourable conditions (Whitcomb, 1929, 1938; Ausland, 1957; van't Sant, 1961; Bohlen, 1967; Jørgensen and Thygesen, 1968). Under laboratory conditions, with water, carbohydrate and protein supplied *ad libitum*, females laid three batches of eggs beginning on the second, sixth and twelfth days following emergence (Brunel, 1979). Estimates of the number of eggs laid per female varied from 5 to 167 (Whitcomb, 1929; Körting, 1940; van't Sant, 1961; Bohlen, 1967), but these estimates were based on small samples. Jørgensen and Thygesen (1968) in a more extensive study of field-caught flies estimated an average of 41 eggs per female. The wide variation in the numbers of eggs observed was probably due to differences in nutritional status of the flies.

In Ontario, Canada, appearance of the first eggs coincides with the appearance of the first blossoms of the choke-cherry (*Prunus virginiana* L.) (Dustan, 1930; Beirne, 1971). In East Anglia, first-generation eggs are first found at the end of May or early June, second-generation eggs first in late July and August, and those of the third generation in early November, but in mild winters eggs have been collected as late as January (Coppock, 1974). Seedling carrots may be attacked any time after the cotyledon stage (Petherbridge and Wright, 1943) and the number of eggs laid increased with the amount of carrot foliage or some factor related to the age of the carrot (Petherbridge, Wright and Davies, 1942). Wakerley (1963) found that egg-laying occurred during bright, warm and dry, and dull, cool and dry weather. Oviposition flights took place during these two weather patterns when temperature, humidity and light intensity were similar, light intensity in the range of 15–17 000 lux being the overriding factor. Wakerley (1963) noted that female flies returned to hedges in the evening despite no change in temperature or humidity and concluded that only light intensity had changed from that required to stimulate oviposition flight. The optimum light intensity for oviposition appears to be within the range 100–1000 lux (Städler, 1975), and is found in the shade of plant foliage (Overbeck, 1978).

Bohlen (1967) described a sequence of behaviour leading to oviposition, in which females landed on carrot leaves and immediately began to run over the upper and lower sides of the leaves extending their ovipositor without it contacting the plant surface. The haustellum was also repeatedly extended, the labella briefly contacting the leaf surface. After at least 20 seconds on the leaf, the females then moved down the leaf petiole to the soil where they oviposited. The choice of oviposition site was determined by tactile stimulation of the ovipositor, roughness of the soil surface, negative photo-taxis and humidity of the substrate. Trichoid sensilla on the cerci of the ovipositor are probably mechanoreceptors although some may possibly func-tion as olfactory or contact chemoreceptors (Behan and Ryan, 1977).

Rhythmic activity

Females have a diel rhythm of activity in which they leave the shelter of the vegetation of field boundaries to oviposit in carrot fields, usually between 16.00 and 20.00 hours (Baker *et al.*, 1942; Petherbridge and Wright, 1943; Fox Wilson, 1945; Williams, 1954; Ausland, 1957; van't Sant, 1961; Städler, 1975). In the laboratory, a similar diel rhythm was observed. At 15 h light, 9 h dark (LD 15:9) there was a low level of activity for 8 h after 'lights-on', increasing to a high level for the period 11–14 h after lights-on (Bohlen, 1967). In LD 12:12, however, peak feeding occurred within 1 h after lights-on, peak oviposition 10–12 h later, and mating, which occurred throughout the day, increased slightly in the latter half of the light phase (Wainhouse, 1977). In actographs in LD 12:12, peak activity of females occurred 10–12 h after lights-on, but in the absence of external time cues (constant dim light), the rhythm persisted with a period of 22.5 h, and was, therefore, presumed to be endogenously controlled (Wainhouse, 1977). In continuous bright light, activity became arrhythmic (Städler, 1975; Wainhouse, 1977). Wainhouse (1977) suggested that reports of activity occurring throughout the day on dull, overcast days may be an artefact from the use of yellow traps, the attractiveness of which to carrot flies varies with the quality of light falling on them. In an acoustic actograph, the presence of carrot plants reduced afternoon activity of females, probably through aggregative effects of the host-plant odour (Brunel, 1977).

Control

NATURAL

Abiotic factors

Few studies on the carrot fly have determined lethal thresholds for weather factors. The interaction of temperature, humidity and wind, and the varying susceptibilities of each stage of the carrot fly is complex.

High egg mortality was recorded above 26°C (Whitcomb, 1938; van't Sant, 1961; Overbeck, 1978), and at low (unspecified) humidity (Whitcomb, 1938; Ausland, 1957; van't Sant, 1961). Early larval mortality was one of two key factors for carrot fly in England (Burn, 1984), and was attributed to unusually low humidity (54% r.h. at the soil surface), low soil moisture (< 10% of field capacity) and high temperature ($\geq 35°C$ immediately below the soil surface). Another example of extreme adverse conditions was flooding of the 8000 ha Bradford Marsh in Ontario, Canada for 2–3 weeks in October, 1954 due to hurricane 'Hazel' which killed all larvae and most pupae (Backs, 1957). The number of adults captured in sweep nets in 1955 was only 11.5% of the 1951–54 average, and was attributed largely to invasion from adjacent upland farms. No larvae or pupae were found in early or late carrots throughout the marsh in 1955, due to the reduced number of flies and possibly also to hot, dry weather. *P. rosae* did not return to pest status until 1961 (Beirne, 1971).

Hanson and Webster (1941) suggested that winter mortality was the greatest factor in population reduction. Third-instar larvae collected from soil at a depth of 5 cm, and a soil temperature of $-3°C$, suffered 60% mortality within 6 days of being returned to 19°C (Labeyrie, 1956). Temperatures above 23°C increased larval and pupal mortality (Städler, 1970).

Biotic factors

The two most important parasitoids attacking the carrot fly are *Chorebus* (= *Dacnusa*) *gracilis* (Nees) (Hymenoptera: Braconidae) and *Basalys* (= *Loxotropa*) *tritoma* (Thoms.) (Hymenoptera: Diapriidae) (Körting, 1940; Wright and Ashby, 1946b; Wright, Geering and Ashby, 1947; van't Sant, 1961; Jørgensen and Thygesen, 1968). Levels of parasitism ranging from 4% to 63% by *C. gracilis*, the more significant of the two parasites, have been recorded (Savzdarg, 1927; Körting, 1940; Barnes, 1942; Wright and Ashby, 1946b; van't Sant, 1961), but this did not prevent extensive damage to carrots because parasitized larvae completed feeding and pupated before the parasitoids emerged (Wright and Ashby, 1946b). Naton (1968) described a rearing method for *C. gracilis*, but considered that releases would be of minimal benefit, as it is a poor searcher during the cool damp weather conditions favourable to *P. rosae*. Nixon's key to the Dacnusinae (Nixon, 1944) can be used to identify *C. gracilis*. Up to 14% parasitism by *B. tritoma* has been recorded (Körting, 1940; Wright and Ashby, 1946b; Wright, Geering and Ashby, 1947). A rearing method was developed for *B. tritoma* (Maybee, 1956) to facilitate releases. *C. gracilis* and *B. tritoma*, obtained from England, were released in large numbers in several carrot-growing areas across Canada from 1949 to 1954 (Maybee, 1954). However, both species failed to establish, for unknown reasons. Several other braconid parasitoids of minor importance have been recorded: *Aphaereta minuta* (Nees), *Dapsilarthra* (= *Adelura*) *apii* (Curt.), *Alysia apii* (Curt.) and two species of *Dacnusa* (Curtis, 1860; Smith, 1921; Smith and Gardner, 1922; Jegen, 1932; Balachowsky and Mesnil, 1936). Burn (1984) recorded parasitism by *Kleidotoma psiloides* Westwood (Hymenoptera: Eucoilidae) which may be the unidentified *Kleidotoma* species identified by Wright, Geering and Ashby (1947) and Sachtleben (1954) recorded *Kleidotoma* (= *Rhynchacis*) *nigra* (Hartig) (Hymenoptera: Eucoilidae). Parasitism by *Aleochara sparsa* Heer (Coleoptera: Staphylinidae) is sporadic and irregular with levels of up to 5% (Wright and Ashby, 1946b; Wright, Geering and Ashby, 1947).

Burn (1982) studied losses of marked *P. rosae* eggs due to predation. Peak egg loss in spring (23–42%) coincided with maximum numbers of Aleocharinae (Coleoptera: Staphylinidae) caught in pitfall traps, whereas peak egg loss in autumn (27–60%) coincided with maximum numbers of *Trechus quadristriatus* (Shrank) (Coleoptera: Carabidae). Egg loss in mid-summer was smaller (13–18%). The damaged remains of marked eggs bore the characteristic feeding patterns made by both groups. Larger carabids were probably not involved in egg predation, as the small size of the carrot fly egg may be outside their detectable prey-size range.

A blackening bacterial disease of larvae and pupae (Barnes, 1942; Wright and Ashby, 1946b) caused 15% pupal mortality (Labeyrie, 1956). Two nematode parasites of larvae and pupae also were reported by Wright and Ashby (1946b) but parasitism was generally below 1%; Labeyrie (1956), however, recorded levels of 70% of which 3% was by mermithids in larvae. The fungus *Empusa* sp. *(=Entomophthora)* can cause high losses of adults in some years, especially in late summer (van't Sant, 1961). Eilenberg and Philipsen (1986) reported three species of Entomophthorales on carrot flies: *Entomophthora muscae* (C.) Fres. (a complex of species with campanulate primary spores); *Conidiobolus apiculatus* (Thax.) Remaud. and Kell., and *Erynia* species. *E. muscae* was the most common, causing up to three epizo-otics per year and infecting up to 60% of flies; it was considered to be a potential agent for biological control (Eilenberg, 1983). Despite their occasional effectiveness in other studies, biotic agents were not found to be key mortality factors by Burn (1984).

APPLIED

Cultural

Several cultural methods have reduced damage to umbelliferous crops by *P. rosae*. It has long been known that first-generation damage can be reduced by delaying sowing (Henderson, 1814; Fletcher, 1904; Glasgow and Gaines, 1929; Whitcomb, 1938; Petherbridge, Wright and Davies, 1942; van't Sant, 1961) but yields may suffer due to the shorter season (Petherbridge, Wright and Davies, 1942; Scott, 1959). As damage to carrots left in the field increases with time, they should be harvested and stored as early as possible (Petherbridge, Wright and Davies, 1942; Petherbridge and Wright, 1943). van't Sant and Brader (1972) found that early sowing allowed early harvest and avoidance of second-generation damage in Holland. The cost of con-structing storage facilities has, however, prevented this practice in the UK. Severely damaged carrots should not be ploughed in but should be harvested and destroyed, otherwise they will provide a source of infestation in the following year (Petherbridge, Wright and Ashby, 1945). Spatial separation of carrot crops in successive years can be highly effective (Whitcomb, 1938; Watkins and Miner, 1943; Petherbridge, Wright and Ashby, 1945; Dabrowski and Legutowska, 1976). However, the minimum distance is far from clear and most carrot growers are unlikely to have large enough acreages to accomplish this. Once the relationship between the distribution of adults in field surrounds and high levels of damage in the adjacent crop had been established (Baker *et al.*, 1942; Wright and Ashby, 1946a), it was recommended that carrots should be grown in large open fields (Wright and Ashby, 1946a; van't Sant and Brader, 1972), and the herb layer in field surrounds removed (Wright and Ashby, 1946a).

Results of intercropping with onions have been inconsistent. Petherbridge, Wright and Davies (1942), van't Sant (1961) and Anthony (1978) were unable to demonstrate any effect on carrot fly damage. Whitcomb (1938) and van Poeteren (1939) achieved, and Hills (1972) claimed, reductions in larval damage; however, details of the experiments are lacking. Uvah and Coaker (1984) obtained maximum reduction in damage by carrot fly larvae when onions were young. This method was effective against the first generation only, and became ineffective once the onions had begun to bulb. Frequent watering was suggested by Ormerod (1881) to keep the 'ground fairly compacted against the fly'. Sabatino (1976) irrigated a fennel crop infested with *P. rosae* larvae, every second day, reducing the larval population below economic level. Spreading grass cuttings between rows reduced infestation in one experiment (Petherbridge, Wright and Davies, 1942) but had no effect other than to reduce yield in another (Coppock, Maskell and Gair, 1975).

Plant resistance

Reports from the late 1800s provide conflicting results from tests for resistance to carrot fly in different carrot cultivars (cvs). This literature was reviewed by Ellis, Wheatley and Hardman (1978). In the early 1970s a systematic effort was mounted by several research teams to identify resistant varieties and the factors responsible for resistance, and to select for increased resistance. As a result, 'modest' sources of resistance were discovered in some cvs (Ellis and Hardman, 1980).

Nieuwhof (1977) and Dabrowski and Legutowska (1977) were unable to identify varietal resistance in the carrot cvs tested. De Ponti *et al.* (1981) tested 200 cvs and accessions and about 70 accessions of wild *Daucus* species. Two generations of line selection resulted in an increase of resistance over the parental generation of about 20%; however, the resistance of wild *Daucus* accessions was low, and was therefore discarded. Sazonova (1982) identified five relatively resistant cvs from screening 474 carrot cvs.

From testing 84 cvs, six including Sytan were identified as consistently less susceptible than most of the others (Ellis, 1977). Ellis, Freeman and Hardman (1978) found that incidence and severity of damage was linearly related to root length, longer roots having more damage than shorter ones. Nevertheless, roots of a given length of some cvs were more susceptible to attack than others. There was a consistent difference in the relative susceptibilities of Royal Chantenay (least) and Speed's Norfolk Giant (most) (Ellis, Wheatley and Hardman, 1978). Although correlated with plant size, allowances for foliage and root size failed to account for all of the difference. In tests of 84 cvs, the Nantes group was less severely damaged than the Chantenay and Autumn King groups (Ellis *et al.*, 1980), the Nantes carrot, Sytan, being the least susceptible. Tests of eight cvs representing a range of resistance were conducted for 2 years at 12 sites in five European countries (Ellis and Hardman, 1981). Again, Sytan proved the most resistant. A 5-year comparison of Sytan and Danvers, which represented the extremes of resistance and susceptibility, showed a reduction in second-generation damage on Sytan compared with that on Danvers of 40–67% (Ellis, Freeman and Hardman, 1984). A programme of selection within the partially resistant Sytan was begun in 1980 (Ellis *et al.*, 1984). Crosses were also made between Sytan and other resistant carrot groups. Ellis *et al.* (1985) advocated two approaches in selection for increased resistance. The first was to cross Sytan with other cvs, and from a preliminary biochemical screening (Cole, 1985), reduce the number of promising lines to about 100 for field screening. The second was to cross cultivated cvs with *D. capillifolius*, a species identified by Hardman and Ellis (1982) as not supporting an appreciable carrot fly population.

Several researchers have provided clues as to the nature of resistance to carrot fly. Ellis, Wheatley and Hardman (1978) compared two cvs with different levels of susceptibility. Both were grown in quadrats or interplanted and there was no indication that the presence or absence of choice affected their relative susceptibilities. Furthermore, it indicated that testing cvs under situations where the carrot fly can choose its host was a valid approach to ranking cvs by their susceptibility to attack. In another study, different cvs were artificially infested with carrot fly eggs (Guerin, Gfeller and Städler, 1981). Differences in susceptibility remained, indicating resistance by means of antibiosis. It was suggested that differences in attack by first-instar larvae could vary with availability of side roots, tissue texture or the presence of suitable host cues. Toxins, feeding deterrents, or the production of defensive compounds were also postulated. Visser and de Ponti (1983), noting differences in volatiles from roots and from the air over foliage, suggested that they might have a role in future screening for resistance. Headspace vapour and steam distillate from the roots of resistant and susceptible cvs were compared by gas–liquid chromatography (Guerin and Ryan, 1984). The presence of both non-preference resistance by ovipositing females and root resistance to larvae, was confirmed in some cvs. Root resistance, however, was considered the crucial prerequisite for success. Headspace vapours over Sytan (resistant) foliage evoked a significantly higher electroantennogram response and oviposition than those over Danvers (susceptible); however, flies contacting foliage preferred Danvers (Guerin and Städler, 1984). Although their findings suggest that olfactory and contact chemostimuli may be involved, they do not fully account for the preferences observed. Cole (1985) found that higher concentrations of chlorogenic acid in carrot roots correlated with increased susceptibility to larval damage. Chlorogenic acid may be essential in the production of insect cuticle.

Insecticidal

Some of the early chemicals were used as repellents and included spirits of tar, asphalt paper mulch, creosote-treated string stretched between rows and naphthalene dust (Anonymous, 1842; McIn, 1843; Curtis, 1860; Whitcomb, 1929; Hanson and Webster, 1941; Petherbridge and Wright, 1943; Fox Wilson, 1945). These methods were only partly effective and crops required frequent retreatment. Early insecticides were applied against adults. Derris powder, although expensive, was effective in reducing damage to carrots (Whitcomb, 1929, 1938; Petherbridge, Wright and Davies, 1942) as was mercurous chloride dust (Glasgow, 1929). In the early 1940s, a poison bait of 2.5% molasses and 0.8% sodium fluoride was applied to dyke-sides and hedges surrounding fields; it was generally effective in reducing root damage, but frequent retreatment was necessary (Petherbridge, Wright and Davies, 1942).

These early methods were rendered obsolete by synthetic organic insecticides in the mid-1940s. DDT applied to carrot foliage, was highly effective against adults (Wright and Ashby, 1945). Aimed principally at larval control, gamma BHC applied as a dust or seed dressing was effective (Newton, Satchell and Shaw, 1946; Wright, 1951; Shaw, 1952; Shaw and McDonald, 1955; Stone, 1956) but tainted the carrots and was replaced by dieldrin (Wright and Bowe, 1957). Resistance to organochlorine insecticides by the carrot fly first appeared in Washington State, USA in 1957 (Howitt and Cole, 1959), then Canada (Niemczyk and Harris, 1962; Finlayson, Fulton and Noble, 1964), France (Missonier *et al.*, 1964), England (Gostick and Baker, 1968; Wright and Coaker, 1968) and Holland (Brown, 1971). Subsequently, the change-over was made to the less-persistent organophosphorous (OP) compounds (Brunel, 1971b), principally phorate and chlorfenvinphos in the UK, and carbamates (Bachmann and Legge, 1968) such as carbofuran in Canada. Carrot crops treated with a granular formulation of phorate incorporated into the soil at sowing would normally be protected for 15–20 weeks in organic soils and 25–30 weeks in mineral soils (Wheatley and Wright, 1970; Wheatley, 1971). The short residue life of this treatment exposed carrots to second- and third-generation damage, especially in areas where carrots were left in the ground over winter (Wheatley, 1969).

Proper placement is necessary to ensure full insecticidal activity. 'Broadcast' applicators which distribute granules over a certain swath width, have been replaced by precision applicators. 'In-furrow' or 'bow-wave' (Makepeace, 1965) applicators generally place granules within 2 cm of the soil surface (Thompson *et al.*, 1986), and 'vertical-band' applicators place granules in a fairly uniform vertical band extending from the soil surface to a depth of about 15 cm (Whitehead, Tite and Bromilow, 1981). Thompson *et al.* (1986), were able to demonstrate increased efficacy from 'vertical-band' as opposed to 'bow-wave' applicators. Soil insecticides act against the carrot fly by being picked up and translocated to feeding sites on the roots at concentrations sufficient to kill larvae (Wheatley and Hardman, 1967). Supplementary insecticidal sprays such as diazinon for the control of the second and third generations (Maskel and Gair, 1973) were reported to be most effective when directed at the crown of the carrot plant and surrounding soil for larval control (Mowat and Martin, 1984); flies on carrot foliage, however, may also be affected (Maskell and Gair, 1973). Indications that phorate was showing reduced effectiveness against first- and second-generation larval attack in England (Maskell and Gair, 1973) led to the suggestion that flies were developing resistance (Wheatley and Percivall, 1974; Thompson and Harris, 1982) but this has not yet been confirmed. Recent evidence suggests that the erratic performance of OP (e.g. fensulfothion) and carbamate (e.g. carbofuran) insecticides may be due in part to the development in soil of microbial populations capable of enhanced degradation of these insecticides (Suett, 1986; Harris *et al.*, 1987).

Supervised control

In recent years, monitoring programmes have been established to reduce applications of insecticides against carrot fly. Trapping efficiency is so low, however, that economic thresholds are essentially based on presence or absence of captured flies. Using yellow sticky traps to monitor adults in carrot crops in the Bradford Marsh, Ontario, 0–5 applications of insecticides were made in contrast to the 5–6 applications applied in routine programmes (Stevenson, 1981c). It was proposed that recommendations based on trap counts could be made for groups of farms located within uniform environs. Getzin and Archer (1983) in Washington State applied larvicide sprays to seedling carrots based on a sticky trap threshold of one fly per trap per 7-day exposure period. A spray threshold of two flies per yellow water trap in the course of 3 consecutive days was established in Poland (Legutowska, 1984a). It was also estimated that chemical control in carrots was economically justified for infestations causing damage to over 1% of roots (Legutowska, 1984b). Insecticidal applications were reduced by 44% compared with previous practices, using thresholds of 0.2–0.5 flies per yellow sticky trap per day in British Columbia (Judd, Vernon and Borden, 1985b). Traps were placed 2–4 m into each field parallel to the perimeter at *c.* 50 m intervals. As the abundance of *P. rosae* varied widely among farms and fields, the monitoring of individual fields was thought to be superior to monitoring regionally. In Ontario in 1986, individual carrot fields were monitored and a spray threshold of 0.1 flies per yellow sticky trap per day was used (M-R. McDonald, personal communication).

Researchers in Denmark have studied the relationship between yellow sticky trap catches and damage in an effort to increase the reliability of treatment recommendations. Münster-Swendsen (1983) found that the number of flies caught on traps was related to activity, density and trap-efficiency. Counts were higher on certain days and in some traps than others. Esbjerg *et al.* (1983) found that the number of flies caught was influenced by wind, rain and trap height. Nielsen (1983) counted eggs and pupae in carrot fields to establish the numbers per sample and the number of samples necessary to obtain acceptable sampling precision. A relationship was found between trap catch and the occurrence of eggs and pupae; however, host-plant age (and therefore susceptibility to oviposition) and exposure to wind had to be taken into account (Esbjerg *et al.*, 1983). The uneven distribution of flies in fields has also made the establishment of thresholds difficult (Philipsen, 1986).

Integrated pest management

Before the term 'integrated pest management' was coined, some researchers recommended a combination of more than one approach to the management of *P. rosae*. A number of possible components of an integrated pest management system for the control of carrot fly on carrots are listed in *Table* 2.

Savzdarg (1927) suggested that the judicious choice of a resistant variety and time of sowing would prevent injury to carrots. Whitcomb (1938) advised scheduling dates of sowing to avoid first-generation damage, and the avoidance of ploughing-in badly mined carrots. Petherbridge, Wright and Ashby (1945) and Wright and Ashby (1946a) also recommended the disposal of damaged carrots, growing early carrots as far away as possible from maincrop carrots, and the reduction of shelter in field surrounds. More recently, Dabrowski and Legutowska (1976) found that crop infestation was significantly influenced by several factors including crop rotation, reduction of shelter, time of sowing, disposal of damaged crops and time of harvest. Bleasdale (1981) proposed that co-operation among neighbouring growers would permit rotations of carrot crops far removed from the previous year's crops, thus breaking the continuity of the pest.

Ellis *et al.* (1987) combined partial resistance in cv. Sytan with specific sowing and lifting times to obtain satisfactory yields of marketable carrots. In another study it was found that one-third the dose of insecticide was necessary to provide effective control of carrot fly on the resistant variety Sytan versus the highly susceptible cv. Danvers (Ellis *et al.*, 1984). Freuler, Fischer and Bertuchoz (1982b) developed pest monitoring beyond simple detection of carrot fly, also allowing for the effects of time of year and time of harvest in determining thresholds. By monitoring individual fields, Judd, Vernon and Borden (1985b) were able to identify fields and sectors within these fields appropriate for early harvest, to minimize what would otherwise have developed into serious damage.

Table 2. Components of an integrated pest management system for the control of the carrot fly on carrots

Time of harvest
Proximity to source of infestation (spatial isolation)
Nature of shelter constituting field surrounds
Time of drilling
Resistant varieties
Disposal of damaged carrots
Management of field surrounds
Insecticides applied according to trap catch thresholds

A system for predicting probable damage levels to carrots was developed, with a view to helping growers avoid high-risk fields (Anonymous, 1983b). Predictions were based on a number of factors including time of harvest, proximity to source of infestation, nature of shelter constituting field surrounds and time of drilling.

Future prospects for carrot fly control

A number of aspects relating to the control of carrot fly could benefit from additional studies. Improvements to monitoring systems, capable of being used by growers, to identify the need and timing of insecticide treatments, would include the development of more efficient traps that are also selective for adult carrot flies.

Further development of trap catch/crop damage relationships would advance current insecticide treatment recommendations beyond what are essentially presence or absence thresholds. Insecticide use could also be reduced through integration with resistant cultivars, sowing and harvesting times. Classification of insecticide efficacy against adult and larval stages is also needed. Further work may also be rewarding on developing the ecological knowledge necessary to identify crops at risk, with reference to shelter provided for adults in field surrounds and the proximity of new crops to sources of flies.

Acknowledgements

We thank our colleagues, who have brought so many papers to our attention, laying the groundwork for this review; A. J. Burn, D. R. Hartley, O. T. Jones, S. F. Nottingham, I. I. I. Uvah and D. Wainhouse.

References

ANONYMOUS (1842). Spirits of tar and carrots. *Gardeners' Chronicle* **2**, 365.

ANONYMOUS (1957). *Distribution Maps of Insect Pests—Series A, Number 84*. Commonwealth Institute of Entomology, London.

ANONYMOUS (1983a). *Carrot fly. Booklet of the Ministry of Agriculture, Fisheries and Food, Leaflet 68*. MAFF Publications, Alnwick, Northumberland.

ANONYMOUS (1983b). Carrot fly control—damage can be reduced. *Farmers Weekly Supplement–Arable* **99** (15), 37–38.

ANTHONY, R. W. V. (1978). *Studies on the Effect of Intercropping Onions and Carrots on the Carrot Fly* (Psila rosae F.). MSc thesis, Imperial College of Science and Technology, London, UK.

ASHBY, D. G. AND WRIGHT, D. W. (1946). The immature stages of the carrot fly. *Transactions of the Royal Entomological Society of London* **97**, 355–379.

AUSLAND, O. (1957). Gulrotflua (*Psila rosae* Fabr.), Undersøkelser over dens biologi i Norge. *Melding fra Statens Plantevern, Number 14*.

BACHMANN, F. AND LEGGE, J. B. (1968). Insecticidal properties of a new group of carbamates. *Journal of the Science of Food and Agriculture (Supplementary Issue)* **19**, 39–43.

BACKS, R. H. (1957). Note on effect of flooding on the carrot rust fly in the Holland Marsh area of Ontario. *Canadian Entomologist* **89**, 89–90.

Baker, F. T., Ketteringham, I. E., Bray, S. P. V. and White, J. H. (1942). Observations on the biology of the carrot fly (*Psila rosae* Fab.): assembling and oviposition. *Annals of Applied Biology* **29**, 115–125.

Balachowsky, A. and Mesnil, L. (1936). *Les Insectes Nuisibles aux Plantes Cultivées. Leurs Moeurs, Leur Destruction. Volume II*, pp. 1372–1374. Busson, Paris.

Barnes, H. F. (1942). Studies of fluctuations in insect populations. IX. The carrotfly (*Psila rosae*) in 1936–41. *Journal of Animal Ecology* **11**, 69–81.

Behan, M. and Ryan, M. F. (1977). Sensory receptors on the ovipositor of the carrot fly (*Psila rosae* (F.))(Diptera: Psilidae) and the cabbage root fly (*Delia brassicae* (Wiedemann)) (Diptera: Anthomyiidae). *Bulletin of Entomological Research* **67**, 383–389.

Biernaux, J. (1968). Observations sur l'hibernation de *Psila rosae* F. *Bulletin des Recherches Agronomiques de Gembloux* **3**, 241–248.

Biernaux, J. (1972). Les étapes du développement nymphal de *Psila rosae* F. (Diptera – Psilidae): éléments de détermination du cycle évolutif de l'insecte. *Bulletin des Recherches Agronomiques de Gembloux* **7**, 3–15.

Biernaux, J. and Seutin, E. (1969). Cycle biologique de la mouche de la carotte en 1968 et résultats d'un premier essai de lutte chimique. *Mededelingen van de Rijksfaculteit der Landbouwwetenschappen te Gent* **34**, 587–597.

Bierne, B. P. (1971). Pest insects of annual crop plants in Canada. I. Lepidoptera II. Diptera III. Coleoptera. *Memoirs of the Entomological Society of Canada* **78**, 63–65.

Bleasdale, J. K. A. (1981). Carrot fly. Strategy for control. *Grower* **95** (23), 26–27.

Bohlen, E. (1967). Untersuchungen zum Verhalten der Möhrenfliege. *Psila rosae* Fab. (Dipt. Psilidae), im Eiablagefunktionskreis. *Zeitschrift für angewandte Entomologie* **59**, 325–360.

Bouché, P. F. (1833). *Naturgeschichte der schädlichen und nützlichen Garten-Insekten und die bewährtesten Mittel zur Verteiligung der ersteren*, pp. 132–134. Nicolaischen, Berlin.

Brooks, A. R. (1949). *The Identification of the Commoner Root Maggots of Garden Crops in Canada*. Dominion Entomological Laboratory, Saskatoon.

Brown, A. W. A. (1971). Pest resistance to pesticides. In *Pesticides in the Environment* (R. White-Stevens, Ed.), volume 1, part II, p. 469. Marcel Dekker, Inc., New York.

Brunel, E. (1968). Étude du développement nymphal de *Psila rosae* Fab. (Diptères Psilidés) en conditions naturelles et expérimentales: quiescence et diapause. *Comptes rendus des séances de la Société de biologie* **162**, 2223–2228.

Brunel, E. (1971a). Influence de la plante hôte (espèce et stade végétatif) sur les captures de *Psila rosae* Fab. (Diptera, Psilidae) au moyen de pièges jaunes. *Mededelingen van de Faculteit der Landbouwwetenschappen der Rijksuniversiteit te Gent* **36**, 241–249.

Brunel, E. (1971b). Comparaison de l'action des produits insecticides organophosphorés utilisés contre la mouche de la carotte (*Psila rosae* Fab.) de 1963 à 1970. *Mededelingen van de Faculteit der Landbouwwetenschappen der Rijksuniversiteit te Gent* **36**, 950–960.

Brunel, E. (1977). Étude de l'attraction périodique de femelles de *Psila rosae* Fabr. par la plante-hôte et influence de la végétation environnante. *Colloques Internationaux du Centre National de la Recherche Scientifique, Number 265. Comportement des Insectes et Milieu Trophique*, pp. 373–389.

Brunel, E. (1979). Etude de l'ovogenèse de *Psila rosae* Fab. (Diptère Psilidés): role de la température, de l'alimentation et de la plante hôte. *Annales de Zoologie—Ecologie Animale* **11**, 227–246.

Brunel, E. and Blot, Y. (1975). Rôle de la couverture végétale sur les captures de *Psila rosae* Fabr. (Diptère Psilidé) au moyen de piège jaune. *Sciences Agronomiques Rennes, 1975*, 91–96.

BRUNEL, E. AND LANGOUET, L. (1970). Influence de caractéristiques optiques du milieu sur les adultes de *Psila rosae* Fab. (Diptères Psilidés): attractivité de surfaces colorées, rhythme journalier d'activité. *Comptes rendus des séances de la Société de biologie* **164**, 1638–1644.

BRUNEL, E. AND RABASSE, J. M. (1975). Influence de la forme et de la dimension de pièges à eau colorés en jaune sur les captures d'insectes dans une culture de carotte. Cas particulier des Diptères. *Annales de Zoologie—Ecologie Animale* **7**, 345–364.

BRUNEL, E., FREULER, J. AND DUCHESNE, J. (1981). Détermination des variations de la réflectance du feuillage de quelques ombellifères en fonction de l'âge et de la variété: conséquences sur le comportement de *Psila rosae*. *Signatures Spectrales d'Objets en Télédétection, Avignon, 8–11 September 1981*, 107–115.

BURN, A. J. (1980). *The Natural Mortality of the Carrot Fly* (Psila rosae *F.*). PhD thesis, University of Cambridge, UK.

BURN, A. J. (1982). The role of predator searching efficiency in carrot fly egg loss. *Annals of Applied Biology* **101**, 154–159.

BURN, A. J. (1984). Life tables for the carrot fly, *Psila rosae*. *Journal of Applied Ecology* **21**, 891–902.

BURN, A. J. AND COAKER, T. H. (1981). Diapause and overwintering of the carrot fly, *Psila rosae* (F.) (Diptera: Psilidae). *Bulletin of Entomological Research* **71**, 583–590.

CHAMBERLAIN, R., SKILLMAN, E. E. AND STEWART, J. H. (1937). The control of the carrot fly (*Psila rosae*) in Northern Ireland. *Journal of the Ministry of Agriculture, Northern Ireland* **5**, 39–51.

COAKER, T. H. AND WHEATLEY, G. A. (1962). Insect pests of vegetable crops: current research at Wellesbourne. *NAAS Quarterly Review Number 56*, 153–158.

COAKER, T. H. AND WHEATLEY, G. A. (1970). Crop loss assessment methods. Host: *Daucus carota* (carrot); Organism: *Psila rosae* (carrot fly). In *FAO Manual on the Evaluation and Prevention of Losses by Pests, Diseases, and Weeds* (L. Chiarappa, Ed.), Section 3. Methods; 3.3. Special Methods, Serial Number 10. FAO, Rome.

COCKBILL, G. F., HENDERSON, V. E., ROSS, D. M. AND STAPLEY, J. H. (1945). Wireworm populations in relation to crop production. *Annals of Applied Biology* **32**, 148.

COLE, R. A. (1985). Relationship between the concentration of chlorogenic acid in carrot roots and the incidence of carrot fly larval damage. *Annals of Applied Biology* **106**, 211–217.

COLLIN, J. E. (1944). The British species of Psilidae (Diptera). *Entomologist's Monthly Magazine* **80**, 214–224.

COLLINGWOOD, C. A. AND CROXALL, H. E. (1954). Carrot fly injury and 'canker' of parsnips in the West Midlands. *Plant Pathology* **3**, 99–103.

COPPOCK, L. J. (1974). Notes on the biology of carrot fly in Eastern England. *Plant Pathology* **23**, 93–100.

COPPOCK, L. J., MASKELL, F. E. AND GAIR, R. (1975). Attempts at cultural control of carrot fly damage to carrots in East Anglia. *Plant Pathology* **24**, 97–101.

CROSBY, C. R. AND LEONARD, M. D. (1918). *Manual of Vegetable-Garden Insects*, pp. 181–185. Macmillan, New York.

CURTIS, J. (1860). *Farm Insects: Being the Natural History and Economy of the Insects Injurious to the Field Crops of Great Britain and Ireland, and also those which infest Barns and Granaries. With Suggestions for their Destruction*, pp. 402–407. Blackie and Son, Glasgow.

DABROWSKI, Z. T. AND LEGUTOWSKA, H. (1976). Wplyw polozenia plantacji i agrotechniki na wystepowanie polyśnicy marchwianki (*Psila rosae* F.). *Wiadomości Ekologiczne* **22**, 265–277.

DABROWSKI, Z. T. AND LEGUTOWSKA, H. (1977). Three years experiment on the damage of some carrot cultivars made by *Psila rosae*. *Annual Plant Resistance to Insects Newsletter Volume 3*, p. 69. Purdue University, West Lafayette.

Dustan, A. G. (1930). Control of the onion maggot and the carrot rust fly. *Ontario Department of Agriculture. Twenty-fifth Annual Report of the Vegetable Growers' Association, 1929*, 47–52.

Eilenberg, J. (1983). Entomophthora-svampe på gulerodsfluen (*Psila rosae* F.). Muligheder for anvendelse af Entomophthora-svampe til biologisk bekaempelse af skadedyr. *Tidsskrift for Planteavl* **87**, 399–406.

Eilenberg, J. and Philipsen, H. (1986). *Entomophthorales* on the carrot fly (*Psila rosae* F.): Field prevalence during two seasons. In *Abstracts of the First International Congress of Dipterology, Budapest, 17–24 August, 1986* (B. Darvas and L. Lapp, Eds), p. 66. Hungarian Academy of Sciences, Budapest.

Ellis, P. R. (1977). The search for resistance to *Delia brassicae* in crucifers and *Psila rosae* in carrots. In *Breeding for Resistance to Insects and Mites. International Organisation for Biological Control of Noxious Animals and Weeds. West Palaearctic Regional Section Bulletin 1977* (3), 7–11.

Ellis, P. R. and Hardman, J. A. (1980). Carrot Fly. A new way to resist attack. *Grower* **93** (20), 55–58.

Ellis, P. R. and Hardman, J. A. (1981). The consistency of the resistance of eight carrot cultivars to carrot fly attack at several centres in Europe. *Annals of Applied Biology* **98**, 491–497.

Ellis, P. R., Freeman, G. H. and Hardman, J. A. (1978). Carrot fly maggot damage to different carrot cultivars in relation to root length. *Journal of Horticultural Science* **53**, 267–274.

Ellis, P. R., Freeman, G. H. and Hardman, J. A. (1984). Differences in the relative resistance of two carrot cultivars to carrot fly attack over five seasons. *Annals of Applied Biology* **105**, 557–564.

Ellis, P. R., Wheatley, G. A. and Hardman, J. A. (1978). Preliminary studies of carrot susceptibility to carrot fly attack. *Annals of Applied Biology* **88**, 159–170.

Ellis, P. R., Hardman, J. A., Jackson, J. C. and Dowker, B. D. (1980). Screening of carrots for their susceptibility to carrot fly attack. *Journal of the National Institute of Agricultural Botany* **15**, 294–302.

Ellis, P. R., Hardman, J. A., Dowker, B. D. and Horobin, J. F. (1984). Progress in the studies of resistance to carrot fly (*Psila rosae*) in carrots. In *Breeding for Resistance to Insects and Mites, International Organisation for Biological Control of Noxious Animals and Weeds. West Palaearctic Regional Section Bulletin 1984*, **VII** (4), 47–48.

Ellis, P. R., Dowker, B. D., Freeman, G. H. and Hardman, J. A. (1985). Problems in field selection for resistance to carrot fly (*Psila rosae*) in carrot cv. Long Chantenay. *Annals of Applied Biology* **106**, 349–356.

Ellis, P. R., Hardman, J. A., Cole, R. A. and Phelps, K. (1987). The complementary effects of plant resistance and the choice of sowing and harvest times in reducing carrot fly (*Psila rosae*) damage to carrots. *Annals of Applied Biology*, in press.

Encobet, J. A. (1911). Datos para el conocimiento de la distribución geográfica de los dípteros de España. *Memorias de la Real Sociedad Española de Historia Natural* **7**, 147, 229.

Esbjerg, P., Jørgensen, J., Nielsen, J. K., Philipsen, H., Zethner, O. and Øgaard, L. (1983). Integreret bekaempelse af skadedyr med gulerødder, gulerodsfluen (*Psila rosae* F., Dipt. Psilidae) og ageruglen (*Agrotis segetum* Schiff., Lep., Noctuidae) som afgrøde-skadedyr model. *Tidsskrift for Planteavl* **87**, 303–355.

Fabricius, J. C. (1794). *Entomologia Systematica Emendata et Aucta. Secundum Classes, Ordines, Genera, Species; Adjectis Synonimis, Locis, Observationibus, Descriptionibus, Volume 4*, p. 356. Proft, Hafniae.

Fabricius, J. C. (1805). *Systema Antilatorum, Secundum Ordines, Genera, Species; Adjectis Synonimis, Locis, Observationibus, Descriptionibus*, p. 319. Carolum

Reichard, Brunsvigae.

FALLÉN, C. F. (1820). *Diptera Sveciae Volume 2*, pp. 8–10. Berlingianis, Lundae.

FINLAYSON, D. G., FULTON, H. G. AND NOBLE, M. D. (1964). Experiments against carrot rust fly (*Psila rosae* (F.)) resistant to cyclodiene organochlorine insecticides. *Proceedings of the Entomological Society of British Columbia* **61**, 13–20.

FLETCHER, J. (1886). Carrot Fly—(*Psila rosae*, Fab.). *Canada Department of Agriculture. Report of the Entomologist, 1885*, 15.

FLETCHER, J. (1904). Insects injurious to Ontario crops in 1903. The carrot rust-fly (*Psila rosae*, Fab.). *Thirty-fourth Annual Report of the Entomological Society of Ontario for 1903*, 66.

FOX WILSON, G. (1945). Investigations on the control of carrot fly (*Psila rosae* F.) in gardens. *Annals of Applied Biology* **32**, 265–276.

FREULER, J., FISCHER, S. AND BERTUCHOZ, P. (1982a). La mouche de la carotte, *Psila rosae* Fab. (*Diptera, Psilidae*). II. Mise au point d'un piège. *Revue suisse de viticulture, d'arboriculture et d'horticulture* **14**, 137–142.

FREULER, J., FISCHER, S. AND BERTUCHOZ, P. (1982b). La mouche de la carotte, *Psila rosae* Fab. (*Diptera, Psilidae*). III. Avertissement et seuil de tolérance. *Revue suisse de viticulture, d'arboriculture et d'horticulture* **14**, 275–279.

GERSDORF, E. AND KRAFT, A. VON (1965). Die 'falsche Möhrenfliege' auch in Deutschland. *Gesunde Pflanzen* **17**, 49–52.

GETZIN, L. W. (1982). Seasonal activity and geographical distribution of the carrot rust fly (Diptera: Psilidae) in western Washington. *Journal of Economic Entomology* **75**, 1029–1033.

GETZIN, L. W. AND ARCHER, T. E. (1983). Carrot rust fly (Diptera: Psilidae) control: monitoring adults to schedule post-plant-emergence, soil-applied larvicides. *Journal of Economic Entomology* **76**, 558–562.

GLASGOW, H. (1929). Mercury salts as soil insecticides. *Journal of Economic Entomology* **22**, 335–340.

GLASGOW, H. AND GAINES, J.G . (1929). The carrot rust fly problem in New York. *Journal of Economic Entomology* **22**, 412–416.

GOEDEWAAGEN, M. A. J. (1926). Der Anbau des Kümmels in den Niederlanden. Eine zusammenfassende Darstellung über die Kümmelpflanze, ihre Kultur und ihre wirtschaftliche Bedeutung. *Heil- und Gewürz Pflanzen* **9**, 1–16.

GORHAM, R. P. (1934). Control of the carrot rust fly, *Psila rosae* Fab. *Twenty-Fifth and Twenty-Sixth Annual Report of the Quebec Society for the Protection of Plants, 1932–1934*, 90–96.

GOSTICK, K. G. AND BAKER, P. M. (1968). Dieldrin resistant carrot fly in England. *Plant Pathology* **17**, 182–183.

GUERIN, P. M. AND RYAN, M. F. (1980). Insecticidal effect of trans-2-nonenal, a constituent of carrot root. *Experientia* **36**, 1387–1388.

GUERIN, P. M. AND RYAN, M. F. (1984). Relationship between root volatiles of some carrot cultivars and their resistance to the carrot fly, *Psila rosae*. *Entomologia experimentalis et applicata* **36**, 217–224.

GUERIN, P. M. AND STÄDLER, E. (1980). Carrot fly olfaction: behavioural and electrophysiological investigations. In *Olfaction and Taste VII. Proceedings of the 7th International Symposium on Olfaction and Taste and of the 4th Congress of the European Chemoreception Research Organisation* (H. van der Starre, Ed.), p. 95. IRL Press, London.

GUERIN, P. M. AND STÄDLER, E. (1982). Host odour perception in three phytophagous Diptera—a comparative study. In *Proceedings of the 5th International Symposium on Insect–Plant Relationships* (J.H. Visser and A. K. Minks, Eds), pp. 95–105. Pudoc, Wageningen.

GUERIN, P. M. AND STÄDLER, E. (1984). Carrot fly cultivar preferences: some influencing factors. *Ecological Entomology* **9**, 413–420.

GUERIN, P. M. AND VISSER, J. H. (1980). Electroantennogram responses of the carrot fly, *Psila rosae*, to volatile plant components. *Physiological Entomology* **5**,

111–119.

Guerin, P. M., Gfeller, F. and Städler, E. (1981). Carrot resistance to the carrot fly—contributing factors. In *Breeding for Resistance to Insects and Mites. International Organisation for Biological Control of Noxious Animals and Weeds. West Palaearctic Regional Section Bulletin 1981*, **IV** (1), pp. 63–65.

Guerin, P. M., Städler, E. and Buser, H. R. (1983). Identification of host plant attractants for the carrot fly, *Psila rosae. Journal of Chemical Ecology* **9**, 843–861.

Hanson, A. J. and Webster, R. L. (1941). *The Carrot Rust Fly. Bulletin of the State College of Washington Agricultural Experiment Station, Number 405.*

Hardman, J. A. and Ellis, P. R. (1982). An investigation of the host range of the carrot fly. *Annals of Applied Biology* **100**, 1–9.

Hardman, J. A., Ellis, P. R. and Stanley, E. A. (1985). *Bibliography of the Carrot Fly* Psila rosae *(F.)*. Vegetable Research Trust, National Vegetable Research Station, Wellesbourne.

Harris, C. R., Chapman, R. A., Morris, R. F. and Stevenson, A. B. (1987). Enhanced soil microbial degradation of carbofuran and fensulfothion—a factor contributing to the decline in effectiveness of some soil insect control programs in Canada. *Journal of Economic Entomology*, in press.

Hassan, S. A. (1971). Forecasting of damage and determining the appropriate time to control the carrot miner fly *Napomyza carotae* (Diptera: Agromyzidae). *Zeitschrift für angewandte Entomologie* **68**, 68–73.

Hendel, F. (1917). Beiträge zur Kenntnis der acalyptraten Musciden. *Deutsche entomologische Zeitschrift* **31**, 33–47.

Henderson, W. (1814). A preventive against the worms infesting the roots of carrots in light early soils *Memoirs of the Caledonian Horticultural Society* **1**, 200–201.

Hennig, W. (1941). Werden alle Möhrenfliegen-Schäden durch *Chamaepsila rosae* F. verursacht? *Arbeiten über physiologische und angewandte Entomologie aus Berlin-Dahlem* **8**, 36–38.

Hills, L. D. (1965). *Perfumes against Pests*. Henry Doubleday Research Association, Braintree.

Hills, L. D. (1972). *Pest Control Without Poisons*, pp. 19–20. Henry Doubleday Research Association, Braintree.

Hinton, H. E. (1962). The fine structure and biology of the egg-shell of the wheat bulb fly, *Leptohylemyia coarctata. Quarterly Journal of Microscopical Science* **103**, 243–251.

Hinton, H. E. and Cole, S. (1965). The structure of the egg-shell of the cabbage root fly, *Erioischia brassicae. Annals of Applied Biology* **56**, 1–6.

Hinton, W. L. (1971). *The Economics of Carrot Production and Marketing in Britain: a Commodity Study. Occasional Papers No. 14.* R. I. Severs Ltd, Cambridge.

Howitt, A. J. and Cole, S. G. (1959). Chemical control of the carrot rust fly, *Psila rosae* (F.), in western Washington. *Journal of Economic Entomology* **52**, 963–966.

Jegen, G. (1932). Dipteren Zweiflüger. In *Handbuch der Pflanzenkrankheiten, Volume 5. Tierische Schädlinge an Nutzpflanzen, Part 2*, (L. Reh, Ed.), p. 17. Verlagsbuchhandlung Paul Parey, Berlin.

Jones, O. T. (1975). Damage to carrots by larvae of *Napomyza carotae* Spencer (Diptera: Agromyzidae). *Plant Pathology* **24**, 62.

Jones, O. T. (1977). *Host Plant Finding and Ecology of* Psila rosae *(F.) Larvae*. PhD thesis, University of Cambridge, UK.

Jones, O. T. (1979). The responses of carrot fly larvae, *Psila rosae*, to components of their physical environment. *Ecological Entomology* **4**, 327–334.

Jones, O. T. and Coaker, T. H. (1977). Oriented responses of carrot fly larvae, *Psila rosae*, to plant odours, carbon dioxide and carrot root volatiles. *Physiological Entomology* **2**, 189–197.

Jones, O. T. and Coaker, T. H. (1980). Dispersive movement of carrot fly (*Psila rosae*) larvae and factors affecting it. *Annals of Applied Biology* **94**, 143–152.

Jørgensen, J. and Thygesen, T. (1968). Gulerodsfluen, *Psila rosae* F. *Tidsskrift for*

Planteavl **72**, 1–25.

JUDD, G. J. R. AND VERNON, R. S. (1985). Seasonal activity of adult carrot rust flies, *Psila rosae* (Diptera: Psilidae), in the lower Fraser Valley, British Columbia. *Canadian Entomologist* **117**, 375–381.

JUDD, G. J. R., VERNON, R. S. AND BORDEN, J. H. (1985a). Monitoring program for *Psila rosae* (F.) (Diptera: Psilidae) in southwestern British Columbia. *Journal of Economic Entomology* **78**, 471–476.

JUDD, G. J. R., VERNON, R. S. AND BORDEN, J. H. (1985b). Commercial implementation of a monitoring program for *Psila rosae* (F.) (Diptera: Psilidae) in southwestern British Columbia. *Journal of Economic Entomology* **78**, 477–481.

KIRBY, W. AND SPENCE, W. (1815). *An Introduction to Entomology: or Elements of the Natural History of Insects: with plates, volume 1*, p.184. Longman *et al.*, London.

KÖRTING, A. (1940). Zur Biologie und Bekämpfung der Möhrenfliege (*Psila rosae* F.) in Mitteldeutschland. *Arbeiten über physiologische und angewandte Entomologie aus Berlin* **7**, 209–232, 269–285.

LABEYRIE, V. (1956). Sur quelques facteurs réglant la durée du cycle et l'intensité des attaques de la mouche de la carotte (*Psila rosae* Faber) dans le Bordelais. *Revue de Zoologie agricole et appliquée* **55**, 22–24.

LADELL, W. R. S. (1936). A new apparatus for separating insects and other arthropods from the soil. *Annals of Applied Biology* **23**, 862–879.

LEGUTOWSKA, H. (1984a). Control of carrot rust fly (*Psila rosae* Fabr.). *Annals of Warsaw Agricultural University–SGGW–AR, Horticulture* **12**, 75–79.

LEGUTOWSKA, H. (1984b). Occurrence of the carrot rust fly (*Psila rosae* Fabr.) in some regions of carrot cultivation and economic losses caused by this pest. *Annals of Warsaw Agricultural University–SGGW–AR, Horticulture* **12**, 69–74.

MACLEOD, G. F. (1929). *Vegetable Garden Insects. Circular of the Pennsylvania Agricultural Experiment Station Number 122*, p. 25.

MCCLANAHAN, R. J. AND NIEMCZYK, H. D. (1963). Continuous rearing of the carrot rust fly, *Psila rosae* (Fab.). *Canadian Entomologist* **95**, 827–830.

MCIN, C. (1843). Spirits of Tar. *Gardeners' Chronicle* **3**, 5.

MCLEOD, D. G. R., WHISTLECRAFT, J. W. AND HARRIS, C. R. (1985). An improved rearing procedure for the carrot rust fly (Diptera: Psilidae) with observations on life history and conditions controlling diapause induction and termination. *Canadian Entomologist* **117**, 1017–1024.

MAJOR, J. (1829). *A Treatise on the Insects Most Prevalent on Fruit Trees, and Garden Produce Giving an Account of the Different States They Pass Through, the Depradations They Commit, and Recipes for Their Destruction, Including the Recipes of Various Authors, With Remarks on Their Utility; Also, a Few Hints on the Causes and Treatment of Mildew and Canker on Fruit Trees and Cucumbers, &c. &c. &c.*, p. 199. Longman *et al.*, London.

MAKEPEACE, R. J. (1965). The application of granular pesticides. *Proceedings of the 3rd British Insecticide and Fungicide Conference for 1965* **1**, 389–396.

MASKELL, F. E. AND GAIR, R. (1973). Experiments on the chemical control of carrot fly in carrots in East Anglia in 1968–72. *Proceedings of the 7th British Insecticide and Fungicide Conference for 1973* **2**, 513–524.

MAYBEE, G. E. (1954). Introduction into Canada of parasites of the carrot rust fly, *Psila rosae* (F.) (Diptera: Psilidae). *84th Annual Report of the Entomological Society of Ontario for 1953*, 58–62.

MAYBEE, G. E. (1956). Observations, life-history, habits, immature stages, and rearing of *Loxotropa tritoma* (Thoms.) (Hymenoptera: Proctotrupoidea), a parasite of the carrot rust fly, *Psila rosae* (F.) (Diptera: Psilidae). *86th Annual Report of the Entomological Society of Ontario for 1955*, 53–58.

MEIGEN, J. W. (1826). *Systematische Beschreibung der bekannten Europäischen zweiflugeligen Insekten, Volume 5*, p. 358. Schultzschen Buchhandlung, Hamm.

MILES, M. (1956). Carrot fly as a pest of lettuce. *Plant Pathology* **5**, 152.

Missonier, J. and Boulle, N. (1964). Remarques sur la biologie de la mouche de la carotte. Détermination des dates de vol piégeage des adultes. *Phytoma* **16**, 31–34.

Missonnier, J., Arnoux, J., Portier, G., Oudinet, R. and André, M. (1964). Problèmes de protection contre les mouches du chou, de l'oignon et de la carotte, en rapport avec le développement de lignées résistantes vis-à-vis des produits insecticides organo-halogénés. *Phytiatrie–Phytopharmacie* **13**, 15–32.

Mowat, D. J. and Martin, S. J. (1984). The control of carrot fly, *Psila rosae* (F.), by midseason directed insecticide sprays. *Crop Research* **24**, 119–127.

Muggeridge, J. (1933). Some recently recorded exotic insect pests. *New Zealand Journal of Agriculture* **47**, 221–228.

Münster-Swendsen, M. (1983). Prøvetagning af skadedyr. Planlaegning og bedømmelse i forbindelse med integreret bekaempelse. *Tidsskrift for Planteavl* **87**, 365–369.

Naton, E. (1968). Zur Biologie von *Dacnusa gracilis* Nees (*Hym.*, *Braconidae*), einem Parasiten der Echten Möhrenfliege, *Psila rosae* Fabr. *Zeitschrift für angewandte Entomologie* **62**, 78–82.

Newton, H. C. F., Satchell, J. E. and Shaw, M. W. (1946). Carrot fly control. *Nature* **158**, 417.

Nielsen, J. K. (1983). Bestemmelse af aeg- og puppeantal af gulerodsfluen (*Psila rosae* F.) ved indsamling af jordprøver. *Tidsskrift for Planteavl* **87**, 379–387.

Niemczyk, H. D. and Harris, C. R. (1962). Evidence of carrot rust fly resistance to aldrin and heptachlor in Canada. *Journal of Economic Entomology* **55**, 560.

Nieuwhof, M. (1977). Some preliminary data on research into carrot fly (*Psila rosae*) resistance. In *Breeding for Resistance to Insects and Mites, International Organisation for Biological Control of Noxious Animals and Weeds. West Palaearctic Regional Section Bulletin 1977*, **III**. 13–16.

Nixon, G. E. J. (1944). A revision of the European Dacnusini (Hym., Braconidae, Dacnusinae). *Entomologist's Monthly Magazine* **80**, 140–151.

Nottingham, S. F. (1987). Effects of non-host plant odours on the anemotactic response to host plant odour in female cabbage root fly, *Delia radicum*, and carrot fly, *Psila rosae*. *Chemical Ecology* **13**, 1313–1318.

Ormerod, E. A. (1881). *A Manual of Injurious Insects, with Methods of Prevention and Remedy for their Attacks to Food Crops, Forest Trees, and Fruit, and with Short Introduction to Entomology*, pp. 45–53. Sonnenschein and Allen, London.

Osborne, P. (1961). Comparative external morphology of *Psila rosae* (F.) and *P. nigricornis* Mg. (Dipt., Psilidae) third instar larvae and puparia. *Entomologist's Monthly Magazine* **97**, 124–127.

Overbeck, H. (1978). Untersuchungen zum Eiablage- und Befallsverhalten der Möhrenfliege, *Psila rosae*, F. (Diptera: Psilidae), im Hinblick auf eine modifizierte chemische Bekämpfung. *Mitteilungen aus der Biologischen Bundesanstalt für Land- und Forstwirtschaft Berlin-Dahlem Number 183*. Kommissionsverlag Paul Parey, Berlin.

Petherbridge, F. R. and Wright, D. W. (1943). Further investigations on the biology and control of the carrot fly (*Psila rosae* F.). *Annals of Applied Biology* **30**, 348–358.

Petherbridge, F. R., Wright, D. W. and Ashby, D. G. (1945). The biology and control of the carrot fly. *Annals of Applied Biology* **32**, 262–264.

Petherbridge, F. R., Wright, D. W. and Davies, P. G. (1942). Investigations on the biology and control of the carrot fly (*Psila rosae* F.). *Annals of Applied Biology* **29**, 380–392.

Pettit, R. H. (1931). Carrot rust-fly found in Michigan. *Quarterly Bulletin of the Michigan Agricultural Experiment Station* **13**, 119–121.

Philipsen, H. (1986). Monitoring carrot flies (*Psila rosae* F.) with yellow sticky traps. In *Abstracts of the First International Congress of Dipterology/Budapest, 17–24 August, 1986.* (B. Darvas and L. Lapp, Eds), p. 188. Hungarian Academy of Sciences, Budapest.

PONTI, O. M. B. DE, FRERIKS, J. C., STEENHUIS, M. AND INGGAMER, H. (1981). Improving the resistance of carrot and onion to respectively carrot fly and onion fly by recurrent selection. In *Breeding for Resistance to Insects and Mites, International Organisation for Biological Control of Noxious Animals and Weeds. West Palaearctic Regional Section Bulletin 1981*, **IV** (1), 59–62.

ROBERT, J., FISCHER, S. AND FREULER, J. (1985). Les ravageurs de la carotte: description et importance relative des dégâts observés. *Revue suisse de viticulture, d'arboriculture et d'horticulture* **17**, 351–356.

ROEBUCK, A. (1945). The carrot fly in the Midlands. *Annals of Applied Biology* **32**, 264–265.

RYAN, M. F. AND BEHAN, M. (1973). The sensory receptors of the carrot fly larva, *Psila rosae* (F.) (Dipt., Psilidae). *Bulletin of Entomological Research* **62**, 545–548.

RYAN, M. F. AND GUERIN, P. M. (1982). Behavioural responses of the carrot fly larva, *Psila rosae*, to carrot root volatiles. *Physiological Entomology* **7**, 315–324.

RYAN, M. F., GUERIN, P. M. AND BEHAN, M. (1978). Possible roles for naturally occurring chemicals in the biological control of carrot fly. In *Proceedings of Symposium on Biological Control (February 17 and 18, 1977) Royal Irish Academy, Dublin*, pp. 130–143. Dundalgan Press Ltd, Dundalk.

SABATINO, A. (1976). Sulla comparsa della *Psila rosae* F. (Dipt. Psilidae) nelle coltivazioni di finocchio in Puglia. *Entomologica* **12**, 131–134.

SACHTLEBEN, H. (1954). Parasiten der Möhrenfliege, *Psila rosae* Fabr. *Beiträge zur Entomologie* **4**, 219–220.

SALT, G. AND HOLLICK, F. S. J. (1944). Studies of wire worm populations. I. A census of wire worms in pasture. *Annals of Applied Biology* **31**, 52–64.

SAVZDARG, E. E. (1927). [The carrot fly (*Psila rosae* F.) and its control.] *Défense des Plantes* **4**, 238–242.

SAZONOVA, L. V. (1982). [Field evaluation of the resistance of carrot to *Psila rosae* F.]. *Byulleten' Vsesoyuznogo Ordena Lenina i Ordena Druzhby Narodov Nauchnoissledovatel'skogo Instituta Rastenievodstva Imeni N.I. Vavilova, Number 120*, 44–48.

SCOTT, H. E. (1952). *The Biology and Control of the Carrot Rust Fly*, Psila rosae (*Fab.*). PhD thesis, Cornell University, USA.

SCOTT, H. E. (1958). Notes on the history and distribution of the carrot rust fly, *Psila rosae* (F.) (Diptera: Psilidae). *Proceedings of the 10th International Congress of Entomology, Montreal, 1956* **1**, 753–756.

SCOTT, H. E. (1959). The effect of planting dates on infestation of the carrot rust fly, (*Psila rosae*) (F.) (Psilidae: Diptera). *Journal of the Elisha Mitchell Scientific Society* **75**, 148–150.

SHAW, M. W. (1952). Carrot fly control. *Scottish Agriculture* **32**, 152–158.

SHAW, M. W. AND MCDONALD, I. (1955). Comparison of aldrin and BHC for the control of carrot fly. *Plant Pathology* **4**, 11–13.

SHAW, M. W., ALLAN, R. M. AND INKSON, R. H. E. (1961). Trials on carrot fly control in north-east Scotland 1956–58. *Plant Pathology* **10**, 110–115.

SMITH, K. M. (1921). The bionomics of the carrot fly. Some further methods of control. *Fruit-Grower, Fruiterer, Florist and Market Gardener* **52**, 955–958, 993–994.

SMITH, K. M. AND GARDNER, J. C. M. (1922). *Insect Pests of the Horticulturalist: Their Nature and Control. Volume I: Onion, Carrot and Celery Flies*, pp. 34, 72. Benn Brothers, London.

SOUTHWOOD, T. R. E. (1978). *Ecological Methods with Particular Reference to the Study of Insect Populations*, pp. 149, 304. Chapman and Hall, London.

SPENCER, K. A. (1966). A clarification of the genus *Napomyza* Westwood (Diptera: Agromyzidae). *Proceedings of the Royal Entomological Society, London (B)* **35**, 29–40.

STÄDLER, E. (1970). Beitrag zur Kenntnis der Diapause bei der Möhrenfliege (*Psila rosae* Fabr., Diptera: Psilidae). *Bulletin de la Société entomologique suisse* **43**, 17–37.

Städler, E. (1971a). Über die Orientierung und das Wirtswahlverhalten der Möhren-fliege, *Psila rosae* F. (Diptera: Psilidae). I. Larven. *Zeitschrift für angewandte Entomologie* **69**, 425–438.

Städler, E. (1971b). An improved mass-rearing method of the carrot rust fly, *Psila rosae* (Diptera: Psilidae). *Canadian Entomologist* **103**, 1033–1038.

Städler, E. (1972). Über die Orientierung und das Wirtswahlverhalten der Möhren-fliege, *Psila rosae* F. (Diptera: Psilidae). II. Imagines. *Zeitschrift für angewandte Entomologie* **70**, 29–61.

Städler, E. (1975). Täglicher Aktivitätsrhythmus der Eiablage bei der Möhrenfliege, *Psila rosae* Fab. (Diptera: Psilidae). *Bulletin de la Société entomologique suisse* **48**, 133–139.

Städler, E. (1977). Host selection and chemoreception in the carrot rust fly (*Psila rosae* F., Dipt. Psilidae): Extraction and isolation of oviposition stimulants and their perception by the female. *Comportment des Insectes et Milieu Trophique Colloques Internationaux du Centre National de la Recherche Scientifique, Number 265*, pp. 357–372.

Städler, E. and Buser, H. R. (1984). Defence chemicals in leaf surface wax synerg-istically stimulate oviposition by a phytophagous insect. *Experientia* **40**, 1157–1159.

Stevenson, A. B. (1977). A disposable adhesive trap for monitoring the carrot rust fly. *Proceedings of the Entomological Society of Ontario for 1976* **107**, 65–69.

Stevenson, A. B. (1981a). *Carrot Insects. Fact Sheet of the Ontario Ministry of Agriculture and Food, February 1981*, Agdex 258/605.

Stevenson, A. B. (1981b). Development of the carrot rust fly, *Psila rosae* (Diptera: Psilidae), relative to temperature in the laboratory. *Canadian Entomologist* **113**, 569–574.

Stevenson, A. B. (1981c). Carrot rust fly: monitoring adults to determine whether to apply insecticides. *Journal of Economic Entomology* **74**, 54–57.

Stevenson, A. B. (1983). Seasonal occurrence of carrot rust fly (Diptera: Psilidae) adults in Ontario and its relation to cumulative degree-days. *Environmental Entomology* **12**, 1020–1025.

Stone, L. E. W. (1954). Carrot fly and 'canker' of parsnips in the south-west. *Plant Pathology* **3**, 118–121.

Stone, L. E. W. (1956). Control of carrot fly on carrots and parsnips. *Plant Pathology* **5**, 141–143.

Suett, D. L. (1986). Accelerated degradation of carbofuran in previously treated field soils in the United Kingdom. *Crop Protection* **5**, 165–169.

Thompson, A. R. and Harris, C. R. (1982). An indication of phorate-tolerance in a population of carrot fly from East Anglia. *Tests of Agrochemicals and Cultivars (Annals of Applied Biology* **100**, Supplement) Number 3, 18–19.

Thompson, A. R., Percivall, A. L., Suett, D. L. and Edmonds, G. H. (1986). Effects of depth of placement of granules at drilling on the performance of insecticides against the cabbage root fly (*Delia radicum*) and carrot fly (*Psila rosae*) on root vegetables. *Aspects of Applied Biology* **12**, 247–257.

Toms, A. M. (1972). Crop protection and farm management, 2. *Big Farm Management* **12**, 61–62.

Tullgren, A. (1917). Skadedjur i Sverige Åren 1912–1916. *Meddelande från Centra-lanstalten för försöksväsendet på jordbruksområdet, Number 152*, 86.

Uvah, I. I. I. and Coaker, T. H. (1984). Effect of mixed cropping on some insect pests of carrots and onions. *Entomologia experimentalis et applicata* **36**, 159–167.

van Poeteren, N. (1939). Verslag over de werkzaamheden van den Plantenziekten-kundigen Dienst in het jaar 1938. *Verslagen en mededelingen van den Plantenziek-tenkundigen dienst te Wageningen, Number 93*, pp. 55–56.

van't Sant, L. E. (1961). Levenswijze en bestrijding van de wortelvlieg (*Psila rosae*

F.) in Nederland. *Mededelingen Instituut voor Plantenziektenkundig Onderzoek Number 240.* H. Veenman en Zonen NV, Wageningen.

VAN'T SANT, L. E. AND BRADER, L. (1972). Oekologische waarnemingen als hulpmiddel bij de bescherming van wortelen tegen de aantasting door de wortelvlieg *Psila rosae. Entomologische Berichten* **32**, 187–188.

VISSER, J. H. AND PONTI, O. M. B. DE (1983). Resistance of carrot to the carrot fly, Psila rosae. In *Progress Report of the CEC 'Integrated and Biological Control' Programme 1979–1981.* (R. Cavalloro and A. Piavaux, Eds), pp. 214–224. Commission of the European Communities, Luxemburg.

WAINHOUSE, D. (1975). *The Ecology and Behaviour of the Carrot Fly,* Psila rosae *(F.).* PhD thesis, University of Cambridge, UK.

WAINHOUSE, D. (1977). Rhythmic activity of adult carrot fly, *Psila rosae. Physiological Entomology* **2**, 323–329.

WAINHOUSE, D. AND COAKER, T. H. (1981). The distribution of carrot fly (*Psila rosae*) in relation to the flora of field boundaries. In *Pests, Pathogens and Vegetation* (J.M. Thresh, Ed.), pp. 263–272. Pitman, London.

WAKERLEY, S. B. (1963). Weather and behaviour in carrot fly (*Psila rosae* Fab., Dipt. Psilidae) with particular reference to oviposition. *Entomologia experimentalis et applicata* **6**, 268–278.

WAKERLEY, S. B. (1964). The sensory behaviour of carrot fly (*Psila rosae* Fab., Dipt., Psilidae). *Entomologia experimentalis et applicata* **7**, 167–178.

WATKINS, T. C. AND MINER, F. D. (1943). Flight habits of carrot rust flies suggest possible method of control. *Journal of Economic Entomology* **36**, 586–588.

WEBLEY, D. (1957). A method for estimating the density of frit fly eggs in the field. *Plant Pathology* **6**, 49–51.

WHEATLEY, G. A. (1969). The problem of carrot fly control on carrots. *Proceedings of the 5th British Insecticide and Fungicide Conference for 1969* **1**, 248–254.

WHEATLEY, G. A. (1971). The role of pest control in modern vegetable production. *World Review of Pest Control* **10**, 81–93.

WHEATLEY, G. A. AND FREEMAN, G. H. (1982). A method of using the proportions of undamaged carrots or parsnips to estimate the relative population densities of carrot fly (*Psila rosae*) larvae, and its practical applications. *Annals of Applied Biology* **100**, 229–244.

WHEATLEY, G. A. AND HARDMAN, J. A. (1967). Microplot studies on the control of carrot fly. *Seventeenth Annual Report of the National Vegetable Research Station for 1966*, 63–64.

WHEATLEY, G. A. AND PERCIVALL, A. L. (1974). Possible organophosphorus resistance in carrot fly populations. *24th Annual Report of the National Vegetable Research Station for 1973*, 77–78.

WHEATLEY, G. A. AND WRIGHT, D. W. (1970). Studies of methods of applying insecticides for controlling carrot fly and carrot-willow aphid. *7th International Congress of Plant Protection, Paris 21–25 September 1970*, 397–399.

WHITCOMB, W. D. (1929). Observations on the carrot rust fly (*Psila rosae* Fab.) in Massachusetts. *Journal of Economic Entomology* **22**, 672–675.

WHITCOMB, W. D. (1938). *The Carrot Rust Fly. Bulletin of the Massachusetts Agricultural Experiment Station, Number 352.*

WHITEHEAD, A. G., TITE, D. J. AND BROMILOW, R. H. (1981). Techniques for distributing non-fumigant nematicides in soil to control potato cyst-nematodes, *Globodera rostochiensis* and *G. pallida. Annals of Applied Biology* **97**, 311–321.

WILLIAMS, J. B. (1954). Occurrence of adults of the carrot rust fly, *Psila rosae* (Fab.) (Diptera: Psilidae), on corn foliage at Bradford, Ontario. *Canadian Entomologist* **86**, 414–415.

WRIGHT, D. W. (1951). The control of the carrot fly (*Psila rosae* Fab.) with benzene hexachloride. *First Annual Report of the National Vegetable Research Station for 1950*, 14–20.

WRIGHT, D. W. AND ASHBY, D. G. (1945). The control of the carrot fly (*Psila rosae*,

Fab.) (Diptera) with DDT. *Bulletin of Entomological Research* **36**, 253–268.

Wright, D. W. and Ashby, D. G. (1946a). Bionomics of the carrot fly (*Psila rosae,* F.). I. The infestation and sampling of carrot crops. *Annals of Applied Biology* **33**, 69–77.

Wright, D. W. and Ashby, D. G. (1946b). Bionomics of the carrot fly (*Psila rosae* Fab.). II. Soil populations of carrot fly during autumn, winter and spring. *Annals of Applied Biology* **33**, 263–270.

Wright, D. W. and Bowe, D. D. (1957). Effect of insecticides on crop flavour. *Seventh Annual Report of the National Vegetable Research Station for 1956,* 53–54.

Wright, D. W. and Coaker, T. H. (1968). Development of dieldrin resistance in carrot fly in England. *Plant Pathology* **17**, 178–181.

Wright, D. W., Geering, Q. A. and Ashby, D. G. (1947). The insect parasites of the carrot fly, *Psila rosae,* Fab. *Bulletin of Entomological Research* **37**, 507–529.

Zetterstedt, J. W. (1846). *Diptera Scandinaviae Disposita et Descripta,* volume 6, pp. 2402–2403. Lundbergiana, Lundae.

5
Nosema spp. (Microspora: Microsporida: Nosematidae) of Stored-Product Coleoptera and their Potential as Microbial Control Agents

A. R. KHAN* AND B. J. SELMAN

Department of Agricultural and Environmental Science, University of Newcastle upon Tyne, NE1 7RU, UK

Introduction

The excessive use of conventional chemical insecticides has resulted in a number of serious problems, e.g. resistance to the chemical insecticides, elimination of economically beneficial insects, persistence in the environment, toxicity to humans and wildlife and higher cost of crop production. Many

---

* Present address: Department of Zoology, Rajshahi University, Rajshahi, Bangladesh.

---

of these problems can be eliminated with care and attention to detail. Unfortunately the resistance of pest insects to insecticides remains a great problem. Resistance to one or more pesticides has been reported in at least 477 species of insects and mites (Georghiou and Mellon, 1983; Anonymous, 1986). Many insects and mites are capable of tolerating virtually all pesticides available for their control as a result of cross and multiple resistance (Metcalf, 1980). *Table 1* lists some insecticides to which *Tribolium castaneum* has already become resistant.

It is because of these limitations that alternative pest control methods have been considered. Several non-chemical methods have been tried, and one of these is the use of microbial agents, including protozoa. Many stored-product insects, especially beetles, suffer high mortality as a result of pathogenic disease. McLaughlin (1971) described the unique nature of the habitat provided by the stored-product environment, stating that the insect's food is evenly distributed and within definite boundaries and that insects often populate small foci which gradually expand as their numbers multiply. The pathogen substrate is, therefore, initially localized but dense. This statement is true for virtually all insect pests of stored commodities, and particularly the Coleoptera.

McLaughlin (1971) concluded that protozoa show promise as control agents for stored-product pests and Brooks (1980) believed that protozoa could be useful for controlling these insects. The group Microsporidia includes most of the entomophilic protozoa and it has been estimated that

Table 1.   Resistance of *T. castaneum* to insecticides

Insecticides	Localities	References
Aprocarb	Australia	Champ and Campbell-Brown (1970)
Demeton methyl	Poland	Cichy (1971)
DDT	Africa	Dyte (1970)
	USA	Speirs, Redlinger and Jones (1971)
	Philippines	Santhoy and Morallo-Rejesus (1972)
Lindane	Australia	Champ and Campbell-Brown (1970)
	India	Bhatia and Pradhan (1972)
	Africa	Hindmarsh (1976)
	Africa	Warui (1976)
	West Germany	Rassmann (1978)
	Switzerland	Hoppe (1981)
Malathion	Australia	Champ and Campbell-Brown (1970)
	Australia	Greening (1970)
	West Germany	Rassmann (1978)
	Switzerland	Hoppe (1981)
	USA	Solomon (1985)
	Israel	Navarro *et al.* (1986)
Pyrethrins	USA	Vincent and Lindgren (1967)
Pybuthrin	Poland	Cichy (1971)
Phosphine	India	Kem (1977)
	Australia	Nakita and Winks (1981)
Juvenile hormone mimics	Australia	Dyte (1972)

most insects have at least one microsporidan as a pathogen (Weiser, 1961, 1976).

The important sporozoans infecting stored-products beetles are: *Adelina tribolii* Bhatia, *Farinocystis tribolii* Weiser, *Lymphotropha tribolii* Ashford, *Nosema whitei* Weiser and *N. oryzaephili* Burges, Canning and Hurst. In addition, *N. weiseri* Lipa, *N. ptinidorum* Purrini and *N. transitellae* Kellen, Hoffman and Collier also infect these beetles. Burgess and Weiser (1973), in a survey of pathogens in stored-products insects, recorded *N. whitei* to be the most prevalent species.

This chapter is concerned with the important biological aspects of some species of *Nosema* producing disease in beetles infesting stored commodities, and the potential of these pathogens as control agents.

Morphology and ultrastructure

The morphology of *N. whitei* has been studied in some detail (West, 1960; Lipa, 1968; Milner, 1972a). Milner (1972b) first studied the ultrastructure of most stages of this microsporidan. Fresh schizonts are approximately spherical and the cytoplasm is full of refractile globules and dark, discrete, spherical nuclei. The schizonts, however, are oval or spherical when treated with methanol-giemsa, with blue-staining cytoplasm and red nuclei. The predominant schizogonic stages were mononuclear (26%) and binuclear (73%), although schizonts with up to five nuclei were observed. The only dividing stage of the life cycle was a dividing schizont with karyokinesis complete and cytokinesis in progress. The schizonts measured approximately 2.7–7.0 μm in diameter in stained preparations.

Changes can take place in the surface structure and cytoplasmic organization of some cells, which then transform into sporonts. Unfortunately, little is known of the stimuli which trigger the parasites to enter sporogony but it is possible that an autogamous fusion of nuclei may initiate the process (Canning, 1977).

In a generalized form, sporogony is heralded by a change in the character of the surface membranes of the parasites by the addition of further layers external to the plasmalemma, and the cytoplasm is less homogeneous. Few changes have been recorded in the cytoplasm at the ultrastructural level beyond a limited increase in the endoplasmic reticulum. The sporonts are elongate, oval cells, 5.6 × 3.1 μm, which stain relatively faintly with giemsa and have one or two nuclei. The binucleate sporont developed directly into the sporoblast which was shorter and narrower (4.2 × 2.2 μm).

The immature spores are of the same dimensions as the mature spores and the polar filament develops as a diagonal, densely staining band. Mature *N. whitei* spores show no internal structures under phase-contrast microscopy. After giemsa staining, the polar filament is an intense blue with the thick wall a lighter blue. Hydrolyzed spores have two discrete, central nuclei, often very close together, giving a bilobed appearance. Microsporidan spores are among the smallest in the protozoa and are unique among them in being Gram positive. The spore size of different *Nosema* species attacking stored-products Coleoptera is given in *Table 2*. The spores have a sub-apical periodic acid-Schiff (PAS)-positive cap: another characteristic feature.

The thick outer coat of the schizonts is in about five equally spaced layers totalling approximately 312 Å, and a thinner, double inner layer of approximately 62 Å. The endoplasmic reticulum has 4–6 parallel double membranes on either side of the nucleus, parallel to the longitudinal axis of the cell. A distinct group of vesicles can be seen which are similar to the 'primitive Golgi-type body' observed by Vávra (1965). No mitochondrial-type structure has been seen in *N. whitei*.

Sporonts are rarely seen with the electron microscope, and the sporoblast is always extremely distorted and electron dense: consequently the internal organization of this stage remains unknown. The outer coat of *N. whitei* spores is multilayered and approximately 376 Å thick. The inner coat has two electron-dense layers each 50 Å thick. The outer spore coat has been shown by cytochemical tests to be composed largely of protein, and the inner coat is chitinous. The polar filament is coiled in the posterior half of the spore and is attached to the spore wall at a projection of the inner spore coat, the polar cap. It is 900 Å in diameter and has two inner and two outer electron-dense concentric rings. Little or no external structure has been observed on *N. whitei* spores by scanning electron microscopy.

The maturation of sporoblasts in microsporida has been reviewed by Canning (1977). Major cytoplasmic reorganization takes place within the sporoblast during its transformation into the spore. There is an increase in endoplasmic reticulum, in irregular or parallel arrays and a progressive differentiation of the characteristic spore organelles. It is believed that an extensive Golgi system, which proliferates in the sporoblast, is responsible, at least in part, for the development of the polar sac, polar filament, polaroblast and posterior vacuole. In the final phase of spore maturation the polaroplast membranes, which appear first as loosely arranged lamellae, become closely piled on one another and several regions may form: (1) a

Table 2. A comparison of the spore size of *Nosema* spp. infecting stored-products beetles (μm)

Beetles	*Nosema* spp.	Unfixed spore size		References
T. castaneum	*N. whitei*	4.0–5.0	× 1.7–2.0	Weiser (1961)
T. castaneum	*N. whitei*	3.8–5.9	× 2.6–3.6	Milner (1972a)
T. castaneum	*N. whitei*	4.39–4.53	× 2.86–2.97	Khan (1981)
T. castaneum	*N. buckleyi* (= *N. whitei*)	4.8–5.7	× 2.8–3.2	Dissanaike (1955)
T. confusum	*N. whitei*	4.0–6.1	× 2.7–3.5	Lipa (1968)
T. confusum	*N. whitei*	4.0–6.0	× 2.3–3.5	Milner (1972a)
T. confusum	*Nosema* sp.	3.75–5.0	× 2.8–3.2	West (1960)
O. surinamensis	*N. oryzaephili*	4.1–4.8	× 2.6–3.3	Burges, Canning and Hurst (1971)
Ptinus brunneus	*N. ptinidorum*	3.6–6.6 mode, 5.4 × 2.4	× 2.3–3.0	Purrini (1983)
Paramyelois transitella	*N. transitellae*	3.62+0.26 × 2.38±0.15		Kellen, Hoffman and Collier (1977)

layer of single membranes; (2) a layer of triple membranes; and (3) a region of smaller vesicles. The filament takes its position as a peripheral coil and the nuclei lie at the centre of the spore surrounded by cytoplasm containing a few cysternae of endoplasmic reticulum and ribosomes, often arranged as flat sheets of helically oriented polyribosomes. During this time, the spore coat is laid down. The electron-dense layer deposited onto the sporont plasmalemma becomes thickened and an electron-lucent layer is interpolated between it and the plasmalemma. These layers later form the exospore and the endospore respectively, which eventually confer resistance to the external environment during transmission.

The morphology and development of *N. oryzaephili* was studied by Burges, Canning and Hurst (1971) and its ultrastructure by Burges, Canning and Hull (1974). The development of *N. oryzaephili* is similar to that of *N. whitei*. However, there are some differences between these two pathogens (*Table 3*).

Lipa (1968) described a new species, *N. weiseri* from the lesser grain borer, *Rhizopertha dominica* Fab. in Poland. The microsporidan was observed using a normal light microscope. The schizonts were 1.5–4.0 μm in diameter, uninucleate and binucleate schizonts were frequent, and the sporonts were elongate and had two nuclei. The spores were ellipsoidal with the polar filament 60 μm long. The ultrastructure and development of

Table 3. Differences between *N. whitei* and *N. oryzaephili*

Characters	*N. whitei*	*N. oryzaephili*
Schizogony	Four-nuclear schizonts rare	Four-nuclear schizonts frequent; schizonts with nuclei distributed one or two at each pole passing through an elongate, then dumb-bell stage and transforming into binucleate daughters
	Nuclei small and evenly staining	Nuclei large with central 'pale sphere'
	Nuclei up to five	Nuclei up to 12
Sporogony	Sporont nuclei small and apical	Sporont nuclei large and central
	Sporonts stain faintly with Giemsa	Sporonts stain intensely with Giemsa
	Sporonts do not divide	Sporonts divide
	Binucleate sporoblasts common; no clearly dividing spore	Sporoblasts rare; clearly dividing spore stage dividing into two sporoblasts
No. of coils of polar filament	11	13
Polar filament length (μm)	112 ± 6.5 (m±SE)	64.1 ± 16.0 (m±SD)
Susceptibility for *T. castaneum*	High	Low
Site of infection	Fat body	Mainly fat body, sometimes nerve fibres and cells

N. transitellae were also found to be similar to those of *N. whitei* and *N. oryzaephili* (Kellen, Hoffman and Collier, 1977).

Taxonomy and systematic position of *Nosema*

The definition of the genus *Nosema* is one of the most difficult problems in microsporidan taxonomy. Kudo (1924) defined this genus as a member of the family Nosematidae in which the sporont develops into a single spore, the type species being *Nosema bombycis* Naegeli (1857). Kudo also defined two related genera. These are *Glugea*, in which the sporant divides to give two spores and the host cell is hypertrophied to form the so-called glugea-cysts, and *Perezia*, which similarly forms two spores from the sporont but shows no cell hypertrophy. This distribution based on the host reaction has been criticized (Dolfein and Reichenow, 1952; Weiser 1961). However, Kudo (1966) has persisted with his original definition. This unsatisfactory situation was reviewed by Kramer (1959, 1965).

The familiar distinctions between *Nosema, Glugea* and *Perezia* have been challenged (Lom and Weiser, 1963; Sprague and Vernick, 1968; Ishihara, 1969). Lom and Weiser (1963) proposed that *Nosema* is a synonym of both *Glugea* and *Perezia*. They further proposed to re-establish the genus *Perezia* for any species in which the ripe spores remain joined by the sporont membrane. This action violates the International Code of Zoological Nomenclature and is invalid. Lom (1970) found that some species may not divide in the sporont stage. However, he did not comment on the taxonomic position of these species.

The conclusions of Lom and Vávra (1963) were accepted by Burges, Canning and Hurst (1971) and Canning and Hulls (1970). Their works describe species where the sporonts divide and, in both cases, the pathogen is placed in the genus *Nosema*. In addition, Burges, Canning and Hurst (1971) stated that if species of microsporida are found that do not divide in the sporont stage, then a new genus should be created for them, although Milner (1970) thought that this was not essential. All the genera in the family Nosematidae can be characterized by the number of sporoblasts per pansporoblast, i.e. the number of times the sporont divides, within the sporont membrane (Milner, 1972a).

Milner (1972a) believed that the sporont of *N. whitei* did not divide and that, therefore, its place in the scheme of Lom and Weiser (1963) is dubious. Thus Milner proposed that the genus *Nosema* should include all species in the family Nosematidae in which the sporont, if it divides, does not form more than two sporoblasts. Unfortunately, this character is very difficult to determine and in some species only a fraction of the sporonts may undergo division (Ishihara, 1969; Lipa and Martignoni, 1960). Clearly a new attempt is needed to find a more suitable basis for defining the genus *Nosema* (Milner, 1972a). In a more recent classification (Weiser, 1977), the genus *Nosema* has been placed in the class Microsporididea of the phylum Microsporidia. Weiser (1977) placed the genera *Perezia* and *Glugea* in the same class.

PRESENT SYSTEMATIC POSITION OF GENUS *NOSEMA*

Phylum: Microspora (Sprague, 1977)
Class: Microsporea (Delphy, 1936)
Order: Microsporida (Balbiani, 1882)
Suborder: Apansporoblastina (Tuzet, Maurand, Fize, Michel and Fenwick, 1971)
Family: Nosematidae (Labbé, 1899)
Genus: *Nosema* (Naegeli, 1857)

Life cycle and generation time

The life cycle of microsporidan pathogens starts with the entrance of the infective germ, the sporoplasm, into the host cell, and ends with the formation of viable spores at the end of the cycle. Infection is by ingestion of spores. According to Milner (1970), *N. whitei* spores hatch almost certainly in the midgut and the sporoplasm is probably injected directly into the fat body. Milner proposed the life cycle shown in *Figure 1* for *N. whitei* in *T. castaneum*.

The generation time of a pathogen is the time taken for the first spore to develop (Kramer, 1965). Unfortunately, it is very difficult to detect the presence of one spore in a host animal and there is considerable variation between insects. It is probably more accurate to determine the average time taken for 50% of the developmental stages to become mature spores (Milner, 1972a). Milner also suggested that the prepatent period was correlated with the generation time. The generation time of *N. whitei* is dependent upon

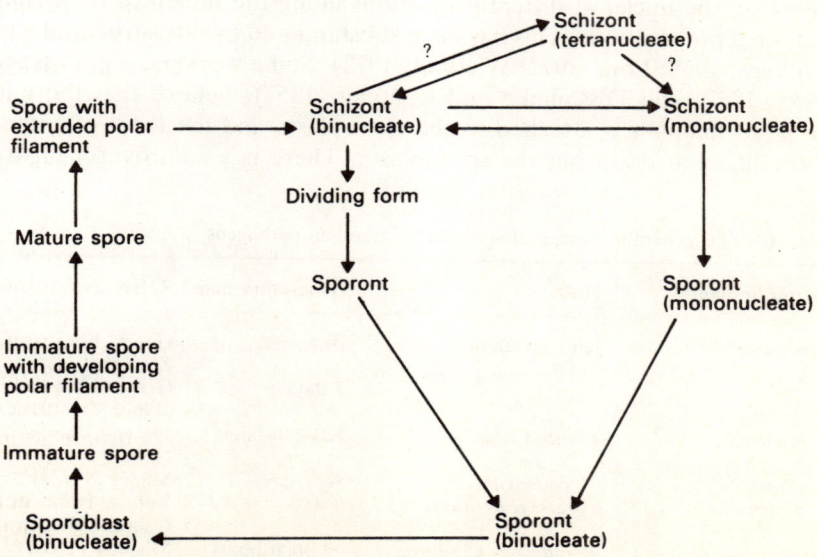

Figure 1.   Life cycle for *N. whitei* in *T. castaneum* (after Milner, 1970); ? = uncertain stage.

dose, temperature and the host insect (Milner, 1972a). Larger doses and higher temperatures reduce the generation time (Maddox, 1968; Milner, 1972a). In addition to temperature and dose, diet may also affect generation time (Kramer, 1965). The generation times of some microsporidan pathogens are shown in *Table 4*. The minimum generation time, under precisely defined conditions in a particular host, may be of taxonomic importance and might reveal strains of the same species of the pathogen with different generation times (Milner, 1972a).

Spore hatching

Infections are initiated in new hosts by the ingestion of mature spores. These hatch in the gut, and the sporoplasms are inoculated into the epithelial cells of the gut wall through the filament. The fresh sporoplasm is a single cell with a nuclear component and some cytoplasm with the usual cytoplasmic organelles and an array of membranes (Ishihara, 1968; Wiedner, 1972; Wiedner and Trager, 1973).

Korke (1916) observed a sporoplasm attached to the end of the filament in *N. ctenocephali* (= *N. pulicis*). He believed the filament conducted the amoebula to a distant part of the tissue and transported the parasite into a fresh area. Morgenthaler (1922) demonstrated the tubular nature of the filament of *N. apis* and the emergence of a fluid mass and a sporoplasm from the open end. Oshima (1937) first put forward the theory that the sporoplasm emerged from the spore through the everted polar filament. This was later confirmed by Gibbs (1953, 1956), Baily (1955), Kramer (1960) and West (1960). It was Kramer (1960) who first produced convincing evidence for the passage of the sporoplasm through the filament, by observing the nuclei at different positions along the lumen of the filaments in stained preparations. This has been substantiated by ultrastructural studies (Ishihara, 1968; Lom, 1972; Wiedner, 1972). Some workers, e.g. Dissanaike (1955, 1957) and Dissanaike and Canning (1957), believe that the end of the polar filament is attached to the sporoplasm and the force of extension of the filament drags out the sporoplasm. There is a controversy regarding

Table 4. The generation times of some microsporidian pathogens

Microsporidians	Hosts	Generation time	References
Nosema apis	*Apis mellifera*	70 hours	Goetz, Eberhardt and Zeutschel (1959)
		7 days	Gray, Cali and Briggs (1969)
N. bombycis	*Bombyx mori*	120–140 hours	Issi and Chervinskaja (1969)
N. whitei	*T. castaneum*	8–17 days	Milner (1972a)
N. transitellae	*Paramyelois transitella*	6 days	Kellen, Hoffman and Collier (1977)
Vairimorpha (*N.*) *necatrix*	*Pseudaletia unipuncta*	99–1000 hours	Maddox (1968)

the nature of the filament. Weiser (1959) and Huger (1960) thought that the filament was solid, while Lom and Vávra (1961), Kudo and Daniels (1963), Weidner (1972) and Weidner and Byrd (1982) believe it is tubular.

Studies by Krieg (1955) and Huger (1960) on the ultrastructure of spores established the basic components of the microsporidian spore, namely the spore coat or wall, polar sac and polar cap, polar filament, polaroplast and sporoplasm.

Ultrastructural studies confirmed that the sporoplasm is a simple cell bounded by a plasmalemma and containing nuclei, membrane and ribosomes (Ishihara, 1968; Wiedner, 1972; Wiedner and Trager, 1973). This invalidated Sprague and Vernick's (1968) hypothesis that the naked genome was inoculated into the host cell, but supported the findings of Sinden and Canning (1974) on the ultrastructure of *N. algerae*. They observed that the two nuclei of the *N. algerae* spore were each surrounded by their own membrane.

Lom and Vávra (1963) proposed from their light-microscopic studies on microsporidan spores that the sporoplasm existed within the spore as a separate entity and was surrounded by a plasmalemma while passing through the filament. However, electron-microscopic studies on spores do not support their view that increased internal pressure resulting from the swelling of the polaroplast and posterior vacuole causes the extrusion of the polar filament (Lom and Vávra, 1963; Lom, 1972; Weidner, 1972). It has been clearly established that the filament is everted during extrusion (Ishihara, 1968; Lom, 1972; Weidner, 1972). However, workers differ considerably in their interpretation.

The 'resting' polar filament consists of three concentric osmiophilic layers, and lies within a membrane-limited cavity in the cytoplasm (Lom, 1972). This is agreed by Canning and Sinden (1973), Gassouma and Ellis (1973) and Sinden and Canning (1974). Lom (1972) further suggested that the two outermost layers were membranous but that the inner osmiophilic core was not. However, Sinden and Canning (1974) found that the core was not membranous but was either amorphous or composed of tubular structures. Consequently, this layer cannot be the outer membrane of the extruded filament as was suggested tentatively by Lom (1972). Sinden and Canning (1974) suggested that the remaining two membranes could invert their positions on extrusion of the filament and that the osmiophilic core was discharged to form part of the glycoprotein coat observed around the extruded filament by Weidner (1972). Weidner described only one membrane bounding the resting filament of *N. michaelis* and proposed that the additional membrane present round the extruded polar filament was derived from the polaroplast. It was suggested that the two membranes may be present in the resting filament and therefore the mechanism of filament extrusion could apply to *N. michaelis* as well as to other species.

A partially everted polar filament after negative staining, as illustrated by Lom (1972), indicates that one or perhaps both of the filament membranes are sealed at the filament tip. It is hard to imagine a means of filament extrusion based on hydrostatic pressure within the spore unless the filament is sealed at its tip. Also, the pressure of a sealed filament provides a simple explanation for the source of the plasmalemma of the sporoplasm. This membrane remains in the spore following the discharge of the sporoplasm, as shown by the electron micrographs of empty spores (Vávra and Undeen, 1970; Weidner, 1972). Thus the sporoplasm cannot be inverted by the plasmalemma of the spore. It has been tentatively suggested that the sporoplasm might acquire a surface membrane by passing through the extruded filament between its two membranes (Weidner, 1972). Alternatively, the micrographs of Ishihara (1968) and Lom (1972) are more indicative that the sporoplasm passes through the central lumen of the tube, considerably distending the tube as it passes through. As Ishihara (1968) has shown a remarkable similarity between the structure of the plasmalemma of the sporoplasm and the membrane of the filament, and Weidner (1972) has shown the two to be continuous, it is reasonable to suggest that the membranes of the sealed tube are distended around the sporoplasm when it reaches the tip of the filament (Sinden and Canning, 1974). Sinden and Canning (1974) proposed, from their ultrastructural studies on *N. algerae* spores, that the positions of the outer membranes of the polar filament are inverted during extrusion and the core becomes a coat to the filament. Some of the cytoplasm, together with the nuclei, passes through the filament and possibly acquires a membrane from the sealed tip of the filament to become the sporoplasm. They suggested that the polaroplast does not participate structurally in extension of the filament.

According to Oshima (1973) spores with dead germs are able to extrude their filaments merely as a result of a change in osmotic pressure. It has been observed that during spore extrusion all the cell contents are suddenly discharged from the spore through the tube (Lom and Vávra, 1963; Weidner, 1972, 1982). There are two stages in microsporidian spore discharge including the sub-second release of the long filament, which is 50–100 μm long (Weidner, 1972; Weidner and Byrd, 1982), and the extrusion of the spore contents through the filament to emerge as an organized cell. It has been observed that the microsporidian spore plasma membrane is discarded with the spore ghost during spore discharge and that the spore polaroplast organelle appears to provide most, if not all, of the newly derived plasmalemma of the extruded sporoplasm (Weidner *et al.*, 1984).

The stimulation of spore hatching in Microsporida has interested many workers. Trager (1937) induced *N. bombycis* spores to hatch in digestive fluid from the silkworm gut, and Oshima (1964) recorded K^+ to be the stimulant in an artificial buffer solution for filament evagination in *N. bombycis*. The extrusion of the polar filaments of *Glugea fumiferanae* was stimulated by chlorides of alkaline metals, and maximum hatching occurred at pH 10.7, at 28°C. Higby *et al.* (1979) used the 'conditioned' culture medium, L-15, normally used for cell culture to induce spore hatching. Undeen (1978) observed the hatching process in some *Nosema* species and found a two-step procedure. Microsporidia from aquatic hosts, e.g. *N. algerae* required pretreatment with water. Those from terrestrial hosts, e.g. *Vairimorpha* (= *N.*) *necatrix* and *heliothis*, were stimulated by a dilute salt solution of high pH. Milner (1972a) employed the wet–dry–wet method of Kramer (1960) for polar filament extrusion in *N. whitei*, a method originally used for the hatching of *Perezia pyraustae* and *N. apis* spores.

Pathogenicity

Milner (1972c) was the first worker to study the quantitative aspects of the pathogenicity of *N. whitei*. The median lethal dose, LD_{50} of *N. whitei* for first-instar *T. castaneum* larvae was 5.4×10^7 spores per gram of flour after 20 days, and 2.2×10^5 after 48 days. The median lethal time, LT_{50} varied from 18.6 days at a dose of 1.0×10^8 to 27.5 days at 8.0×10^5. Milner considered a dose of 10^6 spores per gram of food to be critical. Above this dose insects generally die as larvae, while below they generally emerge as adults. Pupal and adult weights were reduced and fecundity was reduced, although infection had no bearing on the fertility of *T. castaneum* females. No trans-ovum transmission of the pathogen was observed. Khan (1981) and Al-Hafidh (1985) also obtained similar results. Lighter body weight in *T. castaneum* larvae was also recorded by Armstrong (1978a).

Infected *T. castaneum* larvae grow more slowly than the controls and appear normal until infection becomes severe. Severely infected larvae are larger than healthy ones, have swollen bodies, are lighter in colour and move sluggishly (Milner, 1972c; Khan and Selman, 1984). Larvae slowly change in colour after death and eventually become dark brown or black. Fisher and Sanborn (1964) found that *N. whitei*-infected *T. castaneum* larvae grew faster than the controls, and had supernumerary moults. One larva weighed 4.8 mg after 22 days at 27–29°C and 60–65% humidity. This is nearly twice the normal weight. They further reported that infected larvae rarely, if ever, pupated, and that these effects were caused by a juvenile-hormone-like substance produced by *N. whitei* (Fisher and Sanborn, 1962a, 1964). Milner (1972c) found that infected larvae grew larger but never grew faster than the controls. The largest infected larvae weighed 4.18 mg after 41 days at 2×10^7 spores, compared with the heaviest control larvae of 2.5 mg after 17 days. A large proportion of infected larvae emerged as sublethally infected adults if infected at the right age and with the minimum dose. Milner (1972c) concluded that there was no evidence to support the conclusion of Fisher and Sanborn (1962a, 1964) that *N. whitei* produced a juvenile-hormone-like substance.

Larvaeform pupae result from severe infections. They are elongate, narrower, with a spore-laden abdomen, the developing wings and legs are reduced in size, and the characteristic curve of the body is lost (Milner, 1972c; Khan, 1981). They all die at this stage. Severely infected adults have incompletely developed wings, and consequently are unable to close their elytra. The sex-ratio of *T. castaneum* is not significantly altered by nosematosis (Milner, 1970; Khan, 1981; Al-Hafidh, 1985). Adult longevity was reduced by *N. whitei* infection (Milner, 1970; Armstrong, 1982; Khan and Selman, 1988). Adultoids, forms intermediate between pupae and adults, were also recorded but never survived to reproduce.

N. whitei reduced the mating frequency and number of progeny of *T. castaneum* (Armstrong and Bass, 1986). When *T. castaneum* adults were exposed to *N. whitei* after mating, significant differences were recorded between the number of larvae resulting from eggs laid by uninfected and infected females during the second and fourth weeks after exposure.

Burges, Canning and Hurst (1971) found that *N. oryzaephili* was highly infective to *O. surinamensis*. The LD_{50} in young *O. surinamensis* was 2.6 × 10^5 spores g^{-1} of dry food. They found that *T. castaneum* and *T. confusum* were slightly susceptible to *N. oryzaephili*. When many crushed diseased bodies were offered to 2-day-old larvae or mixed with the eggs, 20% of *T. castaneum* and 2% of *T. confusum* adults became diseased.

N. weiseri is highly pathogenic to *Rhizopertha dominica*, and in a highly advanced infection, or in disease-killed insects, the whole fat body is totally destroyed, giving a mortality up to 60% (Lipa, 1968). Kellen, Hoffman and Collier (1977) observed that the mortality of flour beetles was 80% after 30 days when they were reared on food contaminated with *N. transitellae*. *N. whitei* and *N. oryzaephili* infect both the larvae and the adults of beetles, *N. ptinidorum* only infects the adults, while *N. weiseri and N. transitellae* infect the larvae.

Effects of host nutrition

Our knowledge of protozoan nutrition is at present very limited (Lachevalier and Pramer, 1971). We do know that the pathogenicity of *N. whitei* is enhanced by dietary stress of the host. George (1971) demonstrated that supplementation of rice and soya flours with brewer's yeast increases the number of *N. whitei*-parasitized larvae pupating and enables larvae that do not pupate to survive longer. If *N. whitei*-infected *T. castaneum* are supplied with adequate nutrients, e.g. proteins and B-complex vitamins, they are capable of meeting their own metabolic needs and those of the pathogen for a longer period (George, 1971). Host and pathogen appear to compete for stored nutrients in the host's fat body. George (1972) subsequently reared *N. whitei*-infected second-instar *T. castaneum* larvae for 15 days on rice and soya flours and then transferred them to food fortified with brewer's yeast (flour : yeast = 19 : 1). He observed that all infected larvae died when fed on rice flour both unfortified and fortified with yeast, whereas approximately one-third of the uninfected larvae pupated on these diets. George concluded that *N. whitei* infection depleted the nutrient reserve of the fat body in 15 days to such an extent that supplemented diets failed to increase survivorship in diseased larvae.

Armstrong (1978b) investigated the effects of deficient diets on *N. whitei*-infected *T. castaneum*. Beetles grown on vitamin B-deficient diets suffered higher mortality than beetles grown on vitamin-B complete diets. Pupation and longevity were also significantly reduced. The longevity of the host on a vitamin B-complete diet, even when infected, was much longer than both infected and uninfected hosts on deficient diets.

T. castaneum larvae parasitized with *N. whitei* were found to gain much less weight and consume up to twice as much food as healthy larvae. Diseased larvae appeared to be less efficient in converting ingested food to body weight (Armstrong, 1979).

Golemanski and Dukhlinska (1982) in a survey carried out in grain warehouses in Bulgaria, isolated 11 species of Sporozoa and Microsporidia from 13 pest species of insects. The largest number of pathogen species was found in *T. castaneum, T. confusum* and *Tenebrio obscurus*. Infection with *N. whitei, A. tribolii, F. tribolii* and *Mattesia dispora* resulted in the highest level of disease and caused hypertrophy and hyperplasia of the cells of infected larval tissues.

Mechanism of pathogenicity

Most of the pathogenic effects observed in the host–pathogen systems are related directly or indirectly to the depletion of resources in the fat body of the insects. The fat-body cells perform a wide range of functions. These include the metabolism of carbohydrates, lipids and nitrogenous compounds, the synthesis and regulation of blood sugars, the storage of glycogen, fat and protein, and the synthesis of major blood proteins: a diversity that may be unequalled in many other metazoan cells. The fat-body cell can switch its activity pattern in response to nutritional, hormonal and developmental signals to provide for the successive needs of growth, metamorphosis, migration and reproduction. During metamorphosis stored proteins are released and much of the fat body is destroyed through cell lysis. That fraction which survives or is newly formed from undifferentiated cells emerges with renewed efficiency in the adult insect. In the female, the principal function of the fat body, timed to correlate with the reproductive cycle is the production of the precursors of yolk.

Protein synthesis in the fat body can be regulated by both hormonal and non-hormonal signals. Thus, the polysome population in the *Bombyx mori* larval fat body declines during starvation and is readily built up soon after feeding, through a control mechanism that apparently does not depend directly on the intracellular supply of amino acids (Bosquet, 1979). However, little is known about such mechanisms.

It has been suggested that fat-body degeneration may play an important role in the ageing of insects (Miquel, 1971). The fat body is the hub of the insect's metabolism. As the centre controlling metabolism of the haemolymph it determines the physiological activity of other body tissues, and any restrictions on the functions of the fat body limits the physiological capacities of the other tissues (Keeley, 1978). Wyatt (1980) reviewed the role of the fat body in the production of proteins.

Fisher and Sanborn (1962a, 1964) claimed that *N. whitei* produced a juvenile-hormone-like substance, but Milner (1972c) could not confirm their conclusions. Listov (1976, 1977) studied the effects of *N. whitei* and *F. tribolii* on darkling beetles and reported that the hormonal metabolism disorder was probably caused by parasitic destruction of the fat body at metamorphosis, since it was found that fat-body destruction caused morphogenic disruptions. Rabindra, Balasubramanian and Jayaraj (1981) observed a juvenile hormonal activity produced by *F. tribolii* on *T. castaneum*. Ether extracts from the spores of *F. tribolii* when applied to *T. castaneum* pupae and final-instar *Dysdercus cingulatus* produced adultoids. The production of juvenile-hormone-like substances by *N. whitei* probably cannot be ruled out before intensive studies on the biochemistry of the pathogen are carried out.

Issi and Onatskii (1984) concluded that data from the literature and the authors' own results showed that microsporida at the presporulation stage are non-pathogenic to the insect host and even have a stimulating effect at the cellular and organ level. Protein synthesis and energy production increased in the infected cells, and infection at a late instar may have produced heavier pupae. Diseased insects showed a temporarily increased resistance to other pathogens, e.g. bacteria. They added that the stimulating effects ceased with the onset of sporogony and pathogenic effects then started. They believed that this stimulatory effect had evolved within the obligatory host–parasite system and ensured the survival of the host insect.

Disc electrophoresis used to study the effects of *N. whitei* on several metabolic macromolecules of sixth-instar *T. castaneum* larvae showed that nosematosis generally produced a significant decrease in the fractions of the macromolecules (Newton, Armstrong and Reichelderfer, 1983).

It is interesting to note that mitochondria are absent from all stages of the life cycle of *Nosema* spp. and that this is true of all other microsporidans. Intermediary metabolism and endogenous sources of energy are unknown in these animals. However, it has been found that sporoplasms extruded from spores into a tissue culture medium survived better when ATP was added (Weidner and Trager, 1973). This suggests that they rely largely on exogenous sources of energy. It was noted that the developing stages of microsporidia are frequently associated with accumulations of host-cell mitochondria from which their ATP requirements may be derived. Listov (1979) noted reduced respiratory activity and tissue catalase activity in the larvae of flour beetles infected with *N. whitei* and *A. tribolii*. *T. castaneum* larvae with nosematosis do not moult more frequently than healthy insects. George and Townes (1976) suggested that the pathogen, which inhibits the fat body, depletes the reserves of the larvae, inhibiting moulting and causing early pupation.

The effect of starvation on *T. castaneum* larvae both infected with *N. whitei* and uninfected, was to increase larval and pupal mortality, to delay the rate of larval development and to increase the time needed for adult emergence (Armstrong and Newton, 1985). These effects were significantly greater in diseased larvae, which showed a significant decrease in protein content compared to healthy individuals.

Insects, when stressed, are more susceptible to disease. This stress may be produced by a number of environmental factors, e.g. crowding, malnutrition, administration of chemicals and drugs, etc. (Steinhaus, 1956, 1958; Weiser and Lysenko, 1956; Vago, 1959). Some researchers believe that chemical pesticides act as stressors and consequently promote concentration or activation of infectious diseases or make insects more susceptible to microbial toxins. There are two views on the enhanced activity of insecticide–pathogen combinations. These are:

1. that the insecticide makes the insect more susceptible to infection (Zakharcheno, 1959); and
2. that the pathogen weakens the insect so that even though it would not normally die from disease, it cannot in its weakened state tolerate the insecticide (MacBain Cameron, 1963).

However, there is a need for much more intensive investigation of pathogen–chemical insecticide mixtures to determine the exact mode of action. Benz (1971) comprehensively reviewed the integration of chemical insecticides with various pathogen groups prior to 1970.

Khan (1981) first determined the effect of *N. whitei* on *T. castaneum* biology when it was applied in combination with sublethal doses of the organophosphate, pirimiphos methyl. Combined *N. whitei*–pirimiphos methyl doses produced synergistic effects on the mortality of this insect, the effect being greater on larvae than on adults. Combined doses had a significant effect on larval growth and produced a significantly lengthened larval period compared to the controls or the insecticide or the pathogen applied alone (Khan and Selman, 1984). The fecundity and fertility were reduced by *N. whitei*–pirimiphos methyl combinations (Khan, 1981). The emergence of adults was also significantly reduced by the insecticide–pathogen treatment and so was the adult weight (Khan and Selman, 1987). Moreover, the longevity of *T. castaneum* adults was significantly reduced by combined doses (Khan and Selman, 1988).

Al-Hafidh (1985) integrated *N. whitei* with the organophosphates, methacriphos and malathion, the carbamate, carbaryl, and the bacterium, *Bacillus thuringiensis* (Thuricide HP[R] and Dipel[R]), against *T. castaneum*. He observed that combined treatments produced synergistic effects on the mortality of larvae. Infected larvae were significantly more susceptible to malathion than were uninfected larvae of the malathion-resistant strain of the beetle. Growth of *T. castaneum* larvae was significantly affected when *N. whitei* spores were applied in combination with the insecticides and all treatments lengthened the larval period. There was a significantly reduced adult emergence and a significantly greater level of adult deformity by combined treatments as compared to the individual actions of insecticides or the controls. Al-Hafidh also observed a distortion of the 1 : 1 sex-ratio in favour of the males, significantly reduced fecundity and fertility, and a significantly greater adult mortality in *T. castaneum* with combined treatments.

T. confusum and *T. destructor* larvae are resistant to methyl bromide, the former being more resistant (Listov and Nesterov, 1980). When infected with *N. whitei* the resistance of both the species to the fumigant was lowered.

Potentiality of *Nosema* spp. in microbial control

The potentialities of *Nosema* spp. in reducing natural populations of storage beetles have been overshadowed by the use of chemical agents. Their value as successful control agents is discussed below.

HOST RANGE

Pathogens capable of infecting a large number of host species in an ecosystem have a better chance of contracting susceptible hosts, and this, in its turn, increases the pathogen population density and disease prevalence. Protozoan pathogens, once believed to be highly host and tissue specific, are now known to often cross-infect not only genera, but even orders of insects.

According to Weiser (1961) there are at least three barriers which must be satisfied if an infection is to result. These are:

1. the conditions of the gut contents must be favourable for spore germination;
2. the sporoplasm must be able to penetrate the gut and survive the defence mechanisms of the host; and
3. the tissues of the host must be able to support the development of the pathogen.

Thus an insect satisfying all these requirements of a pathogen is a potential host.

Fisher and Sanborn (1962b) attempted to elucidate the factors controlling host range in the Microsporida. They were able to infect many hosts representing several insect orders, however, only in *T. castaneum* and *T. confusum* did *N. whitei* develop unchecked. Milner (1973b) studied the host range of this microsporidan in stored-products insects and found *T. brevicornis*, *T. destructor*, *T. madens* and *O. surinamensis* to be susceptible, in addition to *T. castaneum* and *T. confusum*. The navel orangeworm, *P. transitella* was slightly susceptible to *N. whitei* (Kellen, Hoffmann and Collier, 1977). The host range of *N. oryzaephili* includes *O. mercator*, *T. castaneum*, *T. confusum*, *Lasioderma serricorne* and *Stegobium paniceum* (all beetles) and three species of moths, besides *O. surinamensis*. *N. ptinidorum* infects *P. brunneus* (Coleoptera: Ptinidae). *N. areiseri* infects *R. dominica*. *N. transitellae* was found to be highly pathogenic to *T. castaneum* (Kellen, Hoffmann and Collier, 1977), and the navel orangeworm and other Lepidoptera.

The host range of *Nosema* spp. in stored-products Coleoptera is not as narrow as was previously thought. The effect of host species on the susceptibility to *N. whitei* and a comparison of the pathogenicity of the microsporidan with other pathogens are shown in *Tables 5* and *6*.

VIRULENCE

Greater virulence produces better short-term control of insect populations (Tanada and Fuxa, 1987). However, Anderson (1982) showed by modelling studies, that for the introduction and perhaps for inoculative augmentation, moderately virulent pathogens are better. All the pathogens treated so far are not too virulent and produce chronic and debilitative effects rather than quick mortality.

Table 5. Effect of host species on susceptibility to *N. whitei*

Hosts	LD$_{50}$ (spores/g)	References
T. castaneum	5.4 × 10^7	Milner (1972c)
	3.98 × 10^6	Khan (1981)
	2.41 × 10^6	Al-Hafidh (1985)
T. confusum	7.0 × 10^7	Milner (1973b)
T. anaphe	4.4 × 10^5	Milner (1973b)
O. surinamensis	3.0 × 10^7	Milner (1973b)

Table 6. A comparison of pathogenicity of *N. whitei* with other pathogens

Pathogens	Hosts	LD$_{50}$ (spores/g food)	References
N. whitei	*T. castaneum*	5.4 × 10^7	Milner (1972c)
N. oryzaephili	*O. surinamensis*	2.6 × 10^5	Burges, Canning and Hurst (1971)
F. tribolii	*T. castaneum*	2.4 × 10^6*	Sherlock (1969)
L. tribolii	*T. castaneum*	3.6 × 10^5*	Ashford (1967)
N. locustae	*Melanoplus bivittatus*	9000**	Bucher (1958)

* LD$_{50}$ values multiplied by 8 because there are eight sporozoites in each spore.
** Number of spores ingested/insect.

EFFECT OF PATHOGEN DOSES

The doses of a pathogen applied constitute one of the most important determinants in insect epizootology. This is supported by mathematical modelling (Anderson and May, 1981; Anderson, 1982) and by laboratory determination of median lethal doses, dose-response, field research and studies of natural epizootics (Tanada and Fuxa, 1987). The greater the population density of the pathogen, the greater the chance of contact between pathogens and uninfected hosts. In addition, the effect of pathogen doses on epizootics is interdependent with factors like the susceptibility of the host and the infectivity and persistence of the pathogen in the environment.

PERSISTENCE

Persistence is an important factor in the regulation of host populations by pathogens (Anderson and May, 1980) and it is one of the factors that influences the frequency of transmission (Smith, 1971).

Many attempts have been made to assess the stability of or viability of microsporidan pathogens under various conditions. This subject has been reviewed extensively (Kramer, 1970, 1976; Maddox, 1973, 1977; Brooks, 1980, 1982). Kramer (1970) suggested that the ability to survive varies from species to species and that the spores of some species bound in faeces or in dried cadavers may remain viable for one or more years under room conditions. The longevity of naked dried spores under room conditions usually does not exceed 3–4 months, but may range from approximately 2 weeks to over 1 year. Naked spores in a cold, clear, aqueous medium may survive from 7 to 10 years. Thus, spores are favoured usually by a moist environment, low temperatures and by being protected in faeces or cadavers.

N. whitei, like many other microsporidans, extrudes its polar filament when exposed to wetting agents. However, N. whitei spores will remain viable in flour stored at 4°C for at least 15 months (Milner, 1972d) and can survive temperatures as high as 72°C for 1 hour (Milner, 1970). Burges, Canning and Hurst (1971) reported that spores of N. oryzaephili, mixed with flour and dried, declined in viability after 9 months storage at 2°C. These spores had been suspended in water before mixing with the flour.

HOST POPULATION DENSITY AND DISTRIBUTION

Entomopathogens are almost exclusively host-density dependent (Watanabe, 1987). Increased host density results in increased contact between uninfected and infected hosts and the pathogens. It can also stress the insect population, predisposing it to disease, and making substrates and nutrients in the host more readily available for pathogen growth and reproduction (Fuxa, 1987b). Density dependence is more dominant, or at least more evident, in pathogens that are transmitted horizontally than in those transmitted vertically (Andreadis, 1987). According to Fuxa (1987b), density dependence is an important attribute when a pathogen is recycled or multiplies in a host or is transmitted to new hosts, and a density-dependent pathogen may be at an advantage when it is intended to regulate an insect population continuously.

Armstrong (1980) reared T. castaneum larvae at five levels of crowding (25, 50, 75, 100 and 125 larvae/Petri dish). N. whitei infections were found to be density dependent, and increased mortality while decreasing pupal and adult weight. Al-Hafidh (1985) also obtained similar results.

Transmission of the Nosema spp. of stored-product beetles is horizontal, occurring within and between generations of hosts. Nosema spp., like many other microsporidans, are able to maintain themselves by horizontal transmission alone. These are parasites of the fat body and are transmitted per os. Cadavers are essential for the survival of this category of microsporida. Meiosis is unknown in Nosema, and there is no reason to suspect that species of this genus cannot be transmitted directly (Canning, 1981). Epizootic and enzootic phases of infection would be expected to keep the parasite density within feasible limits.

Steinhaus (1949, 1954) suggested that the tendency of the host to aggregate or disperse may be important to the severity or extent of disease, and that the distribution of disease organisms is likely to be important in control. The dispersion rate of stored-product Coleoptera is high, and the cannibalistic behaviour adds to the transmission of the parasites.

The effect of insect disease is most obvious at high population densities where more contact occurs among individual hosts, and the unfavourable effects of overcrowding are evident (Viktorov, 1971). Horizontal transmission of protozoan infections usually requires high host densities or host aggregation behaviour (May, 1983).

Production of *Nosema* spp.

Microsporidians are obligate parasites requiring living cells for their development. They have been propagated only in laboratory-reared or field-collected insects. In acute infections, the host insects die so rapidly that the replication of the parasite is limited, but in prolonged infections there is abundant production of spores (Weiser, 1963; Kramer, 1968; McLaughlin, 1971; Maddox, 1973). Sublethal infection of early-instar *T. castaneum* larvae results in the production of large, spore-laden final-instar larvae, most of which die before pupation. The mass produciton of *N. whitei* spores would be aided by the use of alternative hosts. Such hosts are *T. anaphe*, which is more susceptible to *N. whitei* than other *Tribolium* spp., and *T. brevicornis* and *T. destructor*, which grow to a larger size and give a higher spore yield.

Large numbers of disease-killed cadavers can be obtained by simply adding infective doses of the pathogen to flour and rearing the larvae of *T. castaneum* or other species on it. Flour beetles of the genus *Tribolium* are cosmopolitan and can be mass reared using simple laboratory techniques and simple rearing media. According to Hostounský and Weiser (1972) the economic production of protozoan spores requires mass-reared insects infected with the lowest possible infective dose. A number of *Nosema* spp. have been mass produced for microbial control programmes (*Table 7*).

Table 7. *Nosema* spp. mass produced for commercial uses (yields in spores/g host)

Species	Hosts	Spore yield	References
N. algerae	*Anopheles stephensi*	8.9×10^5	Undeen and Maddox (1973)
	Heliothis zea	1.8×10^9	
N. eurytremae	*Pieris brassicae*	6.0×10^8	Pilley, Canning and Hammond (1978)
N. fumiferanae	*Choristoneura fumiferana*	1.4×10^8	Wilson (1976)
N. locustae	*Melanoplus bivittatus*	3.9×10^9	Henry and Oma (1981)
N. locustae	*M. bivittatus*	1.6×10^9	Henry (1985)*
		9.1×10^{11}	
N. locustae	*M. differentialis*	3.78×10^9	Henry (1985)*
N. lymantriae	*Lymantria dispar*	3.5×10^7	Ignoffo and Hink (1971)
N. plodiae	*Mamestra brassicae*	9.0×10^8	Hostounský and Weiser (1972)
N. pyrausta	*Ostrinia nubilalis*	8.6×10^7	Raun, York and Brooks (1960)

* Number of spores/insect.

Environmental factors

Environmental factors affect *Nosema* spp., as they do other microsporidans. According to Maddox (1973), temperature is the most important physical factor for microsporidan infection. The pathogenicity of *N. whitei* for *T. castaneum* decreases as the temperature increases above 25°C ($LD_{50} = 4.2 \times 10^6$ spores/g) through 30°C ($LD_{50} = 1.3 \times 10^7$ spores/g) to 35°C ($LD_{50} = 3.2 \times 10^8$ spores/g) (Milner, 1973a). The stress produced by very high or very low temperatures proved harmful to the infected host. Humidity has been observed to produce a significant effect on nosematosis. Milner (1973a) found that the pathogenicity of *N. whitei* for *T. castaneum* decreased consistently from 10 to 70% relative humidity. It is apparent that a low relative humidity stressed the insect and enhanced the pathogenicity of the microsporidan. Milner (1973a) found that any factor which increased the developmental time of the host increased the pathogenicity of *N. whitei*.

EFFECT OF HOST AGE ON INFECTIVE DOSE

Several investigators have observed that a sharp increase in the spore dose is required to infect subsequent instars and adults. Milner (1972c) demonstrated that the difference between the LD_{50} of *N. whitei* in *T. castaneum* was 1.8×10^6 spores/g for first-instar larvae and 1.0×10^{13} spores/g for the fifth-instar larvae.

APPLICATION OF THE ANDERSON AND MAY (1981) MODEL TO *N. WHITEI*

It is interesting to estimate the efficiency of *N. whitei* using the predictions of the model developed by Anderson and May (1981) for the management of *Tribolium* spp. Anderson and May (1981) deduced from a simple population model the conditions that must be satisfied for a pathogen to regulate host population growth. For the eradication of the host, an insect pathogen must be introduced in excess of a critical value A_C determined by the equation:

$$Ac = \frac{\mu r(\alpha + b + \gamma)}{\cup(\alpha - r)}$$

where μ = mortality rate of the free spores; r = intrinsic growth state of the host (birth rate minus death rate); α = disease-induced mortality rate; b = natural mortality rate; γ = recovery rate and $\cup$ = transmission rate.

The transmission rate in Anderson and May's (1981) formula is for horizontal transmission. According to Anderson and May (1980), the progressive nature of microsporidan diseases and lack of a truly effective immune response means that an immune class does not have to be accounted for in the model, and in the case of a highly pathogenic agent the recovery rate may be taken as nil ($\gamma = 0$). Anderson and May (1981) emphasized that pest species with a high population growth rate (large *r*) are difficult to control, and that control efforts are more likely to succeed if they require relatively low rates of introduction of the parasite rather than high rates.

Prospects for future research

The use of pathogenic protozoans alone or in combination with other control agents is still at an early stage. Fortunately, many workers have recently considered the inclusion of entomopathogens into pest-management programmes. This is reflected by the number of recent review articles, e.g. Brooks (1980), Canning (1981, 1982), Henry (1981), Undeen (1982), Davidson and Sweeney (1983), Hazard (1985), Pillai (1985), Flexner, Lighthart and Croft (1986), Maddox (1986), Burges and Pillai (1987) and Fuxa (1987a). The introduction and inoculative augmentation of protozoan pathogens has been successful (Ignoffo, 1985). The difficulty of production and the chronic disease that they cause are disadvantageous in their inoculative augmentation. However, their debilitating, sublethal effects on the host, the large number of species found in every insect order, their frequently high prevalence, and their array of transmission routes are advantageous when they are used in this way. The following guidelines are suggested for future research:

1. Few studies have been made of the host–parasite relationships of *Nosema* spp. of storage insects excepting *N. whitei*. A detailed knowledge is required if they are to be used in pest-management programmes.
2. The development of microbial control in food stores also needs practical trials to provide answers to the following questions:
    (i)   What is their efficacy under normal conditions of commercial storage?
    (ii) Are the pathogens compatible with storage conditions in all types of store?
    (iii) Can they be cost-effective?
    (iv) Are there any unsolicited side-effects?
3. The double infection of insects by protozoan pathogens has been recorded, and cross-infection is also a possibility. These observations need further research.
4. The mass production of *Nosema* spores should be investigated, and cell culture and tissue culture methods should be tried. Methods for the protection of spores from the detrimental effects of the micro- and macro-environment and suitable application techniques need to be found.
5. Much of the variabiilty recorded for many microsporidans is possibly the outcome of strain differences. Conversely, strain differences in host insects may have a significant bearing on the pathogenicity of a microbial agent. This should be determined, though it would be very difficult to do.

6. The effects of behavioural changes in diseased insects should be properly investigated.
7. The genetic manipulation of microsporidan pathogens to amplify their pathogenicity and broaden their host range is a possible future consideration.
8. A combination of pheromones and spores to attract and inoculate stored-products insects has been suggested (Burkholder and Boush, 1974; Sapas, Burkholder and Boush, 1977). The efficiency of pheromone–protozoan combinations in pest control should be evaluated. Bait–pathogen combinations are also important for storage insects.
9. The study of the effects of combining entomopathogens with sublethal insecticide doses to control storage insects is very far from complete. Much more research is needed in this area and a more realistic series of experiments designed to simulate warehouse conditions.
10. The interaction between host nutrition and nosematosis should be properly evaluated.
11. The use of vectors for the transmission of *Nosema* spp. should be considered.

Thus, control agents should be selected for high pathogenicity (large α), good survival (small μ) and high transmission efficiency (large $\cup$). The transmission rate is the probability that a single infective agent (e.g. spore) will be taken up and establish an infection. This is the most difficult value to quantify but if the three elements of the equation can be estimated, it should be possible to predict the effectiveness of a given pathogen as a control agent (Canning, 1981). The depression of pest populations below an economic threshold level, rather than its eradication, is a realistic method in pest-management programmes. For introduction, and perhaps for inoculative augmentation, moderately virulent pathogens are better (Anderson, 1982).

Considering the above it may be concluded that *N. whitei* should not be expected to have a dramatic effect on the host population when used alone. However, it is to be expected that the pathogen may have an important effect on the host population when used in combination with sublethal doses of chemical insecticides, or may be an important check on beetles in an enzootic stage. This is what is to be expected from a microsporidan pathogen that produces chronic, debilitative effects on the host.

Conclusions

Nosema spp. are not expected to be used as fast-acting control agents like some bacteria and viruses or chemical pesticides. They are not expected to replace chemical agents totally but show promise when integrated into the established control system. Storage pests appear to offer good conditions for the development of epizootics and potential for the dispersal and persistence of their pathogens. The increased application of protozoans in pest-control would prove ecologically very advantageous if it also meant the use of reduced levels of chemical control agents. The use of microsporidans in controlling storage pests would be most effective in long-term storage since they produce slow-acting, chronic infections. Besides their use in combination with other pest control strategies, e.g. application of sublethal doses of chemical insecticides, they could furnish better standards of pest management in the stored-product environment.

Acknowledgements

This review was completed during the tenure of a Commonwealth Academic Staff Fellowship awarded to the first author by the Commonwealth Scholarship Commission in the UK. A. R. Khan thanks Rajshahi University, Bangladesh, for granting him study leave.

References

AL-HAFIDH, E. M. T. (1985). *The integration of* Nosema whitei *and some Insecticides on* Tribolium castaneum. Ph.D. Thesis, University of Newcastle upon Tyne.

ANDERSON, R. M. (1982). Theoretical basis for the use of pathogens as biological control agents of pest species. *Parasitology* **84**, 3–33.

ANDERSON, R. M. AND MAY, R. M. (1980). Infectious diseases and population cycles of forest insects. *Science* **210**, 658–661.

ANDERSON, R. M. AND MAY, R. M. (1981). The population dynamics of microparasites and their invertebrate hosts. *Philosophical Transactions of the Royal Society London* **B 291**, 451–524.

ANDREADIS, T. G. (1987). Transmission. In *Epizootiology of Insect Diseases* (J. R. Fuxa and Y. Tanada, Eds), pp. 159–176. John Wiley and Sons, New York.

ANONYMOUS (1986). National Research Council, USA. *Pesticide Resistance: Strategies and Tactics for Management*. National Academy Press, Washington, DC.

ARMSTRONG, E. (1978a). *Nosema whitei*: Body weight changes in larvae of *Tribolium castaneum. Zeitschrift für Parasitenkunde* **56**, 13–15.

ARMSTRONG, E. (1978b). The effects of vitamin deficiencies on the growth and mortality of *Tribolium castaneum* infected with *N. whitei. Journal of Invertebrate Pathology* **31**, 303–306.

ARMSTRONG, E. (1979). *Nosema whitei*: relationship of body weight gains and food consumption of *Tribolium castaneum* larvae. *Zeitschrift für Parasitenkunde* **59**, 27–29.

ARMSTRONG, E. (1980). The effects of crowding on *Nosema whitei* infected and control *Tribolium castaneum. Zeitschrift für Parasitenkunde* **63**, 145–150.

ARMSTRONG, E. (1982). *Nosema whitei*: influence on offspring of *Tribolium castaneum. Journal of Invertebrate Pathology* **39**, 257.

ARMSTRONG, E. AND BASS, L. K. (1986). Effects of infection of *Nosema whitei* on mating frequency and fecundity of *Tribolium castaneum. Journal of Invertebrate Pathology* **47**, 310–316.

ARMSTRONG, E. AND NEWTON, P. B. (1985). The influence of starvation on mortality, development, and protein content in parasitized and unparasitized *Tribolium castaneum. Journal of Invertebrate Pathology* **46**, 103–108.

ASHFORD, R. W. (1967). *A Study of protozoan parasites of stored products Coleoptera*. Ph.D. Thesis, University of London.

BAILY, L. (1955). The infection of the ventriculus of the adult honey bee by *Nosema apis. Parasitology* **45**, 86–94.

BALBIANI, E. G. (1882). Sur les Microsporidies où Psorospermies de l'articule.

Comptes rendus hebdomadaire des Séances de l'Académie des Sciences, Paris **95**, 1168–1171.

BENZ, G. (1971). Synergism of micro-organisms and chemical insecticides. In *Microbial Control of Insects and Mites* (H. D. Burges and N. W. Hussey, Eds), pp. 327–355. Academic Press, New York.

BHATIA, S. K. AND PRADHAN, S. (1972). Studies on resistance to insecticides in *Tribolium castaneum* (Herbst) – V. Cross resistance characteristics of a lindane-resistant strain. *Journal of Stored Products Research* **8**, 89–93.

BOSQUET, G. (1979). Occurrence of an active regulatory mechanism of protein synthesis during starvation and refeeding in *Bombyx mori* larvae. *Biochemistry* **61**, 165–170.

BROOKS, W. M. (1980). Production and efficacy of Protozoa. *Biotechnology and Bioengineering* **22**, 1415–1440.

BROOKS, W. M. (1982). In *Invertebrate Pathology and Microbial Control: Proceedings of the IIIrd International Colloquium on Invertebrate Pathology, University of Sussex, Brighton, 1982*, pp. 336–341.

BUCHER, G. E. (1958). General summary and review of utilization of disease to control insects. *Proceedings of the 10th International Congress of Entomology, Montreal* **4**, 695–701.

BURGES, H. D. AND PILLAI, J. S. (1987). Microbial Bioinsecticides. In *Microbial Technology in the Developing World* (E. J. Dasilva, Y. Domergues, E. J. Nyns and C. Ratledge, Eds), pp. 121–150. Oxford University Press, Oxford.

BURGESS, H. D. AND WEISER, J. (1973). Occurrence of pathogens of the flour beetle *Tribolium castaneum*. *Journal of Invertebrate Pathology* **22**, 464–466.

BURGES, H. D., CANNING, E. U. AND HULLS, I. K. (1974). Ultrastructure of *Nosema oryzaephili* and the taxonomic value of the polar filament. *Journal of Invertebrate Pathology* **23**, 135–139.

BURGES, H. D., CANNING, E. U. AND HURST, J. A. (1971). Morphology, development and pathogenicity of *Nosema oryzaephili* n.sp. in *Oryzaephilus surinamensis* and its host range among granivorous insects. *Journal of Invertebrate Pathology* **17**, 419–432.

BURKHOLDER, W. E. AND BOUSH, G. M. (1974). Pheromones in stored product insect trapping and pathogen dissemination. *Organisation Européenne et Mediterranéenne pour la Protection des Plantes Bulletin* **4**, 455–461.

CANNING, E. U. (1977). Microsporida. In *Parasitic Protozoa* (J. P. Kreier, Ed.), Volume 4, pp. 155–196. Marcel Dekker, New York.

CANNING, E. U. (1981). Insect Control with Protozoa. In *Biological Control in Crop Protection, BARC Symposium Number 5* (G. C. Papavizas, Ed.), pp. 201–216. Allanheld, Osmun, Totowa, NJ.

CANNING, E. U. (1982). An evaluation of protozoal characteristics in relation to biological control of pests. *Parasitology* **84**, 119–149.

CANNING, E. U. AND HULLS, R. H. (1970). A microsporidian infection of *Anopheles gambiae* Giles, from Tanzania, interpretation of its mode of transmission and notes on *Nosema* infections in mosquitoes. *Journal of Protozoology* **17**, 531–539.

CANNING, E. U. AND SINDEN, R. E. (1973). Ultrastructural observations on the development of *Nosema algerae* Vávra and Undeen (Microsporida, Nosematidae) in the mosquito, *Anopheles stephensi* Liston. *Protistologica* **9**, 405–415.

CHAMP, B. R. AND CAMPBELL-BROWN, M. J. (1970). Insecticide resistance in Australian *Tribolium castaneum* (Herbst) (Coleoptera, Tenebrionidae). II. Malathion resistance in eastern Australia. *Journal of Stored Products Research* **6**, 111–131.

CICHY, D. (1971). The role of some ecological factors in the development of pesticide resistance in *Sitophilus oryzae* L. and *Tribolium castaneum* Herbst. *Ekologia polska* **19**, 563–616.

DAVIDSON, E. W. AND SWEENEY, A. W. (1983). Microbial control of vectors: a decade of progress. *Journal of Medical Entomology* **20**, 235–247.

DELPHY, J. (1936). Sous-règne des Protozoaires. In *La Faune de la France et Tableau Synoptiques Illustrés* (R. Perrier, Ed.), Volume 1A, p. 95. Delagrave, Paris.

DISSANAIKE, A. S. (1955). On Protozoa hyperparasitic in helminths, with some observations on *Nosema helminthorum* Moniez, 1887. *Journal of Helminthology* **31**, 47–64.

DISSANAIKE, A. S. (1957). The morphology and life-cycle of *Nosema helminthorum* Mioniez. *Parasitology* **47**, 335–347.

DISSANAIKE, A. S. AND CANNING, E. U. (1957). The mode of emergence of the sporoplasm in Microsporidia and its relation to the structure of the spore. *Parasitology* **47**, 92–99.

DOLFEIN, F. AND REICHENOW, E. (1952). *Lehrbuch der Protozoen Kunde*. Gustar Fischer, Jena.

DYTE, C. E. (1970). Insecticide resistance in stored-product insects with special reference to *Tribolium castaneum*. *Tropical Stored Products Information* **20**, 13–18.

DYTE, C. E. (1972). Laboratory evaluation of susceptible and malathion-resistant strains of *Tribolium castaneum* (Herbst) (Coleoptera, Tenebrionidae). *Journal of Stored Products Research* **8**, 103–109.

FISHER, F. M., JR. AND SANBORN, R. C. (1962a). Production of insect juvenile hormone by a microsporidian parasite, *Nosema*. *Nature* **194**, 1193.

FISHER, F. M., JR. AND SANBORN, R. C. (1962b). Observations on the susceptibility of some insects to *Nosema* (Microsporidia: Sporozoa). *Journal of Parasitology* **48**, 926–932.

FISHER, F. M., JR. AND SANBORN, R. C. (1964). *Nosema* as a source of juvenile hormone in parasitized insects. *Biological Bulletin* **126**, 235–252.

FLEXNER, J. L., LIGHTHART, B. AND CROFT, B. A. (1986). The effects of microbial pesticides on non-target, beneficial arthropods. *Agriculture, Ecosystems and Environment* **16**, 203–254.

FUXA, J. R. (1987a). Ecological considerations for the use of entomopathogens in IPM. *Annual Review of Entomology* **32**, 225–251.

FUXA, J. R. (1987b). Ecological methods. In *Epizootiology of Insect Diseases* (J. R. Fuxa and Y. Tanada, Eds), pp. 23–42. John Wiley and Sons, New York.

GASSOUMA, M. S. S. AND ELLIS, D. S. (1973). The ultrastructure of sporogonic stages and spores of *Thelohania* and *Plistophora* (Microsporida, Nosematidae) from *Simulium ornatum* larvae. *Journal of General Microbiology* **74**, 33–43.

GEORGE, C. R. (1971). Interaction between malnutrition and infection in *Nosema whitei*-infected *Tribolium castaneum*. *Journal of Invertebrate Pathology* **18**, 383–388.

GEORGE, C. R. (1972). Irreversible malnutrition in *Tribolium castaneum* attributable to parasitization by *Nosema whitei*. *Journal of Stored Products Research* **8**, 227–228.

GEORGE, C. R. AND TOWNES, J. (1976). The effects of *Nosema whitei* on moulting in *Tribolium castaneum*. *Journal of Stored Products Research* **12**, 199–200.

GEORGHIOU, G. P. AND MELLON, R. B. (1983). Pesticide resistance in time and space. In *Pest Resistance to Pesticides* (G. P. Georghiou and T. Saito, Eds), pp. 1–46. Plenum Press, New York.

GIBBS, A. J. (1953). *Gurleya* sp. (Microsporidia) found in the gut tissue of *Trachea secalis* (Lepidoptera). *Parasitology* **43**, 143–147.

GIBBS, A. J. (1956). *Perezia* sp. (Fam. Nosematidae) parasitic in the fat-body of *Gonocephalum arenarium* (Coleoptera: Tenebrionidae). *Parasitology* **46**, 48–53.

GOETZ, G., EBERHARDT, F. AND ZEUTSCHEL, B. (1959). *Versuche zur Selbstheilung und Therapie der Nosematose der Honigbiene*. Institut für Bienenkunde der Universität, Bonn.

GOLEMANSKI, V. G. AND DUKHLINSKA, D. D. (1982). Unicellular parasites of insect pests in Bulgaria. I. Composition and distribution of Sporozoa and Microsporidia in storage pests in grain warehouses. *Acta Zoologica Bulgarica* **20**, 26–37.

GRAY, F. H., CALI, A. AND BRIGGS, J. D. (1969). Intracellular stages in the life cycle of the microsporidian *Nosema apis*. *Journal of Invertebrate Pathology* **14**, 391–394.

GREENING, H. G. (1970). Malathion resistance in the rust-red flour beetle *Tribolium castaneum* (Herbst) (Coleoptera: Tenebrionidae). *Journal of the Australian Entomological Society* **9**, 160–162.

HAZARD, E. I. (1985). Microsporidia (Microspora) (Protozoa). *Bulletin, American Mosquito Control Association, No. 6*, pp. 51–55.

HENRY, J. E. (1981). Natural and applied control of insects by protozoa. *Annual Review of Entomology* **26**, 49–73.

HENRY, J. E. (1985). Effect of grasshopper species, cage density, light intensity, and method of inoculation on mass production of *Nosema locustae* (Microsporida: Nosematidae). *Journal of Economic Entomology* **78**, 1245–1250.

HENRY, J. E. AND OMA, E. A. (1981). Pest control by *Nosema locustae*, a pathogen of grasshoppers and crickets. In *Microbial Control of Pests and Plant Diseases 1070–1980* (E. H. D. Burges, Ed.), pp. 573–586. Academic Press, New York.

HIGBY, G. C., CANNING, E. U., PILLEY, B. M. AND BUSH, P. J. (1979). Propagation of *Nosema eurytremae* (Microsporida: Nosematidae) from trematode larvae, in abnormal hosts and in tissue culture. *Parasitology* **78**, 155–170.

HINDMARSH, P. S. (1976). Reduction of post-harvest losses to durable produce in Zambia. *Tropical Stored Products Information* **31**, 13–15.

HOPPE, T. (1981). Pests of stored products in Switzerland: abundance and insecticide resistance. *Mitteilungen der Schweizerischen entomologischen Gesellschaft* **54**, 3–13.

HOSTOUNSKÝ, Z. AND WEISER, J. (1972). Production of spores of *Nosema plodiae* Kellen et Lindergren in *Mamestra brassicae* after different infective dosage. I. *Věstnik Československé zoologické Spolecñosti* **36**, 97–100.

HUGER, A. (1960). Electron microscope study on the cytology of a microsporidian spore by means of ultrathin sections. *Journal of Insect Pathology* **2**, 84–105.

IGNOFFO, C. M. (1985). Manipulating enzootic and epizootic diseases of arthropods. In *Biological Control in Agricultural IPM Systems* (M. A. Hoy and D. C. Herzog, Eds), pp. 243–262. Academic Press, New York.

IGNOFFO, C. M. AND HINK, W. F. (1971). Propagation of arthropod pathogens in living systems. In *Microbial Control of Insects and Mites* (H. D. Burges and N. W. Hussey, Eds), pp. 541–580. Academic Press, New York.

ISHIHARA, R. (1968). Some observations on the fine structure of the sporoplasm discharged from spores of a microsporidian *Nosema bombycis*. *Journal of Invertebrate Pathology* **12**, 245–258.

ISHIHARA, R. (1969). The life cycle of *Nosema bombycis* revealed in tissue cell cultures of *Bombyx mori*. *Journal of Invertebrate Pathology* **14**, 316–320.

ISSI, I. V. AND CHERVINSKAJA, V. P. (1969). On the influence of temperature on the development of *Nosema mesnili* and *Plistophora schubergi* (Microsporidia, Nosematidae). *Zoologicheskiĭ Zhurnal* **48**, 1140–1146.

ISSI, I. V. AND ONATSKII, N. M. (1984). Inter-relationships between microsporidia and insects in early stages of disease. *Protozoologiya (Parazitokhozyainnye otnosheniya) No. 9*, pp. 102–113.

KEELEY, L. L. (1978). Endocrine regulation of fat body development and function. *Annual Review of Entomology* **23**, 329–353.

KELLEN, W. R., HOFFMAN, D. F. AND COLLIER, S. S. (1977). Studies on the biology and ultrastructure of *Nosema transitellae* sp. n. (Microsporidia: Nosematidae) in the navel orangeworm, *Paramyelois transitella* (Lepidoptera: Pyralidae). *Journal of Invertebrate Pathology* **29**, 289–296.

KEM, T. R. (1977). Selection of a strain of *Tribolium castaneum* (Herbst) resistant to phosphine. *Journal of Entomological Research* **1**, 213–217.

KHAN, A. R. (1981). *The Combined Action of Organophosphorus Insecticides and*

Microsporidians on Tribolium castaneum. Ph.D. Thesis, University of Newcastle Upon Tyne.

KHAN, A. R. AND SELMAN, B. J. (1984). Effect of insecticide, microsporidian, and insecticide-microsporidian doses on the growth of *Tribolium castaneum* larvae. *Journal of Invertebrate Pathology* **44**, 230–232.

KHANA, A. R. AND SELMAN, B. J. (1987). Effect of pirimiphos methyl, *Nosema whitei* and pirimiphos methyl–*N. whitei* doses on the growth of *Tribolium castaneum* adults. *Journal of Invertebrate Pathology* **49**, 336–338.

KHAN, A. R. AND SELMAN, B. J. (1988). On the mortality of *Tribolium castaneum* adults from larvae sub-lethally treated with pirimiphos methyl, *Nosema whitei* and pirimiphos methyl–*N. whitei* doses. *Entomophaga* **33**, 53–56.

KORKE, V. T. (1916). On a *Nosema (Nosema pulicis* n. sp.) parasitic in the dog flea (*Ctenocephalus felis*). *Indian Journal of Medical Research* **3**, 725–730.

KRAMER, J. P. (1959). Some relationships between *Perezia pyraustae* and *Pyrausta nubilalis. Journal of Insect Pathology* **1**, 25–33.

KRAMER, J. P. (1960). Observations on the emergence of microsporidian sporoplasm. *Journal of Insect Pathology* **2**, 433–439.

KRAMER, J. P. (1965). Generation time of *Octosporea muscaedomesticae* Flu in adult *Phormia regina* (Meigen) (Diptera: Calliphoridae). *Zeitschrift für Parasitekunde* **25**, 309–313.

KRAMER, J. P. (1968). An octosporeosis of the Black Blowfly, *Phormia regina*: Effect of temperature on the longevity of diseased adults. *Texas Report on Biology and Medicine* **26**, 199–204.

KRAMER, J. P. (1970). Longevity of microsporidian spores with special reference to *Octosporea muscaedomesticae* Flu. *Acta Protozoologica* **8**, 217–224.

KRAMER, J. P. (1976). The extra-corporeal ecology of Microsporidia. In *Comparative Pathobiology*, Vol. 1, *Biology of the Microsporida*, (L. A. Bulla, Jr. and T. C. Cheng, Eds), pp. 127–135. Plenum Press, New York.

KRIEG, A. (1955). Uber Infektionskrankheiten bei Engerlingen von *Melolontha* spec. Unter besonderer Berücksichtigung einer Mikrosporidien-Erkrankung. *Zentralblatt für Bakteriologie, Parasitenkunde, Infektionkrankheiten und Hygiene* **108**, 535–538.

KUDO, R. R. (1924). A biological and taxonomic study of Microsporidia. *Illinois Biological Monographs IX*.

KUDO, R. R. (1966). *Protozoology*. Charles C. Thomas, Springfield, Illinois.

KUDO, R. R. AND DANIELS, E. W. (1963). An electron microscope study of spores of *Thelohania californica. Journal of Protozoology* **10**, 112–120.

LABBÉ, A. (1899). Sporozoa. In *Das Tierreich* (O. Butschli, Ed.), pp. 1–180. Friedlander U. Sohn, Berlin.

LACHEVALIER, H. A. AND PRAMER, D. (1971). *The Microbes*. J. B. Lippincott Co., Philadelphia, PA.

LIPA, J. J. (1968). On two microsporidians; *Nosema whitei* Weiser from *Tribolium confusum* and *Nosema weiseri* sp.n. from *Rhizopertha dominica. Acta Protozoologica* **5**, 375–380.

LIPA, J. J. AND MARTIGNONI, M. E. (1960). *Nosema phryganidine* n.sp., a microsporidian parasite of *Phryganidae californica. Journal of Insect Pathology* **2**, 396–410.

LISTOV, M. V. (1976). Microsporidosis and coccidiosis protozoan diseases of *Tribolium* flour beetles (Coleoptera: Tenebrionidae). *Parazitologiya* **10**, 268–273.

LISTOV, M. V. (1977). The effects of pathogenic protozoa on hormone balance of darkling beetles (Coleoptera: Tenebrionidae). *Entomologicheskoe Obozrenie* **56**, 731–735.

LISTOV, M. V. (1979). The effect of parasitic protozoa on the physiological conditions of tenebrionid beetles. *Parazitologiya* **13**, 429–435.

LISTOV, M. V. AND NESTEROV, V. A. (1980). Resistance of larvae of flour beetles (Coleoptera, Tenebrionidae) to methyl bromide in relation to their infection with microsporidia of the species *Nosema whitei* Weiser, 1953 and coccids of the species *Adelina tribolii* Bhatia, 1937. *Entomologicheskoe Obozrenie* **59**, 725–729.

Lom, J. (1970). Comments on sporoblast development in Microsporidia. *Society of Invertebrate Pathology News sheet* **II** (6), 9–10.

Lom, J. (1972). On the structure of the extruded microsporidian filament. *Zeitschrift für Parasitenkunde* **38**, 200–213.

Lom, J. and Vávra, J. J. (1961). Ultrastructure of the spores of the fish parasite, *Plistophora hyphessobryconis*. *Wiadomósci parazytologiczne* **7**, 828–832.

Lom, J. and Vávra, J. J. (1963). The fine morphology of the spore of microsporidia. *Acta Protozoologica* **1**, 279–284.

Lom, J. and Weiser, J. (1963). Notes on two microsporidian species from *Silurus glanis*, and on the systematic status of the genus *Glugea* Thelohan. *Folia Parasitologica, (Praha)* **16**, 193–200.

MacBain Cameron, J. W. (1963). Factors affecting the use of microbial pathogens in insect control. *Annual Review of Entomology* **8**, 265–286.

McLaughlin, R. E. (1971). Use of protozoans for microbial control of insects. In *Microbial Control of Insects and Mites* (H. D. Burgess and N. W. Hussey, Eds), pp. 151–172. Academic Press, New York.

Maddox, J. V. (1968). Generation time of the microsporidian *Nosema necatrix* in the larvae of the armyworm, *Pseudaletia unipuncta*. *Journal of Invertebrate Pathology* **11**, 90–96.

Maddox, J. V. (1973). The persistence of microsporidia in the environment. *Miscellaneous Publications of the Entomological Society of America* **9**, 99–104.

Maddox, J. V. (1977). Stability of entomopathogenic protozoa. *Miscellaneous Publications of the Entomological Society of America* **10**, 3–18.

Maddox, J. V. (1986). Current status on the use of Microsporidia as biocontrol agents. In *Fundamental and Applied Aspects of Invertebrate Pathology: The Proceedings of the 4th International Colloquium on Insect Pathology* (R. A. Samson, J. N. Vlak and D. Peters, Eds), pp. 518–521.

May, R. M. (1983). Parasitic infections as regulators of animal populations. *American Scientist* **71**, 36–45.

Metcalf, R. L. (1980). Changing role of insecticides in crop protection. *Annual Review of Entomology* **25**, 219–256.

Milner, R. J. (1970). *The Morphology and Pathogenicity of* Nosema whitei *Weiser, a microsporidian pathogen of* Tribolium castaneum *Herbst*. Ph.D. Thesis, University of Newcastle upon Tyne.

Milner, R. J. (1972a). *Nosema whitei*, a microsporidian pathogen of some species of *Tribolium*. I. Morphology, life cycle, and generation time. *Journal of Invertebrate Pathology* **19**, 231–238.

Milner, R. J. (1972b). *Nosema whitei*, a microsporidian pathogen of some species of *Tribolium*. II. Ultrastructure. *Journal of Invertebrate Pathology* **19**, 239–247.

Milner, R. J. (1972c). *Nosema whitei*, a microsporidian pathogen of some species of *Tribolium*. III. Effect on *T. castaneum*. *Journal of Invertebrate Pathology* **19**, 248–255.

Milner, R. J. (1972d). The survival of *Nosema whitei* spores stored at 4°C. *Journal of Invertebrate Pathology* **19**, 256–257.

Milner, R. J. (1973a). *Nosema whitei*, a microsporidian pathogen of some species of *Tribolium*. IV. The effect of temperature, humidity, and larval age on pathogenicity for *T. castaneum*. *Entomophaga* **18**, 305–315.

Milner, R. J. (1973b). *Nosema whitei*, a microsporidian pathogen of some species of *Tribolium*. V. Comparative pathogenicity and host range. *Entomophaga* **18**, 383–390.

Miquel, J. (1971). Aging in male *Drosophila melanogaster* historical, histochemical, and ultrastructural observations. *Advances of Gerontological Research* **3**, 39–71.

MORGENTHALER, O. (1922). The polar filament of *Nosema apis* Zander. *Bee World* **4**, 25.

NAEGELI, K. W. (1857). Über die neue Krankheit der Seidernraupe und verwandte Organismen. *Botanische Zeitung* **40**, 108.

NAKITA, H. AND WINKS, R. G. (1981). Phosphine resistance in immature stages of a laboratory selected strain of *Tribolium castaneum* (Herbst.) (Coleoptera: Tenebrionidae). *Journal of Stored Products Research* **17**, 43–52.

NAVARRO, S., CARMI, Y., KASHANCHI, Y. AND SHAAYA, E. (1986). Malathion resistance of stored products insects in Israel. *Phytoparasitica* **14**, 273–280.

NEWTON, P. B., ARMSTRONG, E. AND REICHELDERFER, C. F. (1983). Some biochemical effects of the microsporidian *Nosema whitei* on the blood components of *Tribolium castaneum* using electrophoresis. *Comparative Biochemistry and Physiology* B **74**, 553–558.

OSHIMA, K. (1937). On the function of the polar filament of *Nosema bombycis*. *Parasitology* **29**, 220–224.

OSHIMA, K. (1964). Effect of potassium ion on filament evagination of the spores of *Nosema bombycis* as studied by the neutralization method. *Annotationes Zoologicae Japonenses* **37**, 102–103.

OSHIMA, K. (1973). Changes of relation between infectivity and filament evagination of debilitated spores of *Nosema bombycis*. *Annotationes Zoologicae Japonenses* **46**, 188–198.

PILLAI, J. S. (1985). Prospects for biological control of disease vectors. *Tropical Biomedicine* **2**, 177–184.

PILLEY, B. M., CANNING, E. U. AND HAMMOND, J. C. (1978). The use of a microinjection procedure for large scale production of the microsporidian, *Nosema eurytremae* in *Pieris brassicae*. *Journal of Invertebrate Pathology* **32**, 355–358.

PURRINI, K. (1983). Über zwei Protozoen-Arten, *Adelina tribolii* Bhatia comb. nov. (Coccidia) und *Nosema ptinidorum* n. sp. (Microsporidia) als Krankheitserreger der vorratsschädlichen Käfer *Ptinus pusillus* Strm., und *P. brunneus* Dft. (Col., Ptinidae). *Zeitschrift für angewandte Entomologie* **95**, 477–482.

RABINDRA, R. J., BALASUBRAMANIAN, M. AND JAYARAJ, S. (1981). The effects of *Farinocystis tribolii* on the growth and development of the flour beetle, *Tribolium castaneum*. *Journal of Invertebrate Pathology* **38**, 345–351.

RASSMANN, W. (1978). Investigations on resistance to malathion and lindane in beetle species infesting stored products in the Federal Republic of Germany. *Anzeiger Schädlingskunde* **51**, 17–20.

RAUN, E. S., YORK, G. T. AND BROOKS, D. L. (1960). Determination of *Perezia pyraustae* infection rates in larvae of the European corn borer. *Journal of Insect Pathology* **2**, 254–258.

SANTHOY, O. AND MORALLO-REJESUS, B. (1972). Toxicity of six organophosphorus insecticides to field-collected DDT-resistant strains of the rice weevil, *Sitophilus oryzae* (L.) and red flour beetle, *Tribolium castaneum* (Herbst). *Philippine Entomologist* **2**, 283–290.

SAPAS, T. J., BURKHOLDER, W. E. AND BOUSH, G. M. (1977). Population suppression of *Trogoderma glabrum* by using pheromone luring for protozoan pathogen dissemination. *Journal of Economic Entomology* **70**, 469–474.

SHERLOCK, P. L. (1969). *A Study of the life cycle and pathogenicity of* Farinocystis tribolii *Weiser*. Honours Thesis, University of Newcastle Upon Tyne.

SINDEN, R. E. AND CANNING, E. U. (1974). The ultra-structure of the spore of *Nosema algerae* (Protozoa, Microsporida) in relation to the hatching mechanisms of microsporidian spores. *Journal of General Microbiology* **85**, 350–357.

SMITH, C. E. G. (1971). The spread and maintenance of infections in vertebrates and arthropods. *Journal of Invertebrate Pathology* **18**, i–xi.

SOLOMON, B. (1985). Gene find may overcome resistance. *Agricultural Research*

Services of U.S. Department of Agriculture **33**, 4.

SPEIRS, R. D., REDLINGER, L. M. AND JONES, R. (1971). DDT-resistant flour beetles from a Georgia peanut seller. *Journal of Economic Entomology* **64**, 1328–1329.

SPRAGUE, V. (1977). Annotated list of species of Microsporidia. In *Comparative Pathobiology*, Volume 2, *Systematics of the Microsporidia* (L. A. Bulla, Jr. and T. C. Cheng, Eds), pp. 31–334. Plenum, New York.

SPRAGUE, V. AND VERNICK, S. H. (1968). Light and electron microscope study of a new species of *Glugea* (Microsporida: Nosematidae) in the 4-spined stickleback *Apeltes quadracus. Journal of Protozoology* **15**, 547–571.

STEINHAUS, E. A. (1949). *Principles of Insect Pathology*. McGraw-Hill, New York.

STEINHAUS, E. A. (1954). The effects of disease on insect populations. *Hilgardia* **23**, 197–261.

STEINHAUS, E. A. (1956). Potentialities for microbial control of insects. *Agriculture and Food Chemistry* **4**, 676–680.

STEINHAUS, E. A. (1958). Stress as a factor in insect disease. *Proceedings of the 10th International Congress of Entomology, Montreal, 1956* **4**, 725–730.

TANADA, Y. AND FUXA, J. R. (1987). The pathogen population. In *Epizootiology of Insect Diseases* (J. R. Fuxa and Y. Tanada, Eds), pp. 113–158. John Wiley and Sons, New York.

TRAGER, W. (1937). The hatching of spores of *Nosema bombycis*, Nageli and the partial development of the organism in tissue cultures. *Journal of Parasitology* **23**, 226–227.

TUZET, O., MAURAND, J., FIZE, A., MICHEL, R. AND FENWICK, B. (1971). Proposition d'un nouveau cadre systematique pour les genres de microsporidies. *Comptes Rendus hebdomadaire des Séances de l'Académie des Sciences, Paris* **272**, 1268–1271.

UNDEEN, A. H. (1978). Spore-hatching process in some *Nosema* species with particular reference to *N. algerae* Vávra and Undeen. *Miscellaneous Publications of the Entomological Society of America* **11**, 29–49.

UNDEEN, A. H. (1982). The production and use of Protozoa for vector control. In *Invertebrate Pathology and Microbial Control: Proceedings of IIIrd International Colloquium on Invertebrate Pathology, University of Sussex, Brighton, 1982*, pp. 382–386.

UNDEEN, A. H. AND MADDOX, J. V. (1973). The infection of non-mosquito hosts by injection with spores of the microsporidian *Nosema algerae. Journal of Invertebrate Pathology* **22**, 258–265.

VAGO, C. (1959). L'enchaînement des maladies chez les insectes. *Annales des epiphyties, 1959*.

VÁVRA, J. (1965). Étude au microscope electronique de la morphologie et du développement de quelques microsporidies. *Comptes Rendus hebdomadaire des Séances de l'Académie des Sciences, Paris* **261**, 3467–3470.

VÁVRA, J. AND UNDEEN, A. H. (1970). *Nosema algerae* n.sp. (Cnidospora, Microsporida) a pathogen in a laboratory colony of *Anopheles stephensi* Liston. *Journal of Protozoology* **17**, 240–249.

VIKTOROV, G. A. (1971). Some general principles of insect population density regulation. *Proceedings of the 13th International Congress of Entomology, Moscow, 1968* **1**, 573–574. Nauka, Leningrad.

VINCENT, L. E. AND LINDGREN, D. L. (1967). Susceptibility of laboratory and field-collected cultures of the confused beetle and red flour beetle to malathion and pyrethrins. *Journal of Economic Entomology* **60**, 1763–1764.

WARUI, C. M. (1976). Insecticide resistance in field strains of *Tribolium castaneum* collected in Mombasa, Kenya. *Kenya Entomologist's Newsletter* **3**, 10–16.

WATANABE, H. (1987). The host population. In *Epizootiology of Insect Diseases* (J. R. Fuxa and Y. Tanada, Eds), pp. 71–112. John Wiley and Sons, New York.

WEIDNER, E. (1972). Ultrastructural studies on microsporidian invasion into cells. *Zeitschrift für Parasitenkunde* **40**, 227–242.

WEIDNER, E. (1982). The microsporidian spore invasion tube. III. Tube extrusion and assembly. *Journal of Cell Biology* **93**, 976–979.

WEIDNER, E. AND BYRD, W. (1982). The microsporidian spore invasion tube. II. The role of calcium in the activation of invasion tube discharge. *Journal of Cell Biology* **93**, 972–975.

WEIDNER, E. AND TRAGER, W. (1973). Adenosine triphosphate in the extracellular survival of an intracellular parasite (*Nosema michaelis* Microsporidia). *Journal of Cell Biology* **57**, 586–591.

WEIDNER, E., BYRD, W., SCARBOROUGH, A., PLESHINGER, J. AND SIBLEY, D. (1984). Microsporidian spore discharge and the transfer of polaroplast organelle membrane into plasma membrane. *Journal of Protozoology* **31**, 195–198.

WEISER, J. (1959). *Nosema laphygmae* n.sp. and the internal structure of the microsporidian spores. *Journal of Insect Pathology* **1**, 52–59.

WEISER, J. (1961). *Die Mikrosporidien als Parasiten der Insekten. Monographie Angewandte Entomologie, 17.* Paul Parey, Hamburg.

WEISER, J. (1963). Sporozoan infections. In *Insect Pathology an Advanced Treatise* (E. A. Steinhaus, Ed.), Volume 2, pp. 291–334. Academic Press, New York.

WEISER, J. (1976). Microsporidia in invertebrates: Host–parasite relations at organismal level. In *Comparative Pathobiology*, Volume 1, *Biology of the Microsporidia* (L. A. Bulla, Jr. and T. C. Cheng, Eds), pp. 163–201. Plenum Press, New York.

WEISER, J. (1977). Contribution to the classification of Microsporidia. *Věstnik Československé zoologické spolecňosti* **41**, 308–320.

WEISER, J. AND LYSENKO, O. (1956). Septikemie bource mouruûového. *Cslka Mikrobiol* **1**, 216–222.

WEST, A. F., JR. (1960). The biology of a species of *Nosema* (Sporozoa: Microsporidia) parasitic in the flour beetle, *Tribolium confusum. Journal of Parasitology* **46**, 747–752.

WILSON, G. G. (1976). A method for mass producing spores of the microsporidian *Nosema fumiferanae* in its host, the spruce budworm, *Choristoneura fumiferana* (Lepidoptera: Tortricidae). *Canadian Entomologist* **108**, 383–386.

WYATT, G. R. (1980). The fat body as a protein factory. In *Insect Biology in the Future* (M. Locke and D. S. Smith, eds), pp. 201–225. Academic Press, New York.

ZAKHARCHENO, N. L. (1959). The effect of introduction of muscavidous fungi and hexachlorane into the soil on the physiological state of beet weevils. *Proceedings of the Ukrainian Academy Agricultural Science Research Institute, Preservation of Plants, 8th meeting*, pp. 57–62.

6
Viruses as Pest-Control Agents

J. B. CARTER

Department of Biology, Liverpool Polytechnic, Byrom Street,
Liverpool L3 3AF, UK

Introduction

Pressures from a variety of sources are causing man to investigate alternatives
to the chemical pesticides which have been used so widely during the past few
decades. Pressures are brought to bear by environmentalists concerned about
the effects of pesticides on wildlife, by pest-control experts concerned about the
effects of these pesticides on parasites and predators of the pests and about the
increasing resistance of the pests to the pesticides, by consumers concerned
about toxic residues in food, and by public health officials concerned about
human poisoning. In Sri Lanka, for example, more people die of pesticide
poisoning than of malaria (Matthews, 1983). Furthermore, research, development and production costs for chemical pesticides have soared, making them
expensive in the developed nations, while in the developing nations, if pesticides
are used at all, farmers select the least expensive—which are usually the most
toxic. 'Biological' control strategies, including the use of pathogens of pests,
attempt to circumvent most of these problems.

Viruses have been used to control mites (Reed, 1981) and rabbits (Fenner,
1983) but this review is concerned principally with the insect-pathogenic
baculoviruses (BVs). Insects are hosts to a wide variety of viruses, including
picornaviruses, parvoviruses and poxviruses. Each of these groups also has
representatives infecting vertebrate animals. Attention has been focused on the
BVs as pesticidal agents because of their lack of similarity to any viruses of
hosts other than invertebrates. There is some logic in this approach, but other
groups of insect-pathogenic viruses should not be ignored. It may be that the
host spectra of other virus groups, such as the cytoplasmic polyhedrosis viruses
and the iridescent viruses, are restricted to invertebrates even though viruses
with similar morphologies and biochemical characteristics infect vertebrates
(and higher plants in the case of iridescent viruses). In fact, a cytoplasmic
polyhedrosis virus is used in Japan against the pine caterpillar, *Dendrolimus*

Abbreviations: BV, baculovirus; DNA, deoxyribonucleic acid; ELISA, enzyme-linked immunosorbent assay; FP,
few polyhedra; GV, granulosis virus; IB, inclusion body; IPM, integrated pest management; LD$_{50}$, median lethal
dose; LT$_{50}$, median lethal time; MNPV, multiple nucleocapsids per virion envelope; MP, many polyhedra; NPV,
nuclear polyhedrosis virus; REN, restriction endonuclease; RNA, ribonucleic acid; SDS-PAGE, sodium dodecyl
sulphate–polyacrylamide gel electrophoresis; SNPV, single nucleocapsid per virion envelope; UV, ultra-violet.

spectabilis (Aizawa, 1976), and an iridescent virus has been tested against leather-jackets (*Tipula* spp. larvae), although with disappointing results (Carter, 1978).

Insect viruses are in use, or are being considered for use, in forestry, horticulture and agriculture, including grassland. Entwistle (1983) listed 31 lepidopteran, 6 hymenopteran and one coleopteran pest species for which control with BVs has been demonstrated to be feasible or highly likely. There is little or no current effort to apply viruses for control of disease vectors or of stored product and timber pests.

If a virus is to be considered seriously as a pest-control agent then detailed knowledge of the virus, its host, and their interactions with the environment must be amassed. Information is required on the structural and biochemical characteristics of the virus, its host spectrum, and median lethal doses (LD_{50}s) and median lethal times (LT_{50}s) for different stages of the host(s). The habits and life cycle of the host, and the mechanisms whereby the virus persists and spreads in the field (epizootiology) must be understood. Techniques for mass production and purification of the virus must be developed and it must be shown to be safe for man and other non-target organisms. This chapter considers all these aspects and discusses the advantages and disadvantages of using insect-pathogenic viruses as pesticides. Industrial aspects are discussed only briefly as they will be considered in detail in a subsequent volume. Recent reviews on the use of viruses as pest-control agents include those of Falcon (1982), Payne (1982) and Entwistle (1983).

The baculoviruses

Only an outline of the structure and replication of BVs is given here. For more detailed accounts the reviews of Harrap and Payne (1979), Granados (1980a) and Kelly (1982) should be consulted.

The BVs have been classified into three subgroups according to whether or not the virions become embedded (occluded) in inclusion bodies (IBs), and, if so, on the size and shape of the IB (Matthews, R.E.F., 1982). Details of the subgroups are presented in *Figure 1* and *Table 1*. The rod-shaped virions are enveloped. Those of the granulosis viruses (GVs) become occluded in capsule-shaped IBs (granules), whereas the nuclear polyhedrosis virus (NPV) IBs (polyhedra) are polyhedral, cuboidal, or 'orange segment-shaped', depending on the virus. The NPVs are subdivided into those in which each occluded virion has a single nucleocapsid per virion envelope (SNPVs) and those in which the occluded virions have multiple nucleocapsids per virion envelope (MNPVs).

STRUCTURE

The virion

BV virions have been found to contain up to 35 proteins (Vlak, 1979), some of which are glycosylated (Dobos and Cochran, 1980) and/or phosphorylated (Tweeten, Bulla and Consigli, 1980). The nucleocapsid consists of a protein capsid containing DNA and further proteins (*Figure 1*). Its dimensions fall

Table 1. Baculovirus subgroups.

Subgroup	Inclusion body dimensions	Number of virions per inclusion body	Single or multiple nucleocapsids per virion envelope	Hosts
Nuclear polyhedrosis viruses	0·8–15 μm diameter	Many		
1. MNPVs			Multiple (1–5 usually; up to 39)	Lepidoptera
2. SNPVs			Single	Lepidoptera, Diptera, Hymenoptera, Trichoptera, Coleoptera, Neuroptera, Crustacea
Granulosis viruses	approx. 200 × 500 nm	1, usually	Single	Lepidoptera
Non-occluded baculoviruses	No inclusion bodies formed		Single	Coleoptera, Diptera (possibly) Mites, Crustacea

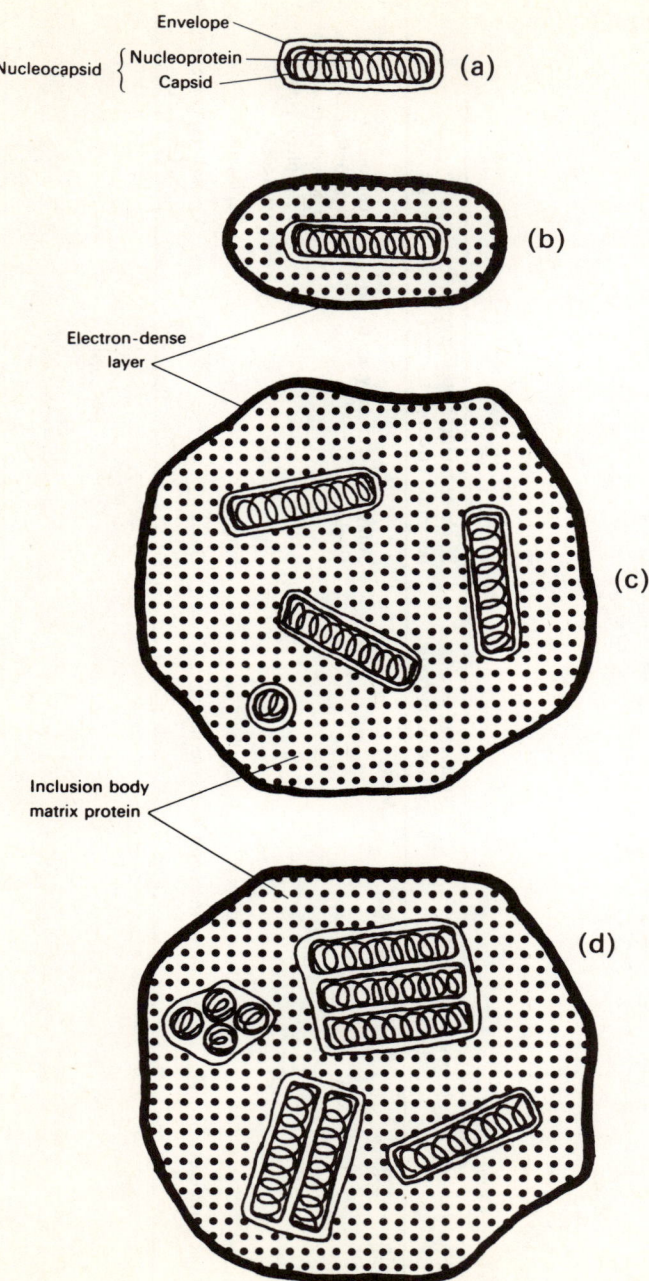

Figure 1. Baculovirus subgroups. (a) Non-occluded baculovirus virion. (b) Granulosis virus inclusion body. (c) Nuclear polyhedrosis virus inclusion body; singly enveloped nucleocapsids (SNPV). (d) Nuclear polyhedrosis virus inclusion body; multiple nucleocapsids per virion envelope (MNPV).

within the ranges 40–140 nm × 250–400 nm. The capsid is constructed from helically arranged subunits (Burley *et al.*, 1982), and structures described as claws and nipples (Kawanishi and Paschke, 1970) or caps (Federici, 1980) are present at its ends.

The double-stranded DNA molecule is a closed circle and is supercoiled. Most molecular weight estimates for BV DNAs fall between 70×10^6 and 120×10^6. Associated with the DNA is a highly basic protein (Tweeten, Bulla and Consigli, 1981) which may play a part in its condensation (Burley *et al.*, 1982).

Many of the larger virion proteins are associated with the lipid-containing membrane which forms the virion envelope. The virions of occluded BVs occur in two forms, each with a distinct envelope: the form which becomes occluded derives its envelope from membrane synthesized within the nucleus, while another form acquires its envelope by budding from the plasma membrane. The latter form normally has only one nucleocapsid per envelope, even if the virus is a MNPV, although Longworth and Singh (1980) observed that a few budded virions of *Epiphyas postvittana* MNPV had two nucleocapsids per virion. The occluded virions are specialized for infection of the host midgut cells, while the budded virions spread the infection to other cells and can readily infect susceptible cell cultures.

Some of the envelope proteins of occluded and budded forms of the same virus are distinct, while others are related (Volkman, 1983). At one end of a budded virion the envelope bears a number of spikes (Summers and Volkman, 1976), which are probably glycoproteins.

The polyamines spermidine and putrescine have been found in some NPVs. In *Heliothis zea* SNPV all of the spermidine and most of the putrescine was shown to be associated with the virion envelope (Elliott and Kelly, 1979).

The inclusion body

IBs are formed by the cytoplasmic polyhedrosis viruses and poxviruses of insects as well as by the occluded BVs. They afford protection to the virions outside the host, often for considerable periods between generations of larvae. Retention of occluded virus infectivity is far superior to that of viruses which do not form IBs, and is a further reason why interest has centred on the occluded viruses as microbial control agents.

The IB matrix is a paracrystalline lattice of protein subunits laid down to form an extremely stable structure which survives putrefaction of the dead host, but is broken down at low and high pH values. Reducing agents enhance the rate of IB dissolution in alkali (Croizier and Meynadier, 1972) and are essential for the dissolution of IBs of poxviruses and of the SNPV of *Tipula paludosa* at pH 10·5 (Bergoin, Guelpa and Meynadier, 1975).

The protein subunits are constructed from monomers, some of which have been reported to be glycosylated and phosphorylated (Kelly, 1981a). Proposals as to how the subunits are formed from the monomers include ionic, hydrophobic and disulphide bonding (Eppstein and Thoma, 1977).

There appears to be a high degree of similarity between the IB monomer proteins of different BVs. Their molecular weights all fall within the range

25 000–33 000. Several serological investigations with antisera have demonstrated relationships within and between the IB proteins of GVs, SNPVs and MNPVs, and these have been confirmed recently using monoclonal antibodies (Roberts and Naser, 1982a; Hohmann and Faulkner, 1983). The latter authors found stronger reactions within BV subgroups than between subgroups. The amino acid sequences of a few IB proteins have been determined and confirm that there is a high degree of similarity between them, especially between lepidopteran NPVs (Rohrmann et al., 1981; Rohrmann, 1982).

Around each IB is a layer of material which appears electron-dense when sections are viewed in the electron microscope. It appears to be more resistant than the IB matrix to alkaline dissolution (Kawanishi, Egawa and Summers, 1972; Green, 1981) and may be composed of carbohydrate (Minion, Coons and Broome, 1979).

IBs from infected insects contain alkaline protease activity. This is displayed when IBs dissolve in alkali (Yamafuji, Yoshihara and Hirayama, 1957) and enhances their rate of dissolution (Summers and Smith, 1975). No such enzyme activity has been detected in IBs from infected cell cultures.

INFECTION OF THE HOST

Most infections are initiated by the ingestion of infective virus. The virions of occluded viruses are released by dissolution of the IBs in the alkali of the midgut. Gut enzymes (Faust and Adams, 1966) and the IB protease may also have roles. Granados and Lawler (1981) found that few *Autographa californica* MNPV IBs remained intact after 15 minutes in the larval midgut (pH 10·4) of the cabbage looper, *Trichoplusia ni*.

The virions must survive the harsh conditions of the midgut while they traverse the peritrophic membrane and attach to the microvilli of midgut cells. The virion envelope fuses with the microvillus membrane, releasing the nucleocapsid(s) into the cell (Granados, 1978).

The SNPVs of the most Diptera and Hymenoptera replicate only in the midgut cells, and IBs are shed into the gut lumen by lysis of infected cells. In the Lepidoptera, however, infection of the midgut is only the preliminary to infection of other tissues. Enveloped nucleocapsids develop and IB protein polymerization may occur in midgut cells, but virions are rarely occluded. Instead they bud into the haemocoel (Harrap, 1970) and are carried in the haemolymph to other susceptible tissues. It has also been suggested (Granados and Lawler, 1981) that some inoculum nucleocapsids may pass straight through the gut cells and bud into the haemocoel.

REPLICATION

BVs replicate in the nucleus, within which the DNA is released from NPV nucleocapsids, while GV nucleocapsids release their DNA into the nucleus via nuclear pores (Granados, 1980a). The infected nucleus hypertrophies and becomes the dominant feature of the cell. A 'virogenic stroma' is formed and nucleocapsids develop at its periphery. Nucleocapsids produced early enter the cytoplasm either by budding through the nuclear envelope (Injac *et al.*, 1971) or via ruptures in it (Adams, Goodwin and Wilcox, 1977). They then leave the cell by budding through a portion of modified plasma membrane which becomes the virion envelope (Hunter, Hoffmann and Collier, 1975). These virions spread the infection to other cells. Later in occluded BV infections most nucleocapsids are retained in the nucleus where they acquire envelopes and become occluded. Occlusion of naked nucleocapsids has never been observed, which suggests that there is an IB protein receptor on the virion envelope.

IBs develop randomly throughout the nucleus, except those of two dipteran SNPVs which develop in intimate association with the inner nuclear membrane (Smith and Xeros, 1954; Stoltz, Pavan and Da Cunha, 1973). The number of IBs produced per cell may vary from a few to several hundred, depending in part on the IB size. The yield per insect depends on many factors, including species and instar. Evans, Lomer and Kelly (1981) found maximum yields of $2 \cdot 7 \times 10^7$ IBs per first-instar larva and $3 \cdot 4 \times 10^9$ IBs per fifth-instar larva for *Mamestra brassicae* MNPV. IBs have been reported to constitute up to 40% of the insect dry weight (Bucher and Turnock, 1983).

Some progress has been made recently in understanding the biochemical events involved in BV replication. Kelly and Lescott (1981) identified four phases of virus protein synthesis in *Spodoptera frugiperda* cell cultures infected with *T. ni* MNPV. The phases were induced in a cascade fashion, with synthesis of one phase blocked if the proteins of the previous phase were rendered non-functional. The early proteins include enzymes such as thymidine kinase (Kelly, 1981b), while the virus structural proteins appear later. Synthesis of the later proteins is probably dependent on virus DNA synthesis, which reaches a high rate.

Nearly all of the late messenger RNA is virus-specific, with approximately 25% of that in *A. californica* MNPV-infected cells specific for IB protein (Adang and Miller, 1982). The control of IB formation is undoubtedly complex; studies with *A. californica* MNPV mutants led Potter and Miller (1980) to suggest that about half of the genome might be involved. Another small protein (molecular weight 10000) is produced late and in large quantities in *A. californica* MNPV-infected cells. It is present in the virion as a minor component, but its function is not known (Smith, Vlak and Summers, 1983).

Insect-cell culture

A brief account of insect-cell culture techniques is relevant because of their value in studies of BV replication (page 380), genetics (page 385) and safety testing (page 388). Furthermore, there are hopes that viruses used as insecticides might be mass produced in cell cultures (page 400). Recent reviews of insect-cell culture include those of Stockdale and Priston (1981), Vaughn and Dougherty (1981) and Grace (1982).

Increasing numbers of insect cell lines and culture media are becoming available (Hink, 1976, 1980). Most of the cell lines are from lepidopteran and dipteran insects; lines from several insect orders, including the Hymenoptera, have not yet been developed.

MNPVs, SNPVs and non-occluded BVs have been replicated in cell cultures.

All attempts to replicate GVs in cell lines have failed so far, but Vago and Bergoin (1963) and Rubinstein, Lawler and Granados (1982) have reported GV replication in primary cell cultures. Replication was incomplete in the latter case. As one of the preferred sites of GV replication is the fat body it will be of interest to see if a GV will replicate in any of the cell lines derived from fat bodies which are now becoming available (Mitsuhashi, 1981).

Susceptible cell cultures are readily infected with budded virions, so the haemolymph of an infected insect or medium from an infected cell culture provides effective inoculum. Virions released from IBs have much lower infectivity for cell cultures.

One of the most widely used cell lines is one derived by Hink (1970) from the ovaries of *T. ni* adults and designated TN-368. It has been used for plaque assays of *A. californica* MNPV (Hink and Strauss, 1977) and *Galleria mellonella* MNPV (Fraser and Hink, 1982). A plaque assay of *H. zea* SNPV in an *H. zea* cell line was described by Yamada and Maramorosch (1981).

A virus which will produce plaques in cell culture can be cloned by picking from single plaques, as carried out by Lee and Miller (1978) for *A. californica* MNPV in a *Spodoptera frugiperda* cell line.

A cell culture, like the whole organism, can harbour inapparent virus infections (Granados, Nguyen and Cato, 1978; Plus, 1978; Heine, Kelly and Avery, 1980). Plus (1980) stressed the importance of initiating cell lines from insects reared from surface-sterilized eggs as a precaution against virus contamination.

Baculovirus characterization and identification

ANALYSIS OF PROTEINS BY SODIUM DODECYL SULPHATE–POLYACRYLAMIDE GEL ELECTROPHORESIS

The technique of sodium dodecyl sulphate–polyacrylamide gel electrophoresis (SDS–PAGE) permits the number of virus proteins to be determined and their molecular weights to be estimated. It provides useful information, but suffers from a number of limitations and should not be used as the sole technique in virus identification (Allaway and Payne, 1983).

SEROLOGY

Serological methods are used to compare different viruses, and to diagnose infection in insects, especially in epizootiological studies. They are also used in safety testing (page 388) where they provide a means of detecting virus or virus components in non-target organisms and of detecting anti-viral antibodies in vertebrates exposed to the virus. Apparently the IB protein, the virion envelope and the nucleocapsid of an occluded BV each bears distinct antigenic determinants.

Prominent among several techniques which have been used is immunodiffusion, which is useful for investigating antigenic relationships, although it lacks sensitivity. The sensitive technique of enzyme-linked immunosorbent assay (ELISA) is becoming widely used. McCarthy and Henchal (1983) used an antiserum against nucleocapsids in an ELISA to detect *A. californica* MNPV virions in larvae and in cell cultures. Brown, Allen and Bignell (1982), investigating the relationships between four MNPVs of *Spodoptera* spp., used an indirect ELISA with enzyme-labelled protein A of *Staphylococcus aureus* in place of enzyme-labelled anti-immunoglobulin.

Monoclonal antibodies are increasing the specificity of serological techniques. Roberts and Naser (1982b) developed hybridomas secreting monoclonal antibodies against the IB protein and against a major virion protein of *A. californica* MNPV. These antibodies were used in several serological methods, and have recently been used in a protein-blotting technique incorporating ELISA (Naser and Miltenburger, 1983). Hohmann and Faulkner (1983) reported the application of a similar technique to investigate BV relationships. Volkman and Falcon (1982) used a monoclonal antibody against the IB protein of *T. ni* SNPV in an ELISA to diagnose infection in larvae. They found that host tissue caused interference, but concluded that the test was sensitive enough to be useful.

RESTRICTION ENDONUCLEASE ANALYSIS OF DNA

For definitive characterization and unequivocal identification of a BV it is preferable to analyse the genome rather than phenotypic characters. One of the techniques that discriminates best between double-stranded DNA viruses is restriction endonuclease (REN) analysis of their nucleic acids. Smith and Summers (1979) could differentiate five *A. californica* MNPV isolates by this technique, whereas the SDS–PAGE protein profiles of the isolates were identical.

BIOASSAYS

Precise bioassay techniques yield important information about the virus–host relationship. This information is vital for selecting virus strains with high infectivity and for estimating suitable rates for field application. Many factors can affect the dose–response relationship and/or the LD_{50} of an insect virus and each of these must be standardized. Larval instar (page 396), larval weight and/or age within instar (Burgerjon *et al.*, 1981; Evans, 1983) diet composition, IB purification technique (Baugher and Yendol, 1981) and incubation temperature (Boucias, Johnson and Allen, 1980) should all be carefully controlled.

Techniques in which a larva consumes only a portion of virus-inoculated diet are less preferable to those in which the whole of the dose is ingested on a leaf disc (Evans, 1981), a small piece of diet (Nordin, 1976), or in a small drop (Klein, 1978; Hughes and Wood, 1981). Laing and Jaques (1980) described a bioassay technique for larvae of boring species such as the codling moth, *Cydia pomonella*.

Estimates of LT_{50}s may also be useful, especially for predicting how rapidly insects will be killed in the field.

HOST RANGE

BV host ranges have been widely investigated but many of the results require confirmation, as viruses which replicated in inoculated hosts were not always identified, often because suitable techniques were not available when these experiments were carried out. Some of the cases of virus replication could have been due to latent virus activation (*see below*) rather than to cross-transmission. In many studies only gross effects of infection (e.g. IB formation, host death) were looked for, although a virus might infect a host sublethally or only some virus functions might be expressed without IB formation. Furthermore, an insect resistant to infection by ingestion of IBs might be susceptible if injected with budded virions.

With these provisos in mind it can be stated tentatively that the MNPVs are the least host-specific of the occluded BVs. *A. californica* MNPV has the widest known host range, infection having been reported in more than 30 insect species and in cell cultures from at least 13 species. Replication of *H. zea* SNPV, on the other hand, appears to be restricted to members of the genus *Heliothis* (Ignoffo and Couch, 1981). Some GVs have been transmitted to other species, e.g. *C. pomonella* GV to five closely related species (Huber, 1982), and *Heliothis armigera* GV to four other species including *T. ni* and two *Spodoptera* species (Hamm, 1982).

When selecting a virus for possible use against more than one pest species it is important to determine the dose–mortality relationship (page 383) for each host. A virus is not likely to control an insect if the LD_{50} is extremely high, as for *Agrotis segetum* GV in *Agrotis exclamationis* larvae in which the LD_{50} for neonate larvae was found to be 1.2×10^6 IBs compared with 1.1×10^4 IBs for the homologous host (Allaway and Payne, 1984).

LATENCY

There have been many reports of insects harbouring 'latent' viruses, especially BVs, but no firm conclusion can be drawn from many of them. The best-substantiated reports concern the development of a homologous NPV in an insect fed with IBs of a heterologous NPV, with both viruses being characterized (Longworth and Cunningham, 1968; Maleki-Milani, 1978; Jurkovičová, 1979). McKinley *et al.* (1981) found that activation of a latent virus was more common than cross-infection after feeding four NPVs to heterologous hosts. Two aspects of their results are particularly interesting: first, there was a straight-line relationship between dose and mortality, i.e. there was no threshold dose of heterologous virus above which activation of homologous virus occurred; second, it appeared that each of the insects in their cultures carried a latent virus.

Because of the phenomenon of latency it is vital that all insect and cell-culture stocks used for virus studies are checked as closely as possible for the presence of latent viruses.

Baculovirus classification and nomenclature

The BV subgroups were described on page 376, and the reader will have gathered that an individual virus is identified by the name of the insect from which it was isolated, e.g. *Gilpinia hercyniae* SNPV, *Pieris rapae* GV. Some insects, e.g. *T. ni*

and the Douglas fir tussock moth, *Orgyia pseudotsugata*, are host to both a
SNPV and a MNPV; regrettably some authors do not specify which type of
virus they have worked with.

The system of naming a BV after an insect host is far from satisfactory because
many, if not most, of the BVs can infect several hosts. The wide host range of
A. californica MNPV has been discussed (page 384), and DNA REN analyses
indicate that this virus, *T. ni* and *G. mellonella* MNPVs (Smith and Summers,
1979) and an NPV from *Diparopsis watersi* (Croizier *et al.*, 1980) are very closely
related. In fact many of the REN pattern differences between these viruses were
no greater than the differences between strains of *A. californica* MNPV.

Sometimes a virus is found to be more infective for another host, e.g. *Pieris
brassicae* GV is more infective for *P. rapae* than for the 'natural' host (Payne,
Tatchell and Williams, 1981), and *M. brassicae* MNPV is more infective for
Plusia gamma than for the 'natural' host (Allaway and Payne, 1984). Clearly, a
more logical approach to BV classification and nomenclature is required.

Baculovirus genetics

There appears to be a multiplicity of genotypes for each of the BVs. A virus
isolated from a single infected larva may contain a variety of genomes, as
demonstrated by the regular presence of submolar fragments of DNA after
REN digestion (e.g. Smith and Summers, 1978; McIntosh and Ignoffo, 1983).
Even when no submolar fragments can be detected in REN analysis, a small
proportion of the genomes may display variability which can be detected in
plaque-purified strains (Smith and Summers, 1980).

There may be differences between virus isolates from members of the same
host species collected from different geographical areas, e.g. isolates of
Spodoptera littoralis MNPV (Kislev and Edelman, 1982), *Neodiprion sertifer*
SNPV (Brown, 1982) and *Lacanobia oleracea* GV (Crook, Brown and Foster,
1982) differed in their DNA REN patterns. Heterogeneity in the genome of a
single 'virus' is also reflected in variability of phenotypic characters. Isolates
may differ serologically, e.g. *A. segetum* MNPV (Allaway and Payne, 1983), in
their SDS–PAGE protein patterns, e.g. *N. sertifer* SNPV (Brown, 1982), and in
biological characteristics of crucial importance in the use of these agents for
pest control. Isolates of *Oryctes rhinoceros* non-occluded BV (Zelazny, 1979),
C. pomonella GV (Harvey and Volkman, 1983) and *A. segetum* MNPV (Allaway
and Payne, 1983) have been shown to differ in LD_{50} for their hosts.

For the reasons just outlined it is preferable that cloned virus strains be used
in all investigations. Viruses which produce plaques in cell culture can be cloned
from single plaques. For those viruses for which no plaque system is available,
the next best approach is to inject groups of insects with serial dilutions of
budded virions, and to select for virus isolation a single infected insect from a
group injected with a dose smaller than the LD_{50}. Green (1981) used the latter
approach with *T. paludosa* SNPV.

A. californica MNPV has been adopted for study by a number of laboratories
and rapid progress is being made in mapping the genome of its dominant
variant. Physical maps have been derived using RENs (Miller and Dawes, 1979;

Vlak, 1980; Cochran and Faulkner, 1983). *Eco*RI digestion yields 24 fragments, 21 of which have been cloned by Lübbert *et al.* (1981). It has been agreed that the map should start at *Eco*RI fragment I, which includes the IB protein gene. Smith and Summers (1982) found that DNAs from several NPVs, GVs and a non-occluded BV had sequences homologous with this fragment.

The locations on the physical map of the genes for several functions, including IB protein, have been found by marker rescue and by using Southern and Northern blotting techniques. The copy-DNA technique has been used to determine the relative amounts of virus messenger-RNA species in the infected cell, and to identify the proteins for which they code (Adang and Miller, 1982; Smith, Vlak and Summers, 1982; Erlandson and Carstens, 1983).

Recombination between MNPV genomes has been demonstrated. Croizier, Godse and Vlak (1980) inoculated *G. mellonella* larvae with MNPVs from *G. mellonella* and *A. californica*, and isolated recombinants. Smith and Summers (1980) plaque-purified recombinants between *A. californica* MNPV and *Rachiplusia ou* MNPV from wild-type *R. ou* MNPV, and suggested that recombination may be important in the evolution of BVs.

The genomes of BVs can now be manipulated using the techniques of genetic engineering. It may soon be possible to construct new virus strains with improved characteristics as microbial control agents.

Epizootiology

A common objective in pest control with a virus is the establishment of an epizootic in a pest population from which the virus is absent or in which it is only enzootic. In order to achieve this it is important that the mechanisms whereby the virus spreads from host to host within a generation and between generations are understood. Some knowledge of how well the virus persists in the field is also necessary.

Insects which feed at plant surfaces become infected with occluded viruses principally by ingesting IBs present on the plant, deposited there from the faeces or the cadavers of infected insects. In order to ensure virus persistence, large quantities of virus are produced, of which only a tiny proportion may be utilized as inoculum. During a SNPV epizootic in the European spruce sawfly, *G. hercyniae*, in Wales it was estimated that more than 10^{14} IBs/hectare were produced, of which only 0·00025% was utilized the following year (Evans and Harrap, 1982).

Virus may be disseminated by the movement of infected larvae, e.g. NPV-infected larvae of the cabbage moth, *M. brassicae* can move several metres in cabbage plots (Evans and Allaway, 1983). NPV-infected larvae of some species, e.g. the gypsy moth, *Lymantria dispar*, (Doane, 1970) and *M. brassicae* (Evans and Allaway, 1983) tend to climb to the tops of plants before they die, thus ensuring maximum contamination of the plants with their virus load. In many lepidopteran species the BV-killed cadaver hangs from the host plant while putrefaction occurs; then the skin bursts, shedding the liquefied contents together with the virus IBs. Soil-dwelling, plant-feeding insects, such as *Tipula* spp. larvae, are less likely to contaminate their food source with infective doses of virus.

The main mode of transmission for two viruses of *Tipula* spp. appears to be by cannibalism (Carter, 1973a, b; Green, 1981).

IBs may also be deposited on plants in the faeces of predators, or they may be transferred from the soil by rain-splash or by the activities of animals.

TRANSMISSION TO THE NEXT GENERATION

In a permanent ecosystem, such as a forest, IBs produced in one generation of larvae may persist on foliage until the next generation has hatched, as Entwistle and Adams (1977) showed for *G. hercyniae* SNPV. Virus may also contaminate the egg surfaces, and this may be ingested by the hatching larvae, as Doane (1975) demonstrated for *L. dispar* NPV.

In an annual crop, on the other hand, virus is transferred to the plants from a reservoir, usually the soil. There have been several investigations into the survival in soil of viruses of brassica pests. David and Gardiner (1967) reported good survival of *P. brassicae* GV in soil for at least two years, and Jaques (1969) found large amounts of an NPV of *T. ni* in soil 231 days after application with little or no evidence of leaching of IBs. Evans (1982), however, found a 98% loss of *M. brassicae* NPV IBs after 52 weeks: nevertheless, with sufficient IBs initially, enough could survive to infect the next generation.

Some larvae which receive small doses of virus, and/or which become infected in a late instar, may survive to produce infected adults which may disperse the virus and transmit it to their progeny. Entwistle (1976) considered that this was an important dispersal mechanism during an epizootic of *G. hercyniae* SNPV.

It has been claimed that some viruses are transmitted within the egg, but this has not yet been unequivocally demonstrated. It has been shown, however, that infected adults can contaminate the egg surface. Hamm and Young (1974) demonstrated transmission of *H. zea* NPV to the next generation in this way.

ROLES OF PARASITES AND PREDATORS

Hymenopteran parasites of insects can act as virus vectors when females oviposit in infected insects and subsequently in uninfected insects. The infective material probably consists of budded virions. Transmission by this mechanism has been shown for several viruses, including *P. rapae* GV (David, 1965), *Heliothis virescens* NPV (Irabagon and Brooks, 1974) and *L. dispar* NPV (Raimo, Reardon and Podgwaite, 1977).

Predators may disperse virus after feeding on infected insects. The following are a few examples of cases in which an infective BV, often in significant amounts, has been demonstrated in the faeces of predators: insects predatory upon *Heliothis punctiger* (Beekman, 1980) and *M. brassicae* (Evans and Allaway, 1983); birds predatory upon *G. hercyniae* (Entwistle, Adams and Evans, 1978) and *Wiseana* spp. (Kalmakoff and Crawford, 1982); and mammals and birds predatory upon *L. dispar* (Lautenschlager and Podgwaite, 1979).

Parasites and predators therefore have important roles in the transmission and dispersal of viruses, in addition to their more direct roles in regulating insect numbers. Integrated pest management (IPM) practices should therefore aim at maximum conservation of these animals.

The normal situation for most virus diseases is an enzootic, occasionally becoming epizootic when the host population density increases. Doane (1976) has described how an NPV epizootic develops in an *L. dispar* population, resulting in a spectacular reduction in population size, which is then likely to remain small for a number of years because of the high level of virus in the environment. Only when this declines is there likely to be a repeat of the cycle of resurgence in insect numbers followed by another epizootic. Briese (1981) proposed that climate, too, might influence the development of GV epizootics in the potato moth, *Phthorimaea operculella*.

Entwistle *et al.* (1983) described the patterns of virus dispersal in *G. hercyniae* SNPV epizootics. The spread of the disease from an initial focus became wave-like and then became random. These authors suggested that other insect viruses, e.g. *O. rhinoceros* non-occluded BV, might follow similar patterns of spread.

Safety

It has been argued (Burges, Croizier and Huber, 1980) that BVs are inherently safe for use as pesticides because man has been exposed to them throughout his evolution and no adverse effects are known. The presence of BVs can be demonstrated on marketed vegetables, some of which are eaten raw. There are, however, a number of potential hazards associated with the mass production and mass application of BVs, and these should be evaluated as fully as possible. It is better to use a pesticide with the confidence that it has passed a series of stringent safety tests than to risk an accident which could set back microbial control for decades.

The viruses which have been most exhaustively tested for safety to date are those registered for use in the US. In a large series of tests on *H. zea* SNPV no adverse effects have been found, except for possible enhancement of simian virus 40-transformation of human amnion cells (McIntosh and Maramorosch, 1973).

The virus, and other materials in the formulation, should be tested for infectivity, toxicity, carcinogenicity, teratogenicity and allergenicity in non-target organisms.

A change in the host specificity of a virus might occur by mutation, or by recombination with another virus or with cellular DNA. BV genomes resemble those of papovaviruses, many of which are oncogenic, in that they are both circular double-stranded DNA molecules. Tests for hybridization between BV and vertebrate virus DNAs, and between BV and cell DNAs, would provide an indication of the likelihood of recombination events.

The safety of humans is the prime concern, and it must be remembered that some highly susceptible individuals, i.e. those with hereditary immunodeficiency, those with acquired immune deficiency syndrome, and those receiving immuno-suppressant therapy, could be exposed to virus-containing sprays and dusts. Persons involved in virus production and field application receive the greatest exposure, especially when the virus is disseminated as a spray.

Perhaps the most likely hazard is an allergic response in the skin or respiratory system. Repeated inhalation might lead to a pulmonary condition similar to farmers' lung disease. One worker involved in *H. zea* SNPV-production is reported to have developed an allergy (Rogoff, 1975).

The welfare of other organisms, including domestic animals, wild mammals, birds, fishes and beneficial insects, must also be safeguarded.

TESTS ON VERTEBRATE ANIMALS

Animals have been inoculated with BV IBs, virions and DNA via a variety of routes. In the vast majority of these tests, e.g. after feeding *H. zea* SNPV IBs to pregnant rats (Ignoffo, Anderson and Woodard, 1973) and after inoculating *O. rhinoceros* non-occluded BV into mice (Gourreau, Kaiser and Monsarrat, 1982), no harmful effects were found. *M. brassicae* NPV IBs and *A. californica* NPV virions were fed to rodents and no chromosomal aberrations were detected (Miltenburger, 1980).

There are two reports of adverse effects in BV-inoculated pigs: Gourreau *et al.* (1979) found an increased rate of liver lesions in pigs inoculated intra-peritoneally with *O. rhinoceros* non-occluded BV; and Döller, Gröner and Straub (1983) found slight temperature increases in piglets fed *M. brassicae* NPV IBs.

G. Döller and co-workers have suggested that an antibody response in an animal is suggestive of virus replication, and have been unable to detect antibodies to IBs and virions in mammals exposed by feeding and inhalation (Döller and Huber, 1983; Döller, Gröner and Straub, 1983). Carey and Harrap (1980), however, found that some rats exposed to *Spodoptera* spp. NPVs developed antibodies to the virions and/or the IB protein and antibody responses have occurred in mice fed IBs (D. L. Knudson, in discussion after Granados, 1980b).

Workers involved in *H. zea* SNPV production (Ignoffo and Couch, 1981) and in field trials with *N. sertifer* SNPV (Entwistle *et al.*, 1978) have been monitored and no antibodies against BV components have been found in their sera.

Care is necessary when interpreting results of serum tests as a number of non-specific reactions have been detected between mammalian sera and IB proteins (Döller, 1980, 1981).

There is evidence both for the survival of BV IBs intact in the mammalian gut, and for their breakdown. Carey and Harrap (1980) recovered infective IBs 21 days after feeding to rats, while Döller, Gröner and Straub (1983) found evidence of IB breakdown in the piglet gut, but were unable to detect infectious virus in the organs.

To test for adverse effects on wildlife the approach of Lautenschlager, Rothenbacher and Podgwaite (1978) could be emulated. These authors monitored a variety of parameters in five species of caged and free-living mammals in a woodland after aerial application of *L. dispar* MNPV; they found no adverse effects. Döller and Enzmann (1982) showed that fish can mount a good antibody response to IB protein, and proposed that tests for immune responses in fish could form part of an environmental monitoring programme.

TESTS ON VERTEBRATE-CELL CULTURES

BVs have been inoculated into a wide variety of vertebrate-cell cultures and in the majority of cases no cytopathic effect occurred and no evidence of virus replication could be found, e.g. *H. zea* SNPV in primate cells (Ignoffo and Rafajko, 1972), *A. californica* MNPV in three mammalian cell lines (Miltenburger, 1980) and *O. rhinoceros* non-occluded BV in mammalian and fish cells (Gourreau, Kaiser and Montsarrat, 1981). Lack of IB production or other cytopathic effect should not be construed as lack of virus replication, but sensitive tests for a range of virus functions should be performed.

BV virions are readily taken up by vertebrate cells in culture. Granados (1980b) reported uptake of *A. californica* MNPV virions into cytoplasmic vacuoles in HeLa and fathead minnow cells, and similar observations were made by Volkman and Goldsmith (1983) and Miltenburger and Reimann (1980). The latter authors (Reimann and Miltenburger, 1983) also found evidence of some nucleocapsids breaking down in the vacuoles, and of others budding out of the cell. They could not detect virus in the cell nuclei, but Tjia, Zu Altenschildesche and Doerfler (1983), using a DNA hybridization technique, found DNA of *A. californica* MNPV in the nuclei of inoculated mammalian cells for at least 24 hours, after which it was rapidly lost. The limit for DNA detection by this technique is one viral genome per 5–10 cells (Miltenburger, 1980), so it is possible that it might have persisted undetected in a few cells. No evidence of transcription of the virus genome could be found.

McIntosh, Maramorosch and Riscoe (1979) found that *A. californica* MNPV virions were taken into cytoplasmic vacuoles in a viper cell line. There was no evidence of virus replication, but the cells grew more slowly, and there was a large increase in the number of C-type particles present in that cell line.

There have been a few reports of BV replication in mammalian cells. The first of these was by Himeno *et al.* (1967) who announced that IBs had developed in human cells inoculated with *Bombyx mori* NPV DNA. Aleshina *et al.* (1973a) subsequently reported replication of *B. mori* NPV in mouse fibroblasts. McIntosh and Shamy (1980) reported evidence of *A. californica* MNPV replication in a Chinese hamster cell line, but no evidence of replication was found by Volkman and Goldsmith (1983) in the same virus-cell system, or by Reimann and Miltenburger (1983) in another Chinese hamster cell line. One further report of a BV-induced change in mammalian cells is of an increase in nuclear size after inoculation with *L. dispar* NPV (Aleshina *et al.*, 1973b).

OTHER COMPONENTS OF BACULOVIRUS PREPARATIONS

Potential hazards from other materials present in a virus formulation must also be assessed. Insect fragments, insect diet and contaminant micro-organisms may be present, depending on the production method. Chemicals may be added to protect the virus from ultra-violet (UV)-light, to enhance adhesion to foliage or to stimulate larval feeding, and some viruses are applied in oil suspensions.

One cause for concern is the possible presence of contaminant viruses. Two small RNA viruses were found in a preparation of *Darna trima* GV (Harrap and Tinsley, 1978), and a small RNA virus has been found in *A. californica* MNPV preparations (Morris, Hess and Pinnock, 1979; Vail *et al.*, 1983). The latter virus has affinities with the mammalian caliciviruses, and is infective for *T. ni* larvae, in which small doses can initiate inapparent infections. *T. ni* larvae are used for *A. californica* MNPV production.

The risks posed by contaminant viruses are still largely unknown, but one small RNA virus (Nodamura virus) isolated from insects is lethal to mice when injected by various routes (Scherer, Verna and Richter, 1968). Until the risks can be shown to be negligible it would seem prudent for any virus which is to be applied as a spray to undergo a purification procedure sufficient, at least, to remove contaminant virions.

Most mass-produced insect virus preparations, however, consist of ground, lyophilized virus-infected larvae, and therefore contain insect material and contaminating micro-organisms. Podgwaite, Bruen and Shapiro (1983) found approximately 10^8–10^9 viable bacteria and fungi per gram of 'Gypchek' (*L. dispar* MNPV). Many of the organisms that they found are opportunistic human pathogens. Padhi and Maramorosch (1983) determined viable bacterial counts in commercial preparations of *H. zea* SNPV. They found 10^5 bacteria per gram in 'Elcar', whereas 'Viron/H' (now discontinued) contained 10^8 bacteria per gram, including *Bacillus cereus* which was pathogenic to silkworm larvae.

Dubois (1976) demonstrated that bacterial contaminants can be destroyed chemically. In the UK, field trials have been carried out with highly purified preparations of *N. sertifer* SNPV (Cunningham and Entwistle, 1981), *P. brassicae* GV (Tatchell and Payne, 1984) and *C. pomonella* GV (Glen and Payne, 1984).

'Gypchek' production also involves the hazard posed by the allergenic, urticarious setae of *L. dispar* larvae. Personnel are protected by filter masks, and a method has been devised for removing the setae during processing (Shapiro *et al.*, 1981).

Other components of BV preparations (e.g. UV-protectants, oils) should be tested for possible hazard, especially for carcinogenicity by inhalation.

REGISTRATION REQUIREMENTS AND GUIDE-LINES FOR SAFETY TESTING

Harrap (1982) has given a comprehensive account of registration requirements for viral (and other microbial) insecticides, and of the guide-lines produced by several national and international bodies for safety testing them. In the UK the controlling body is the Pesticides Safety Precautions Scheme of the Ministry of Agriculture, Fisheries and Food (Papworth, 1980), while in the US it is the Environmental Protection Agency (Rogoff, 1980). Many developing nations lack the facilities and resources for safety testing. Virus preparations which might be of value to those nations could be safety tested in laboratories in the developed nations as a contribution to their overseas aid programmes.

Too few BVs have been exhaustively safety tested to allow conclusions to be drawn about the safety of BVs in general, but the current impression is that these viruses appear to be safe for field use. However, the evidence that IBs can be dissolved in the mammalian gut and that virions can be taken into mammalian cells, together with the reports of replication in mammalian cells and of adverse effects in mammals, mean that several BVs will need to pass stringent safety tests before the group in general receives a blanket seal of approval.

Strategies for pest control with viruses

INTRODUCTION OF VIRUS

Many pests have been introduced into new areas of the world as a result of man's activities. It has been estimated that 30% of the most serious pests in the US are of foreign origin. The pests are often introduced without all of their natural enemies, including viruses. There have been several cases where a virus has subsequently been introduced, either deliberately or by accident, and has provided effective control of the pest. Two examples concern sawflies introduced into North American forests from Europe. In each case the subsequent release of an NPV from Europe initiated epizootics and controlled pests (Bird, 1953; Bird and Elgee, 1957).

This strategy was also applied to the non-occluded BV of the coconut palm rhinoceros beetle, *Oryctes rhinoceros*, which was discovered in Malaysia (Huger, 1966), but was apparently absent from the Pacific islands where *O. rhinoceros* causes serious damage to palms (pages 393 and 396).

SUPPLEMENTATION OF EXISTING DISEASE

Where a virus is present in an insect population, increasing the amount of virus in the environment may lead to a greater proportion of insects becoming infected. One virus application may be sufficient to reduce numbers of a pest to an economically acceptable level, especially in a forest. Cunningham and Entwistle (1981) stated that a single NPV application to young trees is likely to protect them from sawfly damage for their lifetime. In an agricultural situation it may be necessary to use a virus more like a chemical insecticide, with adequate protection provided only by several applications during the lifetime of the crop.

MANIPULATION OF EXISTING DISEASE

In some situations it is possible to increase the level of virus disease in a pest population by the adoption of certain management practices. An example concerns *Wiseana cervinata* which damages pasture in New Zealand. An NPV is widespread and can control this pest, but cultivation of the land buries the virus reservoir beyond the range of the larvae. Kalmakoff and Crawford (1982) therefore recommended oversowing damaged areas of pasture without cultivation. They also recommended the regular movement of stock over pastures to spread the virus.

Techniques for virus dissemination

RELEASE OF INFECTED/CONTAMINATED INSECTS

This dissemination technique has special attraction for viruses which survive poorly outside the host, e.g. *O. rhinoceros* non-occluded BV, which has been introduced into a number of South Pacific islands by releasing infected beetles (Bedford, 1981). Only the mid-gut cells are susceptible in the adult, which may survive for many weeks. The infection has a debilitative effect, however: the beetles stop boring into palms and females stop egg-laying. Monsarrat and Veyrunes (1976) estimated that an infected adult excretes about 300 ng virus per day. Some of this virus is transmitted during mating, and some serves as a source of infection for larvae, in which the infection rapidly becomes systemic and causes death.

Some insect viruses, e.g. *P. brassicae* GV (Tatchell, 1981), are transmitted to the progeny if the ovipositor of the female is contaminated, but this technique has not yet been widely applied to virus dissemination in the field.

SPRAYING

Most viruses are applied in aqueous sprays using equipment developed for spraying chemical insecticides. Morris (1980) and Smith and Bouse (1981) have argued for a research programme to design equipment specifically for the application of viruses and other microbes.

Equipment producing small droplets is preferred. Virus application in droplets with diameters of 100–150 μm usually results in higher insect mortality than in larger droplets (Smith and Bouse, 1981). Entwistle *et al.* (1978) used a micro-droplet machine producing droplets with a mean diameter of 50 μm. Reed and Springett (1971) suggested that *P. operculella* GV might best be disseminated as a mist as the IBs would be more likely to enter the stomata, thereby becoming more accessible to the larvae within the leaves.

Virus dissemination in charged droplets from an electrostatic sprayer means that a larger proportion of IBs adhere to the leaves, especially the undersides which are the sides often favoured by insects, and which provide some protection from sunlight for the virus. A disadvantage of electrostatic sprayers is poorer spray penetration into the plant canopy (Matthews, G.A., 1982).

A modification of spray application was carried out by Hamm and Hare (1982) who introduced NPVs of *H. zea* and *S. frugiperda* on to corn via an overhead irrigation system. Instead of spraying the crop, Young and Yearian (1980) sprayed the soil with an NPV of the soybean looper, *Pseudoplusia includens*, at soybean planting time.

If a virus spreads rapidly, then blanket spraying may be unnecessary. This is the situation with the SNPV of the red-headed pine sawfly, *Neodiprion lecontei*, for which Cunningham (1982) has proposed spot introductions or 'zebra stripe spraying' from aircraft. Spot introductions into glasshouses of the GV of the tomato moth, *Lacanobia oleracea*, were suggested by Crook, Brown and Foster (1982).

BAITS

The application of insecticides in baits has the advantage that less insecticidal material is required, and the disadvantage of increased costs of field application. Baits are especially valuable if insects which have a burrowing or mining habit can be encouraged to spend longer at the plant surface and ingest larger doses of insecticidal material.

Most research into the application of viruses in baits has involved *H. zea* SNPV and baits based on cottonseed and soybean (page 395). Johnson and Lewis (1982) used wheat bran baits to apply two MNPVs to corn.

DIPPING SEEDLINGS

Ignoffo *et al.* (1980) suggested that IBs could be introduced on to cabbages by dipping them in an IB suspension at the time of transplanting.

Viruses undergoing trials and/or in use

Details of viruses registered for use in various countries are given in *Table 2*, and some of them are discussed more fully below.

1. *Heliothis zea* SNPV has been marketed for almost a decade in the US for the control of *H. zea* and *H. virescens* on cotton (Ignoffo and Couch, 1981). Some workers, e.g. Shieh and Bohmfalk (1980), have found it to be an effective

Table 2. Viruses registered for use.

Virus	Used on	Country	Product name
Heliothis zea SNPV	Cotton and other crops	US	Elcar
	Cotton, sorghum	Australia	
Orgyia pseudotsugata MNPV	Fir trees	US	TM Biocontrol-1
		Canada (temporary registration)	Virtuss
Lymantria dispar MNPV	Deciduous trees	US	Gypchek
		USSR	Virin–ENSh
Autographa californica MNPV	Several crops	US (experimental use permit)	SAN 404
Neodiprion sertifer SNPV	Pine trees	US	Neochek S
		USSR	Virin–Diprion
		Finland	none
Neodiprion lecontei SNPV	Pine trees	Canada (temporary registration)	Lecontvirus
Dendrolimus spectabilis cytoplasmic polyhedrosis virus	Pine trees	Japan	Matsukemin

insecticide, while others, e.g. Pfrimmer (1979), have obtained variable and sometimes disappointing results.

Much effort has been expended in attempts to achieve more consistent results. Some of this effort has involved the development of baits, and two in particular have been tested: 'Coax' based on cottonseed and 'Gustol' based on soybean. Many workers (e.g. Hostetter *et al.*, 1982, and Potter and Watson, 1983a) have shown in laboratory and field tests that applying the virus in a bait increases larval mortality. Some of the increased mortality may not be virus-induced, however, as treatment of cotton with 'Coax' alone results in increased mortality (Henry, 1982). This has been attributed to larvae spending longer at the surface before tunnelling into the bolls, thereby extending their exposure to parasites and predators.

Smith, Hostetter and Ignoffo (1978, 1979) compared different formulations, application rates, types of spray nozzle and nozzle pressures. They found that the efficiency of application was affected by nozzle type and droplet size.

H. zea SNPV can also control *Heliothis* spp. on other crops. Ignoffo *et al.* (1978) found that it reduced *H. zea* populations on soybeans by 92–100%, and Smith and Hostetter (1982) reported better control of *H. zea* on soybean and cabbage than on cotton. In Australia *H. zea* SNPV is undergoing tests for its ability to protect navybeans from *Heliothis* spp. (R. E. Teakle, personal communication).

2. *Autographa californica* MNPV, originally isolated from the alfalfa looper, is considered to have a potential commercial value because of its wide host range. It has been reported that it can control *T. ni* as effectively as chemicals on cabbage (Hostetter *et al.*, 1979) and lettuce (Vail, Seay and Debolt, 1980), and it is being assessed as an alternative to *Orgyia pseudotsugata* MNPV for the control of the Douglas-fir tussock moth, *O. pseudotsugata*. Although the latter virus can control its host effectively, the high cost of its production and its limited market mean that there is no commercial interest in it (Martignoni, Steltzer and Iwai, 1982).

3. *Neodiprion sertifer* SNPV has been extensively tested against its host, the European pine sawfly, in Eastern and Western Europe and in North America (Cunningham and Entwistle, 1981). Entwistle *et al.* (in press) have induced high larval mortality in pine forests in Scotland with applications of 5×10^9 to 2×10^{10} IBs/hectare. These quantities of virus can be produced in 20–50 larvae. This remarkable efficiency is attributed to the high larval susceptibility to this virus and to its rapid spread.

4. *Neodiprion lecontei* SNPV has shown promise in trials in Canada. Its host has been controlled with applications of 5×10^9 to 8×10^9 IBs/hectare (Cunningham, 1982).

5. *Galleria mellonella* MNPV was shown by Dougherty, Cantwell and Kuchinski (1982) to control wax moth larvae effectively in bee-hives. A non-hazardous insecticide is especially important for this pest of honeycomb.

6.   *Panolis flammea* NPV has shown very promising results against its host, the pine beauty moth, which is a pest of lodgepole pines in Scotland (P. F. Entwistle, personal communication).

7.   *Heliothis armigera* SNPV was shown to provide control of its host on sorghum in Botswana by Roome (1975) and is still under investigation in that country where *H. armigera* is a pest of many crops (Flattery, 1983).

8.   *Choristoneura occidentalis* NPV and GV have shown promise for the control of their host, the western spruce budworm on Douglas fir. The impact of the NPV on the population size was still detectable one year after spraying (Shepherd, Gray and Cunningham, 1982), and a GV application rate of only 25 'larval equivalents'/acre resulted in 56% mortality (Cunningham, Kaupp and McPhee, 1983).

9.   *Pieris brassicae* GV has been demonstrated by a number of workers, including Kelsey (1958), to provide control of larvae of the small cabbage white butterfly, *Pieris rapae*. Tatchell and Payne (1984) recently found that a spray containing 10^8 IBs/ml reduced the larval population by more than 90%. The virus is rapidly inactivated in the field, however, and regular spraying would be necessary to maintain satisfactory control.

10.   *Cydia pomonella* GV has been tested in many countries for control of the codling moth in orchards. Huber and Dickler (1977) reported that four sprays resulted in good control, but there was no persistence of the disease into the next season. Much virus is probably removed from the orchard on the surface of the apples. Trials carried out by Glen and Payne (1984) led them to conclude that the use of *C. pomonella* GV effectively reduces the more severe forms of fruit damage, but the quantities of virus required to control less severe forms of damage would probably be uneconomic.

11.   *Oryctes rhinoceros* non-occluded BV has been introduced into a number of South Pacific islands. In Tonga it was still infecting 84% of the beetle population after seven years (Young and Longworth, 1981). Control of the rhinoceros beetle has led to a revival of the copra industry in Western Samoa (Marschall and Ioane, 1982).

Timing of field applications

The timing of field applications of virus can be crucial in determining the level of pest control achieved. Significant pest damage is not usually noticed until the larvae are in the later instars, when larger doses of virus are necessary to infect them. This, coupled with the fact that most insect viruses kill their hosts more slowly than chemical insecticides, means that for many pests the virus must be applied before crop damage appears. Pest forecasting systems can be used to indicate when pest numbers are approaching damaging levels.

Increases in LD_{50} of 10^4-fold to 10^6-fold from early to late larval instars have been found for a number of lepidopteran BVs, including *P. brassicae* GV

(Payne, Tatchell and Williams, 1981) and *Mamestra configurata* MNPV (Bucher and Turnock, 1983), and LT_{50}s are often longer in later instars. In some cases the increased LD_{50} may be offset by the increased food consumption of larger larvae, as in *H. armigera* where the first three larval instars have a similar probability of becoming infected with *H. zea* SNPV in the field (R. E. Teakle and J. M. Jensen, personal communication).

Increases in resistance to NPVs in the later instars of sawflies appear to be small compared with those in the Lepidoptera. For *N. sertifer* and *G. hercyniae* the increases in LD_{50} are about tenfold to fiftyfold from the first to the fifth larval instar (Entwistle, Adams and Evans, unpublished work cited in Evans and Harrap, 1982; Entwistle *et al.*, in press). This means that these pests can be controlled if virus is applied after the first instar. Infection of the larvae when they are larger means that more virus is produced and is available to infect the next generation. This approach is more applicable in stable ecosystems, such as forests and pastures where some pest damage can be tolerated, than in annual crops.

Virus persistence in the field

A rapid loss of infective virus from plant surfaces can usually be detected after field application, which could be a physical loss of IBs from plants and/or a loss of infectivity in virus on the plants.

There have been many studies of rates of infectivity loss, but, as pointed out by Richards and Payne (1982), most of them have started with amounts of virus giving 100% mortality in bioassays and have not therefore achieved their objectives. These authors outlined a sound experimental approach which they applied to measure survival of infectivity of a *Pieris* sp. GV on cabbage in the UK. They found that the half-life varied from 0·35 day in June to 1·0 day in October. They showed, using ^{32}P-labelled IBs, that the IBs had not been lost from the cabbages.

If a virus can be protected from inactivation then smaller amounts need be applied and/or the timing of application becomes less critical. The extent to which virus inactivation can make timing of application critical was shown by Potter and Watson (1983b). If they sprayed *H. zea* SNPV against *H. virescens* just after the eggs were laid, 15% of the larvae died, whereas if they sprayed just before the eggs hatched, 80% died.

INACTIVATION BY ULTRA-VIOLET LIGHT

The main factor causing infectivity loss in the field appears to be the UV component of sunlight. Attempts are made to protect some IB preparations by adding a UV-absorbing substance. A polyflavinoid marketed as 'Shade' has been used with *H. zea* SNPV, and increased virus persistence and/or mortality of larvae has resulted. 'Shade' is incorporated into *L. dispar* MNPV preparations (Lewis, 1981), and has been shown to act as a UV-protectant for this virus in a laboratory test, although more protection was afforded by the feeding stimulant 'Coax' (Shapiro, Poh Agin and Bell, 1983).

INACTIVATION BY COTTON LEAF SECRETIONS

Cotton leaf secretions have a high pH due to substances secreted by epidermal glands (Elleman and Entwistle, 1982), and there is some evidence that IBs on the leaf surface can be affected (Andrews and Sikorowski, 1973). Richards (MSc thesis cited in Richards and Payne, 1982) found that an unbuffered suspension of *S. littoralis* NPV was completely inactivated 6 days after application to cotton leaves, whereas infectivity was preserved for much longer if the virus was applied in a phosphate buffer, pH 7. Some workers in the US have applied *H. zea* SNPV to cotton in buffered suspensions, but results have varied in different areas. Further investigations are necessary to determine whether there are advantages to be gained by applying viruses to cotton in buffered suspensions.

Virus production

Viral insecticides are currently produced in the host insect which is either collected in the field or reared in an insectary. For some species more than 10^8 larvae per year are produced. High standards of hygiene are vital to reduce the risk of infection by pathogens which could decimate the insect stocks and contaminate the product.

The production of *H. zea* SNPV ('Elcar') has been described by Ignoffo and Anderson (1979) and Ignoffo and Couch (1981). *H. zea* larvae are reared on a semi-synthetic diet in the wells of plastic trays. Each larva yields about 3.5×10^9 IBs which are extracted, purified and spray dried. The final product contains 99·6% inert ingredients and is stored at $-20°C$. 'Elcar' is produced in the US by Sandoz Inc. who also produce smaller quantities of *A. californica* MNPV and an NPV of *T. ni* for experimental purposes. Both of these viruses are produced in *T. ni* larvae and the IB preparations are spray-dried (Yearian and Young, 1982).

A process for the mass production of *L. dispar* MNPV ('Gypchek') has been described by Shapiro *et al.* (1981) and Shapiro (1982), in which the IB yield represents a 5600-fold increase over the inoculum.

For the production of sawfly viruses, either field-collected larvae are infected and then maintained on host plant material, or infected larvae are collected in the field (Cunningham and Entwistle, 1981).

Production of viruses in insects is labour-intensive and therefore costly in the developed nations. The most time-consuming stage in 'Elcar' production is the introduction of larvae into the trays, while in 'Gypchek' production it is the removal of the infected larvae from their containers.

OPTIMIZING PRODUCTION

The host insect

The insect species from which a virus was isolated may not be the most susceptible (page 385). Use of a more susceptible host for virus production would mean a smaller inoculum requirement.

Alternative hosts might also be considered for insect species which have a long life cycle, which are small and produce a low yield, or which have allergenic and urticarious setae. Shapiro *et al.* (1982) suggested that *O. pseudotsugata* MNPV might be produced in the saltmarsh caterpillar, *Estigmene acrea*, which is more easily reared than the homologous host. It is important to check that virus produced in an alternative host does not have reduced virulence for the original host.

Insect diet

Diet may affect the growth rate of an insect, its susceptibility to virus infection, and the virus yield. Synthetic or semi-synthetic diets are used for most insects, and cost is an important factor. A diet rich in wheat germ was found to be the most cost effective for *L. dispar* MNPV production, although higher IB yields could be obtained using other, more expensive, diets (Shapiro, Bell and Owens, 1981). Shapiro (1982) found a substitute for agar which was 40% cheaper than agar and resulted in improved growth of *L. dispar* larvae with higher IB yields.

Glen and Payne (1984) increased the yield of *C. pomonella* GV by incorporating into the diet a juvenile hormone analogue (methoprene) which resulted in larger larvae.

Insect stage and virus dosage

The lower IB doses necessary to infect younger larvae must be balanced against the early deaths of these larvae with smaller IB yields. The optimum dosage must be determined. If it is too low, many larvae will not become infected, whereas if it is too high, inoculum will be wasted and larval growth will be retarded resulting in suboptimal yields.

Incubation environment

Temperature affects the rate of insect growth, the rate of virus replication and the virus yield. The optimum temperature for each of these may not be the same and it is necessary to determine the optimum for yield. Relative humidity and photoperiod must also be maintained at their optima.

Preservation of virus infectivity

Conditions which destroy infectivity (e.g. increased temperature, extremes of pH) must be avoided during harvesting, purification, formulation and storage of virus.

QUALITY CONTROL

Each batch of virus must be carefully bioassayed (page 383) and tested for the presence of harmful contaminants, especially human pathogens. Morris, Vail and Collier (1981) suggested that quality control procedures should include tests for contaminants such as small RNA viruses.

Future prospects

USE OF VIRAL INSECTICIDES IN INTEGRATED PEST MANAGEMENT

The relatively narrow host spectra of insect viruses may be environmentally attractive, but mean that markets for viral insecticides are restricted and that a virus alone is unlikely to afford protection against all the pests in a particular ecosystem. For example, if *C. pomonella* GV is used for codling moth control in orchards, other lepidopteran pests, especially tortrix moths, may resurge. On the other hand, use of the virus has the advantage that parasites and predators of the fruit-tree red spider mite, *Panonychus ulmi*, are not killed, so damaging numbers of this pest are not reached, which may occur if an organophosphorus insecticide is used to control codling moth (Glen *et al.*, 1984).

A virus may form a useful component of an IPM programme in which pests are controlled by husbandry practices, chemicals and biological agents. The most widely used microbial control agent is *Bacillus thuringiensis*, most strains of which have wide spectra of activity against lepidopteran insects. In fact, the existence of this microbial insecticide is one of the factors limiting the development of viral insecticides, although for some pests the two might be used together. *B. thuringiensis*, together with *P. rapae* GV and *A. californica* MNPV, have been reported to control *P. rapae* and *T. ni* on cabbage almost as effectively as chemical insecticides (Sears, Jaques and Laing, 1983).

IPM on cotton might include the use of *H. zea* SNPV and *B. thuringiensis* or chlordimeform against *Heliothis* spp., *A. californica* MNPV and *B. thuringiensis* against the cotton leaf-perforator, *Bucculatrix thurberiella* (Bell and Romine, 1982), and diflubenzuron against the cotton boll weevil, *Anthonomus grandis* (Bull *et al.*, 1979).

VIRUS PRODUCTION IN CELL CULTURE

Several laboratories are attempting to develop reliable and economic cell-culture systems for the mass production of insect viruses as alternatives to production in insects, which has a number of associated problems (pages 398–399), and because a high-purity product is more feasible from cell cultures. Much progress has been made, but several problems remain to be solved. An outline of insect-cell culture techniques was given on pages 381–382. The possible application of those techniques to virus production will now be discussed.

Production systems

Insect cells can be grown in fermenters of the type used for vaccine production and Vaughn (1981) has suggested that the slack periods of such plants could be used for insect-virus production. When *A. californica* MNPV was produced in TN-368 cells in fermenters 2–3 litres in volume it was found (Hink and Strauss, 1980) that more vigorous aeration was required than in small volumes, and this resulted in foaming and cell damage. Antifoam was added and the concentration of methylcellulose, already present to inhibit cell clumping, was increased to protect the cells.

Attempts have been made to avoid the stresses imposed on cells in traditional fermenters by using alternative systems. Miltenburger and David (1980) blew air through silicone rubber tubing coiled inside a fermenter. Oxygen diffused through the silicone rubber into the medium. Hilwig and Alapatt (1981) and Vaughn and Dougherty (1981) have worked on roller bottle systems, but Stockdale and Priston (1981) believe that they are too bulky and labour-intensive for adoption by industry. Vaughn and Dougherty (1981) are also developing a 'perfusion culture system': this consists of vessels containing coils which provide a large surface area for cell attachment; pH and oxygen concentration are adjusted outside the vessel. Pollard and Khosrovi (1978) presented a design for a continuous-flow tubular fermenter.

Optimizing production

Some of the factors which can affect IB yield were investigated by Gardiner, Priston and Stockdale (1976) for *A. californica* MNPV in TN-368 cells. They found an optimum temperature of 27°C, an optimum pH range of 5·5–6·5 and an optimum osmotic pressure range of 250–500 milliosmoles. For the same virus–cell system Hink (1982) reported production of 10^8 IBs/ml medium and suggested that this must be increased twentyfold before the system becomes economic.

The IB yield can be affected by the growth phase of the cells at the time of virus inoculation (Lynn and Hink, 1978), and by their concentration. The cell concentration giving maximum IB yield per ml of medium was higher than that at which the maximum number of IBs per cell was produced (Hink, Strauss and Ramoska, 1977; Stockdale and Gardiner, 1977). The latter authors suggested that the reduced IB production at higher cell densities might be due to depletion of a vital precursor. Wood, Johnston and Burand (1982) reported a 98% reduction in virus production in high-density attached cultures compared with low-density cultures. Inhibition of virus production did not occur unless there was cell-to-cell contact. Further investigations are necessary into the mechanisms of, and ways of overcoming, inhibition of IB production at high cell concentrations.

Virus strains and cell strains should be selected to give a high-yielding system. Cells should be cloned and the clones screened for desirable properties, e.g. more rapid growth rate (McIntosh and Rechtoris, 1974). The quality of IBs produced in a cell line must also be checked. Lynn and Hink (1980) found that *A. californica* MNPV IBs produced in cells from four insect species were less infective for *T. ni* larvae than IBs produced in *T. ni* cells.

Most insect-cell culture media are expensive, principally because most of them contain foetal bovine serum. Dougherty, Cantwell and Kuchinski (1982) calculated that, for *G. mellonella* MNPV production in cell culture, half the cost, including labour, was for serum. Serum-free media are now being developed and have been reported for a *S. frugiperda* cell line with replication of *A. californica* MNPV (Wilkie, Stockdale and Pirt, 1980), an *L. dispar* cell line with replication of *L. dispar* NPV (Goodwin and Adams, 1980) and for several other cell lines (Mitsuhashi, 1982). Weiss *et al.* (1981) reduced the cost of their medium for *S. frugiperda* cells by omitting antibiotics. They encountered no contamination problems.

Changes in virus on passage

Several studies, e.g. Faulkner and Henderson (1972), have demonstrated that IBs produced during the first few passages in cell culture are as infective as IBs produced in insects. Upon repeated passage, however, the quality and yield of IBs have been found to decline. Hirumi, Hirumi and McIntosh (1975) reported that passage of *A. californica* MNPV in *T. ni* cells led to the production of aberrant virions and a reduction in IB yield. After 40 passages the yield had dropped a hundredfold, with IBs developing in only 4% or less of the cells (McIntosh, Shamy and Ilsley, 1979). MacKinnon *et al.* (1974) found a reduction in average yield of *T. ni* MNPV IBs from 28 per cell initially to 2·5 per cell after 50 passages, with extensive production of abnormal capsids. Knudson and Harrap (1976) found that passage of *S. frugiperda* NPV in *S. frugiperda* cells led to the production of IBs containing few or no virions, and Yamada, Sherman and Maramorosch (1982) reported reduced yields of *H. zea* SNPV IBs after 20 passages in *H. zea* cells.

Hink and Strauss (1976) described two plaque morphologies after passage of *A. californica* MNPV *in vitro*. In one type of plaque there were between 81 and 352 IBs per nucleus, while in the other there were only 2–13 IBs per nucleus. The plaque types were named many-polyhedra (MP) and few-polyhedra (FP) plaques respectively. The FP plaques became increasingly dominant on passage. IBs from MP plaques contained normal virions (multiple nucleocapsids per virion) and were much more infective for *T. ni* larvae than IBs from FP plaques, which either contained only a few virions (each with only a single nucleocapsid per virion) or appeared devoid of virions. Similar phenomena have been reported for *T. ni* MNPV (Potter, Faulkner and MacKinnon, 1976), *G. mellonella* MNPV (Fraser and Hink, 1982) and for *H. zea* SNPV (M. J. Fraser and W. J. McCarthy, unpublished, in Fraser, Smith and Summers, 1983).

Most FP forms are genetically stable. They have a selective advantage *in vitro* as FP-infected cells produce higher titres of budded virions than MP-infected cells (Potter, Jaques and Faulkner, 1978). Wood (1980) suggested that FP forms might be deletion mutants of MP forms, but several FP forms have been found to contain insertions of host DNA (Miller and Miller, 1982; Fraser, Smith and Summers, 1983).

One way of avoiding the FP form becoming dominant in cell cultures would be to return regularly for inoculum to haemolymph from insects infected by ingestion of IBs. Insect haemolymph, however, is unlikely to supply the quantities of inoculum that an industrial-scale process would demand, so some means will have to be found of preventing the development of FP and other aberrant virus forms *in vitro*.

The debate concerning the degree of purification necessary before a virus is sprayed in the field has not yet been resolved. Of primary concern are the potential hazards posed by the presence of contaminants (pages 390–391). Other considerations are the costs involved and the possible effects of purification on virus infectivity and persistence.

There have been several reports (e.g. Magnoler, 1968; Carner, Hudson and Barnett, 1979; Evans and Harrap, 1982) of laboratory and field tests in which purified IB preparations had lower infectivity and/or poorer environmental persistence than IB preparations contaminated with insect fragments, gut contents and micro-organisms. It is well known that proteins can protect viruses from inactivation, so the contaminants may afford some protection, especially from UV light. *N. sertifer* SNPV, however, controls its host effectively when applied as a highly purified IB preparation (Entwistle *et al.*, in press). It is interesting that *L. dispar* larvae were deterred from feeding on foliage contaminated with decayed cadavers or extracts from healthy larvae, but were not deterred by foliage treated with purified NPV IBs (Capinera, Kirouac and Barbosa, 1976).

The most efficient way of purifying IBs is by some form of gradient centrifugation, but this can make the final product prohibitively expensive in the developed nations, let alone the developing nations. In the future, if viruses are produced in cell cultures they should be free from contaminating micro-organisms and minimal purification should be necessary. In the meantime, tests should be carried out to evaluate the hazards posed by contaminants in virus preparations produced in insects, which should be subjected to rigorous quality control procedures before field application.

NEW VIRUS STRAINS

Ideal attributes in a viral pesticide are high infectivity and high virulence for a broad range of pest species, rapid replication with high yields, and good field persistence with high resistance to UV inactivation. No known virus is endowed with all of these attributes, but progress towards the development of such an agent should be possible by two approaches, i.e. by searching in nature for new virus strains and by the genetic manipulation of existing isolates.

There can be no doubt that the number of insect virus strains isolated to date is only a tiny fraction of the total in nature. Virus strains with desirable properties will undoubtedly be found among future isolates.

A virus might be genetically improved by selecting for desired traits, or by using the techniques of genetic engineering. The former approach was used by Brassel and Benz (1979) who selected a strain of *C. pomonella* GV which was 5·6 times more resistant to UV light than the original isolate, and remained infective for twice as long in the field. Wood *et al.* (1981) induced mutations in *A. californica* MNPV, then isolated a mutant with increased virulence for *T. ni* larvae, demonstrated by a significantly reduced LT_{50}. Among several possibilities for genetically engineering virus strains is the suggestion by Miller, Lingg and Bulla (1983) that the gene for an insect-specific toxin might be incorporated into the virus genome to kill the host more rapidly.

PATENTS

A search by Stockdale (in press) revealed that 13 patents had been filed for processes or formulations involving insect viruses. It is not possible to patent the viruses, however, and this is one of the reasons why viral pesticides have not been developed more rapidly. There are some hopes that this situation may change and that it may become possible to patent an organism if it is the product of a biotechnological process (Crespi, 1980) or if it has undergone genetic manipulation (Kayton, 1983). If these hopes are realized then there will be more incentive for commercial concerns to invest in microbial pesticide development.

DEVELOPMENT OF PEST RESISTANCE

There are not yet any reports of selection of an insect strain with high resistance to a virus, as occurred in the rabbit to myxoma virus (Fenner, 1983). It is a possible outcome, however, if a virus is widely used over a long period. Genetic variability, upon which selection could operate, has been demonstrated in a number of insect species, e.g. varying levels of susceptibility to a GV in the Indian meal moth, *Plodia interpunctella* (Hunter and Hoffmann, 1973) and to an NPV in the light brown apple moth, *Epiphyas postvittana* (Briese *et al.*, 1980).

Some workers have attempted to select for virus resistance. Ignoffo and Allen (1972) failed to select for increased resistance to an NPV in *H. zea* after inoculating 25 generations of larvae with doses at, or greater than, the LD_{50} and breeding from the survivors. Briese and Mende (1983), however, selected for resistance to a GV in *P. operculella* within six generations of insects from the wild, but they were able to select for only a slight increase in resistance in a laboratory strain which was already highly resistant.

CONCLUSION

Advances in BV research are providing greater insight into the viruses themselves, e.g. their genetics, and into their interactions with their hosts, e.g. their epizootiology. This information means that the viruses can be used as pesticides on a more rational basis.

A number of insect viruses are currently used as pesticides and the potential of others has been demonstrated. Entwistle (1983) is optimistic that BVs will become the principal means of regulating lepidopteran and sawfly pests of forests. Viral pesticides can have an important role in Third World countries if they can be produced locally and if it can be shown that they are safe to disseminate in an unpurified or semi-purified state.

The potential for viral insecticides in the agriculture and horticulture of the Developed World is more limited at present. Consumers demand fruit and vegetables free from blemishes, so growers look for products which provide a quick and virtually complete kill of pests. Often this cannot be achieved with viruses, so until there is a change in consumer attitude, chemicals are likely to remain the main tools for pest control. As the integrated pest management approach gains ground, however, viruses of pests should have increasingly useful roles.

Acknowledgements

I am grateful to Drs J. C. Cunningham, P. F. Entwistle, C. C. Payne, H. Stockdale and R. E. Teakle for helpful discussions and for allowing me access to their unpublished work, and to Karen Bernard and Christine Guy for typing the manuscript.

References

ADAMS, J.R., GOODWIN, R.H. AND WILCOX, T.A. (1977). Electron microscopic investigations on invasion and replication of insect baculoviruses *in vivo* and *in vitro*. *Biologie Cellulaire* **28**, 261–268.

ADANG, M.J. AND MILLER, L.K. (1982). Molecular cloning of DNA complementary to mRNA of the baculovirus *Autographa californica* nuclear polyhedrosis virus: location and gene products of RNA transcripts found late in infection. *Journal of Virology* **44**, 782–793.

AIZAWA, K. (1976). Recent development in the production and utilisation of microbial insecticides in Japan. In *Proceedings of the First International Colloquium on Invertebrate Pathology, Kingston, Canada*, (T.A. Angus, P. Faulkner and A. Rosenfield, Eds), pp. 59–63. Society for Invertebrate Pathology, Queen's University at Kingston.

ALESHINA, O.A., EGIAZARYAN, L.A., SOLDATOVA, N.V. AND MARTYNOVA, G.S. (1973a). The infection of culture L of mouse cells with nuclear polyhedrosis virus (first communication. [English abstract.] *Review of Applied Entomology, Series A* **61**, 812–813.

ALESHINA, O.A., SOLDATOVA, N.V., MARTYNOVA, G.S. AND EGIAZARYAN, L.A. (1973b). The caryometrical study of the cytopathogenic action of the nuclear polyhedrosis virus on transplantable cultures of mammalian cells. [English abstract.] *Review of Applied Entomology, Series A* **61**, 812.

ALLAWAY, G.P. AND PAYNE, C.C. (1983). A biochemical and biological comparison of three European isolates of nuclear polyhedrosis viruses from *Agrotis segetum*. *Archives of Virology* **75**, 43–54.

ALLAWAY, G.P. AND PAYNE, C.C. (1984). Host range and virulence of five baculoviruses from lepidopterous hosts. *Annals of Applied Biology*, in press.

ANDREWS, G.L. AND SIKOROWSKI, P.P. (1973). Effects of cotton leaf surfaces on the nuclear polyhedrosis virus of *Heliothis zea* and *Heliothis virescens* (Lepidoptera : Noctuidae). *Journal of Invertebrate Pathology* **22**, 290–291.

BAUGHER, D.G. AND YENDOL, W.G. (1981). Virulence of *Autographa californica* baculovirus preparations fed with different food sources to cabbage loopers. *Journal of Economic Entomology* **74**, 309–313.

BEDFORD, G.O. (1981). Control of the rhinoceros beetle by baculovirus. In *Microbial Control of Pests and Plant Diseases 1970–1980* (H.D. Burges, Ed.), pp. 409–426. Academic Press, London.

BEEKMAN, A.G.B. (1980). The infectivity of polyhedra of nuclear polyhedrosis virus (N.P.V.) after passage through gut of an insect-predator. *Experientia* **36**, 858–859.

BELL, M.R. AND ROMINE, C.L. (1982). Cotton leafperforator (Lepidoptera : Lyonetiidae): effect of two microbial insecticides on field populations. *Journal of Economic Entomology* **75**, 1140–1142.

Bergoin, M., Guelpa, B. and Meynadier, G. (1975). Ultrastructure du virus de la polyédrose nucléaire du Diptère *Tipula paludosa* Meig. *Journal de Microscopie et de Biologie Cellulaire* **23**, 9a–10a.

Bird, F.T. (1953). The use of a virus disease in the biological control of the European pine sawfly, *Neodiprion sertifer* (Geoffr.). *Canadian Entomologist* **85**, 437–446.

Bird, F.T. and Elgee, D.E. (1957). A virus disease and introduced parasites as factors controlling the European spruce sawfly, *Diprion hercyniae* (Htg.) in central New Brunswick. *Canadian Entomologist* **89**, 371–378.

Boucias, D.G., Johnson, D.W. and Allen, G.E. (1980). Effects of host age, virus dosage, and temperature on the infectivity of a nucleopolyhedrosis virus against velvetbean caterpillar, *Anticarsia gemmatalis*, larvae. *Environmental Entomology* **9**, 59–61.

Brassel, J. and Benz, G. (1979). Selection of a strain of the granulosis virus of the codling moth with improved resistance against artificial ultraviolet radiation and sunlight. *Journal of Invertebrate Pathology* **33**, 358–363.

Briese, D.T. (1981). The incidence of parasitism and disease in field populations of the potato moth *Phthorimaea operculella* (Zeller) in Australia. *Journal of the Australian Entomological Society* **20**, 319–326.

Briese, T.D. and Mende, H.A. (1983). Selection for increased resistance to a granulosis virus in the potato moth, *Phthorimaea operculella* (Zeller) (Lepidoptera: Gelechiidae). *Bulletin of Entomological Research* **73**, 1–9.

Briese, D.T., Mende, H.A., Grace, T.D.C. and Geier, P.W. (1980). Resistance to a nuclear polyhedrosis virus in the light-brown apple moth *Epiphyas postvittana* (Lepidoptera: Tortricidae). *Journal of Invertebrate Pathology* **36**, 211–215.

Brown, D.A. (1982). Two naturally occurring nuclear polyhedrosis virus variants of *Neodiprion sertifer* Geoffr. (Hymenoptera: Diprionidae). *Applied and Environmental Microbiology* **43**, 65–69.

Brown, D.A., Allen, C.J. and Bignell, G.N. (1982). The use of a protein A conjugate in an indirect enzyme-linked immunosorbent assay (ELISA) of four closely related baculoviruses from *Spodoptera* species. *Journal of General Virology* **62**, 375–378.

Bucher, G.E. and Turnock, W.J. (1983). Dosage responses of the larval instars of the bertha armyworm, *Mamestra configurata* (Lepidoptera: Noctuidae), to a native nuclear polyhedrosis. *Canadian Entomologist* **115**, 341–349.

Bull, D.L., House, V.S., Ables, J.R. and Morrison, R.K. (1979). Selective methods for managing insect pests of cotton. *Journal of Economic Entomology* **72**, 841–846.

Burgerjon, A., Biache, G., Chaufaux, J. and Petré, Z. (1981). Sensibilité comparée, en fonction de leur âge, des chenilles de *Lymantria dispar*, *Mamestra brassicae* et *Spodoptera littoralis* aux virus de la polyédrose nucléaire. *Entomophaga* **26**, 47–58.

Burges, H.D., Croizier, G. and Huber, J. (1980). A review of safety tests on baculoviruses. *Entomophaga* **25**, 329–340.

Burley, S.K., Miller, A., Harrap, K.A. and Kelly, D.C. (1982). Structure of the *Baculovirus* nucleocapsid. *Virology* **120**, 433–440.

Capinera, J.L., Kirouac, S.P. and Barbosa, P. (1976). Phagodeterrency of cadaver components to gypsy moth larvae, *Lymantria dispar*. *Journal of Invertebrate Pathology* **28**, 277–279.

Carey, D. and Harrap, K.A. (1980). Safety tests on the nuclear polyhedrosis viruses of *Spodoptera littoralis* and *Spodoptera exempta*. In *Invertebrate Systems In Vitro* (E. Kurstak, K. Maramorosch and A. Dübendorfer, Eds), pp. 441–450. Fifth International Conference of Invertebrate Tissue Culture, Rigi-Kaltbad, 1979. Elsevier/North-Holland, Amsterdam.

Carner, G.R., Hudson, J.S. and Barnett, O.W. (1979). The infectivity of a nuclear polyhedrosis virus of the velvetbean caterpillar for eight noctuid hosts. *Journal of Invertebrate Pathology* **33**, 211–216.

Carter, J.B. (1973a). The mode of transmission of *Tipula* iridescent virus I. Source of infection. *Journal of Invertebrate Pathology* **21**, 123–130.

Carter, J.B. (1973b). The mode of transmission of *Tipula* iridescent virus II. Route of infection. *Journal of Invertebrate Pathology* **21**, 136–143.

CARTER, J.B. (1978). Field trials with *Tipula* iridescent virus against *Tipula* spp. larvae in grassland. *Entomophaga* **23**, 169–174.

COCHRAN, M.A. AND FAULKNER, P. (1983). Location of homologous DNA sequences interspersed at five regions in the baculovirus *Ac*NPV genome. *Journal of Virology* **45**, 961–970.

CRESPI, S. (1980). Patenting nature's secrets and protecting microbiologists' interests. *Nature* **284**, 590–591.

CROIZIER, G. AND MEYNADIER, G. (1972). Les protéines des corps d'inclusion des *Baculovirus* 1. Etude de leur solubilisation. *Entomophaga* **17**, 231–239.

CROIZIER, G., GODSE, D. AND VLAK, J. (1980). Sélection de types viraux dans les infections doubles à *Baculovirus* chez les larves de Lépidoptère. *Comptes rendus des séances de l'Académie des sciences, Série D* **290**, 579–582.

CROIZIER, G., AMARGIER, A., GODSE, D.-B., JACQUEMARD, P. AND DUTHOIT, J.-L. (1980). Un virus de polyédrose nucléaire découvert chez le lépidoptère Noctuidae *Diparopsis watersi* (Roth.) nouveau variant du *Baculovirus* d'*Autographa californica* (Speyer). *Coton et Fibres Tropicales* **35**, 415–423.

CROOK, N.E., BROWN, J.D. AND FOSTER, G.N. (1982). Isolation and characterization of a granulosis virus from the tomato moth, *Lacanobia oleracea*, and its potential as a control agent. *Journal of Invertebrate Pathology* **40**, 221–227.

CUNNINGHAM, J.C. (1982). Field trials with baculoviruses: control of forest insect pests. In *Microbial and Viral Pesticides* (E. Kurstak, Ed.), pp. 335–386. Marcel Dekker, New York.

CUNNINGHAM, J.C. AND ENTWISTLE, P.F. (1981). Control of sawflies by baculovirus. In *Microbial Control of Pests and Plant Diseases 1970–1980* (H.D. Burges, Ed.), pp. 379–407. Academic Press, London.

CUNNINGHAM, J.C., KAUPP, W.J. AND MCPHEE, J.R. (1983). Ground spray trials with two baculoviruses on western spruce budworm. *Canadian Forestry Service Research Notes* **3**, 10–11.

DAVID, W.A.L. (1965). The granulosis virus of *Pieris brassicae* L. in relation to natural limitation and ecological control. *Annals of Applied Biology* **56**, 331–334.

DAVID, W.A.L. AND GARDINER, B.O.C. (1967). The persistence of a granulosis virus of *Pieris brassicae* in soil and sand. *Journal of Invertebrate Pathology* **9**, 342–347.

DOANE, C.C. (1970). Primary pathogens and their role in the development of an epizootic in the gypsy moth. *Journal of Invertebrate Pathology* **15**, 21–33.

DOANE, C.C. (1975). Infectious sources of nuclear polyhedrosis virus persisting in natural habitats of the gypsy moth. *Environmental Entomology* **4**, 392–394.

DOANE, C.C. (1976). Epizootiology of diseases of the gypsy moth. In *Proceedings of the First International Colloquium on Invertebrate Pathology, Kingston, Canada*, (T.A. Angus, P. Faulkner and A. Rosenfield, Eds), pp. 161–165. Society for Invertebrate Pathology, Queen's University at Kingston.

DOBOS, P. AND COCHRAN, M.A. (1980). Protein synthesis in cells infected by *Autographa californica* nuclear polyhedrosis virus (*Ac*-NPV): the effect of cytosine arabinoside. *Virology* **103**, 446–464.

DÖLLER, G. (1980). Solid phase radioimmunoassay for the detection of polyhedrin antibodies. In *Safety Aspects of Baculoviruses as Biological Insecticides* (H.G. Miltenburger, Ed.), pp. 203–210. Symposium Proceedings, Jülich, 1978, Bundesministerium für Forschung und Technologie, Bonn, Federal Republic of Germany.

DÖLLER G. (1981). Unspecific interaction between granulosis virus and mammalian immunoglobulins. *Naturwissenschaften* **68**, 573–574.

DÖLLER, G. AND ENZMANN, P.-J. (1982). Induction of baculovirus specific antibodies in rainbow trout and carp. *Bulletin of the European Association of Fish Pathologists* **2**, 53–55.

DÖLLER, G. AND HUBER, J. (1983). Sicherheitsstudie zur Prüfung einer Vermehrung des Granulosevirus aus *Laspeyresia pomonella* in Säugern. *Zeitschrift für angewandte Entomologie* **95**, 64–69.

DÖLLER, G., GRÖNER, A. AND STRAUB, O.C. (1983). Safety evaluation of nuclear polyhedrosis virus replication in pigs. *Applied and Environmental Microbiology* **45**, 1229–1233.

DOUGHERTY, E.M., CANTWELL, G.E. AND KUCHINSKI, M. (1982). Biological control of the greater wax moth (Lepidoptera: Pyralidae), utilising in vivo- and in vitro-propagated baculovirus. *Journal of Economic Entomology* **75**, 675–679.

DUBOIS, N. (1976). Effectiveness of chemically decontaminated *Neodiprion sertifer* polyhedral inclusion body suspensions. *Journal of Economic Entomology* **69**, 93–95.

ELLEMAN, C.J. AND ENTWISTLE, P.F. (1982). A study of glands on cotton responsible for the high pH and cation concentration of the leaf surface. *Annals of Applied Biology* **100**, 553–558.

ELLIOTT, R.M. AND KELLY, D.C. (1979). Compartmentalization of the polyamines contained by a nuclear polyhedrosis virus from *Heliothis zea*. *Microbiologica* **2**, 409–413.

ENTWISTLE, P.F. (1976). The development of an epizootic of a nuclear polyhedrosis virus disease in European spruce sawfly, *Gilpinia hercyniae*. In *Proceedings of the First International Colloquium on Invertebrate Pathology, Kingston, Canada* (T.A. Angus, P. Faulkner and A. Rosenfield, Eds), pp. 184–188. Society for Invertebrate Pathology, Queen's University at Kingston.

ENTWISTLE, P.F. (1983). Viruses for insect pest control. *Span* **26**, 59–62.

ENTWISTLE, P.F. AND ADAMS, P.H.W. (1977). Prolonged retention of infectivity in the nuclear polyhedrosis virus of *Gilpinia hercyniae* (Hymenoptera: Diprionidae) on foliage of spruce species. *Journal of Invertebrate Pathology* **29**, 392–394.

ENTWISTLE, P.F., ADAMS, P.H.W. AND EVANS, H.F. (1978). Epizootiology of a nuclear polyhedrosis virus in European spruce sawfly (*Gilpinia hercyniae*): the rate of passage of infective virus through the gut of birds during cage tests. *Journal of Invertebrate Pathology* **31**, 307–312.

ENTWISTLE, P.F., ADAMS, P.H.W., EVANS, H.F. AND RIVERS, C.F. (1983). Epizootiology of a nuclear polyhedrosis virus (Baculoviridae) in European spruce sawfly (*Gilpinia hercyniae*): spread of disease from small epicentres in comparison with spread of baculovirus diseases in other hosts. *Journal of Applied Ecology* **20**, 473–487.

ENTWISTLE, P.F., EVANS, H.F., HARRAP, K.A. AND ROBERTSON, J.S. (1978). *Field Trials on the Control of Pine Sawfly* (Neodiprion sertifer) *using Purified Nuclear Polyhedrosis Virus. First series 1977, Technical Report No. 1*. Unit of Invertebrate Virology, Oxford, UK.

ENTWISTLE, P.F., EVANS, H.F., HARRAP, K.A. AND ROBERTSON, J.S. (in press). Control of European pine sawfly (*Neodiprion sertifer*) (Geoffr.) with its nuclear polyhedrosis virus in Scotland. In *Population Dynamics of Forest Pests* (D. Bevan, Ed.), Proceedings of IUFRO Meeting, Dornoch, Scotland, 1980.

EPPSTEIN, D.A. AND THOMA, J.A. (1977). Characterization and serology of the matrix protein from a nuclear-polyhedrosis virus of *Trichoplusia ni* before and after degradation by an endogenous proteinase. *Biochemical Journal* **167**, 321–332.

ERLANDSON, M.A. AND CARSTENS, E.B. (1983). Mapping early transcription products of *Autographa californica* nuclear polyhedrosis virus. *Virology* **126**, 398–402.

EVANS, H.F. (1981). Quantitative assessment of the relationships between dosage and response of the nuclear polyhedrosis virus of *Mamestra brassicae*. *Journal of Invertebrate Pathology* **37**, 101–109.

EVANS, H.F. (1982). The ecology of *Mamestra brassicae* NPV in soil. In *Invertebrate Pathology and Microbial Control, Proceedings of the Third International Colloquium on Invertebrate Pathology, Brighton, UK* (C.C. Payne and H.D. Burges, Eds), pp. 307–312. Society for Invertebrate Pathology, Glasshouse Crops Research Institute, Littlehampton.

EVANS, H.F. (1983). The influence of larval maturation on responses of *Mamestra brassicae* L. (Lepidoptera: Noctuidae) to nuclear polyhedrosis virus infection. *Archives of Virology* **75**, 163–170.

EVANS, H.F. AND ALLAWAY, G.P. (1983). Dynamics of baculovirus growth and dispersal in *Mamestra brassicae* L. (Lepidoptera: Noctuidae) larval populations introduced into small cabbage plots. *Applied and Environmental Microbiology* **45**, 493–501.

EVANS, H.F. AND HARRAP, K.A. (1982). Persistence of insect viruses. In *Virus Persistence, 33rd Symposium of the Society for General Microbiology* (B.W.J. Mahy, A.C. Minson and G.K. Darby, Eds), pp. 57–96. Cambridge University Press.

EVANS, H.F., LOMER, C.J. AND KELLY, D.C. (1981). Growth of nuclear polyhedrosis virus in larvae of the cabbage moth, *Mamestra brassicae* L. *Archives of Virology* **70**, 207–214.

FALCON, L.A. (1982). Use of pathogenic viruses as agents for the biological control of insect pests. In *Population Biology of Infectious Diseases* (R.M. Anderson and R.M. May, Eds), pp. 191–210. Springer-Verlag, Berlin.

FAULKNER, P. AND HENDERSON, J.F. (1972). Serial passage of a nuclear polyhedrosis disease virus of the cabbage looper (*Trichoplusia ni*) in a continuous tissue culture cell line. *Virology* **50**, 920–924.

FAUST, R.M. AND ADAMS, J.R. (1966). The silicon content of nuclear and cytoplasmic viral inclusion bodies causing polyhedrosis in Lepidoptera. *Journal of Invertebrate Pathology* **8**, 526–530.

FEDERICI, B.A. (1980). Mosquito baculovirus: sequence of morphogenesis and ultrastructure of the virion. *Virology* **100**, 1–9.

FENNER, F. (1983). Biological control, as exemplified by smallpox eradication and myxomatosis. *Proceedings of the Royal Society of London, B* **218**, 259–285.

FLATTERY, K.E. (1983). Bioassay of a purified nuclear polyhedrosis virus against *Heliothis armigera*. *Annals of Applied Biology* **102**, 301–304.

FRASER, M.J. AND HINK, W.F. (1982). The isolation and characterization of the MP and FP plaque variants of *Galleria mellonella* nuclear polyhedrosis virus. *Virology* **117**, 366–378.

FRASER, M.J., SMITH, G.E. AND SUMMERS, M.D. (1983). Acquisition of host cell DNA sequences by baculoviruses: relationship between host DNA insertions and FP mutants of *Autographa californica* and *Galleria mellonella* nuclear polyhedrosis viruses. *Journal of Virology* **47**, 287–300.

GARDINER, G.R., PRISTON, R.A.J. AND STOCKDALE, H. (1976). Studies on the production of baculoviruses in insect tissue culture. In *Proceedings of the First International Colloquium on Invertebrate Pathology, Kingston, Canada* (T. A. Angus, P. Faulkner and A. Rosenfield, Eds), pp. 99–103. Society for Invertebrate Pathology, Queen's University at Kingston.

GLEN, D.M. AND PAYNE, C.C. (1984). Production and field evaluation of codling moth granulosis virus against *Cydia pomonella* in the United Kingdom. *Annals of Applied Biology* **104**, in press.

GLEN, D.M., WILTSHIRE, C.W., MILSOM, N.F. AND BRAIN, P. (1984). Codling moth granulosis virus: effects of its use on other orchard fauna. *Annals of Applied Biology*, in press.

GOODWIN, R.H. AND ADAMS, J.R. (1980). Liposome incorporation of factors permitting serial passage of insect viruses in Lepidopteran cells grown in serum-free medium. *In Vitro* **16**, 222.

GOURREAU, J.M., KAISER, C. AND MONTSARRAT, P. (1981). Étude de l'action pathogène éventuelle du baculovirus d'*Oryctes* sur cultures cellulaires de vertébrés en lignée continue. *Annales de Virologie* (*Institut Pasteur*) **132E**, 347–355.

GOURREAU, J.M., KAISER, C. AND MONSARRAT, P. (1982). Study of the possible pathogenic action of the *Oryctes* baculovirus in the white mouse. *Annales de Virologie* (*Institut Pasteur*) **133E**, 423–428.

GOURREAU, J.-M., KAISER, C., LAHELLEC, M., CHEVRIER, L. AND MONSARRAT, P. (1979). Étude de l'action pathogène éventuelle du *Baculovirus* d'*Oryctes* pour le porc. *Entomophaga* **24**, 213–219.

GRACE, T.D.C. (1982). Development of insect cell culture. In *Invertebrate Cell Culture Applications* (K. Maramorosch and J. Mitsuhashi, Eds), pp. 1–8. Academic Press, London.

GRANADOS, R.R. (1978). Early events in the infection of *Heliothis zea* midgut cells by a baculovirus. *Virology* **90**, 170–174.

GRANADOS, R.R. (1980a). Infectivity and mode of action of baculoviruses. *Biotechnology and Bioengineering* **22**, 1377–1405.

GRANADOS, R.R. (1980b). Replication phenomena of insect viruses *in vivo* and *in vitro*. In *Safety Aspects of Baculoviruses as Biological Insecticides, Symposium Proceedings, Jülich, 1978* (H.G. Miltenburger, Ed.), pp. 163–184. Bundesministerium für Forschung und Technologie, Bonn, Federal Republic of Germany.

GRANADOS, R.R. AND LAWLER, K.A. (1981). *In vivo* pathway of *Autographa californica* baculovirus invasion and infection. *Virology* **108**, 297–308.

GRANADOS, R.R., NGUYEN, T. AND CATO, B. (1978). An insect cell line persistently infected with a baculovirus-like particle. *Intervirology* **10**, 309–317.

GREEN, E.I. (1981). *Interactions Between a Baculovirus and its Host*, Tipula paludosa (*Meigen*). PhD thesis, Liverpool Polytechnic.

HAMM, J.J. (1982). Extension of the host range for a granulosis virus from *Heliothis armiger* from South Africa. *Environmental Entomology* **11**, 159–160.

HAMM, J.J. AND HARE, W.W. (1982). Application of entomopathogens in irrigation water for control of fall armyworms and corn earworms (Lepidoptera: Noctuidae) on corn. *Journal of Economic Entomology* **75**, 1074–1079.

HAMM, J.J. AND YOUNG, J.R. (1974). Mode of transmission of nuclear-polyhedrosis virus to progeny of adult *Heliothis zea*. *Journal of Invertebrate Pathology* **24**, 70–81.

HARRAP, K.A. (1970). Cell infection by a nuclear polyhedrosis virus. *Virology* **42**, 311–318.

HARRAP, K.A. (1982). Assessment of the human and ecological hazards of microbial insecticides. *Parasitology* **84**, 269–296.

HARRAP, K.A. AND PAYNE, C.C. (1979). The structural properties and identification of insect viruses. *Advances in Virus Research* **25**, 273–355.

HARRAP, K.A. AND TINSLEY, T.W. (1978). The international scope of invertebrate virus research in controlling pests. In *Viral Pesticides: Present Knowledge and Potential Effects on Public and Environmental Health, Environmental Protection Agency Symposium, Myrtle Beach, South Carolina, US, 1977* (M.D. Summers and C.Y. Kawanishi, Eds), pp. 27–42. EPA Health Effects Research Laboratory, Research Triangle Park.

HARVEY, J.P. AND VOLKMAN, L.E. (1983). Biochemical and biological variation of *Cydia pomenella* (codling moth) granulosis virus. *Virology* **124**, 21–34.

HEINE, C.W., KELLY, D.C. AND AVERY, R.J. (1980). The detection of intracellular retrovirus-like entities in *Drosophila melanogaster* cell cultures. *Journal of General Virology* **49**, 385–395.

HENRY, J.E. (1982). Use of baits in microbial control of insects. In *Invertebrate Pathology and Microbial Control, Proceedings of the Third International Colloquium on Invertebrate Pathology, Brighton, UK* (C.C. Payne and H.D. Burges, Eds), pp. 45–48. Society for Invertebrate Pathology, Glasshouse Crops Research Institute, Littlehampton.

HILWIG, I. AND ALAPATT, F. (1981). Insect cells lines in suspension, cultivated in roller bottles. *Zeitschrift für angewandte Entomologie* **91**, 1–7.

HIMENO, M., SAKAI, F., ONODERA, K., NAKAI, H., FUKUDA, T. AND KAWADE, Y. (1967). Formation of nuclear polyhedral bodies and nuclear polyhedrosis virus of silkworm in mammalian cells infected with viral DNA. *Virology* **33**, 507–512.

HINK, W.F. (1970). Established insect cell line from the cabbage looper, *Trichoplusia ni*. *Nature* **226**, 466–467.

HINK, W.F. (1976). A compilation of invertebrate cell lines and culture media. In *Invertebrate Tissue Culture Research Applications* (K. Maramorosch, Ed.), pp. 319–369. Academic Press, New York.

HINK, W.F. (1980). The 1979 compilation of invertebrate cell lines and culture media. In *Invertebrate Systems In Vitro, Fifth International Conference on Invertebrate Tissue Culture, Rigi-Kaltbad, 1979* (E. Kurstak, K. Maramorosch and A. Dübendorfer, Eds), pp. 553–578. Elsevier/North-Holland, Amsterdam.

HINK, W.F. (1982). Production of *Autographa californica* nuclear polyhedrosis virus in cells from large-scale suspension cultures. In *Microbial and Viral Pesticides* (E. Kurstak, Ed.), pp. 493–506. Marcel Dekker, New York.

HINK, W.F. AND STRAUSS, E. (1976). Replication and passage of alfalfa looper nuclear polyhedrosis virus plaque variants in cloned cell cultures and larval stages of four host species. *Journal of Invertebrate Pathology* **27**, 49–55.

HINK, W.F. AND STRAUSS, E.M. (1977). An improved technique for plaque assay of *Autographa californica* nuclear polyhedrosis virus on TN–368 cells. *Journal of Invertebrate Pathology* **29**, 390–391.

HINK, W.F. AND STRAUSS, E.M. (1980). Semi-continuous culture of the TN–368 cell line in fermentors with virus production in harvested cells. In *Invertebrate Systems In Vitro, Fifth International Conference on Invertebrate Tissue Culture, Rigi-Kaltbad, 1979* (E. Kurstak, K. Maramorosch and A. Dübendorfer, Eds), pp. 27–33. Elsevier/North-Holland, Amsterdam.

HINK, W.F., STRAUSS, E.M. AND RAMOSKA, W.A. (1977). Propagation of *Autographa californica* nuclear polyhedrosis virus in cell culture: methods for infecting cells. *Journal of Invertebrate Pathology* **30**, 185–191.

HIRUMI, H., HIRUMI, K. AND McINTOSH, A.H. (1975). Morphogenesis of a nuclear polyhedrosis virus of the alfalfa looper in a continuous cabbage looper cell line. *Annals of the New York Academy of Sciences* **266**, 302–326.

HOHMANN, A.W. AND FAULKNER, P. (1983). Monoclonal antibodies to baculovirus structural proteins: determination of specificities by Western blot analysis. *Virology* **125**, 432–444.

HOSTETTER, D.L., BIEVER, K.D., HEIMPEL, A.M. AND IGNOFFO, C.M. (1979). Efficacy of the nuclear polyhedrosis virus of the alfalfa looper against cabbage looper larvae on cabbage in Missouri. *Journal of Economic Entomology* **72**, 371–373.

HOSTETTER, D.L., SMITH, D.B., PINNELL, R.E., IGNOFFO, C.M. AND McKIBBEN, G.H. (1982). Laboratory evaluation of adjuvants for use with *Baculovirus heliothis* virus. *Journal of Economic Entomology* **75**, 1114–1119.

HUBER, J. (1982). The baculoviruses of *Cydia pomonella* and other tortricids. In *Invertebrate Pathology and Microbial Control, Proceedings of the Third International Colloquium on Invertebrate Pathology, Brighton, UK* (C.C. Payne and H.D. Burges, Eds), pp. 119–124. Society for Invertebrate Pathology, Glasshouse Crops Research Institute, Littlehampton.

HUBER, J. AND DICKLER, E. (1977). Codling moth granulosis virus: its efficiency in the field in comparison with organophosphorus insecticides. *Journal of Economic Entomology* **70**, 557–561.

HUGER, A.M. (1966). A virus disease of the Indian rhinoceros beetle, *Oryctes rhinoceros* (Linnaeus), caused by a new type of insect virus, *Rhabdionvirus oryctes* gen.n., sp.n. *Journal of Invertebrate Pathology* **8**, 38–51.

HUGHES, P.R. AND WOOD, H.A. (1981). A synchronous peroral technique for the bioassay of insect viruses. *Journal of Invertebrate Pathology* **37**, 154–159.

HUNTER, D.K. AND HOFFMANN, D.F. (1973). Susceptibility of two strains of Indian meal moth to a granulosis virus. *Journal of Invertebrate Pathology* **21**, 114–115.

HUNTER, D.K., HOFFMANN, D.F. AND COLLIER, S.J. (1975). Observations on a granulosis virus of the potato tuberworm, *Phthorimaea operculella*. *Journal of Invertebrate Pathology* **26**, 397–400.

IGNOFFO, C.M. AND ALLEN, G.E. (1972). Selection for resistance to a nucleopolyhedrosis virus in laboratory populations of the cotton bollworm, *Heliothis zea*. *Journal of Invertebrate Pathology* **20**, 187–192.

IGNOFFO, C.M. AND ANDERSON, R.F. (1979). Bioinsecticides. In *Microbial Technology* (H.J. Peppler and D. Perlman, Eds), Volume 1, 2nd edn, pp. 1–28. Academic Press, New York.

IGNOFFO, C.M. AND COUCH, T.L. (1981). The nucleopolyhedrosis virus of *Heliothis* species as a microbial insecticide. In *Microbial Control of Pests and Plant Diseases 1970–1980* (H.D. Burges, Ed.), pp. 329–362. Academic Press, London.

IGNOFFO, C.M. AND RAFAJKO, R.R. (1972). *In vitro* attempts to infect primate cells with the nucleopolyhedrosis virus of *Heliothis*. *Journal of Invertebrate Pathology* **20**, 321–325.

IGNOFFO, C.M., ANDERSON, R.F. AND WOODARD, G. (1973). Teratogenic potential in rats fed the nuclear polyhedrosis virus of *Heliothis*. *Environmental Entomology* **2**, 337–338.

IGNOFFO, C.M., GARCIA, C., HOSTETTER, D.L. AND PINNELL, R.E. (1980). Transplanting: a method of introducing an insect virus into an ecosystem. *Environmental Entomology* **9**, 153–154.

IGNOFFO, C.M., HOSTETTER, D.L., BIEVER, K.D., GARCIA, C., THOMAS, G.D., DICKERSON, W.A. AND PINNELL, R. (1978). Evaluation of an entomopathogenic bacterium, fungus and virus for control of *Heliothis zea* on soybeans. *Journal of Economic Entomology* **71**, 165–168.

INJAC, M., VAGO, C., DUTHOIT, J.-L. AND VEYRUNES, J.-C. (1971). Libération (release) des virions dans les polyédroses nucléaires. *Comptes rendus des séances de l'Académie des sciences, Série D* **273**, 439–441.

IRABAGON, T.A. AND BROOKS, W.M. (1974). Interaction of *Campoletis sonorensis* and a nuclear polyhedrosis virus in larvae of *Heliothis virescens*. *Journal of Economic Entomology* **67**, 229–231.

JAQUES, R.P. (1969). Leaching of the nuclear polyhedrosis virus of *Trichoplusia ni* from soil. *Journal of Invertebrate Pathology* **13**, 256–263.

JOHNSON, T.B. AND LEWIS, L.C. (1982). Evaluation of *Rachiplusia ou* and *Autographa californica* nuclear polyhedrosis viruses in suppressing black cutworm damage to seedling corn in greenhouse and field. *Journal of Economic Entomology* **75**, 401–404.

JURKOVÍČOVÁ, M. (1979). Activation of latent virus infections in larvae of *Adoxophyes orana* (Lepidoptera: Tortricidae) and *Barathra brassicae* (Lepidoptera: Noctuidae) by foreign polyhedra. *Journal of Invertebrate Pathology* **34**, 213–223.

KALMAKOFF, J. AND CRAWFORD, A.M. (1982). Enzootic virus control of *Wiseana* spp. in the pasture environment. In *Microbial and Viral Pesticides* (E. Kurstak, Ed.), pp. 435–448. Marcel Dekker, New York.

KAWANISHI, C.Y. AND PASCHKE, J.D. (1970). The relationship of buffer pH and ionic strength on the yield of virions and nucleocapsids obtained by the dissolution of Rachiplusia ou nuclear polyhedra. In *Proceedings of the Fourth International Colloquium on Insect Pathology, Maryland, US* (A.M. Heimpel, Ed.), pp. 127–146. Society for Invertebrate Pathology, Agricultural Research Service, Beltsville, Maryland, USA.

KAWANISHI, C.Y., EGAWA, K. AND SUMMERS, M.D. (1972). Solubilization of *Trichoplusia ni* granulosis virus proteinic crystal. II. Ultrastructure. *Journal of Invertebrate Pathology* **20**, 95–100.

KAYTON, I. (1983). Does copyright law apply to genetically engineered cells? *Trends in Biotechnology* **1**, 2–3.

KELLY, D.C. (1981a). Baculovirus replication: electron microscopy of the sequence of infection of *Trichoplusia ni* nuclear polyhedrosis virus in *Spodoptera frugiperda* cells. *Journal of General Virology* **52**, 209–219.

KELLY, D.C. (1981b). Baculovirus replication: stimulation of thymidine kinase and DNA polymerase activities in *Spodoptera frugiperda* cells infected with *Trichoplusia ni* nuclear polyhedrosis virus. *Journal of General Virology* **52**, 313–319.

KELLY, D.C. (1982). Baculovirus replication. *Journal of General Virology* **63**, 1–13.

KELLY, D.C. AND LESCOTT, T. (1981). Baculovirus replication: protein synthesis in *Spodoptera frugiperda* cells infected with *Trichoplusia ni* nuclear polyhedrosis virus. *Microbiologica* **4**, 35–57.

KELSEY, J.M. (1958). Control of *Pieris rapae* by granulosis viruses. *New Zealand Journal of Agricultural Research* **1**, 778–782.

KISLEV, N. AND EDELMAN, M. (1982). DNA restriction-pattern differences from geographic isolates of *Spodoptera littoralis* nuclear polyhedrosis virus. *Virology* **119**, 219–222.

KLEIN, M. (1978). An improved peroral administration technique for bioassay of nucleo-polyhedrosis viruses against Egyptian cotton worm, *Spodoptera littoralis*. *Journal of Invertebrate Pathology* **31**, 134–136.

KNUDSON, D.L. AND HARRAP, K.A. (1976). Replication of a nuclear polyhedrosis virus in a continuous cell culture of *Spodoptera frugiperda*: microscopy study of the sequence of events of the virus infection. *Journal of Virology* 17, 254–268.

LAING, D.R. AND JAQUES, R.P. (1980). Codling moth: techniques for rearing larvae and bioassaying granulosis virus. *Journal of Economic Entomology* 73, 851–853.

LAUTENSCHLAGER, R.A. AND PODGWAITE, J.D. (1979). Passage of nucleopolyhedrosis virus by avian and mammalian predators of the gypsy moth, *Lymantria dispar*. *Environmental Entomology* 8, 210–214.

LAUTENSCHLAGER, R.A., ROTHENBACHER, H. AND PODGWAITE, J.D. (1978). Response of small mammals to aerial applications of the nucleopolyhedrosis virus of the gypsy moth, *Lymantria dispar*. *Environmental Entomology* 7, 676–684.

LEE, H.H. AND MILLER, L.K. (1978). Isolation of genotypic variants of *Autographa californica* nuclear polyhedrosis virus. *Journal of Virology* 27, 754–767.

LEWIS, F.B. (1981). Control of the gypsy moth by a baculovirus. In *Microbial Control of Pests and Plant Diseases 1970–1980* (H.D. Burges, Ed.), pp. 363–377. Academic Press, London.

LONGWORTH, J.F. AND CUNNINGHAM, J.C. (1968). The activation of occult nuclear-polyhedrosis viruses by foreign nuclear polyhedra. *Journal of Invertebrate Pathology* 10, 361–367.

LONGWORTH, J.F. AND SINGH, P. (1980). A nuclear polyhedrosis virus of the light brown apple moth, *Epiphyas postvittana* (Lepidoptera: Tortricidae). *Journal of Invertebrate Pathology* 35, 84–87.

LÜBBERT, H., KRUCZEK, I., TJIA, S. AND DOERFLER, W. (1981). The cloned *EcoRI* fragments of *Autographa californica* nuclear polyhedrosis virus DNA. *Gene* 16, 343–345.

LYNN, D.E. AND HINK, W.F. (1978). Infection of synchronized TN–368 cell cultures with alfalfa looper nuclear polyhedrosis virus. *Journal of Invertebrate Pathology* 32, 1–5.

LYNN, D.E. AND HINK, W.F. (1980). Comparison of nuclear polyhedrosis virus replication in five lepidopteran cell lines. *Journal of Invertebrate Pathology* 35, 234–240.

MCCARTHY, W.J. AND HENCHAL, L.S. (1983). Detection of *Autographa californica* baculovirus nonoccluded virions in vitro and in vivo by enzyme-linked immunosorbent assay. *Journal of Invertebrate Pathology* 41, 401–404.

MCINTOSH, A.H. AND IGNOFFO, C.M. (1983). Restriction endonuclease patterns of three baculoviruses isolated from species of *Heliothis*. *Journal of Invertebrate Pathology* 41, 27–32.

MCINTOSH, A.H. AND MARAMOROSCH, K. (1973). Retention of insect virus infectivity in mammalian cell cultures. *Journal of the New York Entomological Society* 81, 175–182.

MCINTOSH, A.H. AND RECHTORIS, C. (1974). Insect cells: colony formation and cloning in agar medium. *In Vitro* 10, 1–5.

MCINTOSH, A.H. AND SHAMY, R. (1980). Biological studies of a baculovirus in a mammalian cell line. *Intervirology* 13, 331–341.

MCINTOSH, A.H., MARAMOROSCH, K. AND RISCOE, R. (1979). *Autographa californica* nuclear polyhedrosis virus (NPV) in a vertebrate cell line: localization by electron microscopy. *Journal of the New York Entomological Society* 87, 55–58.

MCINTOSH, A.H., SHAMY, R. AND ILSLEY, C. (1979). Interference with polyhedral inclusion body (PIB) production in *Trichoplusia ni* cells infected with a high passage strain of *Autographa californica* nuclear polyhedrosis virus (NPV). *Archives of Virology* 60, 353–358.

MCKINLEY, D.J., BROWN, D.A., PAYNE, C.C. AND HARRAP, K.A. (1981). Cross-infectivity and activation studies with four baculoviruses. *Entomophaga* 26, 79–90.

MACKINNON, E.A., HENDERSON, J.F., STOLTZ, D.B. AND FAULKNER, P. (1974). Morphogenesis of nuclear polyhedrosis virus under conditions of prolonged passage *in vitro*. *Journal of Ultrastructure Research* 49, 419–435.

MAGNOLER, A. (1968). The differing effectiveness of purified and non purified suspensions of the nuclear-polyhedrosis virus of *Porthetria dispar*. *Journal of Invertebrate Pathology* **11**, 326–328.

MALEKI-MILANI, H. (1978). Influence de passages répétés du virus de la polyèdrose nucléaire de *Autographa californica* chez *Spodoptera littoralis* [Lep.: Noctuidae]. *Entomophaga* **23**, 217–224.

MARSCHALL, K.J. AND IOANE, I. (1982). The effect of re-release of *Oryctes rhinoceros* baculovirus in the biological control of rhinoceros beetles in Western Samoa. *Journal of Invertebrate Pathology* **39**, 267–276.

MARTIGNONI, M.E., STELZER, M.J. AND IWAI, P.J. (1982). *Baculovirus* of *Autographa californica* (Lepidoptera: Noctuidae): a candidate biological control agent for Douglas-fir tussock moth (Lepidoptera: Lymantriidae). *Journal of Economic Entomology* **75**, 1120–1124.

MATTHEWS, G.A. (1982). Prospects of better deposition of microbial pesticides using electrostatic sprayers. In *Invertebrate Pathology and Microbial Control, Proceedings of the Third International Colloquium on Invertebrate Pathology, Brighton, UK* (C.C. Payne and H.D. Burges, Eds), pp. 55–59. Society for Invertebrate Pathology, Glasshouse Crops Research Institute, Littlehampton.

MATTHEWS, G.A. (1983). Can we control insect pests? *New Scientist* **98**, 368–372.

MATTHEWS, R.E.F. (1982). Classification and nomenclature of viruses. *Intervirology* **17**, 1–199.

MILLER, D.W. AND MILLER, L.K. (1982). A virus mutant with an insertion of a *copia*-like transposable element. *Nature* **299**, 562–564.

MILLER, L.K. AND DAWES, K. (1979). Physical map of the DNA genome of *Autographa californica* nuclear polyhedrosis virus. *Journal of Virology* **29**, 1044–1055.

MILLER, L.K., LINGG, A.J. AND BULLA, L.A. (1983). Bacterial, viral and fungal insecticides. *Science* **219**, 715–721.

MILTENBURGER, H.G. (1980). Viral pesticides: hazard evaluation for non-target organisms and safety testing. In *Environmental Protection and Biological Forms of Control of Pest Organisms* (B. Lundholm and M. Stackerud, Eds), *Ecological Bulletins* **31**, 57–74. Swedish Natural Science Research Council, Stockholm.

MILTENBURGER, H.G. AND DAVID, P. (1980). Mass production of insect cells in suspension. In *Proceedings of the Third General Meeting of the European Society of Animal Cell Technology 1979, Developments in Biological Standardization* volume 46, pp. 183–186. S. Karger, Basel.

MILTENBURGER, H.G. AND REIMANN, R. (1980). Viral pesticides: biohazard evaluation on the cytogenetic level. In *Proceedings of the Third General Meeting of the European Society of Animal Cell Technology 1979, Developments in Biological Standardization* volume 46, pp. 217–222. S. Karger, Basel.

MINION, F.C., COONS, L.B. AND BROOME, J.R. (1979). Characterization of the polyhedral envelope of the nuclear polyhedrosis virus of *Heliothis virescens*. *Journal of Invertebrate Pathology* **34**, 303–307.

MITSUHASHI, J. (1981). Establishment and some characteristics of a continuous cell line derived from fat bodys of the cabbage armyworm (Lepidoptera, Noctuidae). *Development, Growth and Differentiation* **23**, 63–72.

MITSUHASHI, J. (1982). Media for insect cell cultures. *Advances in Cell Culture* **2**, 133–196.

MONSARRAT, P. AND VEYRUNES, J.C. (1976). Evidence of *Oryctes* virus in adult feces and new data for virus characterization. *Journal of Invertebrate Pathology* **27**, 387–389.

MORRIS, O.N. (1980). Entomopathogenic viruses: strategies for use in forest insect pest management. *Canadian Entomologist* **112**, 573–584.

MORRIS, T.J., HESS, R.T. AND PINNOCK, D.E. (1979). Physicochemical characterization of a small RNA virus associated with baculovirus infection in *Trichoplusia ni*. *Intervirology* **11**, 238–247.

MORRIS, T.J., VAIL, P.V. AND COLLIER, S.S. (1981). An RNA virus in *Autographa californica* nuclear polyhedrosis preparations: detection and identification. *Journal of Invertebrate Pathology* **38**, 201–208.

NASER, W.L. AND MILTENBURGER, H.G. (1983). Rapid baculovirus detection, identification, and serological classification by Western blotting—ELISA using a monoclonal antibody. *Journal of General Virology* **64**, 639–647.

NORDIN, G.L. (1976). Microsporidian bioassay technique for third-instar *Pseudaletia unipuncta* larvae. *Journal of Invertebrate Pathology* **27**, 397–398.

PADHI, S.B. AND MARAMOROSCH, K. (1983). *Heliothis zea* baculovirus and *Bombyx mori:* safety considerations. *Applied Entomology and Zoology* **18**, 136–138.

PAPWORTH, D.S. (1980). Registration requirements in the UK for bacteria, fungi and viruses used as pesticides. In *Environmental Protection and Biological Forms of Control of Pest Organisms* (B. Lundholm and M. Stackerud, Eds), *Ecological Bulletins* **31**, 135–143. Swedish Natural Science Research Council, Stockholm.

PAYNE, C.C. (1982). Insect viruses as control agents. *Parasitology* **84**, 35–77.

PAYNE, C.C., TATCHELL, G.M. AND WILLIAMS, C.F. (1981). The comparative susceptibilities of *Pieris brassicae* and *P. rapae* to a granulosis virus from *P. brassicae*. *Journal of Invertebrate Pathology* **38**, 273–280.

PFRIMMER, T.R. (1979). *Heliothis* spp.: control on cotton with pyrethroids, carbamates, organophosphates, and biological insecticides. *Journal of Economic Entomology* **72**, 593–598.

PLUS, N. (1978). Endogenous viruses of *Drosophila melanogaster* cell lines: their frequency, identification and origin. *In Vitro* **14**, 1015–1021.

PLUS, N. (1980). Further studies on the origin of the endogenous viruses of *Drosophila melanogaster* cell lines. In *Invertebrate Systems In Vitro, Fifth International Conference on Invertebrate Tissue Culture, Rigi-Kaltbad, 1979* (E. Kurstak, K. Maramorosch and A. Dübendorfer, Eds), pp. 435–439. Elsevier/North-Holland, Amsterdam.

PODGWAITE, J.D., BRUEN, R.B. AND SHAPIRO, M. (1983). Microorganisms associated with production lots of the nucleopolyhedrosis virus of the gypsy moth, *Lymantria-dispar* [*Lep.: Lymantriidae*]. *Entomophaga* **28**, 9–16.

POLLARD, R. AND KHOSROVI, B. (1978). Reactor design for fermentation of fragile tissue cells. *Process Biochemistry* **13**, 31–37.

POTTER, K.N. AND MILLER, L.K. (1980). Correlating genetic mutations of a baculovirus with the physical map of the DNA genome. In *Animal Virus Genetics, ICN–UCLA Symposia on Molecular and Cellular Biology* (B.N. Fields, R. Jaenisch and C.F. Fox, Eds), volume 18, pp. 71–80. Academic Press, New York.

POTTER, K.N., FAULKNER, P. AND MACKINNON, E.A. (1976). Strain selection during serial passage of *Trichoplusia ni* nuclear polyhedrosis virus. *Journal of Virology* **18**, 1040–1050.

POTTER, K.N., JAQUES, R.P. AND FAULKNER, P. (1978). Modification of *Trichoplusia ni* nuclear polyhedrosis virus passaged *in vivo*. *Intervirology* **9**, 76–85.

POTTER, M.F. AND WATSON, T.F. (1983a). Laboratory and greenhouse performance of *Baculovirus heliothis*, combined with feeding stimulants for control of neonate tobacco budworm. *Protection Ecology* **5**, 161–165.

POTTER, M.F. AND WATSON, T.F. (1983b). Timing of nuclear polyhedrosis virus-bait spray combinations for control of egg and larval stages of tobacco budworm (Lepidoptera: Noctuidae). *Journal of Economic Entomology* **76**, 446–448.

RAIMO, B., REARDON, R.C. AND PODGWAITE, J.D. (1977). Vectoring gypsy moth nuclear polyhedrosis virus by *Apanteles melanoscelus* [*Hym.: Braconidae*]. *Entomophaga* **22**, 207–215.

REED, D.K. (1981). Control of mites by non-occluded viruses. In *Microbial Control of Pests and Plant Diseases 1970–1980* (H.D. Burges, Ed.), pp. 427–432. Academic Press, London.

REED, E.M. AND SPRINGETT, B.P. (1971). Large-scale field testing of a granulosis virus for the control of the potato moth (*Phthorimaea operculella* (Zell.) (Lep., Gelechiidae)). *Bulletin of Entomological Research* **61**, 223–233.

REIMANN, R. AND MILTENBURGER, H.G. (1983). Cytogenetic studies in mammalian cells after treatment with insect pathogenic viruses [*Baculoviridae*]. II *In vitro* studies with mammalian cell lines. *Entomophaga* **28**, 33–44.

RICHARDS, M.G. AND PAYNE, C.C. (1982). Persistence of baculoviruses on leaf surfaces. In *Invertebrate Pathology and Microbial Control, Proceedings of the Third International Colloquium on Invertebrate Pathology, Brighton, UK* (C.C. Payne and H.D. Burges, Eds), pp. 296–301. Society for Invertebrate Pathology, Glasshouse Crops Research Institute, Littlehampton.

ROBERTS, P.L. AND NASER, W. (1982a). Characterization of monoclonal antibodies to the *Autographa californica* nuclear polyhedrosis virus. *Virology* **122**, 424–430.

ROBERTS, P.L. AND NASER, W. (1982b). Preparation of monoclonal antibodies to a baculovirus. *FEMS Microbiology Letters* **14**, 79–83.

ROGOFF, M.H. (1975). Exposure of humans to nuclear polyhedrosis virus during industrial production. In *Baculoviruses for Insect Pest Control: Safety Considerations, EPA–USDA Symposium, Bethesda, Maryland, USA* (M. Summers, R. Engler, L.A. Falcon and P.V. Vail, Eds), pp. 102–105. American Society for Microbiology, Washington D.C.

ROGOFF, M.H. (1980). Testing requirements for registering biological pesticides in the United States—current status. In *Environmental Protection and Biological Forms of Control of Pest Organisms* (B. Lundholm and M. Stackerud, Eds), *Ecological Bulletins* **31**, 111–134. Swedish Natural Science Research Council, Stockholm.

ROHRMANN, G.F. (1982). Genetic characterization and sequence analysis of polyhedrins. In *Invertebrate Pathology and Microbial Control, Proceedings of the Third International Colloquium on Invertebrate Pathology, Brighton, UK* (C.C. Payne and H.D. Burges, Eds), pp. 226–232. Society for Invertebrate Pathology, Glasshouse Crops Research Institute, Littlehampton.

ROHRMANN, G.F., PEARSON, M.N., BAILEY, T.J., BECKER, R.R. AND BEAUDREAU, G.S. (1981). N-terminal polyhedrin sequences and occluded *Baculovirus* evolution. *Journal of Molecular Evolution* **17**, 329–333.

ROOME, R.E. (1975). Field trials with a nuclear polyhedrosis virus and *Bacillus thuringiensis* against larvae of *Heliothis armigera* on sorghum and cotton in Botswana. *Bulletin of Entomological Research* **65**, 507–514.

RUBINSTEIN, R., LAWLER, K.A. AND GRANADOS, R.R. (1982). Use of primary fat body cultures for the study of baculovirus replication. *Journal of Invertebrate Pathology* **40**, 266–273.

SCHERER, W.F., VERNA, J.E. AND RICHTER, G.W. (1968). Nodamura virus, an ether- and chloroform-resistant arbovirus from Japan. *American Journal of Tropical Medicine and Hygiene* **17**, 120–128.

SEARS, M.K., JAQUES, R.P. AND LAING, J.E. (1983). Utilization of action thresholds for microbial and chemical control of lepidopterous pests (Lepidoptera: Noctuidae, Pieridae) on cabbage. *Journal of Economic Entomology* **76**, 368–374.

SHAPIRO, M. (1982). In vivo mass production of insect viruses for use as pesticides. In *Microbial and Viral Pesticides* (E. Kurstak, Ed.), pp. 463–492. Marcel Dekker, New York.

SHAPIRO, M., BELL, R.A. AND OWENS, C.D. (1981). Evaluation of various artificial diets for in vivo production of the gypsy moth nucleopolyhedrosis virus. *Journal of Economic Entomology* **74**, 110–111.

SHAPIRO, M., POH AGIN, P. AND BELL, R.A. (1983). Ultraviolet protectants of the gypsy moth (Lepidoptera: Lymantriidae) nucleopolyhedrosis virus. *Environmental Entomology* **12**, 982–985.

SHAPIRO, M., MARTIGNONI, M.E., CUNNINGHAM, J.C. AND GOODWIN, R.H. (1982). Potential use of the saltmarsh caterpillar as a production host for nucleopolyhedrosis viruses. *Journal of Economic Entomology* **75**, 69–71.

SHAPIRO, M., OWENS, C.D., BELL, R.A. AND WOOD, H.A. (1981). Simplified, efficient system for in vivo mass production of gypsy moth nucleopolyhedrosis virus. *Journal of Economic Entomology* **74**, 341–343.

SHEPHERD, R.F., GRAY, T.G. AND CUNNINGHAM, J.C. (1982). Effects of nuclear polyhedrosis virus and *Bacillus thuringiensis* on western spruce budworm (Lepidoptera:

Tortricidae) 1 and 2 years after aerial application. *Canadian Entomologist* **114**, 281–282.

SHIEH, T.R. AND BOHMFALK, G.T. (1980). Production and efficacy of baculoviruses. *Biotechnology and Bioengineering* **22**, 1357–1375.

SMITH, D.B. AND BOUSE, L.F. (1981). Machinery and factors that affect the application of pathogens. In *Microbial Control of Pests and Plant Diseases 1970–1980* (H.D. Burges, Ed.), pp. 635–653. Academic Press, London.

SMITH, D.B. AND HOSTETTER, D.L. (1982). Laboratory and field evaluations of pathogen-adjuvant treatments. *Journal of Economic Entomology* **75**, 472–476.

SMITH, D.B., HOSTETTER, D.L. AND IGNOFFO, C.M. (1978). Formulation and equipment effects on application of a viral (*Baculovirus heliothis*) insecticide. *Journal of Economic Entomology* **71**, 814–817.

SMITH, D.B., HOSTETTER, D.L. AND IGNOFFO, C.M. (1979). Nozzle size-pressure and concentration combinations for *Heliothis zea* control with an aqueous suspension of polyvinyl alcohol and *Baculovirus heliothis*. *Journal of Economic Entomology* **72**, 920–923.

SMITH, G.E. AND SUMMERS, M.D. (1978). Analysis of baculovirus genomes with restriction endonucleases. *Virology* **89**, 517–527.

SMITH, G.E. AND SUMMERS, M.D. (1979). Restriction maps of five *Autographa californica* MNPV variants, *Trichoplusia ni* MNPV and *Galleria mellonella* MNPV DNAs with endonucleases *Sma*I, *Kpn*I, *Bam*HI, *Sac*I, *Xho*I, and *Eco*RI. *Journal of Virology* **30**, 828–838.

SMITH, G.E. AND SUMMERS, M.D. (1980). Restriction map of *Rachiplusia ou* and *Rachiplusia ou–Autographa californica* baculovirus recombinants. *Journal of Virology* **33**, 311–319.

SMITH, G.E. AND SUMMERS, M.D. (1982). DNA homology among subgroup A, B and C baculoviruses. *Virology* **123**, 393–406.

SMITH, G.E., VLAK, J.M. AND SUMMERS, M.D. (1982). In vitro translation of *Autographa californica* nuclear polyhedrosis virus early and late mRNAs. *Journal of Virology* **44**, 199–208.

SMITH, G.E., VLAK, J.M. AND SUMMERS, M.D. (1983). Physical analysis of *Autographa californica* nuclear polyhedrosis virus transcripts for polyhedrin and 10,000-molecular-weight protein. *Journal of Virology* **45**, 215–225.

SMITH, K.M. AND XEROS, N. (1954). An unusual virus disease of a dipterous larva. *Nature* **173**, 866–867.

STOCKDALE, H. (in press). Microbial insecticides. In *The Practice of Biotechnology: Commodity Products, Comprehensive Biotechnology, Vol. 2*, Pergamon Press, Oxford.

STOCKDALE, H. AND GARDINER, G.R. (1977). The influence of the condition of cells and medium on production of polyhedra of *Autographa californica* nuclear polyhedrosis virus in vitro. *Journal of Invertebrate Pathology* **30**, 330–336.

STOCKDALE, H. AND PRISTON, R.A.J. (1981). Production of insect viruses in cell culture. In *Microbial Control of Pests and Plant Diseases 1970–1980* (H.D. Burges, Ed.), pp. 313–328. Academic Press, London.

STOLTZ, D.B., PAVAN, C. AND DA CUNHA, A.B. (1973). Nuclear polyhedrosis virus: a possible example of *de novo* intranuclear membrane morphogenesis. *Journal of General Virology* **19**, 145–150.

SUMMERS, M.D. AND SMITH, G.E. (1975). *Trichoplusia ni* granulosis virus granulin: a phenol-soluble phosphorylated protein. *Journal of Virology* **16**, 1108–1116.

SUMMERS, M.D. AND VOLKMAN, L.E. (1976). Comparison of biophysical and morphological properties of occluded and extracellular nonoccluded baculovirus from in vivo and in vitro host systems. *Journal of Virology* **17**, 962–972.

TATCHELL, G.M. (1981). The transmission of a granulosis virus following the contamination of *Pieris brassicae* adults. *Journal of Invertebrate Pathology* **37**, 210–213.

TATCHELL, G.M. AND PAYNE, C.C. (1984). Field evaluation of a granulosis virus for control of *Pieris rapae* in the United Kingdom. *Entomophaga*, in press.

TJIA, S.T., ZU ALTENSCHILDESCHE, G.M. AND DOERFLER, W. (1983). Autographa californica nuclear polyhedrosis virus (AcNPV) DNA does not persist in mass cultures of mammalian cells. *Virology* **125**, 107–117.

TWEETEN, K.A., BULLA, L.A. AND CONSIGLI, R.A. (1980). Structural polypeptides of the granulosis virus of *Plodia interpunctella*. *Journal of Virology* **33**, 877–886.

TWEETEN, K.A., BULLA, L.A. AND CONSIGLI, R.A. (1981). Applied and molecular aspects of insect granulosis viruses. *Microbiological Reviews* **45**, 379–408.

VAGO, C. AND BERGOIN, M. (1963). Dévelopment des virus à corps d'inclusion du Lepidoptère *Lymantria dispar* en cultures cellulaires. *Entomophaga* **8**, 253–261.

VAIL, P.V., SEAY, R.E. AND DEBOLT, J. (1980). Microbial and chemical control of the cabbage looper on fall lettuce. *Journal of Economic Entomology* **73**, 72–75.

VAIL, P.V., MORRIS, T.J., COLLIER, S.S. AND MACKEY, B. (1983). An RNA virus in *Autographa californica* nuclear polyhedrosis virus preparations: incidence and influence on baculovirus activity. *Journal of Invertebrate Pathology* **41**, 171–178.

VAUGHN, J.L. (1981). Insect cells for insect virus production. *Advances in Cell Culture* **1**, 281–295.

VAUGHN, J.L. AND DOUGHERTY, E.M. (1981). Recent progress in *in vitro* studies of baculo-viruses. In *Beltsville Symposia in Agricultural Research*, [5] *Biological Control in Crop Production* (G.C. Papavizas, Ed.), pp. 249–258. Allanheld, Osmun, New Jersey.

VLAK, J.M. (1979). The proteins of nonoccluded *Autographa californica* nuclear poly-hedrosis virus produced in an established cell line of *Spodoptera frugiperda*. *Journal of Invertebrate Pathology* **34**, 110–118.

VLAK, J.M. (1980). Mapping of *Bam*HI and *Sma*I DNA restriction sites on the genome of the nuclear polyhedrosis virus of the alfalfa looper, *Autographa californica*. *Journal of Invertebrate Pathology* **36**, 409–414.

VOLKMAN, L.E. (1983). Occluded and budded *Autographa californica* nuclear polyhedrosis virus: immunological relatedness of structural proteins. *Journal of Virology* **46**, 221–229.

VOLKMAN, L.E. AND FALCON, L.A. (1982). Use of monoclonal antibody in an enzyme-linked immunosorbent assay to detect the presence of *Trichoplusia ni* (Lepidoptera: Noctuidae) S nuclear polyhedrosis virus polyhedrin in *T. ni* larvae. *Journal of Economic Entomology* **75**, 868–871.

VOLKMAN, L.E. AND GOLDSMITH, P.A. (1983). In vitro survey of *Autographa californica* nuclear polyhedrosis virus interaction with nontarget vertebrate host cells. *Applied and Environmental Microbiology* **45**, 1085–1093.

WEISS, S.A., SMITH, G.C., KALTER, S.S. AND VAUGHN, J.L. (1981). Improved method for the production of insect cell cultures in large volume. *In Vitro* **17**, 495–502.

WILKIE, G.E.I., STOCKDALE, H. AND PIRT, S.V. (1980). Chemically-defined media for production of insect cells and viruses *in vitro*. In *Proceedings of the Third General Meeting of the European Society of Animal Cell Technology 1979, Develop-ments in Biological Standardization*, Vol. 46, pp. 29–37. S. Karger, Basel.

WOOD, H.A. (1980). Isolation and replication of an occlusion body-deficient mutant of the *Autographa californica* nuclear polyhedrosis virus. *Virology* **105**, 338–344.

WOOD, H.A., JOHNSTON, L.B. AND BURAND, J.P. (1982). Inhibition of *Autographa californica* nuclear polyhedrosis virus replication in high-density *Trichoplusia ni* cell cultures. *Virology* **119**, 245–254.

WOOD, H.A., HUGHES, P.R., JOHNSTON, L.B. AND LANGRIDGE, W.H.R. (1981). Increased virulence of *Autographa californica* nuclear polyhedrosis virus by mutagenesis. *Journal of Invertebrate Pathology* **38**, 236–241.

YAMADA, K. AND MARAMOROSCH, K. (1981). Plaque assay of *Heliothis zea* baculovirus employing a mixed agarose overlay. *Archives of Virology* **67**, 187–189.

YAMADA, K., SHERMAN, K.E. AND MARAMOROSCH, K. (1982). Serial passage of *Heliothis zea* singly embedded nuclear polyhedrosis virus in a homologous cell line. *Journal of Invertebrate Pathology* **39**, 185–191.

YAMAFUJI, K., YOSHIHARA, F. AND HIRAYAMA, K. (1957). Protease and desoxyribonuclease in viral polyhedral crystal. *Enzymologia* **19**, 53–58.

YEARIAN, W.C. AND YOUNG, S.Y. (1982). Control of insect pests of agricultural importance by viral insecticides. In *Microbial and Viral Pesticides* (E. Kurstak, Ed.), pp. 387–423. Marcel Dekker, New York.

YOUNG, E.C. AND LONGWORTH, J.F. (1981). The epizootiology of the baculovirus of the coconut palm rhinoceros beetle (*Oryctes rhinoceros*) in Tonga. *Journal of Invertebrate Pathology* **38**, 362–369.

YOUNG, S.Y. AND YEARIAN, W.C. (1980). Soil application of *Pseudoplusia* NPV: persistence and incidence of infection in soybean looper caged on soybean. *Environmental Entomology* **8**, 860–864.

ZELAZNY, B. (1979). Virulence of the baculovirus of *Oryctes rhinoceros* from ten locations in the Philippines and in Western Samoa. *Journal of Invertebrate Pathology* **33**, 106–107.

7

Recombinant DNA Technology and Genetic Control of Pest Insects

A.F. COCKBURN*, A.J. HOWELLS† AND M.J. WHITTEN

Division of Entomology, CSIRO, GPO Box 1700, Canberra, ACT 2601, Australia

Introduction

The successful control of the screwworm *Cochliomyia hominivorax* in North America with radiation-sterilized males (Baumhover *et al.*, 1955) has stimulated interest in a range of genetic approaches to the control of pest insects (Waterhouse, LaChance and Whitten, 1974; for recent review *see* Whitten, 1984). This approach has been successfully extended to a few other species (LaChance, 1979). More sophisticated forms of genetic control, including homozygous translocations, meiotic drive, cytoplasmic incompatibility, hybrid sterility, compound chromosomes, sex-linked rearrangements coupled with conditional lethals, have also been developed, but their application has been restricted to a small number of pest species for which adequate genetic knowledge exists to isolate and characterize these phenomena.

The development and evaluation of many of the approaches to genetic control listed above have depended upon the extensive knowledge of the genetics of *Drosophila melanogaster*. This species, which has become the best-understood eukaryote in cytogenetic and genetic terms over the past six decades, now offers new potential to applied entomology because of recent advances in our understanding of its molecular biology.

This paper focuses on those recent advances in the molecular biology of *D. melanogaster* which have some bearing on the development of novel methods of controlling pest insects. We describe the methods currently available for cloning genes and for reinserting cloned genes into germ-line chromosomes. We then discuss how such techniques might be used in genetic control programmes. Whereas a serious limitation to applying classic genetic control

Abbreviations: ACE, acetylcholinesterase; ADH, alcohol dehydrogenase; bp, base pairs; kb, kilobases; SIRM, sterile insect release method; XDH, xanthine dehydrogenase.

*Present address: Department of Biology, University of California, San Diego, La Jolla, California 92093, USA.
†And Department of Biochemistry, The Faculties, Australian National University, Canberra, ACT 2601, Australia.

theory has been the absence of basic genetic information on most pest species, one of the attractive features of the molecular approach is that it may be applicable to pest species where little or no formal genetic information is available. For example, genes from *D. melanogaster* (or from other donors) might be transferred into particular pest species. Alternatively, cloned genes from *D. melanogaster* could be used as probes to isolate homologous genes from the pest insect, which could then be either modified by *in vitro* mutation or inserted at a new genomic site to create a specific genetic variant. In this way, geneticists should be able to exploit the enormous library of *D. melanogaster* mutants as a source of genetic variation for other species of insects.

Cloning and isolation of insect genes

In this section we concentrate on how specific genes can be cloned and identified, because the utility of the genetic control methods we will discuss later depends on the ease with which genetically characterized genes can be obtained. Descriptions of basic techniques such as the use of restriction enzymes to cleave DNA molecules at specific sequences, DNA ligase to join DNA molecules together, DNA hybridization to identify specific sequences, and other relevant techniques, can be found in basic texts on molecular biology. We recommend Old and Primrose (1981) for a general discussion of recombinant DNA technology, and Wu (1979) and Maniatis, Fritsch and Sambrook (1982) for specific experimental details.

CLONING VECTORS

In order to isolate a particular segment of DNA from any insect species, a library of cloned DNA sequences must be prepared. Such libraries are propagated in bacterial hosts, generally *Escherichia coli*. The essential step in the construction of such libraries is to attach the DNA segments to vector DNA molecules that can replicate independently in the bacterium. Three types of cloning vectors can be used in the construction of libraries: plasmids, bacteriophage, and cosmids.

Plasmids

Plasmids are small, circular DNA molecules which exist naturally within bacterial cells. Those used as cloning vectors have been selected by virtue of having the following features: capacity for self-replication, single target sites for cutting by certain restriction enzymes (cloning sites), and antibiotic-resistance genes which can be used to select cells containing the plasmid. One of the most commonly used plasmid vectors is pBR322 (Bolivar and Backman, 1979), the structure of which is shown in *Figure 1*. Naturally occurring plasmids can be very large, up to several hundred kilobases, but the plasmids used in library construction are generally much shorter (only a few kilobases), because there are technical difficulties in working with larger plasmids. Plasmids are seldom used to clone pieces of DNA longer than 10 kb. Once recombinant

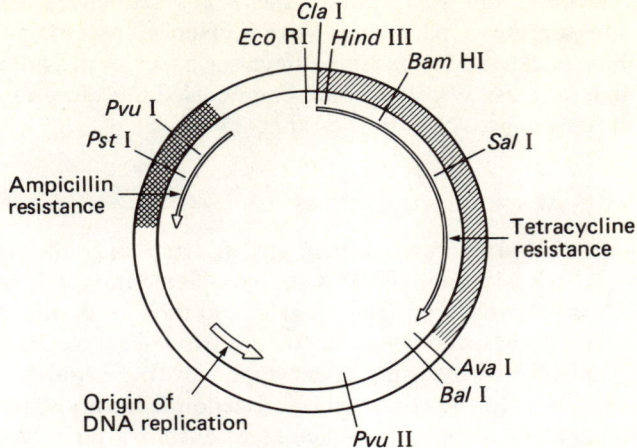

Figure 1.   The genome of pBR322. The double circle represents circular double-stranded DNA and shows the positions of the antibiotic-resistance genes, the origin of replication, and the restriction enzyme sites that are commonly used as insertional cloning sites for cloning DNA fragments (*see* Old and Primrose, 1981).

plasmids consisting of the cloning vector plus passenger DNA fragment have been constructed *in vitro*, they are introduced by DNA transformation into the bacterial host strain for propagation.

Bacteriophage

Bacteriophage cloning vectors are derived from the λ bacteriophage. As with the plasmid vectors, they have been designed specifically to accept passenger DNA sequences. Most of the bacteriophage vectors are replacement vectors, in which a central region of the linear phage DNA is cut out and replaced with a passenger DNA fragment. These recombinant molecules are packaged *in vitro* using bacteriophage proteins to give infectious phage particles. The highly efficient process of infection can therefore be used to introduce the genetic material into the host cell. The need to package the recombinant DNA into the phage head puts strict limits on the length of the passenger DNA fragment — the recombinant DNA can be packaged only if the total length of the molecule is between 39 and 52 kb. A variety of λ replacement vectors is available and can be used to clone fragments of between 8 and 20 kb. Lambda insertional vectors, which contain a single cloning site, are also available. These accept small passenger fragments up to about 8 kb.

Cosmids

Cosmids, which are constructed from plasmids and are similar in size to the plasmid cloning vectors, are self-replicating minichromosomes that convey antibiotic resistance on the host cell, have insertional cloning sites for passenger molecules, and also contain the λ *cos* sequence, which is the recognition site for the λ packaging system. Hence, recombinant cosmid molecules can be packaged *in vitro* into infectious particles provided that they are long enough (at least 39 kb); cosmids can be used therefore to clone very long pieces of passenger DNA (in the range 35–45 kb).

PREPARATION OF PASSENGER DNA MOLECULES

The passenger DNA molecules which are inserted into the vectors are generally of two types: genomic DNA or complementary DNA (cDNA). Genomic DNA is prepared from an organism and then randomly fragmented to give segments of the proper length for insertion. Fragmentation can be achieved mechanically (by shearing) or enzymatically (by partial digestion with a restriction enzyme). In the construction of genomic libraries, the aim is to propagate sufficient recombinant molecules to ensure a high probability of finding any genomic sequence represented in the library. Genomic libraries are usually constructed in either bacteriophage or cosmid vectors; these will accept longer insert fragments than plasmids, so fewer recombinant molecules are required to get a complete library. The procedures that we employed in the construction of a genomic library from the Australian sheep blowfly *Lucilia cuprina* are shown in *Figure 2*.

Complementary DNA (cDNA) is copied from messenger RNA. The synthesis is done using the enzyme RNA-dependent DNA polymerase ('reverse transcriptase'), which uses RNA as a template for DNA synthesis.

Figure 2, *(Opposite).* Construction of a library of genomic DNA from *L. cuprina* in a bacteriophage λ-derived cloning vector.
(i) Preparation of *L. cuprina* DNA: High-molecular-weight DNA is isolated and partially digested with the restriction enzyme *Sau* 3A to an average size of 15 000 bp. The DNA is then treated with phosphatase to remove 5′-terminal phosphates. This prevents the subsequent religation of fragments and ensures that only one fragment can be inserted per recombinant phage.
(ii) Preparation of λ DNA: A detailed description of the cloning vector EMBL 3A is given in Frischauf *et al.* (1983). One feature of this vector is that polylinkers of 16 bp, containing cutting sites for *Sal* I, *Bam* HI and *Eco* RI, have been inserted at each end of the non-essential central region of the phage DNA; the *Bam* HI sites are used as the cloning sites. (For simplification the *Sal* I sites are not shown in our diagrams). EMBL 3A DNA is treated with DNA ligase to circularize the molecules, so protecting the *cos* sites at each end of the linear DNA. The non-essential region is cut out with the restriction enzyme *Eco* RI and the fragments then treated with phosphatase so that they cannot be religated to give intact vector. The mixture is then cut with the restriction enzyme *Bam* HI, which generates on the vector fragments the same overlapping end sequence as *Sau* 3A generates on the *L. cuprina* DNA. Thus the two types of fragments can be ligated together. The *Eco* RI–*Bam* HI fragments (9 bp) are too small to precipitate with ethanol and can therefore be eliminated.
(iii) Library construction: The *L. cuprina Sau* 3A-generated fragments are mixed with the EMBL 3A *Bam* HI-treated vector fragments and incubated with DNA ligase. The ligation reaction requires a phosphate group on at least one of the two ends being ligated together, so the genomic DNA fragments can join to the vector but cannot self-ligate. The non-essential region of EMBL 3A cannot reinsert because it has *Eco* RI overlapping end sequences, which do not match the *Bam* HI ends on the vector fragment. The ligation reaction produces high-molecular-weight concatameric DNA molecules consisting of alternating vector and insert (*L. cuprina*) DNA sequences. The ligated mixture is mixed with structural proteins *in vitro*; during the packaging reaction, which gives rise to infectious phage particles, the concatamers are cut at adjacent *cos* sites to give correctly sized recombinant DNA molecules.

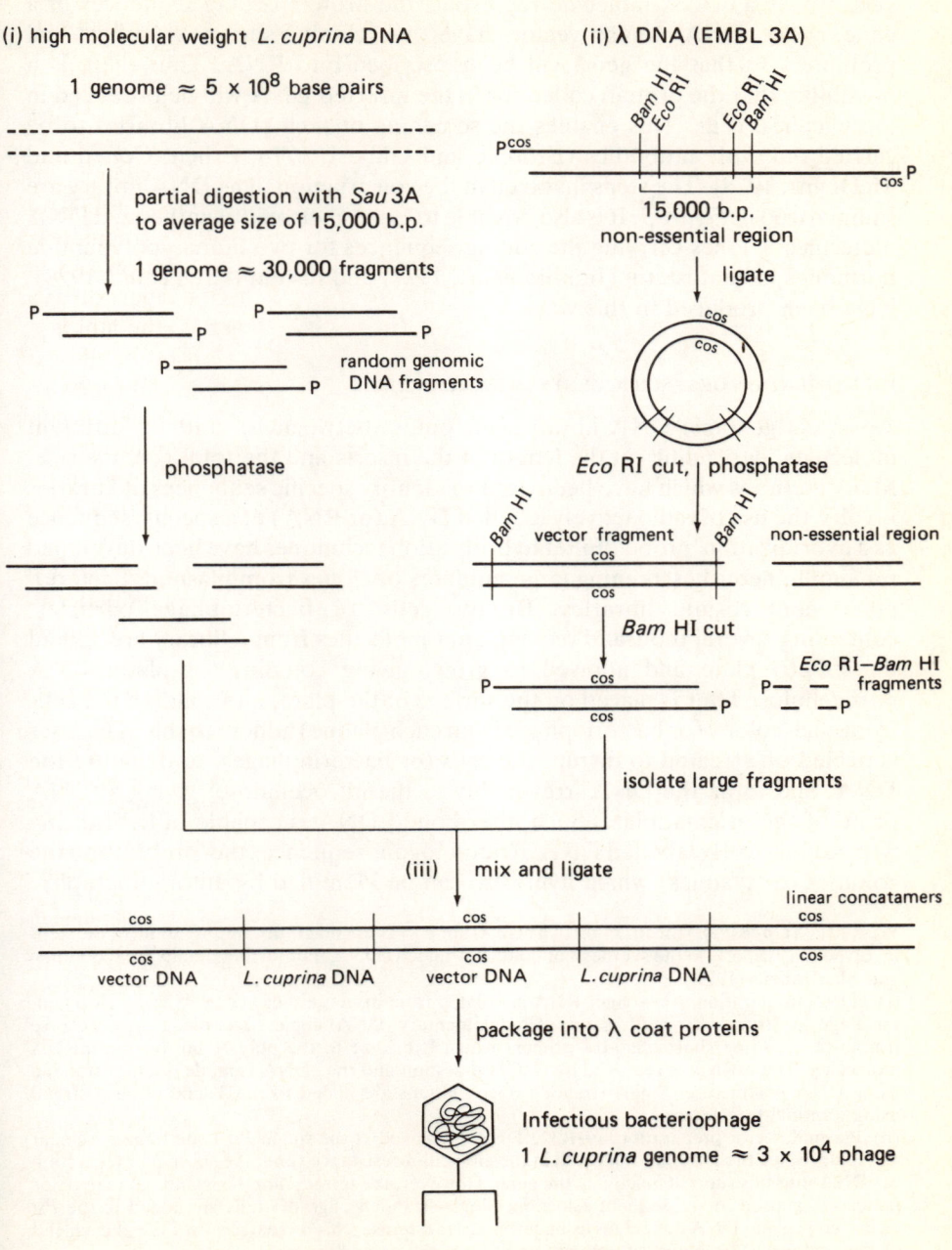

(i) high molecular weight *L. cuprina* DNA

1 genome ≈ 5×10^8 base pairs

partial digestion with *Sau* 3A
to average size of 15,000 b.p.

1 genome ≈ 30,000 fragments

random genomic
DNA fragments

phosphatase

(ii) λ DNA (EMBL 3A)

Bam HI
Eco RI
Eco RI
Bam HI

P cos
cos P

15,000 b.p.
non-essential region

ligate

cos
cos

Eco RI cut, phosphatase

Bam HI
vector fragment
cos
Bam HI
non-essential region
cos

Bam HI cut

P cos
cos P

Eco RI–*Bam* HI
fragments
P— —
— —P

isolate large fragments

(iii) mix and ligate

linear concatamers
cos
cos
cos
cos
cos
cos
vector DNA *L. cuprina* DNA vector DNA *L. cuprina* DNA

package into λ coat proteins

Infectious bacteriophage
1 *L. cuprina* genome ≈ 3×10^4 phage

Since the average messenger RNA molecule is only about 1 kb long, cDNA libraries can be constructed using either plasmid or insertional bacteriophage vectors. As a cDNA molecule represents the protein-coding sequences of a gene, some cDNA cloning vectors have their cloning sites next to bacterial promoters so that the gene will be transcribed into RNA. Thus there is a possibility that the protein coded for in the inserted DNA will be produced in the bacterial cells. This enables the screening of such cDNA libraries to be carried out with antibodies (Broome and Gilbert, 1978; Erlich, Cohen and McDevitt, 1978). The steps involved in the construction of a cDNA library are summarized in *Figure 3*. It is also possible to clone chemically synthesized DNA molecules. Clones carrying the coding sequences for two human polypeptide hormones, somatostatin (Itakura *et al.*, 1977) and insulin (Crea *et al.*, 1978), have been produced in this way.

IDENTIFICATION OF SPECIFIC GENES IN DNA LIBRARIES

An insect genomic DNA library will contain between 10^4 and 10^6 different molecules, depending on the length of the inserts and the total genome size. Many methods which have been used to identify specific sequences in libraries involve the use of radioactively labelled DNA (or RNA) of a specific sequence as a hybridization 'probe'. Filter hybridization techniques have been developed for simultaneously screening large numbers of clones from plasmid, bacteriophage and cosmid libraries. Briefly, cells, or bacteriophage particles, containing several thousand recombinant molecules from a library are spread on a Petri plate and allowed to grow, giving colonies (or plaques). A nitrocellulose filter is placed on the surface of the plate, and some of the cells from each colony (or bacteriophage from each plaque) adhere to this. The filter is peeled off, treated to disrupt the cells (or bacteriophage), to denature the DNA, and to fix the DNA irreversibly to the nitrocellulose, giving a 'DNA print' of the original plate. Such filter-bound DNA is capable of hybridizing with radioactively labelled DNA of homologous sequence (the 'probe') and the colonies (or plaques) which hybridize can be identified by autoradiography.

Figure 3, *(Opposite)*. Construction of a cDNA library. These diagrams simplify some of the steps involved, particularly in the synthesis of double-stranded cDNA. For further details see Goodman and MacDonald (1979).
(i) cDNA preparation: Messenger RNA is isolated from insects either at a specific developmental stage or from a specific tissue. Complementary DNA copies are made with reverse transcriptase, using short oligo-dT primers which hybridize to the poly A tail on the mRNA molecules. The mRNA is removed by NaOH digestion and the cDNA is made double-stranded using DNA polymerase. Single-stranded oligo-dC tails are added to the 3' end of each strand using terminal transferase.
(ii) Plasmid vector preparation: pBR322 DNA is cleaved at the single *Pst* I site to give a linear molecule. Since the *Pst* I site is located in the ampicillin-resistance gene, the subsequent insertion of cDNA into this site will inactivate the gene. However, the tetracycline-resistance gene remains intact and is used for subsequent selection. Single-stranded oligo-dG tails are added to the *Pst* I-cleaved plasmid DNA molecules using terminal transferase (this recreates a *Pst* I site at each end of the vector and simplifies subsequent analysis of the cloned inserts).
(iii) Library construction: The two DNAs are mixed in approximately equal molar ratios and ligated (DNA ligase). The oligo-dC and oligo-dG tails hybridize to each other and ensure that each plasmid can have only one insert. The ligated DNA is used to transform CaCl$_2$-treated competent *E. coli* cells and the mixture is plated on tetracycline agar to select for cells which have received the tetracycline-resistance gene of pBR322.

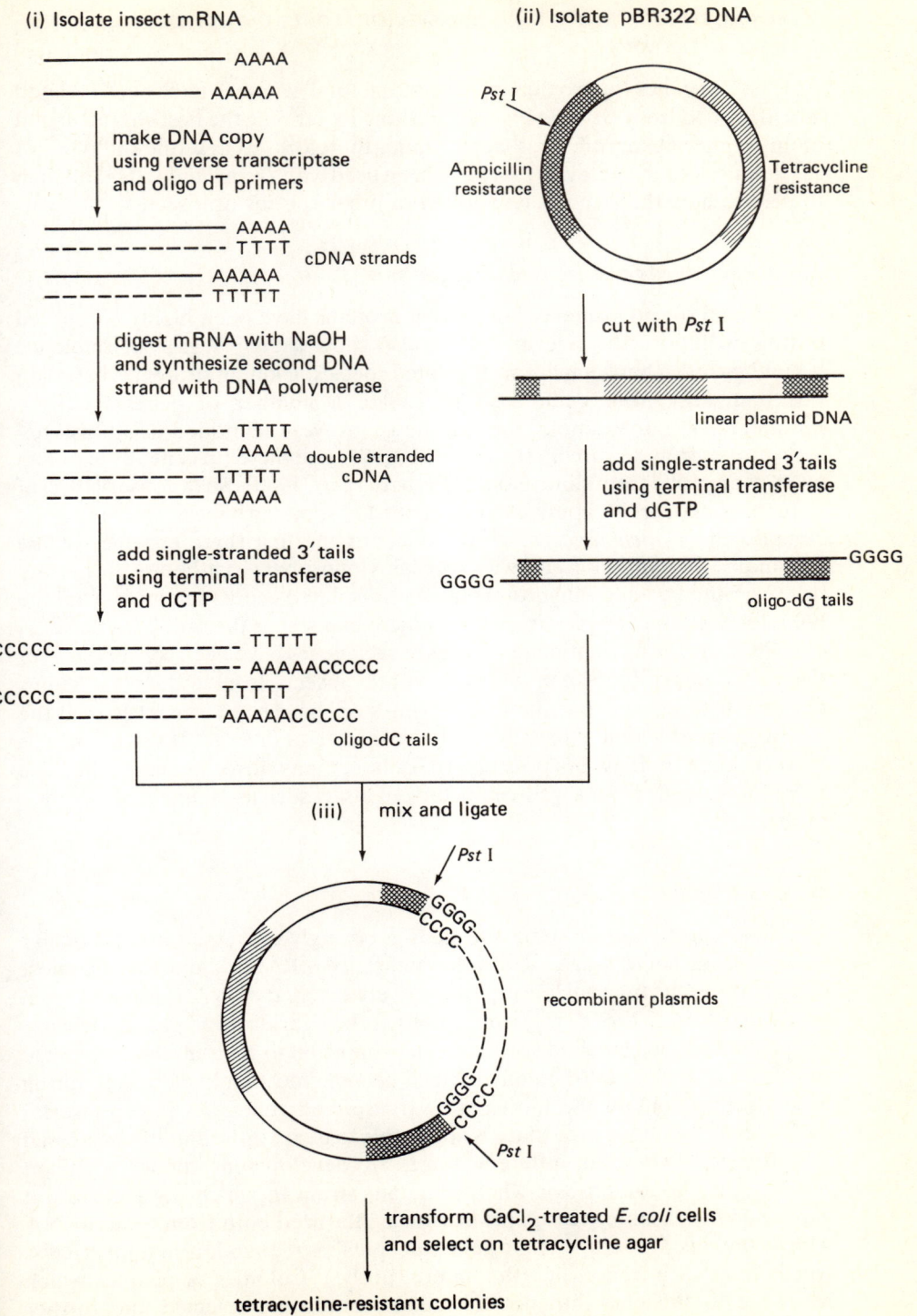

(i) Isolate insect mRNA

make DNA copy using reverse transcriptase and oligo dT primers

cDNA strands

digest mRNA with NaOH and synthesize second DNA strand with DNA polymerase

double stranded cDNA

add single-stranded 3′ tails using terminal transferase and dCTP

oligo-dC tails

(ii) Isolate pBR322 DNA

Pst I

Ampicillin resistance

Tetracycline resistance

cut with *Pst* I

linear plasmid DNA

add single-stranded 3′ tails using terminal transferase and dGTP

oligo-dG tails

(iii) mix and ligate

Pst I

recombinant plasmids

Pst I

transform CaCl₂-treated *E. coli* cells and select on tetracycline agar

tetracycline-resistant colonies

SCREENING TECHNIQUES FOR ISOLATING SPECIFIC CLONED GENES FROM
D. MELANOGASTER

At least 200 genes (approximately 2% of the total genome) have been isolated and identified from *D. melanogaster*, making its genome the best characterized of any higher eukaryote. It is worth while, therefore, to describe the types of successful cloning strategies that have been used with *D. melanogaster* and then to discuss how these might be applied to other species of insect.

The use of cloned genes from other organisms

As the amino-acid sequences of certain proteins have been highly conserved during evolution, the relevant genes also show a high degree of sequence homology, even between distantly related species. Cloning strategies based on this homology have been used to isolate a number of genes from *D. melanogaster*. For example, the histone genes were identified using histone-gene probes from sea urchin (Lifton *et al.*, 1977) and the actin genes by use of an actin-gene probe from slime mould (Fyrberg *et al.*, 1980). Such conservation of sequences appears to apply even to genes for some enzymes. Thus the gene complex in *D. melanogaster* which codes for the first three enzymes of the pyrimidine biosynthetic pathway (carbamylphosphate synthetase, aspartate transcarbamylase and dihydro-orotase) was isolated using sequences from the homologous gene complex from the Chinese hamster as the probe (Segraves *et al.*, 1983). As the nucleotide sequences of such genes have been conserved over the evolutionary time-span which separates insects from rodents, it seems reasonable to suggest that many insect genes will have been conserved over the relatively shorter time it has taken for the Insecta to diverge. If this surmise is correct, then it may be possible to isolate many insect genes using the homologous gene from a well-characterized insect such as *D. melanogaster* as a probe.

cDNA screening

For those cases where a particular gene is strongly expressed in a particular tissue during some stage of development, its mRNA is one of the most abundant in that tissue. Hence, radioactively labelled cDNA (prepared from total mRNA from that tissue) can be used as the hybridization probe. Because the probe contains labelled sequences in proportion to their abundance in the mRNA, those plaques or colonies which become most highly labelled during hybridization contain the most highly transcribed genes. A more powerful variation of the above procedure is applicable to genes differentially expressed in different tissues, at different stages of development, under different physiological conditions, or when a mutant strain in which the gene is not expressed is available. cDNA probes are synthesized both from total mRNA which contains transcripts of the gene (the 'plus' probe) and from total mRNA which lacks such transcripts (the 'minus' probe). Colonies or plaques which hybridize to the plus, but not the minus, probe are selected and further characterized to identify the gene of interest. Genes that are present at frequencies of as low as 10^{-4} of the total cDNA in the plus probe can be isolated. This strategy was used to obtain the genes coding for the heat-shock proteins (Livak *et al.*, 1978), for larval cuticle proteins (Snyder, Hirsh and Davidson, 1981), and the enzyme dopa decarboxylase (Hirsh and Davidson, 1981).

Chromosome walking

Because genomic DNA libraries are usually prepared from randomly fragmented chromosomal DNA, a specific sequence will be carried in the library in several overlapping inserts. Therefore, once any single-copy sequence has been isolated it can be used as the starting point for a 'walk' along the chromosome. Sequences from the ends of the original clone are used as probes to isolate clones carrying overlapping inserts. The ends of these can be used in turn to isolate further clones and so on. The procedure is illustrated in *Figure 4*.

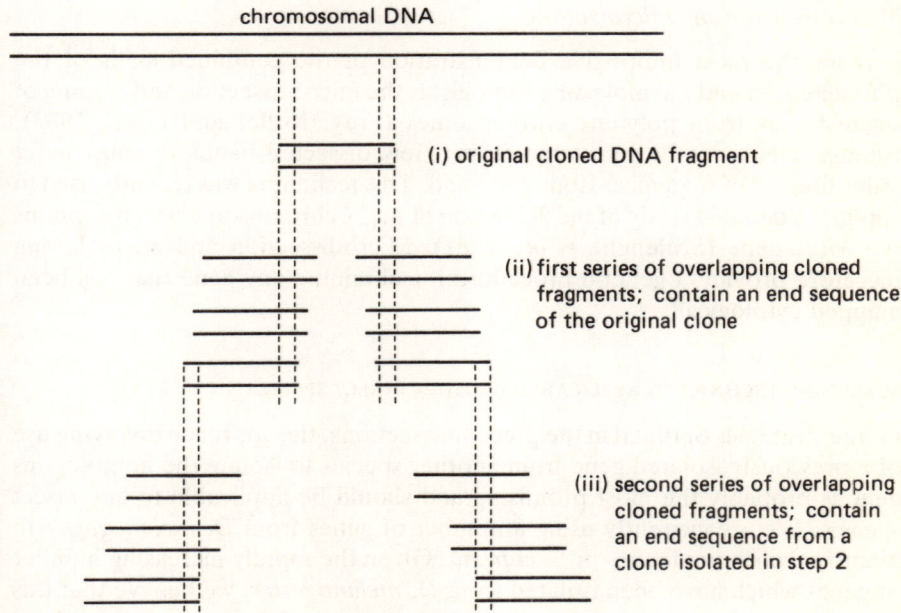

chromosomal DNA

(i) original cloned DNA fragment

(ii) first series of overlapping cloned fragments; contain an end sequence of the original clone

(iii) second series of overlapping cloned fragments; contain an end sequence from a clone isolated in step 2

Figure 4. Chromosome walking.
(i) Short pieces of terminal sequence are isolated from each end of the original cloned DNA fragment that serves as the starting point of the 'walk'.
(ii) Each of these terminal pieces is used as a hybridization probe to select overlapping clones from the library.
(iii) Short fragments from the outermost end of the first series of clones are used to select a second series of overlapping clones. This process can be continued indefinitely, until the end of the chromosome or a region of repetitive sequences is reached. (It is not possible to 'walk' through repeated sequence DNA). *See* Bender, Spierer and Hogness (1983) for a detailed discussion of this technique and its refinements.

The significance of chromosome walking in relation to gene cloning is that it is possible to walk from one previously cloned gene to another that maps genetically nearby on the chromosome. The rate of walking is slow: when using a bacteriophage library, each step gives an advance of about 10 kb and an average polytene chromosome band in *D. melanogaster* contains about 30 kb of DNA (there are about 5000 bands in the genome). However, by the use of chromosome deletions and rearrangements which make it possible to 'jump' from one region of the genome to another, the rate of progress of a 'walk' can be greatly accelerated. These procedures are particularly powerful in *D. melanogaster* because the cytogenetic locations within the chromosomes of many genes are known and because numerous chromosome deletions and rearrangements have been characterized. Furthermore, the chromosomal position of cloned genes can be determined by hybridizing radioactively labelled DNA to polytene chromosome squashes *in situ* (Wensink *et al.*, 1974). Such an approach was used to isolate the *bithorax* complex, the *rosy*$^+$ gene, and the gene for acetylcholinesterase (*Ace*$^+$) in one 'walk' on the third chromosome (Bender, Spierer and Hogness, 1983).

Microdissection and microcloning

Perhaps the most impressive demonstration of the combined skills of the cytogeneticist and the molecular biologist is the microdissection and cloning of single bands from polytene chromosomes (Frey, Koller and Lezzi, 1982). Enough DNA can be recovered from a few dissected bands to construct a 'mini-library' of sequences from that band. This technique was recently used to conduct a detailed study of the 3C region of the X chromosome which contains the *white* gene (Scalenghe *et al.*, 1981). Microdissection and microcloning therefore provide a general procedure for obtaining any gene that has been mapped cytologically.

SCREENING TECHNIQUES APPLICABLE TO OTHER INSECT SPECIES

Of the strategies outlined in the preceding sections, the approach involving use of a previously isolated gene from another species to isolate the homologous gene is probably the most promising and should be applicable to any insect species. We are currently using a number of genes from *D. melanogaster* to identify their homologues in *L. cuprina*. Given the rapidly increasing number of genes which have been isolated from *D. melanogaster*, we believe that this should be the first approach to evaluate when attempting to clone a gene from any other insect species. The cDNA strategies are also general techniques that should be applicable to any insect species.

Chromosome walking will probably not be widely applied in other species although it may be useful in some particular cases. Its general usefulness in *D. melanogaster* is dependent upon the availability of detailed genetic and cytogenetic information on the location of genes and on the presence of polytene chromosomes, which permit cloned genes to be localized precisely by *in situ* hybridization. It is also greatly aided in this species by the large number of cloned genes already available (which provide numerous starting points for walks) and by the availability of a large collection of chromosome rearrangements. No other insect species has such an array of advantageous features.

Chromosomal microdissection is also likely to have fairly restricted application to other species. It can be used only in species that have polytene chromosomes (which does include dipteran pests, such as *L. cuprina* and many mosquito and blackfly species). A second prerequisite for this procedure to be useful is that the genes of interest have been cytologically mapped on the polytene chromosomes. There are very few genes (in species other than *D. melanogaster*) that meet this criterion at present. In *L. cuprina*, some 11 mutations potentially useful in a genetic control programme have each been localized to one or a few polytene bands (Foster *et al.*, 1980, 1981; G.G. Foster, personal communication).

A number of other strategies have been used to identify genes from *D. melanogaster* and other organisms. Among these are screening of cDNA clones using antibodies directed against a specific protein (Broome and Gilbert, 1978; Erlich, Cohen and McDevitt, 1978), complementation of mutants in yeast (Henikoff *et al.*, 1981), and use of transposable elements as mutators (Bingham, Levis and Rubin, 1981). These techniques may have specialized applications, but probably will not be widely used to isolate insect genes.

There is, however, one further approach which has been used successfully with a range of organisms (although not yet, to our knowledge, with insects). This involves utilizing the amino-acid sequences of purified proteins to derive probable mRNA sequences and then chemically to synthesize regions of complementary DNA sequence. Reasonably short synthetic complementary oligonucleotide sequences (10–20 nucleotides), when added to preparations of mRNA, act as primers for the production of specific cDNA probes. For example, in the cloning of the rat gene for the polypeptide hormone relaxin, a synthetic 11-nucleotide primer, complementary to a region of predicted mRNA sequence, permitted the preparation of a labelled cDNA probe which specifically hybridized to DNA from clones carrying the relaxin gene sequences (Hudson *et al.*, 1981). Other examples illustrating the use of amino-acid sequences, derived mRNA sequences and synthetic complementary oligonucleotides in the production of specific probes include human leukocyte interferon (Goeddel *et al.*, 1980) and mouse β-nerve growth factor (Scott *et al.*, 1983). As techniques are now available for determining the amino-acid sequence of small (microgram) amounts of purified protein or polypeptide (Hunkapiller and Hood, 1980), this provides a further general approach to the cloning of genes. It is applicable to any gene (from any organism), the protein product of which can be purified.

In concluding this section, we wish to emphasize, firstly, that it is now a relatively straightforward task to isolate genes from *D. melanogaster* and, secondly, that it may often be possible to use the cloned *D. melanogaster* gene to identify its homologues in genomic DNA libraries prepared from other insect species. Furthermore, in those cases where the above approach is unsuccessful, a variety of other strategies is available. Any insect gene can be isolated if something is known about its expression into RNA or protein, its genetic location, if a mutant is available or if its protein product can be purified and partially sequenced. In the next section we will discuss how isolated genes can be reintroduced into insect chromosomes.

Transformation of *D. melanogaster* with cloned DNA

Although several methods are available for introducing DNA into individual cells in culture, it is much more difficult to transform intact multicellular organisms, such as insects, particularly if one wishes to obtain germ-line transformation. However, successful germ-line transformation of *D. melanogaster* was achieved recently by taking advantage of the P transposable element. We will therefore begin this section by briefly describing the general properties of transposable elements.

GENERAL PROPERTIES OF TRANSPOSABLE ELEMENTS

The molecular structures and properties of prokaryotic transposable elements (transposons) are well established (*see* reviews by Calos and Miller, 1980; Kleckner, 1981). They are discrete sequence elements of DNA that can exist in many different locations in the bacterial chromosome or plasmids. A variety of elements is known, varying in length from simple structures of about 0·7 kb to large composite structures of at least 40 kb. Their characteristic feature is that they can transpose from one site in a chromosome to another. In doing so they cause mutations by inactivation of the genes into which they insert. The frequency of transposition varies greatly between transposons, from about 10^{-4} to 10^{-7} per cell division. A general feature of their structure is that they have the same nucleotide sequence at each end, either in inverted or direct orientation. These terminal repeats, which are quite short (10–50 bp), are recognition sequences for the enzymes ('transposases') that are involved in mobilizing the element. Most prokaryotic transposons are transcribed and presumably code for one or more transposases, and often carry other genes such as ones conferring antibiotic resistance.

Transposons which are similar in structure and properties to those in prokaryotes have been found in many eukaryotes (Calos and Miller, 1980). In *D. melanogaster*, transposons represent about 75% of the moderately repeated DNA (about 10% of the total genome) (Spradling and Rubin, 1981). These elements may fall into as many as 100 families of related sequences. The different families do not share any sequence homology but do have similarities in overall structure. Like the prokaryotic transposons, the frequency of transposition of those in *D. melanogaster* varies greatly, from those which transpose very rarely (copia-like elements) to those, such as the P element, which can transpose at very high frequencies under certain genetic conditions. When transposition does occur, mutations can follow due to the insertional inactivation of the gene at the target site. Naturally occurring mutant alleles of *D. melanogaster* genes are often the consequence of the presence of a transposable element within the gene (Spradling and Rubin, 1981). For example, of the 13 mutant alleles of *white* investigated by Zachar and Bingham

(1982), seven were found to contain a transposon within the gene. In the case of *bithorax*, every mutant allele so far investigated has been found to be due either to a chromosomal rearrangement or to the insertion of a transposable element (Bender, Spierer and Hogness, 1983). The structure of the best-characterized transposon of *D. melanogaster*, copia, is shown in *Figure 5a*.

THE P TRANSPOSABLE ELEMENT AND HYBRID DYSGENESIS

The P element of *D. melanogaster* was originally characterized genetically because it induces a syndrome known as hybrid dysgenesis (Kidwell, 1983). The symptoms of hybrid dysgenesis include a greatly increased mutation rate (due mainly to the insertion of the element into genes), increased recombination in the male germ line, increased germ line chromosome breakage, and gonadal dysgenesis. The syndrome is seen only when a male carrying the elements (generally from a wild strain) is mated to a female lacking them (generally from a laboratory strain). Their progeny exhibit the hybrid dysgenic phenotype, and are normally identified because many of them are sterile. The progeny of the reciprocal cross (females carrying elements mated to males lacking them) are completely normal. It has been proposed that these P^+ females carrying the elements transmit a repressor to their progeny via the egg cytoplasm, which represses both maternal and paternal elements (Engels, 1979a).

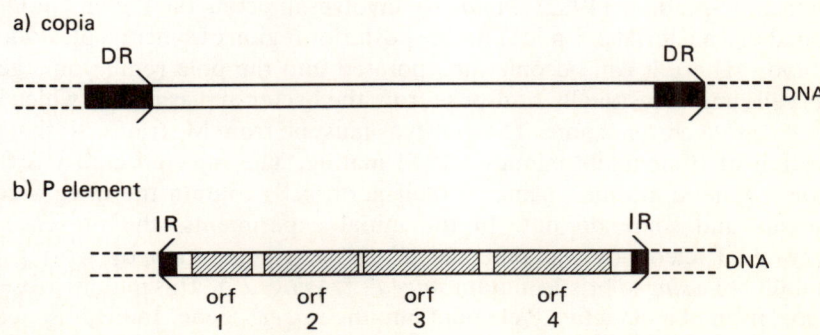

Figure 5. Structures of transposons from *D. melanogaster*.
(a) The copia element. Total length, approx. 5000 bp. Flanked by 276-bp direct repeats (DR). Number of proteins encoded is unknown (*see* Calos and Miller, 1980).
(b) The P element. Total length, 2907 bp. Flanked by 31-bp inverted repeats (IR). Contains four open reading frames (orf) and may encode four proteins (*see* O'Hare and Rubin, 1983).

At least two families of transposable elements cause hybrid dysgenesis in *D. melanogaster*. These are called 'P' (for paternal) and 'I' (for inducer); strains lacking the elements are called 'M' (for maternal) and 'R' (for reactive) respectively. These two elements do not interact and seem to be parts of independent systems. The dysgenic phenotypes are slightly different: P–M dysgenesis affects both males and females, while I–R dysgenesis affects only females. Both P and I elements have been cloned and there are no apparent sequence similarities between the two.

In order to clone the P-element sequences, Rubin, Kidwell and Bingham (1982) selected a number of mutants of the *white* locus that had been caused by insertion of P elements during hybrid dysgenesis. The mutant *white* genes were found to have homologous inserts of different sizes. When the DNA of one of these inserts was used as a hybridization probe, it was found that the DNA of P$^+$ flies contain about 50 copies per genome of variously sized P-element sequences. The longest element (2·9 kb) has been completely sequenced (O'Hare and Rubin, 1983) and the general features of its organization are given in *Figure 5b*. It has 31-base-pair inverted repeats at its ends, causes an 8-base-pair sequence at the site of insertion to be duplicated when it transposes and, within the internal sequences, contains four long non-overlapping open reading frames, suggesting that it could code for four proteins (O'Hare and Rubin, 1983). The shorter elements were found to be deleted versions of the longest one.

Part of the I element has also been cloned by using a similar strategy. It has proved to be unstable when cloned in bacteria, and has not been well characterized to date (Bucheton *et al.*, 1982).

TRANSFORMATION USING P-ELEMENT DNA

The procedure for germ-line transformation of *D. melanogaster* developed by Rubin and Spradling (1982) (*Figure 6*), involves injecting the P-element DNA (cloned in a bacterial plasmid) into the posterior region of syncitial blastoderm embryos. There it can become incorporated into the pole (embryonic germ line) cells and occasionally transpose from the bacterial plasmid (in which it is injected) to a chromosome. The embryos must be from M strains, so that the injection of P elements mimics a P–M mating. The injected embryos then become genetic mosaics: some of their germ cells contain the integrated P elements and some do not. In the initial experiments, the presence of integrated P elements was detected in the next generation by scoring the instability of a *singed* bristle mutant *singed*^{weak} (*singed*^{w}). This mutant arose by the insertion of a defective P element into the *singed*$^+$ gene. In the absence of an intact P element within the genome the mutant is stable, but in the presence of an intact P element it becomes unstable and changes to a more extreme allele (*singed*^{e}) or to wild type at a high frequency (Engels, 1979b). The destabilization of the *singed*^{w} allele is believed to be due to the mobilization of the defective P element which can generate either a small deletion (*singed*^{e}) or

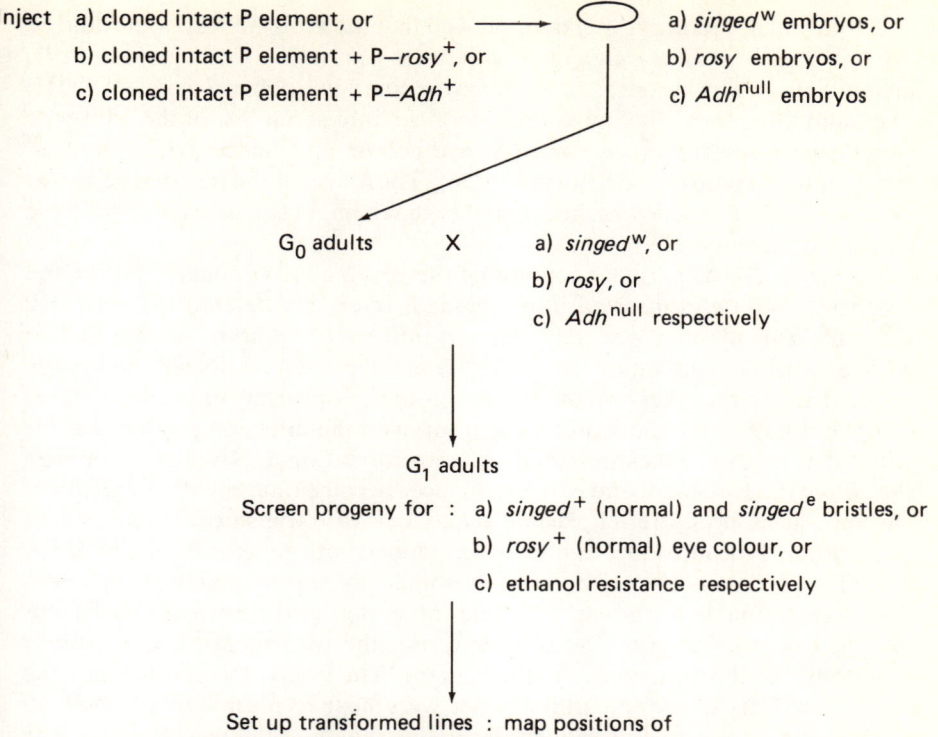

Inject
a) cloned intact P element, or
b) cloned intact P element + P–*rosy*$^+$, or
c) cloned intact P element + P–*Adh*$^+$

a) *singed*^{w} embryos, or
b) *rosy* embryos, or
c) *Adh*^{null} embryos

G$_0$ adults   X
a) *singed*^{w}, or
b) *rosy*, or
c) *Adh*^{null} respectively

G$_1$ adults

Screen progeny for :
a) *singed*$^+$ (normal) and *singed*^{e} bristles, or
b) *rosy*$^+$ (normal) eye colour, or
c) ethanol resistance respectively

Set up transformed lines : map positions of
integrated genes both genetically and by *in situ*
hybridization using polytene chromosomes

Figure 6. Transformation of *D. melanogaster* using P-element DNA. Cloned P-element DNA is injected into early (less than 3-hour) embryos (of different genotypes) either (a) by itself or (b) and (c) in conjunction with the DNA of a P-element plasmid carrying a marker gene. P-*rosy*$^+$ carries the *rosy*$^+$ (xanthine dehydrogenase) gene (Rubin and Spradling, 1982) and P-*Adh*$^+$ carries the *Adh*$^+$ (alcohol dehydrogenase) gene (Goldberg, Posakony and Maniatis, 1983).
The adults which develop from the injected eggs are mated with non-injected adults from the appropriate genetic stock. The progeny of these crosses are screened for evidence of transformation as follows:
(a) the integration and expression of an intact P element destabilizes the *singed*^{w} (weak) bristle mutant, giving rise to *singed*$^+$ (normal) and *singed*^{e} (extreme) progeny; (b) the integration and expression of the *rosy*$^+$ gene changes the mutant *rosy* eye colour giving *rosy*$^+$ (normal) progeny; (c) the integration and expression of the *Adh*$^+$ gene renders the recipients more resistant to ethanol.
Transformed lines are established from individual transformed flies for further analysis (*see* Rubin and Spradling, 1982).

restore the normal gene. Spradling and Rubin (1982) found that about half of the flies which developed from the surviving injected embryos produced *singed*^{e} or wild-type progeny, indicating that these had been transformed with P element.

Rubin and Spradling (1982) then showed that the P-element system could be used to transform *D. melanogaster* with other genes. The *rosy*⁺ gene from *D. melanogaster*, which codes for the enzyme xanthine dehydrogenase (XDH) (Yen and Glassman, 1965) was used because only about 5% of the wild-type activity is necessary to give normal eye colour and because it is not cell autonomous (Hadorn and Schwink, 1956). Thus, even if the inserted gene was expressed poorly, there was the possibility that it might still give a recognizable change in phenotype.

An 8·1 kb DNA fragment containing the *rosy*⁺ gene (Bender, Spierer and Hogness, 1983) was inserted into a cloned, internally deleted (defective) P element. This plasmid was then injected into *rosy*⁻ embryos in conjunction with a plasmid containing an intact P element. The injected flies were backcrossed to *rosy* flies and the eye colour of their progeny examined (*Figure 6*). About half of the individuals which survived the injection produced some wild-type progeny, indicating successful transformation. DNA analysis showed that while the P element and *rosy*⁺ sequences became integrated into recipient chromosomes, no bacterial plasmid sequences were transferred. The sites of integration of the introduced genes were mapped both genetically and by DNA hybridization *in situ* to polytene chromosomes. Of approximately 30 independent transformants analysed, none had integrated at the normal *rosy*⁺ locus and no two had integrated at the same site; the sites seem to be relatively randomly distributed throughout the genome. The levels of expression and the tissue specificity of the integrated genes were near to normal in all cases, so neither the chromosomal sequences at the regions of integration nor the presence of the flanking P-element sequences interfere with transcription of the integrated *rosy*⁺ gene (Spradling and Rubin, 1983).

Several other genes have been reintroduced into *D. melanogaster* using this system, including alcohol dehydrogenase (*Adh*⁺) (Goldberg, Posakony and Maniatis, 1983), dopa decarboxylase (*Ddc*⁺) (Scholnick, Morgan and Hirsh, 1983), *white*⁺ (Hazelrigg, Levis and Rubin, 1984), and some of the chorion protein genes (A.R. Spradling, personal communication). In all cases the genes seem to be normally expressed, even though the lengths of the DNA fragments used in the construction of the transformation plasmids are only a few thousand base pairs longer than the transcription units in some cases. Thus these genes do not seem to require long adjacent regulatory regions to function, an important point if transformation is to be useful for introducing new genes into insects. The largest transformation plasmid used so far (containing the *rosy*⁺ and *white*⁺ genes) carried an insert of about 20 kb (Hazelrigg, Levis and Rubin, 1984). The frequency of transformation obtained with this construct was only about one-tenth of that obtained with the *rosy*⁺ gene alone (8·1 kb insert), suggesting that larger plasmids transform much more poorly than smaller ones. Thus it is important that the fragments to be inserted into the vectors be as small as possible, especially if several genes are to be inserted into a single P-element vector.

Prospects for DNA transformation of other insects

There are at least two major technical problems to be overcome in developing transformation systems for other species. Firstly, suitable transformation vectors need to be developed; secondly, methods for selecting transformants have to be established.

TRANSPOSABLE ELEMENT VECTORS

The P element from *D. melanogaster* codes for its own transposase system. Consequently, it seems possible that it might function as a transposable element in other species. If such is the case, the P-element-derived transformation vectors developed for *D. melanogaster* might be generally useful in insect transformation. Successful transformation using such vectors has apparently been obtained with one species of insect in addition to *D. melanogaster* so far — *D. hawaiensis*, a Hawaiian picturewing species of *Drosophila* (M. Brennan, personal communication).

In the event that P-element vectors do not prove to be useful, then it may be necessary to attempt to develop specific vectors for each insect pest under consideration. Given the widespread occurrence of transposable elements in nature (they have been characterized in bacteria, yeast, nematodes, maize, fruitflies and mice), it seems probable that they exist in the genomes of most organisms. The cloned DNA of the transposable elements of any organism would be the obvious starting point for the development of a transformation vector for that organism.

OTHER APPROACHES TO TRANSFORMATION

An alternative approach to obtaining germ-line transformation of insects would be to attempt to transform insect cells in culture and then to insert such cells into developing embryos. For example, it might be possible to remove germ-line cells, transform them *in vitro* and then transplant them back into recipient embryos. Such cell transplantations have already been achieved with embryos from *D. melanogaster* (Van Deusen, 1976; Zalocar, 1981).

Many different types of vectors have been developed to transform eukaryotic cells in culture. In yeast, for example, both plasmids and minichromosomes have been used. These types of vector replicate within the host cell but cannot integrate into the genome (Hinnen, Hicks and Fink, 1978; Hicks, Hinnen and Fink, 1978). In yeast, plasmids are lost during meiosis (Stinchcomb, Struhl and Davis, 1979). It therefore seems unlikely that non-integrative plasmids will be useful as transformation vectors with insect cells.

Minichromosome vectors differ from plasmids in that they contain centromeric sequences (Clarke and Carbon, 1980; Murray and Szostak, 1983). These are much more stable during mitosis and meiosis than plasmids and so do not require continuous selection. Such vectors are not yet available for cells of higher eukaryotes but might be developed in the future. Various viral transformation vectors have been used in mammalian cells. Minichromosomes based on the monkey virus SV40 are the most commonly used (Elder, Spritz and Weissman, 1981). Insect viruses could be investigated for this purpose, but a major problem with virus-based vectors is the danger of the vector spreading by infection to other, non-pest, species. Of course, it might be possible to genetically engineer certain insect viruses to make them non-pathogenic. However, a discussion of the properties of insect viruses and of the prospects for manipulating their genomes will not be attempted in this review. The use of viruses as pest-control agents is discussed by Carter (1984) in volume 1 of this series.

Mammalian-cell transformation is often done without a specific vector. Under certain conditions tissue-culture cells will take up added DNA and incorporate it into high-molecular-weight complexes, which are then inserted into the genome (Perucho, Hanahan and Wigler, 1980). The frequency of transformation is usually low, so selectable markers need to be available to monitor transformation. However, much higher frequencies of transformation can be obtained by the microinjection of DNA directly into cells (Capecchi, 1980; Gordon *et al.*, 1980).

IDENTIFICATION OF TRANSFORMANTS

The identification of transformants in the P-element transformation experiments in *D. melanogaster* took advantage both of detailed knowledge about hybrid dysgenesis and of the availability of certain mutants of this species. Although neither the mutants nor the other detailed genetic information are readily available in other insects, there are a number of alternative possibilities for monitoring successful transformation.

DNA hybridization

The most reliable (but tedious) method is to use DNA hybridization. Progeny of injected insects can be individually ground up, their nucleic acids extracted, bound to a filter support, and hybridized with radioactive vector DNA. An advantage of this approach is that transformation (when it occurs) would be detected irrespective of whether any marker gene carried by the vector was expressed. (The only genes which would have to be expressed would be those necessary for vector transposition). In addition, it is applicable to any species regardless of the availability of mutant stocks or biochemical information. If the transformation frequency in other species is near that achieved with *D. melanogaster* (e.g. 1–10%), then DNA hybridization will be a practical monitoring system; in fact it would be easier to use this method than to spend time isolating mutants and constructing suitable recipient stocks. However, at present there are no data on which to base predictions about either the transformation frequencies or the optimal transformation conditions for other species.

Visible markers

The first genetic marker used to detect P-element transformation was the ability of an intact P element to destabilize the *singed*[w] mutation. Mutants of this type are only available in *D. melanogaster* at present, and the chances of finding defective P-element induced mutants in other species appears to be small. The first marker actually introduced with the P element into *D. melanogaster* was the xanthine dehydrogenase gene, *rosy*$^+$. In *D. melanogaster* this was a good choice of marker because it affects eye colour and is, therefore, easy to score (Yen and Glassman, 1965); in addition, it can be selected because *rosy* flies are sensitive to added purine in their diet (Glassman, 1965). However, XDH-mutants are not known in other insects. Because XDH affects the level of the drosopterin eye pigments (which are unique to species of *Drosophila*), its absence would not cause an easily recognizable eye-colour phenotype in other insects (Summers, Howells and Pyliotis, 1982). Nevertheless, it may be possible to identify XDH-mutants in other species, either on the basis of their purine sensitivity or by screening for reduced enzyme activity.

Another eye-colour gene which has been introduced into *D. melanogaster* using the P-element system is the *white*$^+$ gene. Although this is not as attractive as *rosy*$^+$ as a marker in *D. melanogaster* as it is cell autonomous, and its gene product is uncharacterized, there are equivalent mutants in many species of insects: *Musca domestica* (Hiraga, 1964; Milani, 1975); *Calliphora erythrocephala* (Langer, 1967); *L. cuprina* (Foster *et al.*, 1981), and *Ephestia kuhniella* (Caspari and Gottlieb, 1975). Because the *white*$^+$ gene of *D. melanogaster* appears to contain a very complex control region (Judd, 1976), it is possible that it will not function in other species. However, it may be possible to isolate the *white*$^+$ gene from the species to be transformed, using the *D. melanogaster* gene as a probe, and to insert that into the vector as the marker. A clone which cross-hybridizes with sequences in the *D. melanogaster white*$^+$ gene has been isolated from a *L. cuprina* genomic DNA library (A. Vacek and A.J. Howells, unpublished work) and is currently being characterized. If it carries the *white*$^+$ gene of *L. cuprina*, we plan to test it as a transformation marker in this species.

Selectable markers

The alcohol dehydrogenase gene (Adh^+) is effective as a selectable marker in P-element transformation of *D. melanogaster* (Goldberg, Posakony and Maniatis, 1983). There are excellent schemes for selecting both Adh^+ individuals (using ethanol; Vigue and Sofer, 1975) and Adh^- individuals (using 1-pentyne-3-ol; O'Donnell *et al.*, 1974). Hence it may be possible to select the necessary Adh^- recipient strains, as well as being able to select for Adh^+ transformants, in other species. The *D. melanogaster* Adh^+ gene is extremely active, so it may be possible to use this gene in other species. Alternatively, if necessary, the gene could be isolated from each species to overcome specificity problems. Since ADH converts ethanol into acetaldehyde, which is also toxic, the insect must have an acetaldehyde detoxification system. In *D. melanogaster* ADH itself detoxifies acetaldehyde (Heinstra *et al.*, 1983) by converting it into acetate.

The insecticide-resistance genes constitute a large group of potentially selectable markers. As yet, no insecticide-resistance gene has been cloned, but the acetylcholinesterase (Ace^+) gene has been implicated in organophosphorus-pesticide resistance in some arthropods, e.g. the cattle tick *Boophilus microplus* (Stone, Nolan and Schuntner, 1976). This gene has been cloned from *D. melanogaster* and is being used as a probe to isolate and characterize resistance alleles at the *Ace* locus in the cattle tick (J. Nolan and P. Riddles, personal communication). In time, other resistance genes will probably be isolated and characterized, and could then be used as markers. An advantage of using such genes in transformation is that they do not require a special recipient strain, since the resistance is a novel function (although in the case of many serious pests this is unfortunately no longer true). Obviously, careful consideration would be necessary before pesticide resistance is introduced into a pest species simply as a marker of successful transformation.

Bacterial antibiotic-resistance genes could also be tried in insects as selectable markers. Kanamycin-resistance (kan^R) genes from *E. coli* have been used (after being spliced to a eukaryotic promoter) as transformation markers in yeast (Jimenez and Davies, 1980), in mammalian tissue-culture cells (Southern and Berg, 1982; Colbere-Garapin *et al.*, 1981), and in the slime mould (Hirth, Edwards and Firtel, 1982). It may be necessary to splice the kan^R gene to an efficient insect promoter sequence. Efficient promoters from *D. melanogaster* are already available and similar sequences can undoubtedly be obtained from genes of other insects.

We conclude this section by noting that the prospects for manipulating the genomes of insects by introducing cloned DNA appear to be good. Several approaches are available for inserting the cloned DNA into embryos and, as far as transformation markers are concerned, there is a wide variety of cloned genes already available for this purpose. Consequently, in the final section of this paper, we will consider ways in which such genome manipulation might contribute to insect-pest control.

Recombinant DNA techniques and genetic control

The aim of a genetic control programme is to manipulate the hereditary apparatus of individuals in a target population such that a high proportion of individuals in some ensuing generations will not survive. Genetic control therefore requires the laboratory propagation and release of genetically modified individuals which, by mating with residents of the target population, will serve to introduce the modified genetic material into the population. The number of individuals and number of releases required are determined by the mechanism used to spread the genetic modification through the target population. The sterile insect release method (SIRM) relies on swamping the target population with large numbers of released insects, hence the need for large production factories and the development of sophisticated rearing and release technologies. The ultimate success of SIRM is realized when all field females in the release areas are inseminated by males treated with ionizing radiation or chemosterilant such that all gametes produced by these individuals carry one or more dominant lethal conditions. The Y-chromosome–autosomal translocation system (*see below*) also relies on an initial swamping of the target population, and on the effects of the genetic load induced while the released genetic material is being eliminated by natural selection. The application of recombinant DNA techniques to this system of genetic control is discussed in the next section.

Some systems of genetic control involve a transporting mechanism, e.g. meiotic drive or negative heterosis, which are devices for enabling the genetic condition to spread through the target population despite the genetic disability which it bestows on its carrier (Whitten, 1984). The techniques of molecular biology may increase options open to applied entomologists working in this area. These possibilities are considered in the final section.

ENHANCEMENT OF 'CLASSIC' GENETIC CONTROL

It is not intended here to give a comprehensive treatment of this subject but simply to provide one example of classic genetic control, i.e. one which draws upon traditional cytogenetics and chromosome mechanics, and to indicate how its efficacy might be enhanced by use of recombinant DNA techniques.

The use of reciprocal translocations between the Y or male-determining chromosome and some autosomes, as a means of limiting conditional lethals to the female sex, has been outlined by Whitten (1979). This system relies on the absence of crossing over during male meiosis to effect complete linkage between the maleness factor and the conditional lethal loci. In species in which crossing over occurs during male meiosis, crossover suppressors such as a chromosome inversion have been used to secure adequate coupling. For example, this approach has been used in developing sex-killing systems in certain mosquito species (LaChance, 1979). The rationale for the Y–autosome translocation system for inducing high genetic loads in the generations following a period of release is outlined in *Figure 7*. The genetic load in this system derives from two sources. Firstly, the released males give rise to a proportion of inviable offspring as a direct result of the production of aneuploid gametes. All their viable male offspring inherit the rearrangement and consequently perpetuate this source of genetic load. The recessive mutations m_1 to m_6 (*see Figure 7*) yield phenotypes which are lethal under field conditions, but are viable in the laboratory. If these loci assort independently during meiosis in the female progeny of the cross between field females and released males, then the female progeny in the next generation will be inviable, unless they receive the wild-type allele for each of the six loci. If the loci are unlinked, the probability of a female offspring being viable is thus $(1/2)^6$ (*see Figure 7*). It is greater if a degree of linkage exists between the loci, and the consequent

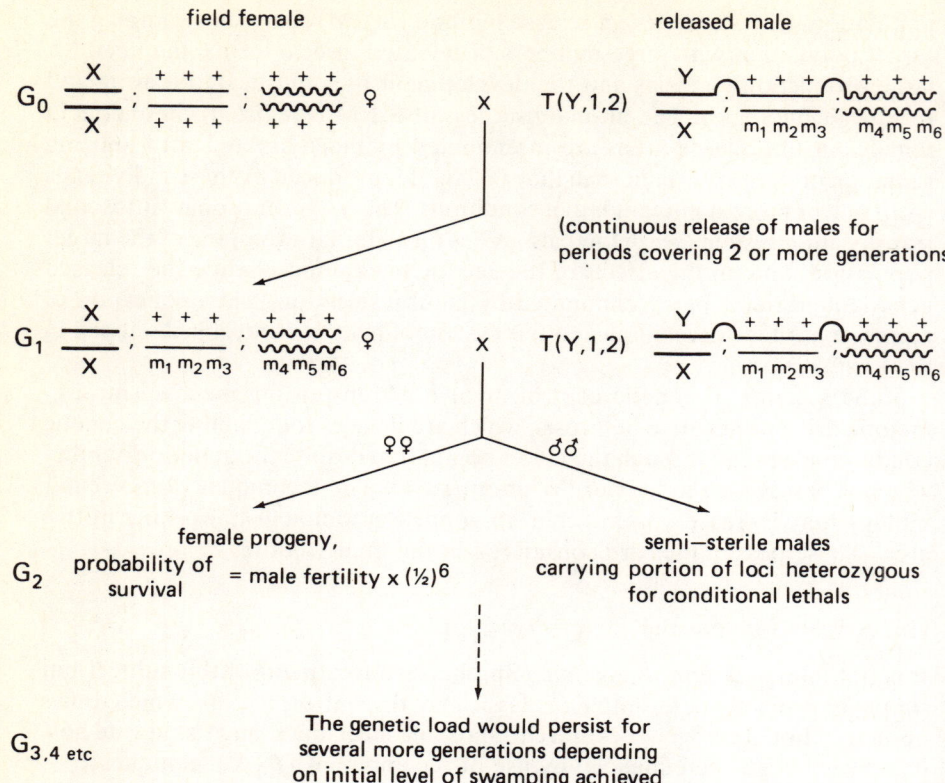

Figure 7. A scheme for using Y–autosome translocations coupled with conditional lethals to control the Australian sheep blowfly, *Lucilia cuprina*. Genetically manipulated males are mass reared and released into a target population, ideally in numbers to ensure that most field females are inseminated by the released males. Females from the released strain are either eliminated during production by a genetic sex killing system (Whitten, 1984), or express mutations which prevent survival in the field. Releases are maintained such that the first-generation daughters (G_1) are largely inseminated by released males. In the following generation (G_2), the only daughters to survive are those not expressing chromosomal duplication/deficiencies which have a wild-type allele (+) at each of the recessive conditional lethal loci (m_1, m_2 etc.).

genetic load is correspondingly lower. The reduced load with linked loci follows from the fact that the conditional lethals are more likely to be coupled, i.e. doubly kill their bearer and therefore be eliminated without additional cost. This system is currently being evaluated in *L. cuprina*.

Two features of the system limit its usefulness. It may be desirable to avoid the male sterility which is necessarily associated with the translocation rearrangements. For example, partially sterile males can create limitations in mass rearing the strain. Secondly, only conditional lethals which are located on the translocated chromosomes can be used. In the case of *L. cuprina* this limits our choice to a maximum of three or six loci depending on whether one or two autosomes are involved in the Y–autosome translocation. This latter argument assumes a chromosomal length of approximately 100 map units, which seems to represent an upper size for many insect chromosomes (e.g. Foster *et al.*, 1981).

Access to cloned genes and a transformation system would permit us to examine the possibility of transferring the wild-type allele of any number of loci to the male-determining chromosome. Alternatively, if the maleness element is itself a discrete entity, it might be cloned and then relocated to another chromosome. In the longer term, it may be feasible to construct a mini-chromosome carrying the maleness factor and the wild-type allele of as many loci as we wish to include in the scheme. It would then suffice to construct multiple marker stocks carrying the desired combination of conditional lethals masked by the wild-type allele on the modified or synthetic Y chromosome. These males would be fully fertile, phenotypically wild type and would be used as a vehicle for introducing a genetic load that would be borne by the female sector of the population. The magnitude of the genetic load depends on the number and location of the genetic loci and on the ratio of wild:released males. The load would decay during ensuing generations, leading to population collapse and extinction in particular cases. Thus the opportunity exists of combining conventional cytogenetic methods with molecular techniques, to induce a change in the genetic information of a target population which is detrimental to the genetic fitness of that population. Sex limitation of conditional lethals represents one application of the molecular techniques that should become available.

DEVELOPMENT OF NOVEL GENETIC CONTROL STRATEGIES

Meiotic drive

As outlined above, genetic control attempts to reduce the genetic fitness of the target population by the injection of inappropriate genetic information into that population. Large releases aimed at swamping the target population may be cost effective in some instances, e.g. SIRM for screwworm. However, it will always remain the dream of geneticists to devise methods which simply require a single seeding of the target population with genetic material that spreads by some infectious mechanism. One possible means of achieving such a spread is to use meiotic drive (Zimmering, Sandler and Nicoletti, 1970), where one chromosome or chromosomal segment enjoys a segregation advantage during meiosis. This transmission advantage ensures its spread through the population despite the fact that it may be propagating some deleterious condition. Little or nothing is yet known about the precise underlying molecular mechanism(s) of meiotic drive and, until more information is available at this level, its application to genetic control is likely to remain limited. Molecular techniques might assist in the elucidation of the mechanisms of meiotic drive systems such as segregation distorter in *D. melanogaster* (Ganetzky, 1977). This knowledge may, in turn, enable the eventual cloning and transfer of such genetic conditions from *D. melanogaster* or facilitate the identification of similar phenomena in pest species.

Transposable elements as control agents

Transposable elements have several properties that might make them useful in genetic control. In particular, when they transpose from one chromosome to another they replicate at the same time, thus increasing their copy number in the genome. Because they can spread in this way they may have a selective advantage over other DNA sequences that are restricted to a single genetic location (Doolittle and Sapienza, 1980; Orgel and Crick, 1980). These authors have pointed out that transposable elements can be considered to be genetic parasites, so using them as control agents is in theory similar to the use of pathogens such as viruses, although it is not thought that transposons can spread by infection (but *see* 'Safety considerations' on page 94).

The P element causes phenomena that reduce the fitness of individuals and could be useful in a genetic control programme: gonadal dysgenesis, an increased mutation rate, and increased chromosome breakage. The sterility of hybrid dysgenic flies is temperature dependent (Kidwell and Novy, 1979). Most flies reared at 18°C are fertile, although not completely so (due to chromosome breakage), but almost all flies reared at 25°C are sterile, because of a failure of the gonads to develop. The mutations generated by P elements (deletions, insertions, translocations) are of the type that completely inactivate genes (and hence cause the loss of the gene product) rather than altering the structure of the gene product (as point mutations might); thus, they are not likely to cause an improvement in fitness. As discussed earlier, the hybrid dysgenesis phenotype is expressed only when the P element becomes derepressed, i.e. when a P male is mated to a M female, because of the absence of the P-element repressor in the cytoplasm of the eggs of such females. Thus, by site-directed mutagenesis of P-element DNA *in vitro*, it might be possible selectively to inactivate the gene for the repressor, leaving the transposase functions unaffected. Such an irrepressible transposon might be very effective in reducing the fitness of populations into which it is introduced.

The critical variables for the spread of a disadvantageous transposon through a population are the rate at which it transposes, the fitness cost to the organism harbouring it, and the initial frequency at which it occurs in the population (Hickey, 1982). Because the initial frequency is limited by the number of individuals that it is practical to release in a genetic control programme, the damage that can be done to a wild population is dependent mainly on the transposition rate of the vector. It would be worth while, therefore, to have available additional high-frequency transposing elements. Elements that transpose rapidly will probably cause an increased mutation rate, so strains of insects that harbour mutator systems, such as those known in *D. melanogaster* (Green, 1976), should be screened for such elements.

Transposable elements for transporting deleterious genes

Mutant genes which have dominant effects on viability or fertility are potentially useful in genetic control. Genes introduced using this system must be dominant, since the P-element vectors randomly introduce genes into the genome and do not replace existing genes. An example of a biochemically well-characterized dominant mutant, which might be introduced into recipients to reduce their fitness, is the defective testes-specific tubulin gene ($B2t^D$) that causes male sterility in *D. melanogaster* (Kemphues *et al.*, 1979). This mutant gene produces a defective tubulin that can copolymerize with normal tubulin, forming aberrant microtubules. The use of highly conserved genes like the tubulin genes increases the probability that they will have the same effects in other species as in *D. melanogaster*. It might be possible to produce similar mutations in genes for other proteins that have a structural role (e.g. actins, myosins, collagens) or are part of multimeric enzyme complexes. An important problem to be overcome when considering the introduction of genes with dominant effects, is the rearing of stocks carrying such genes in the laboratory. It would be desirable to have mutants which express the phenotype only under a certain set of experimental conditions, e.g. temperature-sensitive mutants. Dominant, temperature-sensitive lethal mutants of *D. melanogaster* are known (Suzuki *et al.*, 1976). Unfortunately, few of them have been characterized biochemically and so it will be difficult to isolate the genes involved. However, it may be possible to construct temperature-sensitive dominant genes using recombinant DNA techniques. For example, a temperature-sensitive dominant male-sterility gene could be constructed by fusing the promoter sequence from the 70 kilodalton *D. melanogaster* heat-shock gene (*hsp70*) to the $B2t^D$ tubulin gene. The *hsp70* gene is actively transcribed at 30°C but not at 20°C, and the promoter responsible for this temperature sensitivity has been isolated and sequenced (Corces *et al.*, 1981). Promoters from other types of conditionally activated genes, e.g. those induced by hormones, metabolites, or metal ions, might also be useful in the construction of genes conferring conditional lethality or sterility.

Sex-determining elements

The isolation and use of sex-determining elements to enhance a classic control scheme has been discussed earlier. However, the introduction of additional male-determining genes, as a means of altering the sex ratio of a target population, can also be considered as a genetic control strategy in its own right. A wide variety of sex-determining systems are known in insects (for examples *see* King, 1975), so different genes might have to be isolated for different pest species. With *Musca domestica* (Wagoner, McDonald and Childress, 1974) and some other dipterans (Green, 1980), the maleness element behaves as a single gene, and this may also apply to species which have a typical heterochromatic Y chromosome. For species in which the male-determining element is carried on an autosome and its genetic location is known, e.g. *Chironomus* (Martin *et al.*, 1980), it might be possible to obtain the appropriate DNA by chromosome microdissection followed by microcloning.

SELECTION AGAINST GENETIC CONTROL

The swamping systems outlined earlier depend upon natural selection to exert their controlling influence on target populations. They work because they introduce deleterious genetic information into natural populations, which is lost from the population with the death of its carrier. The genetic load effect is therefore limited in time and amount by the genetic cost of eliminating the carriers of the information.

By way of contrast, systems such as meiotic drive or mobilized transposable elements that would rely on some transmission advantage, will necessarily be opposed by natural selection, should suitable variability exist on which selection can operate. Selection operates to counter meiotic drive systems both in laboratory populations and in natural populations where suppressors of drive are widespread (Lyttle, 1979). A similar reaction in relation to sex determination has occurred in *M. domestica* where the spread of the M maleness factor has been emasculated by an overriding F system. For example, in the Bowhill population in Queensland, the M element is present on three autosomes but the 1:1 sex ratio is still preserved by the existence of yet another sex-determining system, the F system (Ff are female and ff male), which is fully epistatic to the M system (Wagoner, McDonald and Childress, 1974). Thus, where genetic information is spread surreptitiously bringing with it a reduction in Darwinian fitness, its spread may ultimately be thwarted by the generation and spread of some counteracting information to restore normality. As with the development of pesticide resistance, the track record suggests a high probability of resistance developing to the 'infectious vector', but clearly this is not a logical necessity.

SAFETY CONSIDERATIONS

Although genetic engineering is less controversial than it was a few years ago, the use of recombinant DNA technology to control wild insect pests must be preceded by careful analysis of its impact on the environment (Levin, 1979). A potential advantage of the approaches outlined here, over the release of genetically modified insect pathogens such as bacteria or viruses, is that once the novel genetic information becomes stably integrated in an insect genome, there should be no more chance of an inserted gene escaping to cause damage to another species than there is of a naturally occurring insect gene doing so.

The possible use of transposons as genetic control agents must also be evaluated with caution. Transposons share a number of structural features with the mammalian retroviruses and it has been proposed that the two are evolutionarily related (Finnegan, 1983). Certainly, the DNA forms of retroviruses, which integrate stably into mammalian genomes, behave in some ways like transposons. Whether any insect transposon can spread via infection, like retroviruses, will have to be determined carefully. Clearly, transposons must be treated initially as pathogens and tested for infectivity and effects on other insects before the release of any engineered species carrying transposons is undertaken.

Conclusion

Work on the possible impact of recombinant DNA technology on insect-pest control is still in its infancy. It may be several years yet before the correct combinations of transformation vectors and deleterious genes are assembled for any pest species, or before some of the suggestions we have made in this article are tested on a small scale in the laboratory. It is likely that, as molecular genetic research increases our understanding of the structure and functioning of the eukaryotic genome, we will be better able to predict what other sorts of manipulation might ultimately prove useful in controlling insect pests.

References

BAUMHOVER, A.H., GRAHAM, A.J., BITTER, B.A., HOPKINS, D.E., NEW, W.D., DUDLEY, F.H. AND BUSHLAND, R.C. (1955). Screwworm control through release of sterilized flies. *Journal of Economic Entomology* **48**, 462–466.

BENDER, W., SPIERER, P. AND HOGNESS, D.S. (1983). Chromosomal walking and jumping to isolate DNA from the *Ace* and *rosy* loci and the bithorax complex in *Drosophila melanogaster*. *Journal of Molecular Biology* **168**, 17–33.

BINGHAM, P.M., LEVIS, R. AND RUBIN, G.M. (1981). Cloning of DNA sequences from the white locus of *D.melanogaster* by a novel and general method. *Cell* **25**, 693–704.

BOLIVAR, F. AND BACKMAN, K. (1979). Plasmids of *Escherichia coli* as cloning vectors. *Methods in Enzymology* **68**, 245–267.

BROOME, S. AND GILBERT, W. (1978). Immunological screening method to detect specific translation products. *Proceedings of the National Academy of Sciences of the United States of America* **75**, 2746–2749.

BUCHETON, A., WILL, B., SANG, H. AND FINNEGAN, D. (1982). The Inducer-Reactive system of hybrid dysgenesis in *Drosophila melanogaster*. *Heredity* **49**, 138.

CALOS, M.P. AND MILLER, J.H. (1980). Transposable elements. *Cell* **20**, 579–595.

CAPECCHI, M.R. (1980). High efficiency transformation by direct microinjection of DNA into cultured mammalian cells. *Cell* **22**, 479–488.

CARTER, J.B. (1984). Viruses as pest-control agents. In *Biotechnology and Genetic Engineering Reviews* (G.E. Russell, Ed.), volume 1, pp. 375–419. Intercept, Newcastle upon Tyne.

CASPARI, E.W. AND GOTTLIEB, F.J. (1975). The Mediterranean meal moth, *Ephestia kuhniella*. In *Handbook of Genetics* (R.C. King, Ed.), volume 3, pp. 125–147. Plenum Press, New York.

CLARKE, L. AND CARBON, J. (1980). Isolation of a yeast centromere and construction of functional small circular chromosomes. *Nature* **287**, 504–509.

COLBERE-GARAPIN, F., HORODNICEAMU, F., KOURILSKY, P. AND GARAPIN, A.C. (1981). A new dominant hybrid selective marker for higher eucaryotic cells. *Journal of Molecular Biology* **150**, 1–14.

CORCES, V., PELLICER, A., AXEL, R. AND MESELSON, M. (1981). Integration, transcription, and control of a *Drosophila* gene in mouse cells. *Proceedings of the National Academy of Sciences of the United States of America* **78**, 7038–7042.

CREA, R., KRASZEWSKI, A., HIROSE, T. AND ITAKURA, K. (1978). Chemical synthesis of genes for human insulin. *Proceedings of the National Academy of Sciences of the United States of America* **75**, 5765–5769.

DOOLITTLE, W.F. AND SAPIENZA, C. (1980). Selfish genes, the phenotype paradigm and genome evolution. *Nature* **284**, 601–603.

ELDER, J.T., SPRITZ, R.A. AND WEISSMAN, S.M. (1981). Simian virus 40 as a eukaryotic cloning vehicle. *Annual Review of Genetics* **15**, 295–340.

ENGELS, W.R. (1979a). Hybrid dysgenesis in *Drosophila melanogaster*: rules of inheritance of female sterility. *Genetical Research* **33**, 219–236.

ENGELS, W.R. (1979b). Extrachromosomal control of mutability in *Drosophila melanogaster*. *Proceedings of the National Academy of Sciences of the United States of America* **76**, 4011–4015.

ERLICH, H.A., COHEN, S.N. AND McDEVITT, H.O. (1978). A sensitive radioimmunoassay for detecting products translated from cloned DNA fragments. *Cell* **13**, 681–689.

FINNEGAN, D.J. (1983). Retroviruses and transposable elements — which came first? *Nature* **302**, 105–106.

FOSTER, G.G., WHITTEN, M.J., KONOVALOV, C. AND BEDO, D.G. (1980). Cytogenetic studies of *Lucilia cuprina dorsalis* R.-D. (Diptera: Calliphoridae). *Chromosoma* **81**, 151–168.

FOSTER, G.G., WHITTEN, M.J., KONOVALOV, C., ARNOLD, J.T.A. AND MAFFI, G. (1981). Autosomal genetic maps of the Australian Sheep Blowfly, *Lucilia cuprina dorsalis* R.-D. (Diptera: Calliphoridae), and possible correlations with the linkage maps of *Musca domestica* L. and *Drosophila melanogaster* (Mg.). *Genetical Research* **37**, 55–69.

FREY, M., KOLLER, T. AND LEZZI, M. (1982). Isolation of DNA from single microsurgically excised bands of polytene chromosomes of *Chironomus*. *Chromosoma* **84**, 493–503.

FRISCHAUF, A.M., LEHRACH, H., POUSTKA, A. AND MURRAY, N. (1983). Lambda replacement vectors carrying polylinker sequences. *Journal of Molecular Biology* **170**, 827–842.

FYRBERG, E.A., KINDLE, K.L., DAVIDSON, N. AND SODJA, A. (1980). The actin genes of *Drosophila*: a dispersed multigene family. *Cell* **19**, 365–378.

GANETZKY, B. (1977). On the components of segregation distortion in *Drosophila melanogaster*. *Genetics* **86**, 321–365.

GLASSMAN, E. (1965). Genetic regulation of xanthine dehydrogenase. *Federation Proceedings, Federation of American Societies for Experimental Biology* **24**, 1243–1251.

GOEDDEL, D.V. AND 17 OTHER AUTHORS (1980). Human leukocyte interferon produced by *E. coli* is biologically active. *Nature* **287**, 411–416.

GOLDBERG, D.A., POSAKONY, J.W. AND MANIATIS, T. (1983). Correct developmental expression of a cloned alcohol dehydrogenase gene transduced into the *Drosophila* germ line. *Cell* **34**, 59–73.

GOODMAN, H.M. AND MacDONALD, R.J. (1979). Cloning of hormone genes from a mixture of cDNA molecules. *Methods in Enzymology* **68**, 75–90.

GORDON, J.W., SCANGOS, G.A., PLOTKIN, D.J., BARBOSA, J.A. AND RUDDLE, F.H. (1980). Genetic transformation of mouse embryos by microinjection of purified DNA. *Proceedings of the National Academy of Sciences of the United States of America* **77**, 7380–7384.

GREEN, M.M. (1976). Mutable and mutator loci. In *The Genetics and Biology of Drosophila* (M. Ashburner and E. Novitski, Eds), volume 1b, pp. 929–946. Academic Press, London.

GREEN, M.M. (1980). Transposable elements in *Drosophila* and other Diptera. *Annual Review of Genetics* **14**, 109–120.

HADORN, E. AND SCHWINK, I. (1956). A mutant of *Drosophila* without isoxanthopterine which is non-autonomous for the red eye pigments. *Nature* **77**, 940–941.

HAZELRIGG, T., LEVIS, R. AND RUBIN, G.M. (1984). Transformation of *white* locus DNA in *Drosophila*. Dosage compensation, *zeste* interaction, and position effects. *Cell* **36**, 469–481.

HEINSTRA, P., EISSES, K., SCHOONEN, W., ABEN, W., DE WINTER, A., VAN DER HORST, D., VAN MARREWIJK, W., BEENAKKERS, A., SCHARLOO, W. AND THORIG, G. (1983). A dual function of alcohol dehydrogenase in *Drosophila*. *Genetica* **60**, 129–137.

HENIKOFF, S., TATCHELL, K., HALL, B.D. AND NASMITH, K.A. (1981). Isolation of a gene from *Drosophila* by complementation in yeast. *Nature* **289**, 33–37.

HICKEY, D.A. (1982). Selfish DNA: a sexually transmitted parasite. *Genetics* **120**, 33–53.

HICKS, J.B., HINNEN, A. AND FINK, G.R. (1978). Properties of yeast transformation. *Cold Spring Harbor Symposia Quantitative Biology* **43**, 1305–1313.

HINNEN, A., HICKS, J.B. AND FINK, G.R. (1978). Transformation in yeast. *Proceedings of the National Academy of Sciences of the United States of America* **75**, 1929–1934.

HIRAGA, S. (1964). Tryptophan metabolism in eye color mutants of the housefly.

Japanese Journal of Genetics **39**, 240–253.

HIRSH, J. AND DAVIDSON, N. (1981). Isolation and characterization of the dopa decarboxylase gene of *Drosophila melanogaster*. *Molecular and Cellular Biology* **1**, 475–485.

HIRTH, K.-P., EDWARDS, C.A. AND FIRTEL, R.A. (1982). A DNA mediated transformation system for *Dictyostelium discoideum*. *Proceedings of the National Academy of Sciences of the United States of America* **79**, 7356–7360.

HUDSON, P., HALEY, J., CRONK, M., SHINE, J. AND NIALL, H. (1981). Molecular cloning and characterization of cDNA sequences coding for rat relaxin. *Nature* **291**, 127–131.

HUNKAPILLER, M.W. AND HOOD, L.E. (1980). New protein sequenator with increased sensitivity. *Science* **207**, 523–525.

ITAKURA, K., HIROSE, T., CREA, R., RIGGS, A.D., HEYNECKER, H.L., BOLIVAR, F. AND BOYER, H.W. (1977). Expression in *Escherichia coli* of a chemically synthesised gene for the hormone somatostatin. *Science* **198**, 1056–1063.

JIMENEZ, A. AND DAVIES, J. (1980). Expression of a transposable antibiotic resistance element in *Saccharomyces*. *Nature* **287**, 869–871.

JUDD, B.H. (1976). Genetic units of *Drosophila* — complex loci. In *The Genetics and Biology of* Drosophila (E. Novitski and M. Ashburner, Eds), volume 1b, pp. 767–799. Academic Press, London.

KEMPHUES, K.J., RAFF, R.A., KAUFMAN, T.C. AND RAFF, E.C. (1979). Mutation in a structural gene for a beta tubulin specific to testes in *Drosophila melanogaster*. *Proceedings of the National Academy of Sciences of the United States of America* **76**, 3991–3995.

KIDWELL, M.G. (1983). Intraspecific hybrid sterility. In *The Genetics and Biology of* Drosophila (M. Ashburner, H.L. Carson and J.N. Thompson, Jr, Eds), volume 3c, pp. 125–154. Academic Press, London.

KIDWELL, M.G. AND NOVY, J.B. (1979). Hybrid dysgenesis in *Drosophila melanogaster*: sterility resulting from gonadal dysgenesis in the P–M system. *Genetics* **92**, 1127.

KING, R.C. (ED.) (1975). *Handbook of Genetics*, volume 3. Plenum Press, New York.

KLECKNER, N. (1981). Transposable elements in procaryotes. *Annual Review of Genetics* **15**, 341–404.

LACHANCE, L.E. (1979). Genetic strategies affecting the success and economy of the sterile insect release method. In *Genetics in Relation to Insect Management* (M.A. Hoy and J.J. McKelvey, Jr, Eds), pp. 8–18. The Rockefeller Foundation, New York.

LANGER, H. (1967). Uber die pigmentgranula im facettenauge von *Calliphora erythrocephala*. *Zeitschrift für Vergleichende Physiologie* **55**, 354–377.

LEVIN, B.R. (1979). Problems and promise in genetic engineering in its potential applications to insect management. In *Genetics in Relation to Insect Management* (M.A. Hoy and J.J. McKelvey, Jr, Eds), pp. 170–175. Rockefeller Foundation, New York.

LIFTON, R.P., GOLDBERG, M.L., KARP, R.W. AND HOGNESS, D.S. (1977). The organization of the histone genes in *Drosophila melanogaster:* functional and evolutionary implications. *Cold Spring Harbor Symposia Quantitative Biology* **42**, 1047–1051.

LIVAK, K.J., FREUND, R., SCHWEBER, M., WENSINK, P.C. AND MESELSON, M. (1978). Sequence organization and transcription at two heat shock loci in *Drosophila*. *Proceedings of the National Academy of Sciences of the United States of America* **75**, 5613–5617.

LYTTLE, T.W. (1979). Experimental population genetics of meiotic drive systems. II. Accumulation of genetic modifiers of segregation distorter (SD) in laboratory populations. *Genetics* **91**, 339–357.

MANIATIS, T., FRITSCH, E.F. AND SAMBROOK, J. (1982). *Molecular Cloning. A Laboratory Manual*. Cold Spring Harbor Labs., Cold Spring Harbor, NY. 545 pp.

MARTIN, J., KUVANGKADILOK, C., PEART, D.F. AND LEE, B.T.O. (1980). Multiple sex determining regions in a group of related *Chironomus* species (Diptera: Chironomidae). *Heredity* **44**, 367–382.

MILANI, R. (1975). The house fly, *Musca domestica*. In *Handbook of Genetics* (R.C. King, Ed), volume 3, pp. 377–399. Plenum Press, New York.

MURRAY, A.W. AND SZOSTAK, J.W. (1983). Construction of artificial chromosomes in yeast. *Nature* **305**, 189–193.

O'DONNELL, J., GERACE, L., LEISTER, F. AND SOFER, W. (1974). Chemical selection of mutants that affect ADH in *Drosophila*. II. Use of 1-pentyne-3-ol. *Genetics* **79**, 73–83.

O'HARE, K., AND RUBIN, G.M. (1983). Structures of P transposable elements and their sites of insertion and excision in the *Drosophila melanogaster* genome. *Cell* **34**, 25–35.

OLD, R.W. AND PRIMROSE, S.B. (1981). *Principles of Gene Manipulation*, 2nd edn. Blackwell Scientific Publications, London.

ORGEL, L.E. AND CRICK, F.H.C. (1980). Selfish DNA: the ultimate parasite. *Nature* **284**, 604–607.

PERUCHO, M., HANAHAN, D. AND WIGLER, M. (1980). Genetic and physical linkage of exogenous sequences in transformed cells. *Cell* **22**, 309–317.

RUBIN, G.M. AND SPRADLING, A.C. (1982). Genetic transformation of *Drosophila* with transposable element vectors. *Science* **218**, 348–353.

RUBIN, G.M., KIDWELL, M.G. AND BINGHAM, P.M. (1982). The molecular basis of P–M hybrid dysgenesis: the nature of induced mutations. *Cell* **29**, 987–994.

SCALENGHE, F., TURCO, E., EDSTROM, J.E., PIROTTA, V. AND MELLI, M. (1981). Microdissection and cloning of DNA from a specific region of *Drosophila melanogaster* polytene chromosomes. *Chromosoma* **82**, 205–216.

SCHOLNICK, S.B., MORGAN, B.A. AND HIRSH, J. (1983). The cloned dopa decarboxylase gene is developmentally regulated when reintegrated into the *Drosophila* genome. *Cell* **34**, 37–45.

SCOTT, J., SELBY, M., URDEA, M., QUIROGA, M., BELL, G.I. AND RUTTER, W.J. (1983). Isolation and nucleotide sequence of a cDNA encoding the precursor of mouse nerve growth factor. *Nature* **302**, 538–540.

SEGRAVES, W.A., LOUIS, C., SCHEDL, P. AND JARRY, B.P. (1983). Isolation of the rudimentary locus of *Drosophila melanogaster*. *Molecular and General Genetics* **189**, 34–40.

SNYDER, M., HIRSH, J. AND DAVIDSON, N. (1981). The cuticle genes of *Drosophila*: a developmentally regulated gene cluster. *Cell* **25**, 165–177.

SOUTHERN, P.J. AND BERG, P. (1982). Transformation of mammalian cells to antibiotic resistance with a bacterial gene under control of the SV40 early region promotor. *Journal of Molecular and Applied Genetics* **1**, 327–341.

SPRADLING, A.C. AND RUBIN, G.M. (1981). *Drosophila* genome organization: conserved and dynamic aspects. *Annual Review of Genetics* **15**, 219–264.

SPRADLING, A.C. AND RUBIN, G.M. (1982). Transposition of cloned P elements into *Drosophila* germ line chromosomes. *Science* **218**, 341–347.

SPRADLING, A.C. AND RUBIN, G.M. (1983). The effect of chromosomal position on the expression of the *Drosophila* xanthine dehydrogenase gene. *Cell* **34**, 47–57.

STINCHCOMB, D.T., STRUHL, K. AND DAVIS, R.W. (1979). Isolation and characterization of a yeast chromosomal replicator. *Nature* **282**, 39–43.

STONE, B.F., NOLAN, J. AND SCHUNTNER, C.A. (1976). Biochemical genetics of resistance to organophosphorus acaricides in three strains of the cattle tick *Boophilus microplus*. *Australian Journal of Biological Sciences* **29**, 265–279.

SUMMERS, K.M., HOWELLS, A.J. AND PYLIOTIS, N.A. (1982). Biology of eye pigmentation in insects. *Advances in Insect Physiology* **16**, 119–166.

SUZUKI, D.T., KAUFMAN, T., FALK, D. AND THE U.B.C. *DROSOPHILA* RESEARCH GROUP (1976). Conditionally expressed mutations in *Drosophila melanogaster*. In *The Genetics and Biology of* Drosophila (M. Ashburner and E. Novitski, Eds), volume

1a, pp. 206–263. Academic Press, New York.

VAN DEUSEN, E.B. (1976). Sex determination in germ line chimeras of *Drosophila melanogaster*. *Journal of Embryology and Experimental Morphology* **37**, 173–185.

VIGUE, C. AND SOFER, W. (1975). Chemical selection of mutants that affect ADH activity in *Drosophila*. III. Effects of ethanol. *Biochemical Genetics* **14**, 127–135.

WAGONER, D.E., McDONALD, I.C. AND CHILDRESS, D. (1974). The present status of genetic control mechanisms in the house fly, *Musca domestica* L. In *The Use of Genetics in Insect Control* (R. Pal and M.J. Whitten, Eds), pp. 181–197. Elsevier/North-Holland, Amsterdam.

WATERHOUSE, D.F., LaCHANCE, L. AND WHITTEN, M.J. (1974). Use of autocidal methods. In *Theory and Practice of Biological Control* (C. Huffaker and P.S. Messenger, Eds), pp. 637–659. Academic Press, New York.

WENSINK, P.C., FINNEGAN, D.J., DONELSON, J.E. AND HOGNESS, D.S. (1974). A system for mapping DNA sequences in the chromosomes of *Drosophila melanogaster*. *Cell* **3**, 315–325.

WHITTEN, M.J. (1979). The use of genetically selected strains for pest replacement or suppression. In *Genetics in Relation to Insect Management* (M.A. Hoy and J.J. McKelvey, Jr, Eds), pp. 31–41. Rockefeller Foundation, New York.

WHITTEN, M.J. (1984). The theoretical basis of genetic control. In *Comprehensive Insect Physiology, Biochemistry, and Pharmacology* (G.A. Kerkut and L.I. Gilbert, Eds), volume 12, chapter 15. Pergamon Press, New York, in press.

WU, R. (ED.) (1979). Recombinant DNA. *Methods in Enzymology*, volume 68. Academic Press, New York.

YEN, T.T. AND GLASSMAN, E. (1965). Electrophoretic variants of xanthine dehydrogenase in *Drosophila melanogaster*. *Genetics* **52**, 977–981.

ZACHAR, Z. AND BINGHAM, P.M. (1982). Regulation of *white* locus expression: the structure of mutant alleles at the *white* locus of *Drosophila melanogaster*. *Cell* **30**, 529–541.

ZALOCAR, M. (1981). A method for injection and transplantation of nuclei and cells in *Drosophila* eggs. *Experientia* **37**, 1354–1356.

ZIMMERING, S., SANDLER, L. AND NICOLETTI, B. (1970). Mechanisms of meiotic drive. *Annual Review of Genetics* **4**, 409–436.

8

The Biology and Behaviour of Slugs in Relation to Crop Damage and Control

C. M. PORT* AND G. R. PORT†

*Agricultural Development and Advisory Service, Government Buildings, Kenton Bar, Newcastle upon Tyne, NE1 2YA, UK and †Department of Agricultural Biology, The University, Newcastle upon Tyne, NE1 7RU, UK

Introduction
Slug biology
         Systematics and taxonomy—Ecology—Life cycles—Reproduction—
         Activity—Feeding
Slugs as crop pests
         Crops at risk—Damage estimates for potatoes—Damage estimates for
         wheat
Slug control
         Slug population estimation—Cultural control—Chemical control by
         molluscicides—Treatment timing—Seed treatments—Plant extracts—
         Biological control
Future prospects for control
Conclusions
References

Introduction

Slugs (Gastropoda: Pulmonata) occur in many parts of the world, but it is in the cool moist temperate climates that they most frequently attain pest status. They are pests both in terms of their damage to crops and their role as secondary hosts of vertebrate parasites. In this review we discuss the role of slugs as crop pests, especially in British agriculture.

    Runham and Hunter (1970) have provided a good general account of slugs and recently Godan (1983) has produced a comprehensive book on pest slugs and snails although most of the literature cited predates 1979. Stephenson and Bardner (1976) highlighted the importance of slugs in British agriculture. Our aim is to review recent developments in the understanding of slugs, including their pest status, and to concentrate on research that has obvious application or has potential for exploitation in controlling slugs.

Slug biology

SYSTEMATICS AND TAXONOMY

In Europe there are four families in the Order Stylommatophora of the subclass Pulmonata, whose members are commonly known as slugs: Arionidae, Milacidae, Limacidae and Testacellidae. The important crop pests are found in the first three families, the Testacellidae being largely carnivorous. Some important species are listed in *Table 1* and Godan (1983) gives more extensive tables of pest species and their distribution. In the United Kingdom (UK) *Deroceras reticulatum*, the *Arion hortensis* aggregate and *Milax budapestensis* are usually considered to be the most important slug pests of field crops.

Changes in nomenclature and inconsistency among authors in its use make the literature on slugs confusing to follow. The species in *Table 1* are largely those mentioned by Kerney (1976). In recent years some species have been segregated and new species described. Comparison of the form of the genitalia and, more recently, electrophoretic techniques (Evans, 1985) have proved useful taxonomic tools. *Limax pseudoflavus* (Evans, 1978), formerly

Table 1.   Important species of slug

Family	Genus	Species
Arionidae	*Arion*	*A. ater* (L.)
		A. subfuscus (Draparnaud)
		*A. fasciatus** (Nielsson)
		*A. circumscriptus** Johnston
		*A. silvaticus** Lohmander
		A. hortensis† Ferrusac
		A. distinctus† Mabille
		A. intermedius Normand
	Ariolimax	*A. columbianus‡* (Gould)
Milacidae	*Milax*	*M. gagates* (Draparnaud)
		M. sowerbyi (Ferrusac)
		M. budapestensis (Hazay)
Limacidae	*Limax*	*L. maximus* L.
		L. flavus L.
		L. pseudoflavus Evans
		L. marginatus Müller
	Deroceras	*D. reticulatum* (Müller)
		D. agreste (L.)
		D. caruanae (Pollonera)
		D. laeve (Müller)
Veronicellidae		*Vaginulus* spp.§
		Veronicella spp.§

*Often grouped as *Arion fasciatus* aggregate
†Often grouped as *Arion hortensis* aggregate
‡A North American species often used in behavioural studies.
§Species of these genera are sometimes pests in tropical and equatorial regions.

classified as a variant of *L. flavus*, has been found in western England and Ireland and has been the subject of several behavioural and ecological investigations (Cook, 1979a, 1980); however, it is not a major crop pest. *Arion distinctus* is one of the two new species, the other being *Arion owenii*, described by Davies (1977, 1979) from the *Arion hortensis* aggregate. Recent investigations of slug pests from this group have not distinguished between species.

The British slug fauna may be identified using any of several keys: Quick (1960) produced a comprehensive guide which forms the basis for more recent keys; Godan (1983) includes a key to some species, and that by Cameron, Jackson and Eversham (1983) is a useful field key to British slugs. Whereas some species are easily identified by external features, others prove more difficult, especially some of the *Arion* species. *Deroceras reticulatum* may be separated from *D. agreste* only following dissection of the genitalia (Quick, 1960). Few workers specifically distinguish these species, but in Britain confusion of *D. reticulatum* with *D. agreste* is likely to occur only in areas where the latter species is common, that is, in Scotland and northern England (Kerney, 1976).

ECOLOGY

Habitat

Most pest species of slug live in intimate contact with the soil. Given a choice of aggregate sizes, *Deroceras reticulatum* shows distinct preference for a soil structure where it can lie in between the soil aggregates, keeping a large proportion of its body surface in contact with the soil (Duval, 1970; Stephenson, 1975b). Movement through soil is easy for a soft-bodied slug compared with snails with their rigid shells; nevertheless, movement is facilitated by an open soil structure.

Because of their permeable integument, slugs are restricted to habitats where conditions remain moist most of the time, for example 'heavy' soils and dense vegetation (Gould, 1961). Slugs may move to greater depths in the summer as soils dry out, resulting in a corresponding decrease in surface activity. Persistent drought is likely to kill slugs; similarly, activity declines at lower temperatures and extreme cold may prove fatal. Hunter (1966) reported population decline after hard winters. Mellanby (1961) found that temperatures of 5°C immobilized *A. hortensis* agg. and *M. budapestensis*, but that *D. reticulatum* continued to feed actively on cereals at 0.8°C. Kemp and Newell (1987) have shown that *D. reticulatum* has low energy reserves (glycogen content) compared with *A. hortensis* agg. and this may explain why *D. reticulatum* is active at extremes of temperature and humidity which appear to be limiting for other species.

Homing behaviour

Slugs are active at night when conditions are suitable and by day they shelter in resting sites in or on the soil. These sites are often in regular use and slugs show an ability to return to them (homing). Taylor (1907) described homing behaviour in *Limax maximus* and homing is a phenomenon now recorded from many genera of gastropod molluscs including *Ariolimax, Arion, Deroceras* and *Limax* (South, 1965; Newell, 1966; Rollo and Wellington, 1981).

Cook has made an extensive study of homing in *Limax pseudoflavus*. In the laboratory this species can return home without following mucus trails laid on the substrate and it is likely that olfactory cues are important (Cook, 1979b). The optic (posterior) tentacles seem to have an important role in olfaction of air-borne chemicals (Gelperin, 1974). In the field *L. pseudoflavus* usually homes by moving upwind. An individual may not always use the same home and homes are perhaps best regarded as specific resting sites favoured by a species (Duval, 1972; Cook, 1980). If long-range olfaction is not being used, for example if the home lies downwind, then *L. pseudoflavus* follows mucus trails on the substrate, either left by itself or other members of the species (Cook, 1979a, 1980). The anterior tentacles play a greater role in trail-following.

Regulatory factors

As already indicated, extremes of drought and cold may cause direct mortality. Extreme weather conditions also reduce the numbers of suitable resting sites, and competition for the remaining sites intensifies. Inter- and intra-specific aggression has been well documented by Rollo and Wellington (1979). Other factors that influence slug numbers are natural enemies: predators, parasites and pathogens which are likely to have density-dependent effects. Vertebrate predators include birds, badgers, foxes and small mammals (*see*, for example, South, 1980), but none of these specialize on slugs and their impact on slug numbers is likely to be small. Invertebrate natural enemies have been discussed by Stephenson and Knutson (1966) who concluded that few of these animals showed great potential as control agents, although some insects may impose significant control as slug populations have been found to increase after insecticide treatment (Grant *et al.*, 1982).

LIFE CYCLES

Slug life cycles are variable in length and mature slugs may reproduce over a prolonged period. Mating and oviposition occur during periods of weather conditions favouring general activity. As a result, field populations usually comprise mixed age groups. There are recognized peaks of egg laying in most species, but these vary with year and location. The most extensive information on slug life histories has been published for the pest species. The information below is for UK conditions based on Hunter's studies on slugs in Northumberland (Hunter, 1968b, 1978) and Davies' studies on *Arion hortensis* agg. (Davies, 1977, 1979).

Milax budapestensis

This species mates in the autumn, winter and spring, with a peak in October and November. Eggs are found between December and April and many hatch by June. Eggs laid earlier hatch earlier, but low temperatures delay development so that eggs laid over several months in winter may all hatch within a few weeks in May and June. The young slugs grow during the autumn, spring and summer and lay eggs the following winter. Slugs hatching early in April may possibly lay eggs at about one year of age.

Arion hortensis *agg.*

Arion hortensis s.s. matures in early autumn and may breed throughout winter until about April. The breeding season of *A. distinctus* follows that of *A. hortensis* s.s. by a few weeks and is mainly in spring and summer, but may continue throughout the year. *A. distinctus* is less sensitive to the extreme conditions in summer and winter. There is a peak of hatching in late spring due to delay in egg development caused by low temperature. The slugs grow quickly and may lay eggs within a year.

Deroceras reticulatum

This species shows much greater variability in its life history. Hunter (1968b) originally regarded it as having two complete generations a year, with slugs maturing and breeding in the spring and autumn and their offspring maturing and breeding in the autumn and spring respectively. The actual pattern is probably more complex than this (Hunter and Symonds, 1971; Hunter, 1978). As an example, slugs hatching in July (year 1) matured by the following April (year 2) when they laid eggs. These in turn gave rise to a third generation in December (year 2), the intervals between generations being about 9 months. At the same time and place another group of *D. reticulatum* hatched in November (year 1) and laid eggs in August (year 2) and these gave rise to a third generation in April (year 3). This pattern is probably very susceptible to changes in temperature: South (1982) has shown that *D. reticulatum* reared at 5°C takes over a year from hatching to first egg laying, whereas at 18°C they take four and a half months to begin egg laying.

REPRODUCTION

Slugs are all hermaphrodite. The sperm mature slightly before the ova (protandry); in some species such as *Arion ater* this is distinct, but in *D. reticulatum* some eggs are laid before the major phase of spermatogenesis is complete. Self fertilization is possible in many species and Davies (1977) suggests that mating may be unnecessary in *A. intermedius*, but cross fertilization is more probable in most species under normal conditions. Self fertilization is most unlikely in certain species, for example *D. reticulatum*, which even if reared in isolation rarely produce fertile eggs (Runham and Hunter, 1970).

Courtship precedes sperm transfer and is usually a lengthy process, taking several hours in some large species. This behaviour is usually initiated by trail-following in which one slug follows the mucus trail of another, the trail having been laid some time before. When the pursuer contacts the trail maker, courtship may follow. Most studies of trail-following have been as adjuncts to investigations of homing behaviour (pages 257–258).

The period between mating and oviposition varies between species from a few days to several weeks. Eggs are normally laid in crevices in the soil and for *D. reticulatum* it has been shown that eggs are laid preferentially in soil with 75% water content (Carrick, 1942; Arias and Crowell, 1963) and with an aggregate range between 3 and 10 mm (Stephenson, 1975b). By contrast, adult slugs preferred to rest in coarser soil. The number of eggs laid, varies (*Table 2*).

As already stated the rate of egg development is greatly affected by temperature and this may serve to increase the synchrony of hatching of eggs laid at various times over winter. The time taken for young slugs to reach maturity is dependent upon both the quality and quantity of the food available, but it is generally recognized that there are several phases to growth. In the *infantile* phase, growth is rapid; this is followed by a phase of slow growth, the *juvenile* phase, and then finally a phase of little or no growth, the *mature* phase. Many *Arion* species begin to produce eggs only in the mature phase. In *D. reticulatum* and *A. circumscriptus*, oviposition starts in the second phase and the third phase is less distinct (South, 1982). The rate of development is temperature dependent: this, combined with the

Table 2.   Recorded total fecundity of various slug species

Species	Total number of eggs per slug	Reference
Deroceras reticulatum	500	Carrick, 1938
	200	Hunter, 1978
	260	South, 1982
Milax budapestensis	30	Hunter, 1978
Arion hortensis agg.	50	Hunter, 1978
Arion hortensis s.s.	281	Davies, 1977
Arion distinctus	310	Davies, 1977
Arion owenii	183	Davies, 1977
Arion intermedius	302	Davies, 1977
	202	South, 1982

other factors affecting growth rate, means that weight of a slug is a poor indication of its age (Prior, 1983).

ACTIVITY

Feeding and mating occur in weather conditions favourable for slug activity and, because of the dependence of crop damage on slug activity, many studies have been made on this aspect of their biology. Direct observations of slugs in the field have supported the view that weather conditions are very important in regulating activity although no single factor, or even simple combination of factors, has been identified (Barnes and Weil, 1945; White, 1959).

Like many other animals, slugs show a circadian rhythm of activity (Dainton, 1954; Sokolove *et al.*, 1977; Morton, 1979) and are generally active at night. Occasionally they may become active by day if certain combinations of conditions occur. Rollo (1982) considered circadian rhythms to be an invaluable adaptation for slugs enabling them to avoid unfavourable weather conditions, in spite of their slow speed of movement and poor conservation of water. In the field, weather acts on the amplitude of the rhythm; in favourable conditions it is freely expressed, but usually combinations of weather variables impose some constraint.

Light is generally regarded as the environmental factor entraining the circadian rhythm (zeitgeber) (Sokolove *et al.*, 1977), but there are conflicting ideas about the relative importance of light and temperature in inducing daily activity. Some authors consider light to be of overriding importance (Newell, 1968; Lewis, 1969) whereas others regard temperature, or more particularly change in temperature, to be of greatest significance (Dainton, 1954; Karlin, 1961; Dainton and Wright, 1985). Wareing and Bailey (1985) have discussed some of the shortcomings of laboratory investigations on activity and have shown that fluctuating temperatures produce higher levels of activity in *D. reticulatum*.

In assessments of activity in the field, temperature has always been found to be of significance. Webley (1964) studied the numbers of slugs trapped at baits and found that temperature (mean night air temperature and grass minimum temperature) accounted for much of the variation in slug numbers from day to day. However, *Arion hortensis* were not affected by meteorological factors to the same extent as other species. Crawford-Sidebotham (1972a) assessed activity of slugs by night-time searching; again he found temperature and, to a lesser extent, vapour pressure deficit to be important; activity tended to increase with increasing temperature, but to decrease with increasing vapour pressure deficit. Hogan (1985) used slugs caught in traps as a measure of night-time activity and found temperature to be the most useful variable for explaining fluctuations in numbers captured. In this work up to 84% of the variation in numbers of *D. reticulatum* could be explained by multiple regression models.

Rollo (1982) has conducted the most detailed study of this nature, examining activity of *Limax maximus* in field enclosures: his regression models explained 73–87% of the observed variation in activity measured hourly. His approach was novel in that he regarded many of the relationships between activity and weather variables as non-linear and he made allowance for activity being restricted to above a threshold and below a limit for particular variables. Time of day, light intensity, changes in light intensity and surface temperature were all important determinants of activity.

In all these studies relative humidity, vapour pressure deficit or evaporation rate have been considered, but have never been found to have primary importance in regulating slug activity. Nevertheless, it is clear that slug activity is greatest in moist conditions: the observation that slug activity depends on there being a film of moisture over sites of activity (Barnes and Weil, 1945) would bear further investigation.

FEEDING

Most slugs will sample potential foods in their habitat and may exhibit prolonged feeding on certain foods. Food preferences can vary with growth stage, species and habitat. Slugs from the families listed in *Table 1* are all primarily plant feeders.

Range of food

The range of food varies from living plant tissue to decaying organic matter and occasionally animal material. Rottger and Klingauf (1976) have recorded *Deroceras laeve* feeding on the eggs of the mangold fly *Pegomyia betae* (Curtis) (now *Pegomya hyoscyami* (Panzer)) a pest of sugar beet, and Fox and Landis (1973) showed that *D. laeve* would feed on aphids and other insect material. Slugs will also feed on one another: cannibalism of probably moribund individuals may be observed in crowded cultures and rasp marks are often found on slugs, both in cultures and in the wild. Direct observations show that biting may be an early response in many slug encounters.

In laboratory conditions slugs may be maintained and reared on a range of foodstuffs and most workers use carrot, potato and lettuce although even rodent laboratory diet has been found suitable (Crowell, 1979). Experiments on slug feeding often necessitate a synthetic diet and considerable success has been found with agar-based diets (Standen, 1951; Wright, 1973; Whelan, 1982).

In natural habitats, slugs feed on a wide range of the available plant material, usually herbaceous plants and grasses. They may have a substantial impact on the competitive success of plants (Dirzo and Harper, 1980; Cottam, 1985).

Food preferences

Most slugs exhibit a preference for certain food plants (Pallant, 1972) and whereas physical barriers such as the presence of silica may deter slug feeding (Wadham and Wynn Parry, 1981) several studies have shown that these preferences are more usually due to the presence of allelochemicals in the non-preferred plants (Cates, 1975; Rice, Lincoln and Langenheim, 1978; Lincoln and Langenheim, 1979; Gouyon, Fort and Caraux, 1983). The best-known interaction of this type is the avoidance by slugs and snails of clover (*Trifolium repens* L.) and other plants that are cyanogenic, i.e. release cyanide when the leaves are crushed (Jones, 1962, 1966). Varieties that are acyanogenic, being unable to release cyanide, are selectively grazed and are consequently rarer in habitats where slugs and snails are important herbivores (Angseesing and Angseesing, 1973; Angseesing, 1974; Dirzo and Harper, 1982). Slugs rapidly learn to recognize unpalatable food (Gelperin, 1975).

Slugs show a considerable range of preference for different varieties of potato (Gould, 1965; Winfield, Wardlow and Smith, 1967; Hunter, Symonds and Newell, 1968; Pinder, 1974). For example, Maris Piper and Cara are cultivars very susceptible to damage whereas Pentland Dell is much less susceptible (*see Table 4*). Richardson (1979) outlined the procedures for testing new potato varieties in England and Wales: observations are made on the severity of slug attack on maincrop varieties and varieties are rated on a scale 1 (most susceptible) to 9 (least susceptible). Ratings for the 1986 recommended varieties are given by the National Institute of Agricultural Botany (NIAB) (1985).

The basis for the slug's preference has been the subject of several investigations which have produced contrasting results. Atkin (1979) found that varieties with a low total-protein content were preferred: tubers with a relatively high protein content caused a reduction of gut proteolytic activity when consumed by the slug, probably due to trypsin inhibitors in the tuber. Slugs confined on resistant varieties showed much slower growth rates than slugs on susceptible varieties. South (1973) states that potato resistance correlates with structural features of the tuber skin. Storey (1985) showed that preference was also due to variation in starch, glycoalkaloid and phenolic acid content. It is probable that several of these features have a role in determining preference, but further studies are needed.

There has been little work on preferences between cereals although it is generally recognized that barley is not as susceptible to damage as wheat. Duthoit (1964) demonstrated this in laboratory preference tests using seed of both cereals. Barley seed has the grain enclosed within the lemma and palea whereas wheat has a naked seed and A. Young (unpublished data) has shown that removing these structures from barley seed made it more palatable to *Deroceras reticulatum*. As a crop, winter barley is usually sown earlier than winter wheat and will therefore more rapidly pass through the stage when it is susceptible to slug damage.

Among varieties of wheat (*Triticum aestivum* L.) there seem to be no major preferences, but at the University of Newcastle upon Tyne (K. Ashover, unpublished data) differences between species of wheat (*T. aestivum, T. timopheevi* Zhuk., *T. turgidum* L. and *T. durum* Desf.) have been found: the preference appears to be for species with a high total-nitrogen content in the seed. There is obviously a need for further work to investigate feeding preferences in cereals (*Figure 1*).

Figure 1. The field slug, *Deroceras reticulatum*, approaching a wheat seed.

Many of these preference tests are conducted in conditions where the slug has a choice of food. In their simplest form, the tests show that one food is preferred over another. We urgently need more information on the feeding behaviour of slugs as it is not clear whether feeding is equally reduced on non-preferred varieties when the slugs have no choice (Crawford-Sidebotham, 1972b). Consequently, in a field planted with a non-preferred variety, or a variety treated with a repellent (page 286), feeding damage may not be significantly lower.

Slugs as crop pests

CROPS AT RISK

Much of our information on crops at risk is based on data from England and Wales. The most important damage, in monetary terms, to crops in the UK is suffered by potatoes and winter (autumn-sown) wheat. However, a wide range of other UK crops may be affected by slugs, as documented in the Ministry of Agriculture, Fisheries and Food (MAFF) seasonal pest summaries. Field crops at risk include oilseed rape in the seedling stage, peas by contamination of the harvested crop (Wharton and Ensor, 1969) and sugar beet shortly after germination. Similarly, grass is most susceptible to slug attack at, or shortly after, germination; direct-seeded grass is also vulnerable, particularly in permanent pasture (Savage and Thomas, 1985). Clements *et al.* (1985) considered that pasture yield may be reduced by slugs as yield increases after molluscicide use were noted. Slugs may also damage clover seedlings in pastures (Clements and Bentley, 1983).

On vegetable crops slugs may reduce quality, for example by grazing the buttons of Brussels sprouts and celery stalks or by contamination of the produce, for example lettuce (Oakley, 1984). They also damage root vege- tables such as carrots, and have been implicated in the transmission of plant pathogens including the fungus *Mycocentrospora acerina* affecting carrots (Dawkins, Luxton and Bishop, 1985) and the bacterium *Erwinia carotovora* affecting potatoes (Dawkins *et al.*, 1986) and these authors also cite other examples although the economic importance of these, remains to be estab- lished. Slugs may hole strawberry fruits, making them unmarketable. They are not regarded as pests of other fruit crops, but blackcurrants are sometimes affected by snail contamination of the harvested fruit (Stringer and Morgan, 1969). Slugs feature as occasional pests in glasshouses (Foster, 1977) and of ornamental crops such as flower bulbs (Eaton and Tompsett, 1976). In a study of allotment gardens in northern England, slugs were considered to be the most important pests (Atkinson, Gibson and Evans, 1979).

This pattern of crops at risk is reflected throughout northern Europe and Godan (1983) has given a detailed account of damage reported from the Federal Republic of Germany. She also lists crops at risk in other countries, both in Europe and throughout the world. In North America, damage is to maize and many horticultural crops (Wilkinson, 1972; Godan, 1983). Recent work has featured slugs as pests of forage crops (Byers and Bierlein, 1984; Dowling and Linscott, 1983). Similarly, in New Zealand forage crops are considered to be at risk (Charlton, 1978). In Central America slugs of the family Veronicellidae are important pests (Andrews, 1983).

In the past, damage estimates have been made for potatoes and wheat in the UK: here we attempt to update them.

DAMAGE ESTIMATES FOR POTATOES

Slugs do no economic damage to seed potatoes after planting and do not affect plant stand: the economic damage is done by slugs feeding on the maincrop tubers in the period between late summer and harvest. Later- harvested crops are more at risk, particularly in wet autumns; crops harvested in August (early potatoes and most seed crops) are not at risk.

Slugs feeding on tubers reduce actual yield (tuber weight) by very little, but render tubers unacceptable for human consumption. Damaged tubers have to be separated and used for stock feed, involving extra labour and resulting in marketable yield loss. Stephenson and Bardner (1976) discussed estimates of slug damage and the available data are summarized in *Table 3*.

No surveys of slug damage have been made since that of Church, Hampson and Fox (1970); however, Kerr (1984) includes assessments of slug damage, as percentage of total yield at lifting, for four new maincrop varieties tested

Table 3.   Estimates of slug damage to potatoes

Year	Crop unfit for ware (%)	Basis for estimate	Reference
1954	0.29	Potato Marketing Board	
1955	0.16	pre-harvest tuber	
1956	0.12	samples (England and Wales)	Baker and Waines, 1957
1960–64	0.74		Strickland, 1965
1966	0.81	Opinions of ADAS district advisers (England and Wales)	Hunter, 1969a
1965	2.1	Samples of potatoes in	Church, Hampson and
1966	2.6	store (Great Britain)	Fox, 1970

by NIAB at 49–63 sites in England and Wales, 1981–83. Slug damage accounted for 0.4–1.2% of outgrades: these fall within the range of older estimates which vary between 0.12% and 2.6% of the total potato crop. Stephenson and Bardner (1976) placed a monetary value of £0.14–3.17 million on the loss attributable to slug damage at 1974 values (maincrop). They considered this to be an overestimate because, first, the damaged part of the crop would have some value as stock feed or could be used for industrial purposes and, secondly, slug damage can be confused with cutworm damage. Using 1984 and 1985 values for total potato sales (MAFF, 1986), losses attributable to slug damage range from £0.3 million to £7 million.

Since 1974–76, although potato area has declined by 11%, production has increased by 27%. The value of potential loss is likely to be larger today than over 10 years ago, although these estimates can vary a great deal due to large fluctuations in the price of potatoes and variation in slug damage between seasons. An example of the latter is provided by assessments of slug-damaged tubers in NIAB variety trials for cv. Maris Piper of 0.2% in 1983 and 55% in 1984 at the same site (S. P. Kerr, unpublished data). Slug damage varies each season because of weather conditions: for example wet summers followed by wet autumns provide ideal conditions for slug multiplication and damage to crops.

Potential slug damage may have increased over the last ten years because of the larger area growing more susceptible varieties of potato, for example, Maris Piper (*Table 4*). However, actual loss may have been reduced by increased use of molluscicides: for example, in 1977 0.8% of the maincrop in England and Wales was treated and by 1982 this had increased to 11% (Steed *et al.*, 1979; Sly, 1986). The cost of molluscicides must, however, be considered as an indirect loss due to slugs and would amount to £0.25 million at 1985 prices, to which the costs of application and crop damage should also be added.

Table 4. Major maincrop varieties of potato, their susceptibility to slug damage (NIAB, 1985) and % of maincrop in 1974 and 1985. Susceptibility rating 1 (high) to 9 (low)

Maincrop variety	Susceptibility to slugs	Percentage of total maincrop*	
		1985	1974
Cara	2	7.07	0
Desirée	4	15.89	9.25
King Edward	4	3.39	18.63
Kingston	2	0.88	0
Majestic	5†	0	5.28
Maris Piper	1	21.37	11.36
Pentland Crown	4	10.28	25.53
Pentland Dell	6	7.33	11.02
Pentland Hawk	4†	2.92	1.42
Pentland Ivory	6†	0.45	5.40
Pentland Squire	5	7.92	0
Record	4	15.86	9.05
Romano	4	4.59	0
Others		2.35	3.06

* Percentage of total crop after J. Pearson, PMB, (unpublished data)
† Unpublished ratings courtesy of S. P. Kerr, NIAB, (unpublished data)

DAMAGE ESTIMATES FOR WHEAT

Types and cost of damage

Wheat is more susceptible to damage than barley and winter wheat is at greater risk than spring wheat. Winter wheat is vulnerable for a longer period in the cool and moist conditions of autumn and winter which favour slug activity. The most important damage done to winter wheat is by slugs feeding on the germ of seeds shortly after sowing (Gould, 1961; Duthoit, 1964): this hollowing of grain and the severe grazing of newly emerging shoots leads to complete loss of individual plants. Shredding of leaves of older plants in the winter, although not usually considered to be of economic importance, nevertheless sometimes can be so on backward and thin wheat crops. N. French (unpublished data) found that in six out of nine wheat crops moderately to severely damaged by slugs in the autumn (1974–76) (plant density 48–114 plants/m^2), the cost of slug control measures applied in December, January or February was repaid by extra yield.

In wet summers, slugs are sometimes observed feeding on mature plant foliage, especially if the crop has lodged (MAFF seasonal pest summaries). They may even graze the flag leaf, but this damage has never been considered to be of economic importance. *Table 5* gives a summary of damage estimates.

Stephenson and Bardner (1976) regarded the figures provided by Strickland (1965) and Hunter (1969a) as underestimates of potential damage to winter wheat as they are a percentage of total area sown to wheat (winter and spring combined). They estimated the total cost of slug damage (including the

Table 5.   Estimates of factors associated with slug damage to winter wheat

Year	Crop area (%)	Basis of estimate	Reference
1964	2.23	Estimated equivalent grain loss	Strickland, 1965
1966–67	1.11	Redrilled (including part fields)	Hunter, 1969a
	0.49	Treated with molluscicides	
	0.22	Cost (including redrilling, yield loss and control measures)	Stephenson and Bardner, 1976 (based on Hunter, 1969a)
1974	2.00		Chapman, Sly and Cutler, 1977
1977	2.82	Treated with molluscicides	Steed et al., 1979
1982	30.08		Sly, 1986

cost of molluscicide application, the cost of resowing and the cost of reduced yield from sowing spring instead of winter wheat) based on estimates of Agricultural Development and Advisory Service (ADAS) advisers: their figure of 0.22% of the value of the crop was equivalent to £191 700 in 1967, £600 000 in 1974 and £2.69 million in 1985.

Factors affecting damage and damage estimates

Some factors may have led to an increase in the damage by slugs to winter wheat since the early 1970s, but this is offset by the marked increase in the percentage of the crop treated with molluscicide (*Table 5*). Hunter's and Strickland's estimates were based on total area of wheat, because MAFF census figures do not distinguish between winter and spring wheat. The proportion of spring-sown wheat is small and has decreased since these estimates were made, for example from 5.3% in 1974 (Chapman, Sly and Cutler, 1977) to about 3% in 1982 (Sly, 1986): a slightly larger proportion of the crop will therefore be at risk.

Stephenson and Bardner (1976) drew attention to the increase in brassica seed crops, especially oilseed rape, the dense canopy of which provides ideal conditions for slugs. This trend has continued, as shown in *Figure 2*. Most oilseed rape crops are followed by winter wheat which may, as a consequence, be at greater risk from slugs.

It is widely accepted that cultivations lead to a reduction of slug damage (page 273). Where crops are direct-drilled (no cultivation), slug populations are suspected to be larger, but there are few data available (Edwards, 1975; Grant *et al.*, 1982). Another reason for the increased susceptibility of direct-drilled crops is that seeds in the drill slit are more accessible to slugs. When conditions favour closure of the drill slit, slug damage is reduced (Hughes and Gaynor, 1984). It is not clear whether the proportion of direct-drilled wheat has changed in recent years: in 1974 Allen (1975) estimated that 5.6%

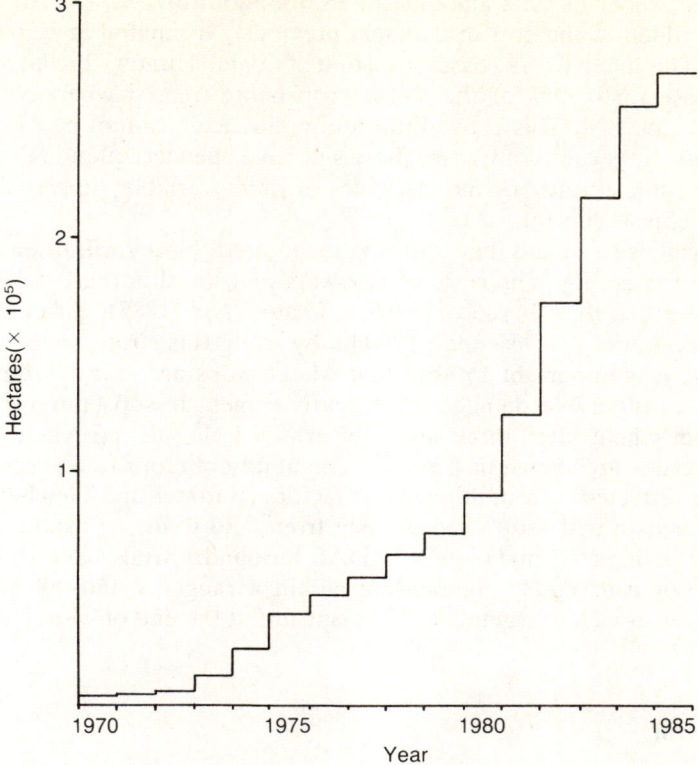

Figure 2. Area of oilseed rape grown in England and Wales 1970–85 (after MAFF census data, © Crown Copyright).

of all winter wheat was direct-drilled; however Chapman, Sly and Cutler (1977) estimated the figure to be 0.8% for the same year. Estimates for 1977 and 1982 were 0.9% and 3% respectively (Steed *et al.*, 1979; Sly, 1986).

There has been a trend for winter wheat to be sown earlier: in 1974, only 22% of winter wheat was sown by mid-October (Chapman, Sly and Cutler, 1977), but in 1977 this had increased to 47% (Steed *et al.*, 1979). The current recommendation, based on several agronomic factors, is to complete sowing by mid-October on heavy soils (MAFF, 1985). This trend may, in part, be due to an increase in wheat crops following oilseed rape which is harvested early, before most cereals. Early-sown crops are less prone to slug damage as the crop passes through the susceptible stage more rapidly and slug activity is likely to be lower in the drier conditions of early autumn.

All these factors will affect the current value of the earlier slug damage estimates and more precise figures would be obtainable only from an extensive survey. However, it is clear from the figures for molluscicide usage in winter wheat crops that the farmer's perception of slugs as pests has changed in recent years. Many farmers and advisers are aware that wheat following oilseed rape is particularly prone to slug damage. The cost of molluscicides applied to wheat in 1982 amounts to £8.6 million (at 1985 prices) and this must be added to the cost of damage, previously estimated at £2.69 million in 1985. The latter figure, based on Hunter's data (Hunter, 1969a), includes an estimate for 0.49% of the wheat crop being treated whereas the 1982 figure was 30.08%. This expenditure on molluscicide cannot be regarded as a total loss as, because of its use, there will have been a reduction in damage. However, the efficacy of molluscicides is quite variable; this is discussed under chemical control.

If damage is so severe that resowing is required, barley will often be sown the following spring. The costs of resowing and the difference in yield and sale price mean that, based on 1986 estimates (Nix, 1985), a farmer would experience a loss just less than £200/ha by using this strategy.

Finally, it is important to note that wheat crops are able to compensate for even considerable damage in the early stages. Jessop (1969) measured yield from wheat after three levels of artificial slug damage in December and his results are shown in *Figure 3*. The ability of crops to compensate for damage is affected by a multiplicity of factors (Bardner and Fletcher, 1974), but similar data to Jessop's can be seen from field trials: an example of one such trial is depicted in *Figure 4*. ADAS husbandry trials have shown that wheat crops can readily compensate within a range of 150–400 plants/m^2 without loss of yield (target, 200–250 plants/m^2 at the end of winter) (MAFF, 1985).

Slug control

SLUG POPULATION ESTIMATION

Direct counts

The need for, and efficiency of, slug-control measures is often based on direct or indirect assessment of the slug population in the field. Slugs are usually nocturnal and by day seek refuge in the soil or under objects on the soil surface: direct counts are therefore feasible only at night. Such methods have been used in several studies including the classic work of Barnes and Weil (1944, 1945), who assessed spatial and temporal variation in slug activity. The drawback of this approach is that it is not only labour intensive, but it is also subject to variation, in that an unknown and probably variable proportion of the slug population in an area is active on any given night.

Damage assessment

As damage is ultimately dependent upon the activity and numbers of slugs, damage assessment has been widely used as an indirect measure of population size. This may entail leaving either normal or novel food in an area (Duthoit, 1961; Glen and Orsman, 1986). More usually, damage suffered by a crop is used to assess the efficacy of different slug-control measures: for potatoes, tuber samples may be scored for severity of damage; for wheat, where loss

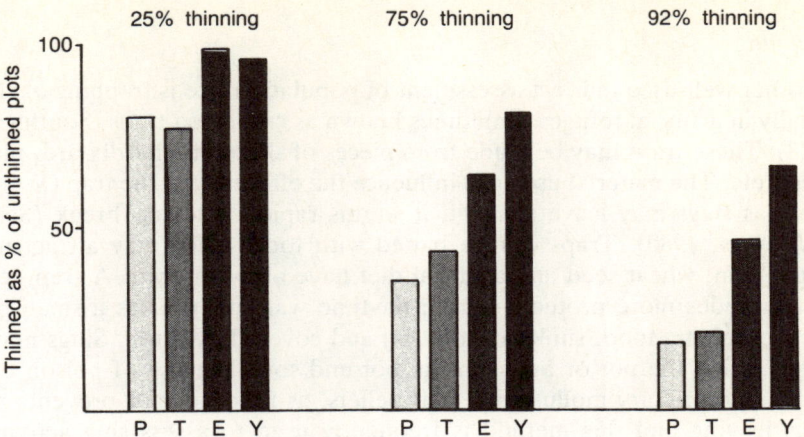

Figure 3. Winter wheat: Effect of artificial thinning in December at three different levels on plants (P) and tillers (T) in March, ears (E) in July and yield (Y) in August (after Jessop, 1969, © Crown Copyright).

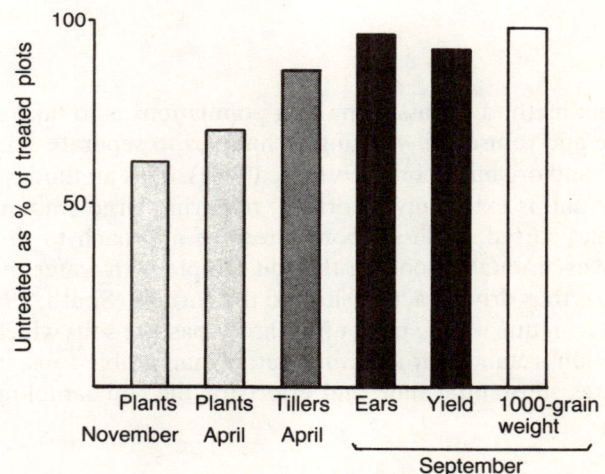

Figure 4. Compensation by winter wheat for slug damage: Performance of untreated crop as a percentage of crop with single treatment of 4% methiocarb pellets (0.22 kg a.i./ha) broadcast soon after sowing.

of plants can be total, plant counts per unit row length or area are used. Wheat plants may be lost to other causes and a more direct assessment of slug activity may be made by digging lengths of drill line and counting the numbers of hollowed grain. This is more laborious and soil conditions may lead to unacceptable variability in assessments made in this way (Gould, 1962).

Trapping

Another well-used indirect assessment of population size is trapping of slugs, usually in artificial refuges sometimes known as cryptozoa traps (Southwood, 1978). These traps may be made from pieces of slate, tile, hardboard, plastic sheet, etc. The material used can influence the efficiency of the trap (Webley, 1963) as slugs may leave a trap if it warms rapidly after daybreak (Schrim and Byers, 1980). Traps can be baited with food which may attract more slugs: bran, wheat seed and artificial diet have all been used. A trap design that provides more protection from daytime warming utilizes a small plant pot, containing food, sunk into the soil and covered by a tile. Slugs may be found inside the pot or between the pot and soil. The use of poison baits, such as proprietary molluscicide bait pellets, as food in traps prevents most slugs leaving and this method is frequently used to assess slug activity in field trials. As traps rely on slug activity they suffer from the drawback of other indirect methods, but they are easy to use. If results are interpreted with caution, traps provide valuable data. These traps can be a useful method for slug-population estimation by farmers and growers.

Soil sampling

The only direct method for assessing slug populations is to take soil samples of known size and to use soil-washing techniques to separate slugs and their eggs from soil and organic debris (Hunter, 1968a). This method is still subject to some error and is extremely laborious, requiring large amounts of soil to be sampled and sorted. A less labour-intensive approach to the extraction of slugs involves gradually flooding the soil sample with water over a period of several days, thus driving active slugs to the surface (South, 1964; Hunter, 1968a). This technique will be most efficient for pasture soils which, following sample collection, retain their structure better than arable soils. The method still necessitates substantial time and effort for the soil sampling.

Marking

Several workers have used marking for estimating slug populations by mark, release and recapture. The methods used include feeding slugs food containing radioisotopes (Francois, Riga and Moens, 1965, 1968), dye, for example agar jelly containing neutral red (South, 1965), and dyeing the skin (Müller and Ohnesorge, 1985); however, the difficulties of marking slugs have prevented full exploitation of this method. Freeze branding has been used to mark larger species individually for behavioural studies (Richter, 1976). Hogan and Steele (1986) have shown that injecting dye, using a dental inoculator, into the slug's skin provides easily distinguishable long-lasting marks with little ill-effect to the slug.

CULTURAL CONTROL

Cultivations

In determining potential crop damage, cultural conditions are of primary importance in addition to population size and effects of weather on activity. However, cultural activities may be manipulated to the advantage of the farmer. Slugs are more of a problem on heavy soils (Gould, 1961), but this may be counteracted, at least in part, by additional cultivations to give a finer and firmer seedbed which will deter slug activity (Stephenson, 1975a,b; Moens, 1983). These additional cultivations will also serve directly to reduce slug numbers, presumably by physical damage (Hunter, 1967; Gould and Webley, 1972; Hogan, 1985).

Conversely, minimal cultivation and direct-drilling do not have these adverse effects on slugs and are usually associated with large amounts of crop residues, resulting in greater risk of slug damage (Edwards, 1975). This is most applicable to cereals (page 268).

Crop residues

Organic matter such as crop residues favours slug populations. This may be important in cereal growing where there is increasing concern about the environmental considerations of straw disposal by burning. In the UK, burning has been the favoured method of reducing straw residues, particularly on heavy land in the predominantly arable eastern areas, the region where most slug damage occurs. It is likely that straw disposal by burning will be increasingly controlled by legislation or voluntary restraint in future. The effects of alternative straw-disposal methods on slug populations have been examined and it has been shown that slug numbers and damage increase where straw is not burnt (Glen, Wiltshire and Milsom, 1984; Smith *et al.*, 1985). In one cereal experiment, slug populations were much smaller on plots burnt after harvest (14 slugs/m^2) compared with plots where straw was incorporated to 10 cm or left on the soil surface (180 slugs/m^2). The proportion of cereal seedlings damaged was largest where straw remained on the surface (47%) and was reduced by incorporation (13%) or burning (5%) of straw. Other observations suggested that the beneficial effects of burning resulted from depriving slugs of food and shelter rather than direct mortality. Although shallow incorporation of straw did not reduce slug numbers, it did reduce crop damage (Glen, Wiltshire and Milsom, 1984).

Some crops, such as cereals, oilseed rape, peas, beans and grass, encourage slugs whereas others, such as potatoes and beet, are not so favourable (Gould, 1961). Hence, on land prone to slugs, choice of crop rotation may be considered as a control.

The susceptibility to slugs of crop species and varieties has already been discussed (pages 262–264). Cultural conditions for a particular crop may be manipulated to reduce the likelihood of slug damage: for example, higher seed rates may be used or the crop may be sown at a time when slug activity is smaller; examples already given include time of sowing wheat (page 269) and winter barley (page 263).

CHEMICAL CONTROL BY MOLLUSCICIDES

A range of poisons, formulated as baits, were available in the early part of this century and the development of chemical slug control is documented by Godan (1983). The first compounds used included calcium salts, Paris Green (copper aceto-arsenite) and copper sulphate (Miles, Wood and Thomas, 1931). The molluscicidal properties of metaldehyde were first discovered and exploited in the 1930s (Gimingham, 1940; Crowell, 1979); today it is the second-commonest molluscicide in the UK, the most widely used being methiocarb (4-methylthio-3,5-xylyl methylcarbamate).

Various inorganic compounds are registered under the Pesticides Safety Precautions Scheme of MAFF for use as molluscicides in the UK, but none are widely used. Considerable attention has been given to copper compounds, which show contact toxicity: this was first documented by Anderson and Taylor (1926) and there is a continuing interest (Ryder and Bowen, 1977a,b; Bowen and Jones, 1985). Other proprietary formulations are based on aluminium sulphate or mixtures of salts, but so far none has shown particular promise (Glen, Milsom and Wiltshire, 1986; Glen and Orsman, 1986).

Numerous chemicals have been screened for use as molluscicides and a few have shown promise (Barry, 1969; Judge, 1969; Judge and Kuhr, 1972; Symonds, 1975; Crowell, 1979), including certain carbamates and herbicides. Some other chemicals have been found useful in combination with metaldehyde (Crowell, 1979). Some insecticides have been shown to have molluscicidal properties, including aldrin (Stephenson, 1959), carbaryl (Ruppel, 1959) and aldicarb (Crowell, 1979), but have subsequently failed to realize this promise. The molluscicidal properties of methiocarb (known in some countries as mesurol) were described by Martin and Forest (1969). Features of methiocarb and metaldehyde are given in *Table 6*.

Mode of action and toxicity

Methiocarb acts as a stomach poison, the slug ingesting the chemical mixed with an attractive bait. It has some contact activity, but this is less important. Conversely, metaldehyde is often more effective as a contact poison. However, it is often formulated as a bait and it is not clear whether the poisoning results primarily from contact or ingestion. The toxicity data for slugs are difficult to obtain. The mucus on the animal's integument may prevent effective contact with a poison applied directly to the skin. The alternative technique for assessing contact toxicity, that of letting the animals

Table 6. Features of metaldehyde and methiocarb relevant to control of slugs

Feature	Metaldehyde	Methiocarb	Reference
Oral toxicity to slugs (LD_{50}):			
D. reticulatum (0.3–0.6 g)	85.2 µg/slug		Henderson, 1969
D. reticulatum (0.4–0.6 g)	45.71 µg/slug	21.88 µg/slug	Hunter and Johnston, 1970
Contact toxicity to slugs: (LC_{50})	42.37 mg/g		Henderson, 1968
Toxicity to mammals: acute oral LD_{50}	dogs 600–1000 mg/kg	male rats 100 mg/kg guinea pigs 40 mg/kg rats 350–400 mg/kg	Worthing and Walker 1983
acute percutaneous LD_{50}	500 mg/kg		Frain, 1982
Common formulations	Bait pellets (4–6% a.i.)	Bait pellets (4% a.i.)	
Pellet density	up to 144/m²	36/m²	
Cost at above rate (1985 prices)	£11.6/ha	£19.2/ha	
Approximate crop area treated (ha) 1982	83 000	582 000	Sly, 1986

move over a treated surface, suffers from the drawback that ingestion of the poison may also occur (Henderson, 1968). To investigate oral toxicity, without confounding the test with contact contamination, it is necessary to inject the chemical into the slug's buccal cavity (Cragg and Vincent, 1952) and Henderson (1969) describes a device for ensuring that test material is not regurgitated.

Slugs, poisoned by methiocarb, move normally for a short time, but then become immobilized as muscle tonus is lost (Godan, 1983). Poisoning by metaldehyde is also characterized by immobility, but the slug is stimulated to produce copious mucus and dehydration is a frequent cause of death. There is no evidence of slugs evolving resistance to either chemical and this is not likely, given the variable efficacy of control and the consequent weak selection pressure.

It is widely reported that slugs of different ages (sizes) show differential susceptibility to molluscicides (Crowell, 1979; Godan, 1983; Kemp and Newell, 1985). Juveniles usually show lower mortality than adults although in some studies the converse is true (Daxl, 1971). The exact causes of this effect are not clear, but may be related to relative size of the slug's foot (Godan, 1983) and/or different patterns of activity in adults and juveniles (Rollo, 1982).

The toxicity of metaldehyde and methiocarb to non-target organisms, especially vertebrates, must also be considered. Methiocarb is an insecticide and may give partial control of pests such as leatherjackets (Diptera: Tipulidae), but it also affects other soil fauna: earthworms may be at risk (Martin and Forrest, 1969; Symonds, 1975) and so may beetles (Martin, Davis and Morris, 1969) although population reductions following methiocarb use have not been detected (Kelly and Curry, 1985).

Vertebrate poisonings are occasionally reported. Many of these involve metaldehyde (for example Maddy, 1975; Longbottom and Gordon, 1979) or less commonly methiocarb (Giles et al., 1984). During the period 1980–83 in England and Wales, 22 incidents involving vertebrate deaths were attributed to poisoning by molluscicide pellets (Fletcher and Stanley, 1981; Stanley and Fletcher, 1982; Fletcher and Hardy, 1983, 1984): three incidents were associated with methiocarb and 19 with metaldehyde. Over half the poisonings involved dogs; other cases included cats, various small mammals and birds. Deaths of many of the dogs had resulted from ingestion of molluscicide pellets which were not securely stored. In some cases poisoned animals were found near areas tested with molluscicide bait pellets: either pellets or poisoned slugs may have been eaten. Fletcher and Hardy (1983) considered that there were surprisingly few incidents of vertebrate poisoning in the UK considering the annual usage of molluscicides. A recent survey in Melbourne, Australia found that 82% (293 cases) of all poisonings seen in veterinary practices were due to metaldehyde (57%) or methiocarb (43%) pellets consumed by dogs and cats. Some dogs were reported to make considerable efforts to obtain pellets and even meticulously to eat scattered individual pellets from treated areas (Studdert, 1985).

Factors affecting efficacy

The commercial forms of methiocarb and most forms of metaldehyde used in agriculture are formulated as pellets composed of a crushed cereal matrix with active ingredient, fungistatic agents, binders, etc. Slugs are attracted to the pellets as a source of food, the object being to induce the slug to feed so that it consumes a lethal dose of the active ingredient. Several workers have investigated potential additives that may increase attraction to, and consumption of, baits (Smith and Boswell, 1970). As slugs are very dependent on chemical and tactile stimuli it is likely that attraction to bait or any food is chemically mediated: chemical attractants may diffuse from a pellet in the soil water or they may be airborne. At present there is little information as to whether slugs do react to food from a distance. Direct observations of behaviour in natural arenas suggest that, often, chance contact with wheat seeds or poison bait is the first step leading to consumption. In laboratory trials, response to contact with plant volatiles has been noted (Pickett and Stephenson, 1980) and both pairs of tentacles and the lips have chemo-receptors (Stephenson, 1979). Chemically mediated anemotaxis (responding to a chemical source by moving upwind) probably plays a part in homing behaviour (Cook, 1980) and may be significant in food finding when the food has a strong odour.

Weather will influence the efficiency of molluscicidal pellets in several ways, the most important being the effect of weather in regulating slug activity in the field (page 261). Secondly, the prevailing weather following poisoning may have an effect on mortality. This is particularly true for metaldehyde, where slugs with a sub-lethal dose for direct mortality may nevertheless be immobilized and stimulated to produce copious mucus. If kept under ideal conditions of high relative humidity (as in recovery chambers) many of these slugs survive; if, however, conditions are dry, the immobilized slugs may suffer lethal dehydration. Thirdly, weather conditions will affect the persistence of the toxic bait in the field. Modern pellet formulations are fairly resistant to weathering, partly as a result of considerable experimental work on formulations (e.g. Webley, 1966). In one investigation, commercial methiocarb pellets were still attractive and potent to slugs after 30 days' weathering (Hogan, 1985).

As contact with pellets is probably not dependent on long-range attraction, it follows that a higher density of pellets will give a greater chance of slug–pellet contact (Webley, 1970). Hunter and Symonds (1970) investigated this and suggested that the ideal distance between pellets should be 10 cm or 100 pellets/m^2. Metaldehyde pellet size was originally larger and baiting points per unit area fewer than for methiocarb pellets, but now most metaldehyde and methiocarb pellets are a similar size and, at typical commercial rates, a larger number of baiting points are obtained with metaldehyde (*Table 6*).

Several different formulations of metaldehyde bait are commercially available in the UK in contrast to one widely available formulation of methiocarb bait for agricultural use. There may be differences in bait composition which influence field performance (Glen and Orsman, 1986). An important feature of formulations of both chemicals is that the presence of the active ingredient acts to reduce consumption of the bait (Wright and Williams, 1980): this means that, as the concentration of active ingredient increases, there will be an optimal point, where most active ingredient is consumed, and at higher concentrations consumption of bait, and active ingredient uptake, will be less. Slugs feeding on poison pellets feed for less time and at a slower rate. This is partly due to a repellency effect of metaldehyde and methiocarb and, in the case of the former, due partly to the onset of paralysis during the meal (Wedgewood and Bailey, 1986). As Symonds (1975) suggested, slugs consuming sub-lethal doses of metaldehyde are more susceptible to subsequent poisoning (Kemp and Newell, 1985), but this does not appear to be true for methiocarb.

Costs of treatments and alternatives

The costs of treatment are given in *Table 6* and to these must be added application costs which, for a single application of molluscicide pellets from a fertilizer spreader, amount to £5/ha (Nix, 1985). If application is to a standing crop then some plants may suffer wheeling damage. It is interesting to compare these costs with those of alternative methods for reducing slug damage. Extra cultivations such as rolling cost £6/ha (Nix, 1985). For cereals a farmer might consider using a higher seed rate to compensate for expected loss to slugs. Increasing the seed rate of wheat by 20% will increase the costs by £8/ha (Nix, 1985).

Changes in molluscicide usage

In recent years there have been large increases in the usage of molluscicides in England and Wales. This is only partly due to a change in crop types. The increased usage on wheat (*Table 5*) suggests that farmers perceive a greater risk of slug damage (page 269). In barley, less than 1% of the crop was treated in 1977 (Steed *et al.*, 1979) and this had increased to 9% in 1982 (Sly, 1986). Most of this was application to the autumn-sown crop and can be explained by the marked swing from spring to winter barley during this time. Winter crops are at greater risk from slugs, but barley is less at risk than wheat (page 263). Similar trends occur in potatoes and may be partly due to the whole crop being more susceptible to slug damage today (page 266). Insurance treatment of this high-value crop is more easily justified.

In 1977 less than 8% of the oilseed rape crop was treated with molluscicide (Steed *et al.*, 1979). In 1982 it was estimated that 93 575 ha, i.e. 54% of the oilseed rape area in England and Wales, was treated with methiocarb pellets and a small area was treated with metaldehyde pellets (Sly, 1986). This would amount to a cost of £1.8 million (chemical alone) at 1985 prices, but at 1982 estimates of usage. Slugs are occasionally considered to be a pest of winter oilseed rape grown on heavier soils, particularly in direct-drilled crops

(Ward *et al.*, 1985). In recent years, winter oilseed rape has occupied most of the oilseed rape area in the UK: however, it is mainly sown before mid-September on fine firm seedbeds and the crop establishes rapidly; these conditions are not conducive to slug activity. Direct-drilling (no cultivation) is a popular method of establishment, in dry seasons on suitable types of heavy land (Ward *et al.*, 1985), although the overall proportion of the crop established in this way is small (J. T. Ward, unpublished data).

Use of poison bait pellets mixed with the seed to control slugs is recommended when direct-drilling on heavy land. Apart from this, treatment is rarely justified as the crop outgrows the effects of all but very serious attacks (Ward *et al.*, 1985). Molluscicide baits at sowing, significantly ($P <$ 0.05) increased plant stand in only one of 16 crops studied in Northumberland and Durham in 1981 and 1982, and plant numbers at establishment were more than adequate on untreated areas at all sites (C. M. Port, unpublished data). Hancock (1986) mentions a recommended use of metaldehyde pellets as a diluent to aid in sowing of the small seed, although Ward *et al.* (1985) state that cereal drills can now handle oilseed rape seed effectively. It would be irrational to use methiocarb for this purpose when cheaper diluents are available.

Farmers probably overestimate the risk of slug damage. Possible reasons are confusion with damage caused by adult cabbage stem flea beetle (*Psylliodes chrysocephala* (L.)) in the autumn (John and Evans, 1984) and the known association of large slug numbers with oilseed rape, borne out by an increased risk of slug damage in following winter wheat crops. These factors are compounded by the lack of experience of many farmers unfamiliar with this crop, which has expanded in popularity only in recent years (*Figure 1*). The level of molluscicide usage on oilseed rape (based on the 1982 estimate) is not justified by risk of slug damage and the usage would be expected to decline once farmers become aware of this.

It is probable that total molluscicide usage has ceased to increase at the rate seen in the 1970s. Recent figures suggest fluctuations in usage from year to year, but no overall trend (H. G. Mannall, unpublished data). *Table 7* shows the estimated use in 1982 by major crop type.

Table 7. Estimated molluscicide usage for England and Wales, 1982. (After Sly, 1986, © Crown Copyright)

Crop type	Area treated (ha)	Treated with methiocarb (%)	Molluscicide used (tonnes)
Grass	7 095	31	4.9
Cereals	550 313	86	174.4
Other arable crops	117 953	92	28.9
Total	675 361		208

Relative efficacy of metaldehyde and methiocarb

It has been widely accepted that methiocarb is a more effective molluscicide than metaldehyde, and this has contributed to its more widespread use. Laboratory studies support this view, but it is the field performance of molluscicides that is of prime importance. In *Table 8* evidence is summarized from experiments comparing metaldehyde and methiocarb in either the laboratory or the field. Laboratory tests tend to measure the direct poisoning effect under controlled conditions, whereas, in the field, factors such as attraction to bait, availability of alternative food, bait distribution, numbers of baiting points, weather during and after baiting and persistence of baits have a great influence on molluscicide performance.

Detailed conclusions from the data in *Table 8* should be made only with further reference to the original work. There is considerable variation in the formulations used, in the methods of assessment and in particular in the rate of chemical application (here standardized to kg a.i./ha) which sometimes bears no relation to recommended pest-management practice. The results from field experiments listed were rarely decisive and, when damage to the crop or test plants was assessed, no difference between the efficacy of metaldehyde and that of methiocarb was found. Where differences occurred, methiocarb was more often rated better than metaldehyde than vice versa (Martin and Forrest, 1969; Frain and Newell, 1983; Glen and Orsman, 1986). However, some of the evidence was based on data for slug mortality obtained either by counting dead slugs on the soil surface or by placing poisoned slugs in recovery chambers before assessing final mortality. Symonds (1975) suggested that this methodology would tend to underestimate the performance of metaldehyde because poisoned slugs, which may recover in chambers, would have had the opportunity to ingest more poison bait in the field. Subsequently, Frain and Newell (1983) recommended that the more accurate method of assessment was to measure the surviving (residual) population. (Methods of population estimation or damage assessment are discussed on pages 270–272.) In some cases mortality was assessed at various times after treatment. We have mostly quoted data for mortality after a period has been allowed for recovery of slugs with sub-lethal poisoning; however, the immediate poisoning effects may be equally important in reducing crop damage. Many of the comparative trials listed in *Table 8* included assessments of slug activity after treatment (Martin and Forrest, 1969; Gould and Webley, 1972; Frain and Newell, 1983; Glen and Orsman, 1986) and/or assessments of damage (Martin and Forrest, 1969; Gould and Webley, 1972; Rayner, 1975; MAFF, 1980; Glen and Orsman, 1986).

Plot size in field experiments

It is important to emphasize here a major problem facing the experimenter conducting field trials—that of choice of plot size. Slug populations are usually aggregated (South, 1965; Airey, 1984), but also relatively mobile.

Small plots may reduce variation from uneven population distribution, but also reduce differences in treatment effects because of inter-plot slug movement. The few studies on slug movement suggest that they disperse at about 1–2 m/day (Hogan, 1985). Inevitably, the final choice will be a compromise but, generally, small plots are adequate where measurements are to be taken soon after treatment. Larger plots are more suitable where long-term data such as yield are required. Barriers may reduce slug movement, but they are laborious to install (Moens *et al.*, 1967; Glen, Milsom and Wiltshire, 1986).

TREATMENT TIMING

The damage controlled by molluscicide application may be affected by the timing of treatments. Treatments may be applied at different times of year, at different stages of the cropping cycle or at different times with respect to weather.

Hunter (1969b, 1978) suggested that treatments should be applied when the pest populations are small, before eggs are laid, the implication being that reduction of the parent population will lead to reduction of the next generation. Although this strategy may seem logical there are no data to support it. In addition, several factors limit the value of this approach: first, there is little information on factors regulating slug populations, but, if density-dependent mortalities occur, populations may recover if there is a long interval between timing of treatment and damage; secondly, slug breeding patterns are variable and generations overlap; thirdly, applications made into a standing crop or crop residue may be less effective, because of the presence of alternative food (Airey, 1986). Furthermore, there is no information on the proportion of a slug population that is killed by a molluscicide application. It may be best to apply treatments as near as possible to the time when damage is done, in order to kill the slugs active at that time.

The value of applications at different times in the cropping cycle has been the subject of several investigations. Rogers-Lewis (1976, 1977) found that, in potato crops, applications at different times were effective only if slugs were active on the surface and this was assessed by means of traps with poison bait. However, results of subsequent trials have not supported this conclusion (Rayner *et al.*, 1978). For wheat crops, Gould and Webley (1972) observed no difference between treatments applied in the stubble of the previous crop, after ploughing or after drilling. In another trial (Rogers-Lewis, 1977), treatments applied 7 weeks before drilling gave no effective control, but treatments nearer the time of drilling were all effective. Our own data (unpublished) support the recommendation that treatments at about the time of drilling wheat are most effective in years when weather follows the normal pattern. Later applications of pellets, after ploughing or further seedbed preparations, are focused on a smaller residual population, reduced already by the effects of cultivations.

Table 8. Summary of laboratory and field experiments comparing metaldehyde and methiocarb for efficiency in slug control. All rates are expressed as weights of active ingredient (a.i.)

Treatment method	Assessment method	Slug species	Result (metaldehyde/methiocarb)*	Reference
A. Laboratory assessments				
Metaldehyde or methiocarb mixed with loose bran (2–4% a.i.). Soil-filled arena.	Mortality after 14 days	Arion ater	20–24%/ 69–85%	Crowell, 1967
Metaldehyde or methiocarb as sprays applied to young pea plants growing in trays (2.24 kg a.i./ha)	Mortality after 7 days; % of plants damaged	Deroceras reticulatum	65%/67% 14.5%/42.5%	Judge, 1969
Metaldehyde (10% a.i.) mixed with loose bran and methiocarb (4% a.i.) pellets in soil-filled arena.	Mortality after 48 hour recovery period	Milax sowerbyi Arion hortensis A. ater D. reticulatum D. caruanae M. budapestensis	27%/58% 12%/41% 18%/48% 22%/48% 45%/56% 26%/32%	Crawford-Sidebotham, 1970
Metaldehyde (5% a.i.) or methiocarb (2% a.i.) mixed with loose bran. Soil-filled arena.	Mortality after 10 days	A. ater	37%/100%	Crowell, 1979
Metaldehyde and methiocarb (2% a.i.) as above.	Mortality after 10 days	A. ater rufus Limax flavus	7%/90–98% 37%/73–79%	
Metaldehyde (6% a.i.) and methiocarb (4% a.i.) pellets	Mortality after first meal and 7-day recovery period	D. reticulatum	18%/46%	Kemp and Newell, 1985
	Mortality of survivors from above after second meal and 7-day recovery period	D. reticulatum	52%/31%	
	Combined results	D. reticulatum	61%/62%	

Treatment	Assessment	Species	Result	Reference
Metaldehyde, several commercial formulations (3.8–20% a.i.) and methiocarb (4% a.i.) pellets. Soil-filled arena.	Mortality after 9-day recovery period	*D. reticulatum*	0–13%/30%	Glen and Orsman (1986)
	% food consumption		NSD	
B. Field assessments				
'Commercially available' metaldehyde pellets and methiocarb (4% a.i.) pellets. 10 g pellets under tile traps.	Mortality after 10-day recovery period	*D. reticulatum*	58%/94%	Martin and Forrest, 1969; Martin, Davis and Morris, 1969
Pellets as above applied to small plots sown with wheat. Metaldehyde rate not specified, methiocarb (0.45 kg a.i./ha)	Counts of dead slugs on surface	Not specified	Significantly more killed by methiocarb	
	Post-emergence damage to plants.		NSD	
Pellets as above applied to field of spring cabbage. Metaldehyde rate not specified, methiocarb (0.14–0.45 kg a.i./ha).	Counts of dead slugs on surface	Not specified	Significantly more killed by methiocarb	
	Assessment of survivors caught in metaldehyde-baited traps		Significantly more survived metaldehyde	
Pellets as above applied to field of winter wheat one week after drilling. Metaldehyde rate not specified, methiocarb (0.22–0.45 kg a.i./ha)	Counts of dead slugs on surface	Not specified	Methiocarb at higher rate significantly better than metaldehyde.	
	Plant counts		NSD	
	Plants with grazing damage		Methiocarb significantly better than metaldehyde.	

cont'd

Table 8. *cont'd*

Treatment method	Assessment method	Slug species	Result (metaldehyde/methiocarb)*	Reference
Metaldehyde (5% a.i. pellets) Methiocarb (4% a.i. pellets)	Mortality of slugs at bait	Not specified	85%/94%	Webley, 1969
Metaldehyde (0.94 kg a.i./ha) and methiocarb (0.45 kg a.i./ha) applied to plots of young peas	Reduction in damage to wheat baits after 12 days	D. reticulatum	58%/94%	Wharton and Ensor, 1969
	Reduction in damage to wheat baits after 5 weeks		25%/51%	
	Number of slugs on vine at harvest		NSD	
	Number of slugs in soil by soil flooding		NSD	
Metaldehyde (3% a.i., 0.94 kg a.i./ha) and methiocarb (4% a.i., 0.22–0.45 kg a.i./ha) on mown lawn.	Mortality after 3-day (metaldehyde) and 7-day (methiocarb) recovery period	D. reticulatum	No consistent difference	Hunter and Symonds, 1970
Metaldehyde (3–6% a.i., 0.94 kg a.i./ha) and methiocarb (4% a.i., 0.22–0.45 kg a.i./ha) pellet treatments to winter wheat	Counts of dead slugs on surface	D. reticulatum predominant	In some trials methiocarb significantly better than metaldehyde	Gould and Webley, 1972
	Assessment of survivors caught in methiocarb baited traps.		NSD	
	Damage to baits of wheat seed		NSD	
	Plant counts		NSD	

Treatment	Measurement	Species	Result	Reference
Metaldehyde (3% a.i., 0.94–1.88 kg a.i./ha) and methiocarb (4% a.i., 0.22–0.90 kg a.i./ha) pellets applied at various times to potatoes.	Proportion of slug-damaged tubers	Many species	NSD	Rayner, 1975
Metaldehyde (3–6% a.i., 0.94 kg a.i./ha) and methiocarb (4% a.i., 0.22 kg a.i./ha) pellets in winter wheat	Mortality after 2-day recovery period	*D. reticulatum*	In 3/4 trials methiocarb killed more slugs than metaldehyde	Symonds, 1975
Metaldehyde (6% a.i., 0.45–0.9 kg a.i./ha) and methiocarb (4% a.i., 0.22 kg a.i./ha) pellets in winter wheat	Damaged grain Plant counts at establishment % plants damaged	*D. reticulatum*	NSD NSD NSD	MAFF, 1980
Metaldehyde (5–10% a.i., 0.25–0.75 kg a.i./ha) in various formulations and methiocarb (4% a.i., 0.12 kg a.i./ha) pellets in grass and clover field	Counts from baited and unbaited traps of poisoned slugs Counts, 2 weeks after treatments, of survivors caught in baited and unbaited traps	*D. reticulatum* predominant	Methiocarb significantly better than metaldehyde NSD	Frain and Newell, 1983
Metaldehyde (3.8–20% a.i., 0.93–5.26 kg a.i./ha) in several commercial formulations and methiocarb (4% a.i., 0.48 kg a.i./ ha) pellets in orchard	Counts of survivors in bran baited traps Damage to test plants	Several species	Methiocarb significantly better than metaldehyde for 4/8 species NSD	Glen and Orsman (1986)

* Results given for metaldehyde first followed by those for methiocarb; NSD = no significant difference

These trials emphasize that molluscicide treatments are more effective when the slugs are active. This activity is usually assessed by using slug traps, but if weather data could be used to predict slug activity (pages 261–262) then it should be feasible to forecast when best to apply control measures; further investigations of this possibility are required.

SEED TREATMENTS

Attempts have been made to protect cereal grains, especially wheat, from slug damage by testing seed treatments containing various chemicals. Seed treatments provide a more convenient and cheaper method of control than baits or sprays. Toxic chemicals are likely to be effective at smaller rates in this form and should therefore be less of an environmental hazard. As the seeds of wheat are most susceptible to slugs (page 267), seed treatments provide an effective control measure in this crop.

Gould (1962) tested various of the molluscicides available at that time, as seed treatments applied to wheat at rates of about 1% a.i./weight of seed. Metaldehyde and Paris Green gave the best results in laboratory tests, but copper oxychloride consistently performed best in field tests in terms of reducing slug damage. However, control was insufficient when slug attack was severe. In addition, the large amounts of active ingredient and sticker applied to seeds created difficulties with drill operation, sometimes reducing plant stand even though germination tests had indicated that phytotoxicity should not be a problem. Symonds (1975) found that thiocarboxime as a seed treatment (rate unspecified) killed *D. reticulatum* and prevented grain hollowing in the laboratory.

Since 1975, a wide range of materials, mainly from chemical groups known to have molluscicidal properties, have been screened in the laboratory, mostly at rates of 0.2% a.i./weight of seed (Scott, Griffiths and Stephenson, 1977; Scott, 1981; Scott *et al.*, 1984). Of these, the carbamate thiocarboxime was the most active in terms of reducing grain damage and killing slugs (*D. reticulatum*). However, its extreme toxicity to vertebrates precluded further development (Scott, Griffiths and Stephenson, 1977). Methiocarb was also effective.

The phenolic herbicides ioxinil and bromoxinil and related compounds previously shown to be molluscicidal (Stephenson, 1967) protected seeds from damage by repelling rather than killing slugs. Various compounds related to nereistoxin (a substance found in marine annelid worms, *Lumbrinereis* spp.), such as thiocyclam and cartap, also protected seeds in this manner. All were phytotoxic to a greater or lesser extent, expecially compounds related to ioxinil. Polymeric forms of thiocyclam were developed which proved effective, but not phytotoxic (Scott *et al.*, 1984). The most promising compounds were tested in field trials and at one site, where a severe slug attack was experienced, seed treatments of methiocarb (0.1% a.i./weight of seed), cartap hydrochloride and thiocyclam polystyrene-hydrosulphonate (both 0.2% a.i./weight of seed) considerably reduced grain hollowing and significantly increased yield of the wheat crop, by up to 50%

(methiocarb) (Scott *et al.*, 1984). The seed treatments were more effective than 4% methiocarb pellets mixed with the seed. However, these particular compounds were considered too toxic to birds and other vertebrates to warrant commercial development as seed treatments. Moens (1983) also demonstrated the effectiveness of methiocarb as a seed treatment to wheat. In tests where *D. reticulatum* were added to plant pots containing soil and wheat seed on or below the surface, methiocarb seed treatment (0.4% a.i./ weight of seed) gave a much greater reduction in grain damage than 4% methiocarb bait pellets.

PLANT EXTRACTS

Techniques for testing molluscicides as seed protectants against slug damage also have potential for screening a wide range of compounds derived from plants for both repellent (Scott, 1982) and attractant properties: Scott *et al.* (1984) reported one such plant extract, geraniol, which repelled slugs. Pickett and Stephenson (1980) devised a bioassay to detect attractive volatile components in plant extracts, which involved slugs following a trail of volatiles laid on moist paper. Volatiles from various whole-plant extracts produced a response and various attractive components were isolated from lettuce, although response was inconsistent. The potential exists, for example, for incorporation of such attractants into poison baits to improve slug control.

Plants with molluscicidal properties have long been recognized and Webbe and Lambert (1983) state that more than a thousand species have been screened for these effects and several have shown promise. Some plant extracts have been used in snail-control trials in attempts to limit the spread of schistosomiasis by control of the parasite's intermediate host. Webbe and Lambert (1983) considered that certain saponins offered the greatest potential as molluscicides, but so far the cultivation of plants on a large scale exclusively for molluscicide production or mollusc control *in situ* has not been attempted.

BIOLOGICAL CONTROL

The manipulation of natural enemies in order to control slug numbers is a possibility that has been considered by some researchers. The groups of natural enemies that affect slugs have been discussed previously (page 258); however, a natural enemy must possess certain specific attributes in order to become a useful biological control agent (Debach, 1974). For example, a degree of host specificity is usually desirable in order to ensure a high rate of parasitism or predation, although a general slug enemy is required for many slug pest populations composed of several species. A high reproductive capacity with respect to the prey is also desirable, together with a good host-searching ability: natural enemies lacking these attributes may need augmentation by mass culture. On these grounds, the most likely candidates as biological control agents would appear to be invertebrate parasites, predators and pathogens. Certain vertebrates, although useful predators of slugs (South, 1980), are less amenable to manipulation.

Parasites and predators

Stephenson and Knutson (1966) listed 46 species of invertebrates associated with slugs and considered that the most important natural enemies were protozoans, brachylaemid flatworms, lungworms, lampyrid beetles and sciomyzid fly larvae. Baronio (1974) also considered that lampyrid and drilid beetles and sciomyzid flies were important natural enemies, the latter two families depending almost exclusively on slugs and snails. Stephenson (1968) outlined the biology of four species of *Tetanocera* (Diptera: Sciomyzidae). *Tetanocera elata* (Fab.) provides a good example of the parasitic and predacious habits of this family: first-instar larvae parasitize only *D. reticulatum* and *D. laeve*, eventually killing the slug and adopting a more predatory life style in later instars when they kill several species of slug. Each larva kills between four and nine slugs (Knutson, Stephenson and Berg, 1965). *T. elata* is common and widely distributed in Europe and North America and probably completes two generations per year. More recently, Trelka and Berg (1977) have studied both the searching and attack behaviour of *Tetanocera* larvae, which can use fresh slime trails to track slugs (as do carnivorous snails: Cook, 1985) and immobilize their prey with a toxic injection. It is probable that larval fireflies (Coleoptera: Lampyridae) also use a paralysing toxin to immobilize their prey (Copeland, 1981).

Several common species of carabid beetles which feed on slugs have also been studied (Stephenson, 1968; Stephenson and Bardner, 1976). It was concluded that although the influence of carabids and sciomyzids on slug populations was unknown, exploitation by manipulating cultural practices was a possibility. Altieri *et al.* (1982) actually demonstrated a significant reduction in slug and snail populations by releasing ground beetles, *Scaphinotus* (*Brennus*) *striatopunctatus* (Chandoir). Orth *et al.* (1975) and Fisher *et al.* (1976) described the potential of a rove beetle, *Staphylinus* (*Ocypus*) *olens* (Müller) (Coleoptera: Staphylinidae), which feeds on slugs and snails, as a biological control agent of brown garden snail (*Helix aspersa* (Müller)), which is a serious pest in California, USA. They considered that culture and mass release of this predator would be feasible for snail control.

Pathogens

Bacteria and other micro-organisms can be found in slugs. Virus-like particles have been observed (David, Taylor and Atkey, 1977), but similar particles have been shown to be composed of galactogen and glycogen and are carbohydrate reserves (Kassanis, Woods and Macfarlane, 1984). Brooks (1968) and Jones (1985) have investigated protozoan parasites: their results show that some parasites may have dramatic effects on slugs. Brooks (1968) studied two ciliate species, one of which, *Tetrahymena rostrata* (Kahl), showed some promise as a control agent for *D. reticulatum*. *T. rostrata* reduced longevity and fecundity of slugs and could be passed to the next generation by transovum transmission. Jones (1985) studied a microsporidian parasite, *Microsporidium novocastriensis* (Jones and Selman, 1985), which appeared to be specific to *D. reticulatum*. The result of infecting slugs was a 40% reduction of fecundity (Jones and Selman, 1984), a 30% reduction in growth and a 23% reduction of longevity (Jones, 1985). Young slugs become infected by feeding on the remains of their eggs, which are often contaminated with *M. novocastriensis* spores, or by grazing on food contaminated by faeces of an infected slug. One remarkable effect of infection was a dramatic reduction of feeding activity, occurring within 24 hours of infection and continuing for many weeks (Jones, 1985). A potential control agent with such useful effects certainly requires fuller investigation.

Future prospects for control

Having discussed which methods of slug control have been considered to date, we now attempt to predict which will be important in the future. Some of these are well-tried and tested approaches, whereas others are more novel in character. In many crops, cultural practices have long been of overriding importance in slug control and they will continue to have a significant role. Cultivations might assume greater importance in future because of the probable decrease in disposal of crop residues by burning.

New molluscicides may become available but, at present, poison bait formulations of metaldehyde and methiocarb are the predominant means of chemical control. There is considerable scope for improving the efficacy of these baits, which are limited by their dependence on slug activity. Short-term forecasting of activity, based on weather, shows some promise but needs further development. Little is known about attraction to, and consumption of, existing bait formulations. Using new techniques for studying behaviour, for example low-light video monitoring (Hogan, 1985) and feeding monitors (Senseman, 1978), methods for improving baits may become apparent and new bait formulations can be assessed. There is scope for improving attractiveness of baits and for inducing slugs to feed more avidly once they have contacted them. The persistence of baits in the field may be improved to increase the chance of effective slug–bait contact. New or existing molluscicides may be useful in formulations other than baits, for example seed treatments and sprays.

The mechanisms of resistance to slug damage in plant species and varieties requires more detailed study to enable exploitation in plant-breeding programmes. Such studies may also reveal new feeding attractants and repellents. Another facet of slug behaviour requiring further study is their remarkable homing ability and trail following, as seen in reproductive behaviour. Pheromones are implicated in these behaviour patterns and they may prove to be of value as in other pest-management programmes, for example for pest monitoring and bait enhancement.

Biological control has so far appeared to be limited in its contribution, yet the potential for exploitation of certain pathogens suggests that they may have a significant role in slug control in future.

Conclusions

Only a few species of slug achieve pest status, but these species vary in their biology and behaviour to an extent that may limit the impact of potential control measures. In many aspects of their biology, such as their life cycles, slugs show great plasticity. Because of their permeable integument, slugs are very susceptible to extremes of weather and they have a refined behavioural repertoire which has contributed to their relatively high degree of success as terrestrial animals (Rollo, 1982).

Worldwide, they are pests of a range of crops, but in Western Europe potato and wheat are at greatest risk. Currently, damage in the UK is estimated to cost £0.14–3.17 million for potatoes and £2.69 million for wheat. Control costs are estimated as £0.25 million for potatoes and £8.6 million for wheat. A substantial amount of molluscicide use, however, for example on oilseed rape, may be needless.

Cultural conditions have been, and will continue to be, a very important means of damage reduction. Chemical control is largely based on baits containing metaldehyde or methiocarb: neither compound shows outstanding efficiency in protecting crops from damage. The increased use of molluscicides in the UK in recent years suggests that farmers have an increased awareness of slugs as pests, and while the relative importance of slugs may be increasing, molluscicide use has increased at a greater rate. There are several approaches which may improve the efficiency of chemical control. The scope for biological control appears to be limited except in the case of certain pathogens.

No single method has achieved outstanding success in the control of slugs. Attempts to control these pests should utilize all possible methods rather than relying solely on one. Integration of the many partial methods may lead to effective control and significant damage reduction. To achieve this goal, more research is required into improving individual techniques, both existing and novel, and into integrating them to obtain effective pest management.

Acknowledgements

We would like to thank our many colleagues, both in Newcastle and elsewhere, who have allowed us to quote their unpublished data, who have brought new papers to our attention or who have commented on this review.

References

AIREY, W. (1984). The distribution of slug damage in a potato crop. *Journal of Molluscan Studies* **50**, 239–240.

AIREY, W. (1986). The influence of an alternative food on the effectiveness of proprietary molluscicide pellets against two species of slugs. *Journal of Molluscan Studies* **52**, in press.

ALLEN, H. P. (1975). ICI Plant Protection Division experience with direct-drilling systems, 1961–1974. *Outlook on Agriculture* **8**, Special Number, 213–215.

ALTIERI, M. A., HAGEN, K. S., TRUJILLO, J. AND CALTAGIRONE, L. E. (1982). Biological control of *Limax maximus* and *Helix aspersa* by indigenous predators in a daisy field in central coastal California. *Acta Œcologica Œcologia applicata* **3**, 387–390.

ANDERSON, A. W. AND TAYLOR, T. H. (1926). *The Slug Pest. Bulletin of the University of Leeds No. 143.*

ANDREWS, K. L. (1983). Slugs of the genus *Vaginulus* as pests of common bean, *Phaseolus vulgaris* in Central America. *Proceedings 10th International Congress of Plant Protection 1983, Vol. 3*, 951.

ANGSEESING, J. P. A. (1974). Selective eating of the acyanogenic form of *Trifolium repens. Heredity* **32**, 73–83.

ANGSEESING, J. P. A. AND ANGSEESING, W. J. (1973). Field observations on the cyanogenesis polymorphism in *Trifolium repens. Heredity* **31**, 276–282.

ARIAS, R. O. AND CROWELL, H. H. (1963). A contribution to the biology of the gray garden slug. *Bulletin of the Southern California Academy of Sciences* **62**, 83–97.

ATKIN, J. C. (1979). *Varietal Susceptibility of Potatoes to Slug Attack.* Unpublished PhD thesis, University of Newcastle upon Tyne.

ATKINSON, H. J., GIBSON, N. H. E. AND EVANS, H. (1979). A study of common crop pests in allotment gardens around Leeds. *Plant Pathology* **28**, 169–177.

BAKER, C. B. M. AND WAINES, R. A. (1957). Wireworm and slug damage to ware potatoes 1954–56. *Plant Pathology* **6**, 115–122.

BARDNER, R. AND FLETCHER, K. E. (1974). Insect infestations and their effects on the growth and yield of field crops: A review. *Bulletin of Entomological Research* **64**, 141–160.

BARNES, H. F. AND WEIL, J. W. (1944). Slugs in gardens: their numbers, activities and distribution. Part 1. *Journal of Animal Ecology* **13**, 140–175.

BARNES, H. F. AND WEIL, J. W. (1945). Slugs in gardens: their numbers, activities and distribution. Part 2. *Journal of Animal Ecology* **14**, 71–105.

BARONIO, P. (1974). Gli insetti nemici dei molluschi gasteropodi. *Bolletino dell'-Instituto di Entomologia della Universita degli studi di Bologna* **32**, 169–187. (English summary only).

BARRY, B. D. (1969). Evaluation of chemicals for control of slugs on field corn in Ohio. *Journal of Economic Entomology* **62**, 1277–1279.

BOWEN, I. D. AND JONES, G. W. (1985). Getting pesticides into cells. *Industrial Biotechnology (Wales)* **5**, 29–32.

BROOKS, W. M. (1968). Tetrahymenid ciliates as parasites of the gray garden slug. *Hilgardia* **39**, 205–276.

BYERS, R. A. AND BIERLEIN, D. L. (1984). Continuous alfalfa: Invertebrate pests during establishment. *Journal of Economic Entomology* **77**, 1500–1503.

CAMERON, R. A. D., JACKSON, N. AND EVERSHAM, B. (1983). A field key to the slugs of the British Isles. *Field Studies* **5**, 807–824.

CARRICK, R. (1938). The life history and development of *Agriolimax agrestis* L., the gray field slug. *Transactions of the Royal Society of Edinburgh* **59**, 563–597.

CARRICK, R. (1942). The grey field slug, *Agriolimax agrestis* L., and its environment. *Annals of Applied Biology* **29**, 43–55.

CATES, R. G. (1975). The interface between slugs and wild ginger. Some evolutionary aspects. *Ecology* **56**, 391–400.

CHAPMAN, P. J., SLY, J. M. A. AND CUTLER, J. R. (1977). *Pesticide Usage Survey Report 11. Arable Farm Crops 1974.* Ministry of Agriculture, Fisheries and Food and Department of Agriculture and Fisheries for Scotland, UK.

CHARLTON, J. F. L. (1978). Slugs as a possible cause of establishment failure in pasture legumes oversown in boxes. *New Zealand Journal of Experimental Agriculture* **6**, 313–317.

CHURCH, B. M., HAMPSON, C. P. AND FOX, W. R. (1970). The quality of stored main-crop potatoes in Great Britain. *Potato Research* **13**, 41–58.

CLEMENTS, R. O. AND BENTLEY, B. R. (1983). The effect of three pesticide treatments on the establishment of white clover (*Trifolium repens*) sown with a seed slotter. *Crop Protection* **2**, 375–378.

CLEMENTS, R. O., GILBEY, J., BENTLEY, B. R., FRENCH, N. AND CRAGG, I. A. (1985). Effects of pesticide combinations on the herbage yield of permanent pasture in England and Wales. *Tests of Agrochemicals and Cultivars No. 6 (Annals of Applied Biology* **106**, Supplement), 126–127.

COOK, A. (1979a). Homing by the slug *Limax pseudoflavus* (Evans). *Animal Behaviour* **27**, 545–552.

COOK, A. (1979b). Homing in the Gastropoda. *Malacologia* **18**, 315–318.

COOK, A. (1980). Field studies of homing in the pulmonate slug *Limax pseudoflavus* (Evans). *Journal of Molluscan Studies* **46**, 100–105.

COOK, A. (1985). Functional aspects of trail following by the carnivorous snail *Euglandina rosea*. *Malacologia* **26**, 173–181.

COPELAND, J. (1981). Effect of larval firefly extracts on molluscan cardiac activity. *Experientia* **37**, 1271–1272.

COTTAM, D. A. (1985). Frequency-dependent grazing by slugs and grasshoppers. *Journal of Ecology* **73**, 925–933.

CRAGG, J. B. AND VINCENT, M. H. (1952). The action of metaldehyde on the slug *Agriolimax reticulatus* (Müller). *Annals of Applied Biology* **39**, 392–406.

CRAWFORD-SIDEBOTHAM, T. J. (1970). Differential susceptibility of species of slugs to metaldehyde/bran and to methiocarb baits. *Oecologia* **5**, 303–324.

CRAWFORD-SIDEBOTHAM, T. J. (1972a). The influence of weather upon the activity of slugs. *Oecologia* **9**, 141–154.

CRAWFORD-SIDEBOTHAM, T. J. (1972b). The role of slugs and snails in the maintenance of the cyanogenesis polymorphisms of *Lotus corniculatus* and *Trifolium repens*. *Heredity* **28**, 405–411.

CROWELL, H. H. (1967). Slug and snail control with experimental poison baits. *Journal of Economic Entomology* **60**, 1048–1049.

CROWELL, H. H. (1979). *Chemical Control of Terrestrial Slugs and Snails. Oregon Agricultural Experimental Station, Station Bulletin No. 628.* Oregon State University, Oregon, USA.

DAINTON, B. H. (1954). The activity of slugs. I. The induction of activity by changing temperatures. *Journal of Experimental Biology* **31**, 165–187.

DAINTON, B. H. AND WRIGHT, J. (1985). Falling temperature stimulates activity in the slug *Arion ater*. *Journal of Experimental Biology* **118**, 439–443.

DAVID, W. A. L., TAYLOR, C. E. AND ATKEY, P. T. (1977). Nonoccluded virus-like particles in the mollusc *Agriolimax reticulatus* (Stylommatophora: Limacinae). *Journal of Invertebrate Pathology* **29**, 242–243.

DAVIES, S. M. (1977). The *Arion hortensis* complex, with notes on *A. intermedius* Normand (Pulmonata: Arionidae). *Journal of Conchology* **29**, 173–187.

DAVIES, S. M. (1979). Segregates of the *Arion hortensis* complex (Pulmonata: Arion-idae), with the description of a new species *Arion owenii*. *Journal of Conchology* **30**, 123–127.

DAWKINS, G., LUXTON, M. AND BISHOP, C. (1985). Transmission of liquorice rot of carrots by slugs. *Journal of Molluscan Studies* **51**, 83–85.

DAWKINS, G., HISLOP, J., LUXTON, M. AND BISHOP, C. (1986). Transmission of bacterial soft rot of potatoes by slugs. *Journal of Molluscan Studies* **52**, 25–29.

DAXL, R. VON (1971). Der einflus von temperatur und relativer luftfeuchte auf die molluskizide wirkung des metaldehyd, isolan und ioxynil auf *Limax flavus* L. und dessen eier. *Zeitschrift für angewandte Entomologie* **67**, 57–87. (English summary only).

DEBACH, P. (1974). *Biological Control by Natural Enemies.* Cambridge University Press, London.

DIRZO, R. AND HARPER, J. L. (1980). Experimental studies on slug–plant interactions. II. The effect of grazing by slugs on high density monocultures of *Capsella bursa-*

pastoris and *Poa annua. Journal of Ecology* **68**, 999–1011.

DIRZO, R. AND HARPER, J. L. (1982). Experimental studies on slug–plant interactions. III. Differences in the acceptability of individual plants of *Trifolium repens* to slugs and snails. *Journal of Ecology* **70**, 101–117.

DOWLING, P. M. AND LINSCOTT, D. L. (1983). Use of pesticides to determine relative importance of pest and disease factors limiting establishment of sod-seeded lucerne. *Grass and Forage Science* **38**, 179–185.

DUTHOIT, C. M. (1961). Assessing the activity of the field slug in cereals. *Plant Pathology* **10**, 165.

DUTHOIT, C. M. G. (1964). Slugs and food preferences. *Plant Pathology* **13**, 73–78.

DUVAL, D. M. (1970). Some aspects of the behaviour of pest species of slugs. *Journal of Conchology* **27**, 163–170.

DUVAL, D. M. (1972). A record of slug movements in late summer. *Journal of Conchology* **27**, 505–508.

EATON, H. J. AND TOMPSETT, A. A. (1976). Avoiding slug damage to lily bulbs at Rosewarne Experimental Horticulture Station, Camborne, Cornwall. In *Lilies 1976 and other Liliaceae*, pp. 63–65. Royal Horticultural Society, London.

EDWARDS, C. A. (1975). Effects of direct drilling on the soil fauna. *Outlook on Agriculture* **8**, Special Number, 243–244.

EVANS, N. J. (1978). *Limax pseudoflavus* Evans: A critical description and comparison with related species. *Irish Naturalists Journal* **19**, 231–236.

EVANS, N. J. (1985). The use of electrophoresis in the separation of two closely related species of terrestrial slugs. *Biochemical Systematics and Ecology* **13**, 325–328.

FISHER, T. W., MOORE, I., LEGNER, E. F. AND ORTH, R. E. (1976). *Ocypus olens*: A predator of brown garden snail. *California Agriculture* **30**, 20–21.

FLETCHER, M. R. AND HARDY, A. R. (1983). *Research and Development Report: Pesticide Science 1982. Reference book of the Ministry of Agriculture, Fisheries and Food No. 252(82)*. HMSO, London.

FLETCHER, M. R. AND HARDY, A. R. (1984). *Research and Development Report: Pesticide Science 1983. Reference book of the Ministry of Agriculture, Fisheries and Food No. 252(83)*. HMSO, London.

FLETCHER, M. R. AND STANLEY, P. I. (1981). *Research and Development Report: Pesticide Science 1980. Reference book of the Ministry of Agriculture, Fisheries and Food No. 252(80)*. HMSO, London.

FOSTER, G. N. (1977). Problems in cucumber crops caused by slugs, cuckoo-spit insect, mushroom cecid, hairy fungus beetle and the house mouse. *Plant Pathology* **26**, 100–101.

FOX, L. AND LANDIS, B. J. (1973). Notes on the predaceous habits of the gray field slug *Deroceras laeve. Environmental Entomology* **2**, 306–307.

FRAIN, J. (1982). Chemical control of molluscs using metaldehyde. *International Pest Control* **6**, 150–151.

FRAIN, J. M. AND NEWELL, P. F. (1983). Testing molluscicides against slugs—the importance of assessing the residual population. *Journal of Molluscan Studies* **49**, 164–173.

FRANCOIS, E., RIGA, E. AND MOENS, R. (1965). Labelling the grey field slug *Agriolimax reticulatus* by means of radionuclides. *Parasitica* **21**, 138–151.

FRANCOIS, E., RIGA, A. AND MOENS, R. (1968). Estimation des populations de *Agriolimax reticulatus* Müller au moyen de la technique de marquage au radiophosphore ^{32}P et recapture. *Parasitica* **24**, 63–78.

GELPERIN, A. (1974). Olfactory basis of homing behaviour in the giant garden slug *Limax maximus. Proceedings of the National Academy of Sciences of the United States of America* **71**, 966–970.

GELPERIN, A. (1975). Rapid food-aversion learning by a terrestrial mollusk. *Science* **189**, 567–570.

GILES, C. J., PYCOCK, J. F., HUMPHREYS, D. J. AND STODULSKI, J. B. J. (1984).

Methiocarb poisoning in a sheep. *Veterinary Record* **114**, 642.

GIMINGHAM, C. T. (1940). Some recent contributions by English workers to the development of methods of insect control. *Annals of Applied Biology* **27**, 161–175.

GLEN, D. M. AND ORSMAN, I. A. (1986). Comparison of molluscicides available to gardeners, based on metaldehyde, methiocarb or aluminium sulphate. *Crop Protection* **5**(6), in press.

GLEN, D. M., MILSOM, N. F. AND WILTSHIRE, C. W. (1986). Evaluation of a mixture containing copper sulphate, aluminium sulphate and borax for control of slug damage to potatoes. *Tests of Agrochemicals and Cultivars No. 7 (Annals of Applied Biology* **108**, Supplement), 26–27.

GLEN, D. M., WILTSHIRE, C. W. AND MILSOM, N. F. (1984). Slugs and straw disposal in winter wheat. *1984 British Crop Protection Conference—Pests and Diseases* 139–144.

GODAN, D. (1983). *Pest Slugs and Snails: Biology and Control*. Springer-Verlag, Berlin.

GOULD, H. J. (1961). Observations on slug damage to winter wheat in East Anglia 1957–1959. *Plant Pathology* **10**, 142–147.

GOULD, H. J. (1962). Tests with seed dressings to control grain hollowing of winter wheat by slugs. *Plant Pathology* **11**, 147–152.

GOULD, H. J. (1965). Observations on the susceptibility of maincrop potato varieties to slug damage. *Plant Pathology* **14**, 109–111.

GOULD, H. J. AND WEBLEY, D. (1972). Field trials for the control of slugs on winter wheat. *Plant Pathology* **21**, 77–82.

GOUYON, P. H., FORT, P. H. AND CARAUX, G. (1983). Selection of seedlings of *Thymus vulgaris* by grazing slugs. *Journal of Ecology* **71**, 299–306.

GRANT, J. F., YEARGAN, K. V., PASS, B. C. AND PARR, J. C. (1982). Invertebrate organisms associated with alfalfa seedling loss in complete-tillage and no-tillage plantings. *Journal of Economic Entomology* **75**, 822–826.

HANCOCK, M. (1986). The use of pesticides during planting and establishment of oilseed rape, potato and sugar beet in the UK. *British Crop Protection Council Monograph No. 33, Symposium on Healthy Planting Material* (D. Rudd-Jones and F. A. Langton, Eds), pp. 147–153. BCPC, Croydon.

HENDERSON, I. F. (1968). Laboratory methods for assessing the toxicity of contact poisons to slugs. *Annals of Applied Biology* **62**, 363–369.

HENDERSON, I. F. (1969). A laboratory method for assessing the toxicity of stomach poisons to slugs. *Annals of Applied Biology* **63**, 167–171.

HOGAN, J. M. (1985). *The Behaviour of the Grey Field Slug*, Deroceras reticulatum (*Müller*) *with Particular Reference to Control in Winter Wheat*. Unpublished PhD thesis, University of Newcastle upon Tyne.

HOGAN, J. M. AND STEELE, G. (1986). Dye-marking slugs. *Journal of Molluscan Studies* **52**, 138–143.

HUGHES, K. A. AND GAYNOR, D. L. (1984). Comparison of Argentine stem weevil and slug damage in maize direct-drilled into pasture or following winter oats. *New Zealand Journal of Experimental Agriculture* **12**, 47–53.

HUNTER, P. J. (1966). The distribution and abundance of slugs on an arable plot in Northumberland. *Journal of Animal Ecology* **35**, 543–557.

HUNTER, P. J. (1967). The effect of cultivations on slugs of arable land. *Plant Pathology* **16**, 153–156.

HUNTER, P. J. (1968a). Studies on slugs of arable ground. I. Sampling methods. *Malacologia* **6**, 369–377.

HUNTER, P. J. (1968b). Studies on slugs of arable ground. II. Life cycles. *Malacologia* **6**, 379–389.

HUNTER, P. J. (1969a). An estimate of the extent of slug damage to wheat and potatoes in England and Wales. *NAAS Quarterly Review* **85**, 31–36.

HUNTER, P. J. (1969b). Slugs and their control. *Proceedings 5th British Insecticide and Fungicide Conference (1969)*, 715–719.

HUNTER, P. J. (1978). Slugs—a study in applied ecology. In *Pulmonates Vol. 2a, Systematics, Evolution and Ecology* (V. Fretter and J. Peake, Eds), pp. 271–286. Academic Press, London.

HUNTER, P. J. AND JOHNSTON, D. L. (1970). Screening carbamates for toxicity against slugs. *Journal of Economic Entomology* **63**, 305–306.

HUNTER, P. J. AND SYMONDS, B. V. (1970). The distribution of bait pellets for slug control. *Annals of Applied Biology* **65**, 1–7.

HUNTER, P. J. AND SYMONDS, B. V. (1971). The leap-frogging slug. *Nature* **229**, 349.

HUNTER, P. J., SYMONDS, B. V. AND NEWELL, P. F. (1968). Potato leaf and stem damage by slugs. *Plant Pathology* **17**, 161–164.

JESSOP, N. H. (1969). The effects of simulated slug damage on the yield of winter wheat. *Plant Pathology* **18**, 172–175.

JOHN, M. E. AND EVANS, E. J. (EDS) (1984). *Control of Pests and Diseases of Oilseed Rape 1984. Booklet of the Ministry of Agriculture, Fisheries and Food No. 2387(84)*. HMSO, London.

JONES, A. A. (1985). *Evaluation of a Microsporidian Parasite of the Grey Field Slug*, Deroceras reticulatum (*Müller*). Unpublished PhD thesis, University of Newcastle upon Tyne.

JONES, A. A. AND SELMAN, B. J. (1984). A possible biological control agent of the grey field slug (*Deroceras reticulatum*). *1984 British Crop Protection Conference— Pests and Diseases*, 261–266.

JONES, A. A. AND SELMAN, B. J. (1985). *Microsporidium novocastriensis* n.sp., a microsporidian parasite of the grey field slug, *Deroceras reticulatum*. *Journal of Protozoology* **32**, 581–586.

JONES, D. A. (1962). Selective feeding of the acyanogenic form of the plant *Lotus corniculatus* L. by various animals. *Nature* **193**, 1109–1110.

JONES, D. A. (1966). On the polymorphism of cyanogenesis in *Lotus corniculatus*. I Selection by animals. *Canadian Journal of Genetics and Cytology* **8**, 556–567.

JUDGE, F. D. (1969). Preliminary screening of candidate molluscicides. *Journal of Economic Entomology* **62**, 1393–1397.

JUDGE, F. D. AND KUHR, R. J. (1972). Laboratory and field screening of granular formulations of candidate molluscicides. *Journal of Economic Entomology* **65**, 242–245.

KARLIN, E. J. (1961). Temperature and light as factors affecting the locomotor activity of slugs. *Nautilus* **74**, 125–130.

KASSANIS, B., WOODS, R. D. AND MACFARLANE, I. (1984). Galactogen, a virus-like particle from slugs. *Annals of Applied Biology* **105**, 587–589.

KELLY, M. AND CURRY, J. P. (1985). Studies on the arthropod fauna of a winter wheat crop and its response to the pesticide methiocarb. *Pedobiologia* **28**, 413–421.

KEMP, N. J. AND NEWELL, P. F. (1985). Laboratory observations on the effectiveness of methiocarb and metaldehyde baits against the slug *Deroceras reticulatum* (Müll.). *Journal of Molluscan Studies* **51**, 228–230.

KEMP, N. J. AND NEWELL, P. F. (1987). The energy reserves, activity and survival of certain agricultural slug pests. *Symposium Papers Vol. 4, IX International Malacological Congress, Edinburgh 1986*, in press.

KERNEY, M. P. (ED.) (1976). *Atlas of the Non-Marine Mollusca of the British Isles*. Institute of Terrestrial Ecology, Cambridge.

KERR, S. P. (1984). Potato varieties completing recommended list trials 1983. *Journal of the National Institute of Agricultural Botany* **16**, 623–628.

KNUTSON, L. V., STEPHENSON, J. W. AND BERG, C. O. (1965). Biology of a slug-killing fly *Tetanocera elata* (Diptera: Sciomyzidae). *Proceedings of the Malacological Society of London* **36**, 213–220.

LEWIS, R. D. (1969). Studies on the locomotor activity of the slug *Arion ater* (Linnaeus). I. Humidity, temperature and light reactions. *Malacologia* **7**, 295–305.

LINCOLN, D. E. AND LANGENHEIM, J. H. (1979). Variation of *Satureja douglasii* monoterpenoids in relation to light intensity and herbivory. *Biochemical Systematics and Ecology* **7**, 289–298.

LONGBOTTOM, G. M. AND GORDON, A. S. M. (1979). Metaldehyde poisoning in a dairy herd. *Veterinary Record* **104**, 454–455.

MADDY, K. T. (1975). Poisoning of dogs with metaldehyde in snail and slug poison bait. *California Veterinarian* **29**, (3) 27–28; (4) 24–25.

MAFF (1980). *Agricultural Science Service. Crop Pests and Diseases 1979. Reference Book of the Ministry of Agriculture, Fisheries and Food, 256.* HMSO, London.

MAFF (1985). *Winter Wheat Husbandry: Recent Developments. Booklet of the Ministry of Agriculture, Fisheries and Food No. 2513.* HMSO, London.

MAFF (1986). *Annual Review of Agriculture 1986.* HMSO, London.

MARTIN, T. J. AND FORREST, J. D. (1969). Development of Draza in Great Britain. *Pflanzenschutz-Nachrichten Bayer* **22**, 205–243.

MARTIN, T. J., DAVIS, M. E. AND MORRIS, D. B. (1969). Development work with methiocarb in Great Britain. *Proceedings 5th British Insecticide and Fungicide Conference (1969)*, 434–441.

MELLANBY, K. (1961). Slugs at low temperatures. *Nature* **189**, 944.

MILES, H. W., WOOD, J. AND THOMAS, I. (1931). On the ecology and control of slugs. *Annals of Applied Biology* **18**, 370–400.

MOENS, R. (1983). Essais sur la protection des grains froment contre l'attaque des limaces. *Revue de l'Agriculture* **36**, 1303–1317.

MOENS, R., FRANCOIS, E., RIGA, A. AND VAN DEN BRUEL, W. E. (1967). A mechanical barrier against terrestrial gasteropods. *Parasitica* **23**, 22–27.

MORTON, B. (1979). The diurnal rhythm and the cycle of feeding and digestion in the slug *Deroceras caruanae*. *Journal of Zoology* **187**, 135–152.

MÜLLER, ST VON AND OHNESORGE, B. (1985). Die verwendung markierter schnecken zur populationsdichteabschatzung und zum studium des migrationsverhaltens von *Arion* sp. *Anzeiger für Schadlingskunde, Pflanzenschutz, Umweltschutz* **58**, 123–126.

NEWELL, P. F. (1966). The nocturnal behaviour of slugs. *Medical and Biological Illustration* **16**, 146–156.

NEWELL, P. F. (1968). The measurement of light and temperature as factors controlling the surface activity of the slug *Agriolimax reticulatus* (Müller). In *The Measurement of Environmental Factors in Terrestrial Ecology. Symposium of the British Ecological Society, 8.* (R. M. Wadsworth, Ed.), pp. 141–147. Blackwell, Oxford.

NIAB (1985). *Farmers Leaflet No. 3, Recommended Varieties of Potato 1986.* National Institute of Agricultural Botany, Cambridge.

NIX, J. (1985). *Farm Management Pocketbook*, 16th edition (1986). Farm Business Unit, Department of Agricultural Economics, Wye College, University of London.

OAKLEY, J. N. (1984). *Slugs and Snails. Leaflet of the Ministry of Agriculture, Fisheries and Food, 115.* HMSO, London.

ORTH, R. E., MOORE, I., FISHER, T. W. AND LEGNER, E. F. (1975). A rove beetle *Ocypus olens* with potential for biological control of the brown garden snail *Helix aspersa* in California including a key to the nearctic species of *Ocypus*. *Canadian Entomologist* **107**, 1111–1116.

PALLANT, D. (1972). The food of the grey field slug *Agriolimax reticulatus* (Müller) on grassland. *Journal of Animal Ecology* **41**, 761–769.

PICKETT, J. A. AND STEPHENSON, J. W. (1980). Plant volatiles and components influencing behaviour of the field slug, *Deroceras reticulatum* (Müll.). *Journal of Chemical Ecology* **6**, 435–444.

PINDER, L. C. V. (1974). The ecology of slugs in potato crops with special reference to the differential susceptibility of potato cultivars to slug damage. *Journal of Applied Ecology* **11**, 439–451.

PRIOR, D. J. (1983). The relationship between age and body size of individuals in isolated clutches of the terrestrial slug *Limax maximus* (Linnaeus, 1858). *Journal of Experimental Zoology* **225**, 321–324.

QUICK, H. F. (1960). British slugs (Pulmonata; Testacellidae, Arionidae, Limacidae). *Bulletin of the British Museum (Natural History) Zoology* **6**, 105–226.

RAYNER, J. M. (1975). Experiments on the control of slugs in potatoes by means of molluscicidal baits. *Plant Pathology* **24**, 167–171.

RAYNER, J. M., BROCK, A. M., FRENCH, N., GOULD, H. J. AND LEWIS, S. C. (1978). Further experiments on the control of slugs in potatoes by means of molluscicidal baits. *Plant Pathology* **27**, 186–193.

RICE, L. R., LINCOLN, D. E. AND LANGENHEIM, J. H. (1978). Palatability of mono-terpenoid compositional types of *Satureja douglasii* to a generalist molluscan herbivore *Ariolimax dolichophallus*. *Biochemical Systematics and Ecology* **6**, 45–53.

RICHARDSON, D. E. (1979). Potato variety testing at NIAB. *Journal of the National Institute of Agricultural Botany* **15**, 92–97.

RICHTER, K. O. (1976). A method for individually marking slugs. *Journal of Molluscan Studies* **42**, 146–151.

ROGERS-LEWIS, D. S. (1976). Use of molluscicides for control of slug damage in maincrop potatoes on silt soils. *Experimental Husbandry* **31**, 125–134.

ROGERS-LEWIS, D. S. (1977). Slug damage in potatoes and winter wheat on silt soils. *Annals of Applied Biology* **87**, 532–535.

ROLLO, C. D. (1982). The regulation of activity in populations of the terrestrial slug *Limax maximus* (Gastropoda: Limacidae). *Researches on Population Ecology* **24**, 1–32.

ROLLO, C. D. AND WELLINGTON, W. G. (1979). Intra- and inter-specific agonistic behaviour among terrestrial slugs (Pulmonata: Stylommatophora). *Canadian Journal of Zoology* **57**, 846–855.

ROLLO, C. D. AND WELLINGTON, W. G. (1981). Environmental orientation by terrestrial mollusca with particular reference to homing behaviour. *Canadian Journal of Zoology* **59**, 225–239.

ROTTGER, U. AND KLINGAUF, F. (1976). *Deroceras laeve* Müll. (Mollusca: Limacidae) ein einrauber von *Pegomyia betae* Curt. (Muscidae). *Anzeiger für Schad-lingskunde, Pflanzenschutz, Umweltschutz* **49**, 49–51.

RUNHAM, N. W. AND HUNTER, P. J. (1970). *Terrestrial Slugs*. Hutchinson University Library, London.

RUPPEL, R. F. (1959). Effectiveness of sevin against the grey garden slug. *Journal of Economic Entomology* **52**, 360.

RYDER, T. A. AND BOWEN, I. D. (1977a). The use of x-ray microanalysis to dem-onstrate the uptake of the molluscicide copper sulphate by slug eggs. *Histo-chemistry* **52**, 55–60.

RYDER, T. A. AND BOWEN, I. D. (1977b). The slug foot as a site of uptake of copper molluscicide. *Journal of Invertebrate Pathology* **30**, 381–386.

SAVAGE, M. J. AND THOMAS, M. R. (1985). *Control of Pests and Diseases of Grass and Forage Crops. Booklet of the Ministry of Agriculture, Fisheries and Food, 2045*. HMSO, London.

SCHRIM, M. AND BYERS, R. A. (1980). A method of sampling three slug species attacking sod-seeded legumes. *Melsheimer Entomological Series* **29**, 9–11.

SCOTT, G. C. (1981). Experimental seed treatments for the control of wheat bulb fly and slugs. *Proceedings 1981 British Crop Protection Conference—Pests and Diseases*, 441–448.

SCOTT, G. C. (1982). *FAO Guidelines for Seed Treatments: Slug Damage to Cereals in Great Britain*. Food and Agriculture Organization, Rome.

SCOTT, G. C., GRIFFITHS, D. C. AND STEPHENSON, J. W. (1977). A laboratory method for testing seed treatments for the control of slugs in cereals. *Proceedings 1977 British Crop Protection Conference—Pests and Diseases*, 129–134.

SCOTT, G. C., PICKETT, J. A., SMITH, M. C., WOODSTOCK, C. M., HARRIS, P. G. W.,

Hammon, R. P. and Koetecha, H. D. (1984). Seed treatments for controlling slugs in winter wheat. *1984 British Crop Protection Conference—Pests and Diseases*, 133–138.

Senseman, D. M. (1978). Short-term control of food intake by the terrestrial slug *Ariolimax*. *Journal of Comparative Physiology* **124**, 37–48.

Sly, J. M. A. (1986). *Pesticide Usage Survey Report 35. Arable Farm Crops and Grass 1982*. Ministry of Agriculture, Fisheries and Food, UK.

Smith, B., Jordan, V., Kendall, D. and Glen, D. (1985). Straw disposal and its effects on pests, diseases and pesticide use. In *Straw Soils and Science*, pp. 20–21; 27. Agricultural and Food Research Council, London.

Smith, F. F. and Boswell, A. L. (1970). New baits and attractants for slugs. *Journal of Economic Entomology* **63**, 1919–1922.

Sokolove, P. G., Beiswanger, C. M., Prior, D. J. and Gelperin, A. (1977). A circadian rhythm in the locomotor behaviour of the giant garden slug *Limax maximus*. *Journal of Experimental Biology* **66**, 47–64.

South, A. (1964). Estimation of slug populations. *Annals of Applied Biology* **53**, 251–258.

South, A. (1965). Biology and ecology of *Agriolimax reticulatus* (Müll.) and other slugs: Spatial distribution. *Journal of Animal Ecology* **34**, 403–417.

South, A. (1973). Degats causes par les limaces en Grande-Bretagne. *Haliotis* **3**, 19–25.

South, A. (1980). A technique for the assessment of predation by birds and mammals on the slug *Deroceras reticulatum* (Müller) (Pulmonata: Limacidae). *Journal of Conchology* **30**, 229–234.

South, A. (1982). A comparison of the life cycles of *Deroceras reticulatum* (Müller) and *Arion intermedius* Normand (Pulmonata: Stylommatophora) at different temperatures under laboratory conditions. *Journal of Molluscan Studies* **48**, 233–244.

Southwood, T. R. E. (1978). *Ecological Methods with Particular Reference to the Study of Insect Populations*, 2nd edition. Chapman and Hall, London.

Standen, O. D. (1951). Some observations upon the maintenance of *Australorbis glabratus* in the laboratory. *Annals of Tropical Medicine and Parasitology* **45**, 80–83.

Stanley, P. I. and Fletcher, M. R. (1982). *Research and Development Report: Pesticide Science 1981. Reference Book of the Ministry of Agriculture, Fisheries and Food No. 252(81)*. HMSO, London.

Steed, J. M., Sly, J. M. A., Tucker, G. G. and Cutler, J. R. (1979). *Pesticide Usage Survey Report 18. Arable Farm Crops 1977*. Ministry of Agriculture, Fisheries and Food and Department of Agriculture and Fisheries for Scotland, UK.

Stephenson, J. W. (1959). Aldrin controlling slug and wireworm damage to potatoes. *Plant Pathology* **8**, 53–54.

Stephenson, J. W. (1967). The molluscicidal properties of ioxynil. In *Rothamsted Experimental Station Report for 1966*, pp. 197–198.

Stephenson, J. W. (1968). A review of the biology and ecology of slugs of agricultural importance. *Proceedings of the Malacological Society of London* **38**, 169–178.

Stephenson, J. W. (1975a). Laboratory observations on the effect of soil compaction on slug damage to winter wheat. *Plant Pathology* **24**, 9–11.

Stephenson, J. W. (1975b). Laboratory observations on the distribution of *Agriolimax reticulatus* (Müll.) in different aggregate fractions of garden loam. *Plant Pathology* **24**, 12–15.

Stephenson, J. W. (1979). The functioning of the sense organs associated with feeding behaviour in *Deroceras reticulatum* (Müll.). *Journal of Molluscan Studies* **45**, 167–171.

Stephenson, J. W. and Bardner, R. (1976). Slugs in agriculture. In *Rothamsted Report for 1976, Part 2*, pp. 169–187.

Stephenson, J. W. and Knutson, L. V. (1966). A resumé of recent studies of invertebrates associated with slugs. *Journal of Economic Entomology* **59**, 356–360.

STOREY, M. (1985). *The Varietal Susceptibility of Potato Crops to Slug Damage*. Unpublished PhD thesis, City of London Polytechnic.

STRICKLAND, A. H. (1965). Pest control and productivity in British agriculture. *Journal of the Royal Society of Arts* **113**, 62–81.

STRINGER, A. AND MORGAN, N. G. (1969). Population and control of snails in blackcurrant plantations. *Proceedings 5th British Insecticide and Fungicide Conference (1969)*, 453–457.

STUDDERT, V. P. (1985). Epidemiological features of snail and slug bait poisoning in dogs and cats. *Australian Veterinary Journal* **62**, 269–271.

SYMONDS, B. V. (1975). Evaluation of potential molluscicides for the control of the field slug *Agriolimax reticulatus* (Müll). *Plant Pathology* **24**, 1–9.

TAYLOR, J. W. (1907). *Monograph of the Land and Freshwater Mollusca of the British Isles; Testacellidae, Limacidae, Arionidae*. Taylor Brothers, Leeds, UK.

TRELKA, D. G. AND BERG, C. O. (1977). Behavioural studies of the slug-killing larvae of two species of *Tetanocera* (Diptera: Sciomyzidae). *Proceedings of the Entomological Society of Washington* **79**, 475–486.

WADHAM, M. D. AND WYNN PARRY, D. (1981). The silicon content of *Oryza sativa* L. and its effect on the grazing behaviour of *Agriolimax reticulatus* Müller. *Annals of Botany* **48**, 399–402.

WARD, J. T., BASFORD, W. D., HAWKINS, J. H. AND HOLLIDAY, J. M. (1985). *Oilseed Rape*. Farming Press Ltd, UK.

WAREING, D. R. AND BAILEY, S. E. R. (1985). The effects of steady and cycling temperatures on the activity of the slug *Deroceras reticulatum*. *Journal of Molluscan Studies* **51**, 257–266.

WEBBE, G. AND LAMBERT, J. D. H. (1983). Plants that kill snails and prospects for disease control. *Nature* **302**, 754.

WEBLEY, D. (1963). Experiments with slug baits in 1959. *Plant Pathology* **12**, 19–20.

WEBLEY, D. (1964). Slug activity in relation to weather. *Annals of Applied Biology* **53**, 407–414.

WEBLEY, D. (1966). Waterproofing of metaldehyde on bran baits for slug control. *Nature* **212**, 320–321.

WEBLEY, D. (1969). A comparison of methiocarb and metaldehyde baits for the control of four species of slugs. *Proceedings 5th British Insecticide and Fungicide Conference (1969)*, 442–444.

WEBLEY, D. (1970). Observations on the effects of distribution and numbers of slug pellets on the catch of slugs. *Annals of Applied Biology* **66**, 347–352.

WEDGEWOOD, M. A. AND BAILEY, S. E. R. (1986). The analysis of single meals in slugs feeding on molluscicidal baits. *Journal of Molluscan Studies*, in press.

WHARTON, A. L. AND ENSOR, H. (1969). The slug problem in peas for processing. *Proceedings 5th British Insecticide and Fungicide Conference (1969)*, 445–442.

WHELAN, J. (1982). An artificial medium for feeding choice experiments with slugs. *Journal of Applied Ecology* **19**, 89–94.

WHITE, A. R. (1959). Observations on slug activity in a Northumberland garden. *Plant Pathology* **8**, 62–68.

WILKINSON, A. T. S. (1972). *Control of Slugs. Publication 1213*. Agriculture Canada, Ottawa.

WINFIELD, A. L., WARDLOW, L. R. AND SMITH, B. F. (1967). Further observations on the susceptibility of maincrop potato cultivars to slug damage. *Plant Pathology* **16**, 136–138.

WORTHING, C. R. AND WALKER, S. B. (1983). *Pesticide Manual*, 7th edition. British Crop Protection Council, Croydon.

WRIGHT, A. A. (1973). Evaluation of a synthetic diet for the rearing of the slug *Arion ater* L. *Comparative Biochemistry and Physiology* **46A**, 593–603.

WRIGHT, A. A. AND WILLIAMS, R. (1980). The effect of molluscicides on the consumption of bait by slugs. *Journal of Molluscan Studies* **46**, 265–281.

9

Ecological and Agricultural Considerations in the Management of Twospotted Spider Mite (*Tetranychus urticae* Koch)

R. L. BRANDENBURG AND G. G. KENNEDY

Department of Entomology, Box 7613, North Carolina State University, Raleigh, North Carolina 27695–7613, USA

Introduction

The pest status of *Tetranychus urticae* is attributable in large part to the particular array of life history attributes it possesses. These represent adaptations for survival in natural ecosystems characterized by a high degree of spatial and temporal variability in habitat and host-plant suitability. In natural ecosystems, populations of *T. urticae* generally consist of widely scattered colonies which are in dynamic equilibrium with their natural enemies and rarely cause severe damage to their host plant. In contrast, in modern agroecosystems, populations of *T. urticae* commonly reach extremely high densities and, if remedial control measures are not applied, will often completely destroy their host plants. There are a number of factors which contribute to population differences between natural and agricultural ecosystems, including the suppression of natural enemies and the stimulation of population growth rates which result from many of our modern agricultural practices. In addition, in modern agroecosystems the patterns of spatial and temporal availability as well as the available biomass of host plants (crops) are vastly different and far more predictable than in natural ecosystems. *T. urticae* possesses many life history attributes which enable it to cope with the high degree of spatial and temporal variation in habitat and host availability and suitability characteristic of natural ecosystems. As a result, this species is highly efficient for exploiting the spatially and temporally predictable superabundance of suitable habitats and host plants provided by modern agriculture.

Our objective in this review is to examine the recent literature pertaining to the ecological and agricultural factors contributing to the pest status and management of *T. urticae*. There is a wealth of literature pertaining to tetranychid mites in general and to *T. urticae* in particular. We have emphasized, in our review, research published since the oustanding reviews of Huffaker, van de Vrie and McMurtry (1969, 1970); McMurtry, Huffaker and van de Vrie (1970), and van de Vrie, McMurtry and Huffaker (1972). For additional information the reader is also referred to the excellent recent volumes by Helle and Sabelis (1985a, b).

Ecological considerations

LIFE HISTORY ATTRIBUTES

Overwintering

T. urticae can overwinter either as diapausing adult females (Veerman, 1985) or as actively reproducing populations (Brandenburg and Kennedy, 1981; Takafuji and Kambayashi, 1984), with the diapause condition occurring where winters are more harsh and overwintering host plants less available. Diapause in spider mites has been reviewed recently by Veerman (1985). The general description of the phenomenon as it relates to *T. urticae*, described below, is taken from that review. Diapause is induced in response to day length (short days) with the response modified by temperature, as is true for many arthropods. The condition of the host plant can also influence the incidence of diapause, but that influence is largely limited to periods when daylength/temperature is approaching a critical threshold. Diapausing *T. urticae* do not feed or oviposit; they pass the winter in sheltered locations such as crevices in bark of trees and shrubs, clods of soil, and in leaf litter, with high levels of mortality occurring among mites overwintering in soil. Termination of diapause also apparently involves a response to photoperiod and temperature which is further modified by the duration of chilling experienced while in diapause. There is extensive intraspecific variation in the diapause response, with the critical daylength for induction of diapause known to vary among populations from different latitudes. Non-diapausing populations often occur in glasshouses and in areas with moderate winter temperatures. Recently, Ignatowicz and Helle (1986) reported that the suppression of diapause in response to short days was inherited as a recessive trait under monogenic control. In areas, such as eastern North Carolina, USA, where winter temperatures are sufficiently mild and suitable host plants are available throughout the winter, populations of *T. urticae* remain active throughout the winter (Brandenburg and Kennedy, 1981; Margolies and Kennedy, 1985). These populations are quite cold tolerant and persist despite exposure to occasional brief periods with ambient temperatures as low as $-15°C$.

A knowledge of the diapause status of overwintering populations in an area is important in understanding the dynamics of colonization of crops in the spring. In the case of *T. urticae* populations on glasshouse cucumbers, the occurrence of diapausing populations is an important determinant of the protocols necessary to achieve effective biological control using the predacious mite *Phytoseiulus persimilis* A. H. (Hussey and Scopes, 1985a). Further, diapausing *T. urticae* have been reported to exhibit higher tolerances to a number of pesticides (Veerman, 1985).

Reproduction

Reproduction by *T. urticae* involves arrhenotokous parthenogenesis with unmated females producing all male progeny and mated females producing both male and female offspring. Although the potential sex ratio is under genetic control (Overmeer and Harrison, 1969; Mitchell, 1972), the actual sex ratio of progency produced by individual females is also dependent upon the amount of sperm a female receives during mating (Overmeer, 1972; Helle and Pijnacker, 1985), host quality (Wrensch and Young, 1983), population density (Wrensch and Young, 1978) and, perhaps, temperature (Hazan, Gerson and Tahori, 1973).

Eggs are deposited on the foliage and upon hatching the developing mites pass through the following stages: larva, protochrysalis, protonymph, deutochrysalis, deutonymph, teliochrysalis and adult. The protochrysalis, deutochrysalis and teliochrysalis stages are inactive. Males emerge before females, guard female teliochrysali and copulate with the adult female immediately upon her emergence. Males typically mate more than once, whereas for females only the first mating is effective (Boudreaux, 1963; Helle, 1967). Additional details of *T. urticae* development and reproductive behaviour can be found in Potter (1979), Everson and Addicott (1982) and Crooker (1985).

As an exploiter of transient resources, *T. urticae* is well adapted to disperse widely to locate new hosts, to exploit newly found hosts through its high reproductive rate, and to generate large numbers of dispersers before a decline in host plant or host patch leads to localized extinction. In an excellent discussion of the reproductive strategies and life history evolution among spider mites, Sabelis (1985a), has summarized the relevant research findings and has argued convincingly that *Tetranychus* species, including *T. urticae*, have been strongly selected for high intrinsic rates of increase (r_m).

Many parameters affect r_m, including age of first reproduction, survivorship, fecundity, duration of the oviposition period, egg viability and various aspects of the sex ratio (Wrensch, 1985). Of these, the factors having the greatest influence on r_m are age of first reproduction and fecundity, with the former being far more important, especially at higher r_ms (Cole, 1954). Lewontin (1965) and Caswell and Hasting (1980) have demonstrated that changes in the rate of development affect r_m more than equal changes in fecundity: this was shown empirically for *T. urticae* by Wrensch and Young (1975). On the basis of an analysis of life history data available for several *Tetranychus* species, Sabelis (1985a) convincingly argued that these species have been subjected to such strong selection for high r_m that their developmental rates approach physiological limits.

A number of factors, both intrinsic (e.g. mite strain, level of inbreeding, colony density, age of population, females' mating status) and extrinsic (e.g. temperature, relative humidity, light, level of predation, intra- and interspecific competition, host plant, pesticides), influence the reproductive parameters affecting r_m (Wrensch, 1985). On the basis of the previous discussion, it is clear that factors affecting developmental time will have the most marked effect on r_m. Factors affecting fecundity and survivorship will also affect r_m, although proportionally greater changes in these parameters, as compared to developmental time, are required to effect comparable changes in r_m. Next to developmental time, fecundity is of greatest importance in determining r_m and it is sensitive to a variety of environmental influences (Wrensch, 1985). Nevertheless, Young and Wrensch (1981) concluded that the ability to produce large numbers of progeny under even marginal conditions is an important reason why spider mites are such successful colonizing organisms, and, one might add, such important pests.

The reproductive parameters for *T. urticae* are strongly influenced by temperature, relative humidity and host plant. As is typical for arthropods, the effects of temperature are manifest primarily on developmental rate. The relationship between developmental rate and temperature is curvilinear with a low-temperature developmental threshold of approximately 10°C (Herbert, 1981). Mori (1961) reported the temperature preference of *T. urticae* to be 13–35°C. However, this range as well as the low-temperature developmental thresholds and optimum temperature for development might be expected to vary to some degree among locally adapted populations, although the general shape of the temperature/development rate curve is not likely to vary much. An understanding of the relationship between temperature and development rate is a prerequisite to the prediction of real population growth rates and to the modelling of predator/prey interactions.

The most severe mite problems are typically experienced under hot and dry conditions. Crop canopy temperatures in drought-stressed plants can be significantly higher than in non-stressed plants, leading to more rapid development strictly attributable to the effects of temperature. The daily mean high temperatures in the canopy of drought-stressed maize plots over a 21-day period was 3.4°C higher than in maize plots receiving adequate moisture, even though all plots were being grown in the same field (P.

Ellsworth, unpublished). Moisture stress may also affect mite population growth independently of temperature through its effects on the physiology of the host plant.

Relative humidity can affect reproductive rate through its direct effects on mite development, survival and reproduction. *T. urticae* develops faster and produces more eggs under conditions of low (25–30%) than high (85–90%) relative humidity (Nickel, 1960). In addition, survival of immature stages, adult longevity and percentage egg hatch are all reduced under conditions of high relative humidity, although percentage egg hatch may be reduced by low relative humidity as well (Boudreaux, 1958; Harrison and Smith, 1961). The effects of low relative humidity on active stages may result from increased feeding rates to replace water loss (Crooker, 1985). Conversely, conditions of high relative humidity may interfere with mite feeding by preventing water loss through the cuticle (Boudreaux, 1958), although excretion of water via the digestive system may alleviate this to a certain extent (McEnroe, 1963). Tulisalo (1974) attempted to exploit the adverse effects of high relative humidity as a means of controlling *T. urticae* on greenhouse cucumber by misting the plants periodically, but found that drowning of mites in water droplets was also an important factor which could be enhanced by adding a surfactant to the water.

Host plants can exert profound effects on the biology of spider mites including *T. urticae* (e.g. van de Vrie, McMurtry and Huffaker, 1972; Jeppson, Keifer and Baker, 1975). These effects can be manifested as differences in developmental rate, survival, reproduction and longevity attributable to differences among plant species, among plant varieties or cultivars, or among plants of the same variety but in different phenological stages of growth or subjected to different environmental regimes. Effects can be mediated through leaf surface structure, cuticle thickness, chemical composition, osmotic pressure of the plant cells, microclimate of the plant canopy and, no doubt, other avenues as well (Jesiotr, Suski and Badowska-Czubik, 1979).

The nutritional status of the host plant as influenced by fertilization has long been known to affect the performance of *T. urticae* and much of the research on this topic has been reviewed previously by van de Vrie, McMurtry and Huffaker (1972) and Suski and Badowska (1975). The effects of nitrogen (N) fertilization and the ratio of nitrogen to phosophorus (P) and potassium (K) has been most studied. In general, increased levels of N fertilization improve host quality. Rodriguez *et al.* (1970) associated mite population increases on strawberry plants with increased levels of foliar nitrogen. Wermelinger, Oertli and Delucchi (1985), studying the performance of *T. urticae* on bean and apple leaf discs from plants grown at different levels of N fertilization, found that nitrogen deficiency increased preimaginal development time and preoviposition period, and decreased adult female weight, fecundity and oviposition rate. On apple leaves, they found that a 50% reduction in leaf N was associated with a tenfold decline in fecundity. Hami and Huffaker (1978) similarly reported that raised N fertilization regimes decreased the doubling time of *T. urticae* populations on strawberry plants. In another study, Mellors and Propts (1983) reported that fertilizer composition had no effect on the number of *T. urticae* per radish plant but that the highest population intensities (number of mites/gram plant foliage) were associated with fertilizer regimes involving a high ratio of N to P and K. Thus, the basis for expressing population size can alter the perceived effects of the fertilizer treatments.

Plant water stress also alters the nutritional status of the host plant (Mellors and Propts, 1983) and tends to enhance susceptibility to spider mites (Hollingsworth and Berry, 1982). Feeding by spider mites can exacerbate plant water stress and lead to an accumulation of soluble leaf carbohydrates which may play a part in enhancing mite reproduction (DeAngelis *et al.*, 1982; DeAngelis, Berry and Krantz, 1983a, b). Although it is clear that plant nutrient status affects spider mite population growth, a general lack of consistency among experimental protocols investigating this phenomenon, combined with the confounding effects of differences in host plant species, phenologies and abiotic environmental conditions, has hindered complete understanding of these effects.

Plant phenology can also profoundly affect host-plant suitability for tetranychid mites, with outbreak populations often occurring coincidentally with flowering and fruiting in the host plant (Huffaker, van de Vrie and McMurtry, 1969). Such an association has been observed between densities of *T. urticae* and host phenology on strawberry (Poe, 1971, 1979), corn (Ehler, 1974; Kattes, 1976; Margolies and Kennedy, 1984), soybean (Cadapan, 1976), chrysanthemum (Poe, 1979) and peanut (Boykin, 1983; Margolies and Kennedy,1984). Although these observations suggest a causal relation to host phenology, only studies with corn, peanut and strawberry have clearly demonstrated that mite populations respond directly to the phenological stage of the host crop (Poe, 1979; Boykin, 1983; Margolies and Kennedy, 1984). On both corn and peanut, Margolies and Kennedy (1984) demonstrated that mite populations on plants in the reproductive stage increased significantly faster than populations on the vegetative stages of the same host. The change from vegetative to reproductive growth involves a redistribution of water, sugars, nitrogenous compounds, organic acids and inorganic ions among plant parts, as well as changes in the form of these metabolites (Noggle and Fritz, 1976). Such changes may account for the enhanced suitability of the host during particular phenological stages. Regardless of the specific mechanisms involved, an understanding of the relationship between plant phenology and suitability for spider mite population growth is important to a prediction of the potential for mite outbreaks.

In addition to the host-plant effects discussed above, recent work by Karban and Carey (1984) and Karban (1986) has demonstrated that both mite feeding and purely mechanical injury to cotton seedlings can initiate a plant response which results in a decrease in the suitability of the plant as a host for *T. urticae*. The effect was not specific to *T. urticae* or even to spider mites, but also affected both *Spodoptera exigua* and the plant-pathogenic fungus *Verticillium* spp. (Karban, 1986). The generality of such induced resistance among plant species and its importance in the establishment of mite infestations and subsequent population growth and decline warrants further investigation.

Other constitutive components of plants, including secondary chemistry and structural characteristics such as leaf trichomes, influence their suitability as hosts for *T. urticae* (e.g. DaCosta and Jones, 1971; Patterson *et al.*, 1975; Gould, 1978; Larson and Berry, 1984; Snyder and Carter, 1984; Carter and Snyder, 1985, 1986; Craig *et al.*, 1986). Host suitability for *T. urticae* can be manipulated through plant breeding to achieve increased levels of host-plant resistance. Such resistance may be manifested as alterations in the levels of fecundity, developmental time, mortality and/or age of first reproduction. Each of these parameters may affect r_m to a different degree. Trichilo and Leigh (1985) have suggested the use of life table statistics to assess resistance, because that approach allows one to identify those particular effects of resistance that have greatest impact on population increase, so that they may be selected for specifically in a plant breeding programme. The labour required to do this, however, is substantial and, in many cases, may be prohibitive. As it is population growth rate and plant damage that we wish to minimize when we use host-plant resistance, a reasonable alternative is to base selection for resistance on the magnitude of population change and damage on each plant entry during the course of the selection experiment. In the absence of detailed information on how a particular resistance influences the mites' reproductive capacity, it is risky to select arbitrarily one parameter such as mortality, fecundity or developmental time as the basis for selection. Various components of the resistance may affect various reproductive parameters of the mite differentially, and it is important to select for those resistance components that have the greatest potential for enhancement through selection and the greatest impact on mite population growth rate. DePonti (1985) has recently reviewed the subject of host-plant resistance to spider mites and the reader should consult that work for a thorough treatment of the topic.

Host adaptations

T. urticae populations show a propensity towards adaptation to particular host plants such that, when populations are transplanted to another host, they often require more than one generation to adapt fully to the new plant (Jesiotr, 1976; Dabbour, 1977). When he switched a *T. urticae* population which had been on rose for over two years to bean, Jesiotr (1979) observed an initial reduction in both R_o (number of female progeny born to a female, on average, during her lifetime) and r_m; six generations on bean were required before the population achieved R_o and r_m levels equal to those of the original population on rose. In a subsequent study, Jesiotr (1980) concluded that the relative ability of different *T. urticae* populations to switch hosts, as measured by the magnitude and direction of changes in reproductive rate, was dependent on the relative food quality of the new host compared with the original host, and on the genetic characteristics of the particular mite population. His results demonstrate not only that several generations are required for some populations to adapt fully to a new host, but also that the specific parameters (e.g. fecundity, mortality, etc.) affected may vary among host-adapted populations.

The host plants on which spider mite populations are found can influence the mites' susceptibility to acaricides (van de Vrie, McMurtry and Huffaker, 1972). Saba (1961), for example, demonstrated large differences in the rate of resistance development to TEPP by different host-adapted populations of *T. urticae*, and Gould, Carroll and Futuyma (1982) demonstrated that *T. urticae* populations selected for survival on a mite-resistant cucumber variety had higher levels (7–14%) of tolerance to several pesticides. Mullin and Croft (1983) reported host-dependent changes in the detoxicative enzymes in *T. urticae*, but found no correlation between the ability to colonize a particular host plant and the magnitude of alteration of detoxicative enzymes associated with the colonization of that plant. Where adaptation to particular host plants is associated with increased levels of tolerance to one or more acaricides, associated alterations in the detoxicative enzymes of the mite may be involved. This whole area and its relevance to spider mite management warrants further investigation.

DYNAMICS OF *T. URTICAE* POPULATIONS

T. urticae is well adapted to exploit suitable host plants rapidly and to disperse in search of new hosts when a given host is depleted through over-exploitation by the mites or natural senescence. The *T. urticae* population in Chowan County, North Carolina, for example, over the course of the growing season exploits a temporal sequence of spatially separate hosts which includes overwintering and early spring weed hosts, maize, peanut and late-season weed hosts (Brandenburg and Kennedy, 1982a; Margolies and Kennedy, 1985; Kennedy and Margolies, 1985a). Understanding the dynamics and pest status on various crops in the Chowan County agroecosystems requires a knowledge of this host sequence and the dispersal that occurs among hosts. The dynamics of a subpopulation in a particular field of a particular crop cannot be fully understood by considering only that subpopulation and disregarding the remaining, much larger, portion of the population which occurs outside the field. In other more temporally stable agroecosystems or communities characterized by perennial hosts (e.g. orchards), the impact of dispersal on the local level population dynamics may not be as significant. When defining a mite population for study or for management, it is important to do so in a manner that recognizes the contribution of dispersal to the mite population dynamics.

Dispersal

Dispersal by spider mites has recently been reviewed by Kennedy and Smitley (1985) and can be considered as movement of individuals away from the mite colony in which they developed. Considered in this way, dispersal includes both intraplant movements, leading to colonization of previously uninfested plant parts, and interplant movements, leading to colonization of previously uninfested plants or host patches.

   T. urticae populations are characterized by cycles of initial colonization by a mated female, followed by rapid population growth and localized (e.g. leaf or plant) exploitation with subsequent dispersal to a new resource. Successful exploitation of temporally and spatially transient resources involves the ability to utilize currently available resources fully, as well as the production of large numbers of potential colonizers.

Prereproductive females in a colony have a tendency to disperse to other uncolonized plant parts, whereas ovipositing females have less tendency to emigrate (Hussey and Parr, 1963). As a result, the infestation spreads throughout the plant or patch, generally progressing upward and outward to the periphery (e.g. Silberman, 1983; Wilson *et al.*, 1983; Sites and Cone, 1985). Presumably, this dispersal by prereproductive females prolongs the supply of food available to the established colony, more fully exploits the available colonization sites on the host or in the patch, and spreads the risk of colony extinction resulting from locally adverse abiotic and biotic factors (Kennedy and Smitley, 1985). When a plant or patch has been fully exploited or begins to decline for other reasons, emigration to other hosts is essential and typically involves both crawling and passive dispersal as aerial plankton. Either form of dispersal carries a high risk of mortality. The probability of one or more dispersers from a colony locating a suitable host plant and successfully founding a new colony is related to both the abundance and distribution of suitable hosts within the dispersal range of the mite and the number of dispersers. Mitchell (1970, 1973) and Wrensch and Young (1978, 1983) have identified a number of adaptations which appear to compensate for the high risks inherent in dispersal. The wide host range of *T. urticae* effectively increases the probability of locating a suitable host. Mating prior to dispersal ensures that dispersing females are capable of producing both male and female offspring when they found a colony. The small size of males (relative to females), which do not disperse, combined with the females' ability to delay one-half of their increase in biomass until they have dispersed, enables larger numbers of dispersers to be produced from a limited amount of host resource. This is further enhanced by an increase in the number of female eggs produced and an increase in the mortality rate of males during development as the host leaf deteriorates.

Crawling is a common means of dispersal throughout portions of a host plant (McEnroe and Dronka, 1971). In dense aggregations of host plants including monocultures characteristic of modern agriculture, crawling over intertwined foliage or over the ground is also important in interplant movement (Brandenburg and Kennedy, 1982a; Margolies and Kennedy, 1985). The factors known to stimulate dispersal by crawling and the behaviours involved have been reviewed by Kennedy and Smitley (1985).

Aerial transport of *T. urticae* is also important in dispersal, both within large host patches as well as among host patches. Although dispersal *per se* is passive, in that the mites cannot control where they are carried on the winds, the initiation of aerial dispersal involves a complex behaviour which facilitates being carried aloft by air currents. Prior to take-off, the mites assume a dispersal posture which involves standing facing away from the light source with their forelegs raised upright above their bodies (Smitley and Kennedy, 1985). The posture is manifest by mites only in the presence of both wind and light, and typically occurs after the mites have concentrated on the upper portions or periphery of their host plants as a result of the positive phototactic response described by Suski and Naegele (1963) and McEnroe and Dronka (1971). Under field conditions in maize, mites assuming the dispersal posture are typically concentrated on the stalk, tassel, around leaf margins, and on the ears, where they face downward towards the ground with forelegs uplifted. In this position, they are readily carried aloft on updraughts. Desiccation, host-plant condition and mite population density are involved in conditioning the dispersal response, although desiccation appears to play the greatest part (Hoy, Groot and van de Baan, 1985; Smitley and Kennedy, 1985). In a recent study of aerial dispersal of *T. urticae* from maize under field conditions, Smitley and Kennedy (1987) found that the incidence of aerial dispersal was positively correlated with mite population density and percentage of leaf area infested, but was negatively correlated with incidence of the fungal pathogen *Neozygites floridana* Weiser and Muma in the population, and with the numbers of hours per week with a greater than 90% relative humidity.

Although aerial dispersal of *T. urticae* frequently involves enormous numbers of mites (Brandenburg and Kennedy, 1982a; Boykin and Campbell, 1984; Hoy, Groot and van de Baan, 1985) and long-distance dispersal certainly occurs, gradients of immigrant mite density which are inversely proportional to distance from the source suggest that the majority of aerial dispersal occurs over relatively short distances (100 m or less) (Brandenburg and Kennedy, 1982a; Margolies and Kennedy, 1985; Miller, Croft and Nelson, 1985). This contention is further reinforced by the observation that fertility barriers between local populations of *T. urticae* separated by as little as 5–10 km are common (de Boer, 1982, 1985). The fertility barriers generally result in incomplete reproductive isolation and involve a number of reproductive events ranging from egg production and fertilization to the fertility of hybrid daughters (Young, Wrensch and Kongchuensin, 1985).

A knowledge of the role of dispersal in the wide-area population dynamics of *T. urticae* is essential to understanding those dynamics, as well as predicting and managing mite outbreaks. It is also important in managing acaricide resistance (Hoy, Groot and van de Baan, 1985; Margolies and Kennedy, 1985; Miller, Croft and Nelson, 1985).

Population development

When a dispersing *T. urticae* reaches a suitable host site, she begins to feed and produce webbing. The webbing essentially defines the colony bounds and the webbed area expands with the colony. Eggs are deposited, and larvae and nymphs develop beneath the web canopy. Webbing serves several functions including protection from rain, wind and some species of general predators, and a reduction of competition from weaker webbing species. In addition, dense webbing may offer some protection against acaricide spray droplets (Gerson, 1985).

As discussed previously, in the absence of pesticides the rate of population growth is affected by temperature, relative humidity and suitability of the host plant, as well as the occurrence and level of intra- and interspecific competition and the abundance and type of natural enemies present. An analysis by Carey (1983), of the age distribution of spider mite populations on cotton in the general absence of predators, revealed that increasing populations have a higher proportion of eggs, a lower proportion of immatures and approximately the same proportion of adults as would be expected with a stable age distribution (SAD). He found that the population age structure began to change at *c.* 5000 mites per plant when the population crested. In cresting populations, the proportion of eggs dropped below the SAD, while the proportions of both immatures and adults increased to above the SAD. These directional changes were intensified in crashing populations. By analysing departures from the stable age distribution and knowing that egg predation was rare, Carey (1983) deduced that the population crest resulted from a reduction in fecundity at increased densities, and the population crash resulted from an intensification of reduced fecundity. This conclusion is consistent with the known effects of population density and resource quality on the life history traits of spider mites (Davis, 1952a; Wrensch and Young, 1975, 1978; Wrensch, 1979).

The apparent crash of local populations may be due to intraspecific competition, destruction of the host plant as a result of mite-feeding injury (or other factors), natural senescence of the host plant, the impact of natural enemies, or the onset of adverse abiotic conditions which kill the mites or induce diapause. The effects of intraspecific competition and a rapid decline in host quality may be manifested as both mortality and dispersal. The factors principally responsible for a population decline will vary from place to place at a given time, as well as from time to time at a given place or on a given host species. In a recent field study of *T. urticae* populations on maize, in which predacious arthropods were eliminated through the application of carbaryl, Smitley, Kennedy and Brooks (1986) found that the local populations declined in association with massive aerial dispersal of the mites or with epizootics of the entomogenous pathogen *Neozygites floridana*. The relative importance of these factors varied from year to year, however, depending on environmental conditions.

NATURAL ENEMIES OF *T. URTICAE*

Natural enemies have an important role in the population ecology of spider mites, including *T. urticae*. That role has been the subject of extensive research and a number of excellent reviews are available (Huffaker, van de Vrie and McMurtry, 1969, 1970; McMurtry, Huffaker and van de Vrie, 1970; van de Vrie, McMurtry and Huffaker, 1972; Hoy, 1982a, b, 1985a; Helle and Sabelis, 1985b). Hence, no attempt is made here to provide a review of the relevant literature; rather we attempt to identify those aspects of the *T. urticae*/natural-enemy interactions that are important in understanding *T. urticae* population dynamics and in exploiting the natural enemies for its control.

Pathogenic fungi

Research on the role of natural enemies in the regulation of tetranychid mite populations has focused primarily on predacious arthropods, but in some environments fungal pathogens appear to be important mortality agents. The entomogenous fungus *Neozygites* (= *Entomophthora* = *Triplosporium*) spp. has frequently been reported to infect large percentages of individuals in *T. urticae* populations (Carner and Canerday, 1968; Smith and Furr, 1975; Humber, Moraes and dos Santos, 1981; Brandenburg and Kennedy, 1981, 1982b; Boykin, Campbell and Beute, 1984; Smitley, Kennedy and Brooks, 1986). Epizootics of *Neozygites* can have devastating effects on populations of *T. urticae*, but apparently require sustained periods of high relative humidity or free moisture. Brandenburg and Kennedy (1982b) observed that epizootics of *Neozygites floridana* were preceded by a 2-day period of high relative humidity and relatively cool (<29°C) summer temperatures. In laboratory studies with this pathogen, conidia production was greatest over a range of temperatures from 15°C to 26°C at 100% relative humidity; no conidia were produced at 32°C or at relative humidities of 85% or lower (Smitley, Brooks and Kennedy, 1986). In the same study, epizootics could be induced in a glasshouse by 14-hour periods of approximately 100% relative humidity per day, but not by 30 minutes of simulated rainfall per day. Thus, the impact of *N. floridana* as a natural factor in population dynamics is limited to areas where relative humidity conditions are permissive. In such areas, however, its impact can be significant, provided that it is not eliminated as a factor by fungicides applied to the host plant (Brandenburg and Kennedy, 1983; Boykin, Campbell and Beute, 1984; Smitley, Kennedy and Brooks, 1986). Indeed, in the corn/peanut agroecosystem of Chowan County, North Carolina, where *T. urticae* populations developing on maize early in the season disperse and colonize peanut late in the season, the occurrence of epizootics of *N. floridana* was found to be a major factor in determining the size of the mite populations that develop on maize. In those years when epizootics of this pathogen decimated mite populations on maize before significant aerial dispersal occurred, relatively few mites dispersed and infestations subsequently experienced in peanuts were low (Kennedy and Smitley, 1985; Smitley and Kennedy, 1987).

Predators

Arthropods representing a large number of families are predacious on spider mites and the reader is referred to Helle and Sabelis (1985b) for a detailed review of recent research on this subject. The greatest attention has been focused on phytoseiid predators because a number of them possess attributes which enable them effectively to control spider mite populations at low densities. Most are able to persist at low prey densities and have an extraordinary ability to recover from starvation and to reproduce after finding a new prey patch (Sabelis, 1985b). In general, phytoseiids require little food for maintenance and most is used in egg production. Some species of phytoseiids will consume other species of mites and pollen when spider mites are scarce or absent (Overmeer, 1985). The rate of prey consumption is linearly related to oviposition, except at very low prey densities (Sabelis, 1985c). Thus, phytoseiids manifest a strong numerical response to prey density. This is extremely important, as phytoseiids do not consume large numbers of prey and the functional response, although well defined in these mites, is not, in itself, adequate to effect control of prey in the absence of mass releases (Sabelis, 1985d). Phytoseiids, like their prey, are exploiters of transient resources and have been subjected to strong selection for high r_m and dispersal ability (Sabelis, 1985b). Some species disperse as aerial plankton and manifest a dispersal behaviour very similar to that of *T. urticae* (Johnson and Croft, 1976, 1981; Hoy, 1982b). Hoy, Groot and van de Baan (1985) estimated that 100–150 million predacious mites dispersed aerially from an 18 ha almond orchard during a single season. This well-developed ability to disperse enables phytoseiids to locate isolated patches of spider mites even though they apparently lack the ability to home in on mite colonies from a distance (Sabelis and Dicke, 1985). The relative dispersal capabilities of the predator are extremely important in the dynamics of predator/prey systems. It is especially important in ephemeral habitats such as annual crops, which must be recolonized each year. This feature explains why most successful biological control programmes involving phytoseiids are in stable perennial crops, or in glasshouse crops where the predators are introduced in the appropriate numbers (Logan, 1982; Hussey and Scopes, 1985a).

Three of the phytoseiid species (*Phytoseiulus persimilis, Metaseuilus occidentalis* and *Amblyseius fallacis*), which are most effective as predators in biological (integrated) control programmes all have high powers of dispersal, a close correlation in distribution with their primary prey species, a high reproductive potential, and at least a moderate degree of specificity for tetranychid mites; they vary in both voracity and ability to survive at low prey density (McMurtry, 1982). Efforts to model predator/prey interactions involving phytoseiids and *T. urticae* have generally shown that the most important factors in determining the dynamics of the predator/prey interactions are the abundance of prey and rate of increase of the prey population, the predator's functional response, and the predator's numerical response (Logan, 1982; Shaw, 1984, 1985; Bernstein, 1985).

Rabbinge and Hoy (1980) utilized a population model to identify critical factors in the control of *T. urticae* through release of *M. occidentalis*. Sensitivity analyses with the model showed the time of release of the predator to be of critical importance. The ability of the predator to control the prey was less sensitive to predator/prey ratio, the frequency of predator releases, or differences in dispersal rate of the predator (over the limited ranges tested). Models of this type can provide guidance for maximizing the impact of phytoseiid predators and the efficiency of spider mite management programmes using phytoseiid predators.

The greatest progress in the use of predators for biological control of *T. urticae* involves the use of *Phytoseiulus persimilis* on greenhouse cucumbers. *P. persimilis* is used for *T. urticae* control on a commercial scale (McMurtry, 1982) on approximately 60% of the greenhouse cucumber acreage in the

Netherlands, 75% in England and 70–75% in Finland, Sweden and Denmark. These programmes are quite complex and are based on a thorough knowledge of the biology and life history of both *T. urticae* and *P. persimilis*, as well as an accurate knowledge of how much spider mite damage can be tolerated before yields are affected (Hussey and Scopes, 1985a, b).

Maintaining the correct predator/prey balance is crucial to successful control. This is often accomplished by releasing *T. urticae* and *P. persimilis* to ensure the correct ratio and distribution throughout the glasshouse. This is essential if the predator is to maintain control during the period when post-diapause *T. urticae* are emerging and invading the crop in the spring (Hussey and Scopes, 1985a). Recently, Havelka and Kindlmann (1984) have used simulation models to define the optimum release rates for both *T. urticae* and *P. persimilis*. The optimum release rates they identified involved a 20-fold decrease in the quantity of *T. urticae* and a three- to fourfold decrease in the quantity of *P. persimilis* released, relative to the generally used release rates.

There are several reasons why biological control of *T. urticae* has been so successful in glasshouse cucumbers: these include the controlled conditions and the limited number of pest species characteristic of the glasshouse environment; the confinement of the predators to the crop; the high-value labour-intensive nature of the crops, and the high cost of chemical controls, combined with the tendency of glasshouse populations of *T. urticae* to develop resistance to acaricides (McMurtry, 1982). Despite the general lack of these characteristics in crops grown outside, phytoseiids and other predators have important roles in integrated control programmes. A number of such programmes are reviewed in Helle and Sabelis (1985b).

One such programme is that on almonds in California which utilizes the release of *M. occidentalis* selected in the laboratory for resistance to insecticides commonly used in almond orchards and the application of reduced rates of certain acaricides, if needed, to establish the appropriate predator/prey ratio. This programme is widely used among almond growers in California and results in a cost savings to growers of US $24–44 per acre annually over conventional control programmes (Roush and Hoy, 1981a, b; Hoy, 1982b, 1985b; Hoy, Groot and van de Baan, 1985; Headley and Hoy, 1986). In other tree fruit crops, phytoseiid predators resistant to organophosphate insecticides, as a result of selection during their long period of use, now occur naturally and are important components of pest management programmes on these crops (Croft, 1977; Asquith *et al.*, 1980).

Agricultural considerations

Effective management of the twospotted spider mite requires a knowledge of those factors directly influencing its abundance. Modern agriculture is characterized by numerous variables that fluctuate from year to year and farm to farm. The literature discussing various agricultural production aspects (i.e. fertility, host plants, pesticide use, etc.) as they affect *T. urticae* is extensive, but few workers have integrated this information into the total agricultural production setting. Jeppson, Keifer and Baker (1975) briefly focused on several agricultural practices that influence spider mite ecology, but the subject has been largely addressed in individual areas of concern. However, there is a great need to take a wide-area view of the agroecosystem as it relates to many crop pests, spider mites included (Rabb, 1978). The success of IPM programmes for *T. urticae* also requires that management programmes should not be restricted by commodity orientation as spider mites are spatially dynamic.

Although pest characteristics may modify agricultural production schemes, most pest problems are a reflection of the system they exploit. Numerous factors act upon spider mites in the field, and several of these are subject to manipulation by humans. For the purpose of this review, four major areas of consideration in agricultural production will be addressed: pesticide interactions; cropping patterns, cultural and production practices; crop damage, economic thresholds and sampling; and pesticide resistance. These areas are not intended to be all-inclusive, but rather represent areas in which significant contributions have been made which, upon integration, can significantly enhance our mite-management abilities. The area of biological control has been the subject of extensive research but is not addressed in this review; several reviews, including those of Helle and Sabelis (1985a, b), have thoroughly covered this important area.

PESTICIDE INTERACTIONS

It could reasonably be argued that pesticides play the most significant part in regulating arthropod populations in most agroecosystems; this is certainly true for those factors over which humans have some influence. At the time of their review, van de Vrie, McMurtry and Huffaker (1972) found over 60 studies of pesticide effects on the development of the twospotted spider mite, and hundreds more on other phytophagous mites. The concern over pesticides and their interactions in agriculture has become an area of great interest to entomologists, acarologists and environmentalists. Interactions associated with various pesticide groups and their implications are of great relevance to pest management programmes. Although the last 30 years have seen a great increase in the use and types of insecticides, crop damage has not significantly decreased (Pimentel *et al.*, 1978). Now, however, the worldwide acceptance of integrated pest management philosophies and an increased awareness of environmental consequences of pesticide use is resulting in a more judicious use of pesticides. In addition, increased costs, development of resistance and resurgence of pests has further promoted more selective use of pesticides (Metcalf, 1980; National Academy of Sciences, 1986).

Even though many broad-spectrum and preventative pesticide treatments have been eliminated in current IPM programmes, the application of only a minimum number of pesticides can still significantly affect twospotted spider mite population dynamics. Despite extensive efforts to eliminate the pesticide treadmill scenario, spider mites continue to be an important pest.

Outbreaks of the twospotted spider mite are associated with the use of various pesticides (Jeppson, Keifer and Baker, 1975). Historically, scientists and agriculturists agreed that these outbreaks were in response to a release from predation. Predators are capable of effectively suppressing spider mite populations in a wide variety of crops. Van de Vrie, McMurtry and Huffaker (1972) suggested at least six effects of pesticides responsible for tetranychid mite population increases: (1) effect on behaviour; (2) direct influence on reproduction; (3) change in plant nutritional quality or physiology; (4) inert residues increasing leaf suitability; (5) release from competition (both intra- and interspecific), and (6) effect on sex ratio and copulation frequency. Three major areas that encompass most of these effects have received particular attention. These include: (1) suppression of natural enemies (Huffaker, van de Vrie and McMurtry, 1970; (2) insecticides stimulating mite dispersal (Huffaker, van de Vrie and McMurtry, 1969), and (3) direct and indirect stimulatory effects (Chaboussou, 1966; Luckey, 1968). These are addressed separately for purposes of review but probably operate simultaneously in complex interactions in the field.

Impact of pesticides on natural enemies of T. urticae

Prior to the 1950s, the twospotted spider mite was rarely a serious pest. The most obvious regulating factor was an abundance of natural enemies important in suppressing spider mite populations (McMurtry, Huffaker and van de Vrie, 1970). During the 1950s and 1960s several accounts began to document a relationship between pesticide use and spider mite outbreaks (Clancy and Pollard, 1952; Davis, 1952b; Klostermeyer and Rasmussen, 1953; Klostermeyer, 1959; Attiah and Boudreaux, 1964a, b; Saini and Cutkomp, 1966). Although the impact on natural enemies was often measured qualitatively rather than quantitatively, the disruptive effects of insecticides were apparent. Early studies focusing on the relationship of pesticide use and spider mite outbreaks were reviewed by Unwin (1971). In addition, Bartlett (1968) listed 59 insecticides and fungicides as being associated with increases in mite or aphid populations. During this time the use of pesticides increased, and more compounds were implicated in spider mite outbreaks on crops that had traditionally suffered few such outbreaks (Leigh, 1985). Spider mite outbreaks soon became closely associated with the use of synthetic organic pesticides (e.g. Duncombe, 1972; Ghobrial and Dittrich, 1980; Smith Meyer, 1981). The effects of pesticides, in general, on natural enemies and subsequent arthropod pest responses has been well documented for a variety of crops (Croft and Brown, 1975; Newsom, 1972). The widespread occurrence of such interactions contributed to the move towards integrated control and selectivity in pesticide use. With the development of the integrated pest management concept in the 1970s, scientists working with spider mites began to consider more carefully the effects of pesticide application on natural enemies.

Our ability to manage spider mites effectively has been further challenged by the development of synthetic pyrethroids. These compounds, in general, are highly toxic to predators, such as phytoseiid mites (Roush and Hoy, 1978; Wong and Chapman, 1979; Aliniazee and Cranham, 1980; Hull and Starner, 1983; Riedl and Hoying, 1983); at the same time, many of these compounds are relatively poor acaricides against pest species (Croft and Hoyt, 1978; Hoyt, Westigard and Burts, 1978; Chapman and Penman, 1979; Hall, 1979; Hoy *et al.*, 1979; Riedl and Hoying, 1980). Studies in California found two strains of *T. urticae* 20–40 times more tolerant to permethrin than three strains of the predacious phytoseiid mite *Metaseuilus occidentalis* (Roush and Hoy, 1978). Synthetic phrethroid residues, even if they do not actually kill them, exhibit significant repellency of the predacious mites *Amblyseius fallacis* Garman and *Typhlodromus occidentalis* (Penman, Chapman and Jenson, 1981; Reidl and Hoying, 1983) either by contact with the residue or by feeding on treated spider mite eggs. Such repellency to predators has also been observed with organophosphate insecticides (Hislop *et al.*, 1981).

As a result, the use of synthetic pyrethroids in various agroecosystems has great potential for disrupting current pest management programmes. This is especially true in orchard systems where Riedl and Hoying (1980) warned of a return to exclusive reliance on pesticides such as that which occurred when organophosphate insecticides were first introduced. Fortunately, many current studies are more closely evaluating new insecticides for their impact on arthropod predators and the potential to induce secondary pest outbreaks (Riedl and Hoying, 1980; Grafton-Cardwell and Hoy, 1983; Duso and Liquori, 1984; Osman, Abo-korah and Ghattas, 1985). Hoy *et al.* (1979) found the use of permethrin in a grape pest management programme in California unacceptable because of its toxicity to several important predators. To minimize adverse effects on predators, the use of reduced rates in the field may offer some promise (Roush and Hoy, 1978). Other studies indicate that synthetic pyrethroids might be incorporated into orchard IPM programmes through techniques to reduce their impact on natural enemies. Increased spray intervals, and combining an organophosphate insecticide with a low rate of a pyrethroid, are two suggestions recommended for Pennsylvania apple growers (Hull and Starner, 1983).

Such studies are not restricted to synthetic pyrethroids. Insecticidal soap can be used in a single application to reduce spider mite populations without eliminating predators (Osborne and Petitt, 1985). Multiple applications, however, while maintaining spider mite abundance at low levels, would virtually eliminate the predator population. Studies in New Zealand on commercial raspberry gardens found several common insecticides (carbaryl, azinphosmethyl and *Bacillus thuringiensis*) compatible with integrated control programmes, whereas methomyl resulted in outbreaks of twospotted spider mite (Charles, Collyer and White, 1985).

Avermectin B_1 is a relatively new insecticide that is highly toxic to *T. urticae* and several other mites; it is a mycelial extract of *Streptomycetes avermitilis*, and probably inhibits signal registration at neuromuscular junctions (Putter *et al.*, 1981). This compound is less toxic to the predacious mite *M. occidentalis* (Nesbitt) than to *T. urticae* (Grafton-Caldwell and Hoy, 1983). Proposed field rates for avermectin B_1, however, would appear to be quite toxic to *M. occidentalis*. Combinations of a light oil and avermectin B_1 might enhance predator survival, but this is also dependent on the survival of some prey (Hoy and Cave, 1985). Reduced rate trials in the field will be necessary for further evaluation.

These discussions have focused on pesticides applied to foliage and the consequences of direct contact with the pesticide during application or with its residues. The use of soil insecticides was also associated with spider mite outbreaks as early as 1953 (Klostermeyer and Rasmussen, 1953). Spider mite outbreaks in maize are commonly associated with the use of certain soil-applied insecticides (Schweissing, 1973; Lesiewicz, 1981; Margolies, Kennedy and van Duyn, 1985). Although a reduction in natural enemies may result from this treatment (Lesiewicz, 1981) other factors to be discussed later may also operate.

Fungicides also have an impact on spider mite abundance, although studies evaluating these interactions have been limited. McMurtry, Huffaker and van de Vrie (1970) reviewed the early studies of fungicides and their impact on predators. Bartlett (1968) listed several fungicides associated with increases in mite or aphid infestations, and Unwin (1971) listed several fungicides reportedly toxic to mite predators. Use of certain fungicides has resulted in *T. urticae* outbreaks on peanuts in North Carolina (Campbell, 1978). The fungicide benomyl, however, also has a low-level ovicidal activity against the twospotted spider mite (Spadafora and Lindquist, 1972; Campbell *et al.*, 1974). More recently, Fungo and Curry (1983) found five fungicides (benomyl, carbendazim, mancozeb, quinomethioate and zineb) to be particularly toxic to adult females and eggs, whereas dinocap was highly toxic only to adult females. However, results of laboratory excised leaf studies such as these do not always correlate with results of field studies (Boykin, Campbell and Beute, 1984), indicating that other factors are involved. In a field study, Rishi and Rather (1983) reported that benomyl and captan were toxic to *T. urticae* and less toxic to the predatory mite *Amblyseius finlandicus* on apple.

Although most discussions of the off-target effects of pesticide use generally concentrate on the destruction of predacious arthropods, fungicides can have effects on other natural enemies such as fungal pathogens. The pathogen *Neozygites floridana* (Weiser and Muna) has a significant role in spider mite population dynamics throughout the south-eastern United States (Carner and Canerday, 1970; Carner, 1976; Campbell, 1978; Brandenburg and Kennedy, 1982b; Smitley, Kennedy and Brooks, 1986). Campbell (1978) first offered evidence that certain pesticides were affecting *N. floridana*. Subsequent studies documented that several fungicides affected either conidial germination or growth of the fungus and contributed to spider mite outbreaks (Brandenburg and Kennedy, 1983; Boykin, Campbell and Beute, 1984; Smitley, Kennedy and Brooks, 1986).

The frequency with which such pesticide interactions are observed in association with spider mite outbreaks cautions us to proceed slowly when incorporating new pesticides into management programmes. Although laboratory studies can assist in revealing potential interactions, field studies are necessary for final evaluation. McMurtry, Huffaker and van de Vrie (1970) warned that laboratory and small-plot trials often produce different results. Field conditions are usually more stressful for the mites (Moriarty, 1969), and various environmental and cropping practices can significantly influence a particular pesticide and its effect. Pyrethroids, for example, are relatively less effective at higher temperatures (Harris and Kinoshita, 1977). Even host plants can influence the susceptibility of spider mites to a pesticide (Neiswander, Rodriguez and Neiswander, 1950).

Our discussion has focused on spider mite outbreaks resulting from pesticides affecting natural enemies of *T. urticae*. However, as early as 1950, Huffaker and Spitzer (1950) determined that DDT caused spider mite outbreaks under conditions that ruled out natural enemy depletion. Klostermeyer and Rasmussen (1953) and Boudreaux (1963) also noted rapid increases in mite populations that could not be explained by lack of predators. These studies provide ample evidence that, while spider mite outbreaks are commonplace, they involve more than simply a reduction in natural enemy activity (Hall, 1979; Boykin and Campbell, 1982; Boykin, Campbell and Beute, 1984). The impact of pesticides on spider mite reproduction, behaviour and host-plant quality has been well established and indicates that spider mite outbreaks following pesticide use are often the result of complex interactions. Many of the early studies on these sublethal effects were reviewed by Moriarty (1969).

The influence of pesticides on spider mite behaviour

One explanation for population increases of spider mites has been pesticide-induced mite dispersal (Huffaker, van de Vrie and McMurtry, 1969). Davis (1952b) first demonstrated that DDT accelerated dispersal of spider mites reared on banana squash. A similar irritant or repellent effect of DDT was observed in the European red mite, *Panonychus ulmi* (Koch) (Pielou, 1960). Later studies, however, were unable consistently to demonstrate enhanced dispersal, and speculated that a reduction in natural enemies aided dispersal (Putman, 1963; Attiah and Boudreaux, 1964a, b).

Studies with synthetic pyrethroids have more consistently demonstrated behavioural responses in *T. urticae*. Hall (1979) observed spider mite outbreaks (*P. ulmi* (Koch)) in the absence of predators. Laboratory tests on *T. urticae* subjected to fenvalerate residues on lima bean leaf discs showed changes in feeding behaviour. Oviposition site selection was also influenced by fenvalerate and permethrin residues on soybean, *Glycine max* (L.) Merrill (Donahue, McPherson and Poe, 1985).

Dispersal appears to be affected most consistently and significantly. Fenvalerate produced a strong repellent effect on twospotted spider mites exposed to residues on leaf discs (Penman, Chapman and Jenson, 1981) as well as an acaricidal effect on phytoseiid mites. Those spider mites remaining on leaf discs showed significantly inhibited oviposition. Whole-plant studies on bean plants produced 'spin-down' and walking away from treated leaf surfaces (Penman and Chapman, 1983) as well as aerial dispersal behaviour (D. C. Margolies and G. G. Kennedy, unpublished work). In addition, studies by Iftner (1982) and Iftner and Hall (1983) demonstrated that the synthetic pyrethroids, fenvalerate and permethrin, produced a repellency and a tendency for mites to move from areas of high synthetic pyrethroid residues to areas of lower (or no) residues. Disruption of feeding and changes in habitat selection and probing behaviour also occurred.

Several organophosphate and carbamate pesticides do not appear to increase spider mite dispersal (Boykin, Campbell and Nelson, 1983; Penman and Chapman, 1983). Other compounds may cause dispersal responses similar to those caused by the synthetic pyrethroids. Franklin and Knowles (1984) reported increased 'walk-off' and 'spin-down' for *T. urticae* on foliage treated with formamidines. It is even possible that other groups of pesticides, such as herbicides, could play an important part in spider mite population increases through repellency (Boller, Janser and Potter, 1985).

The implications of pesticide-induced spider mite dispersal may be very significant. As synthetic pyrethroids are applied to crops at low field rates, it is possible that these sublethal effects frequently occur in the field. Pyrethroids induce mites to seek out areas of lower residues (Penman and Chapman, 1983). The resulting dispersal of spider mites from population foci will cause the populations to become spread out and freed from the constraints imposed by intraspecific competition. Because pyrethroid applications are lethal or repellent to many predators, the widely scattered spider mites can develop colonies without the suppressive effect of natural enemies. The effective use of synthetic pyrethroids in situations where spider mites are present requires more careful attention to application to attain thorough coverage. However, even the most efficient spray application equipment cannot provide perfectly uniform spray deposition. Pesticide-induced dispersal can also significantly alter mite-sampling plans and decision-making confidence (Trumble, 1985).

Trophobiosis and hormoligosis

Spider mite outbreaks resulting from stimulation of spider mite fecundity following the application of pesticides has been widely investigated. Two possible explanations include the direct stimulation of mite fecundity by altering the mite's physiology, a phenomenon referred to as hormoligosis by Townsend and Luckey (1960) and Luckey (1968), or an altering of the host-plant's physiology, termed trophobiosis by Chaboussou (1966). Regardless of the mechanism, an increase in the fecundity of the twospotted spider mite can cause dramatic population increases which exceed the compensatory capabilities of predators.

Most early studies on stimulation of mite fecundity have not sought to separate hormoligosis or trophobiosis as the mechanism involved, but rather evaluated only resultant changes in fecundity due to pesticide treatments or residues. Although changes in mite fecundity resulting from exposure to pesticides is not a new concept, inconsistent spider mite responses following exposure to pesticides plagued studies in the 1960s (Bartlett, 1968).

Changes in host physiology and suitability and associated increases in spider mite populations were reported in the early 1950s (Rodriguez, 1951, 1954). At this same time, Klostermeyer and Rasmussen (1953) found marked increases in spider mite abundance after soil application of DDT. There was also a significant linear increase in mite numbers with increasing amounts of technical DDT applied to the soil. At high dosages, growth of the host plants (potatoes) was affected by the insecticide, suggesting that the increase in mite populations might be due to the effects of the insecticide on plant nutrition and composition. Similar results (apparent trophobiosis) were obtained by Rodriguez, Maynard and Smith (1960) and Rodriguez, Chen and Smith (1960). Rodriguez examined the effects of soil insecticides on apple trees by measuring changes in the plant sugars and resultant effects on mite nutrition. Saini and Cutkomp (1966) observed similar effects on spider mites when green beans (*Phaseolus vulgaris*) were treated with DDT.

In more recent studies, Hoyt, Westigard and Burts (1979) and Hall (1979) speculated that changes in plant physiology following the application of synthetic pyrethroids in pear and apple orchards might, in part, explain spider mite population increases. These studies also suggested a rate response in association with the use of synthetic pyrethroids. The same phenomenon was observed on cotton with a significant interaction between the plant and the pesticides (Leigh and Wynholds, 1980). From studies using organophosphate insecticides, they concluded that spider mite outbreaks in cotton are caused not only by a reduction in natural enemies but also by stimulated reproduction of spider mites early in their adult life. This stimulation appears to be the result of a physiological change in the plant and an increase in its suitability as a host.

In field plots, the application of methyl parathion or one of its breakdown products, phosphoric acid, significantly increased the number of eggs during the first four days and on the second and third days, respectively, compared with untreated plants (Maggi and Leigh, 1983). This response was probably attributable to increased phosphorus levels available to the host plant, although such responses might be reduced when alternative high-level sources of phosphorus, such as fertilizers, are available. The increase measured in this study, although small, could be quite important if it persisted over a growing season.

Several commonly used pesticides appear to influence spider mite fecundity, development and/or longevity directly. This effect of hormoligosis was defined by Luckey (1968) as the phenomenon in which subharmful quantities of many stress agents may be helpful when presented to organisms in suboptimal environments. Bartlett (1968) speculated that such an effect might be occurring in studies where the impact of natural enemies was eliminated. The phenomenon resurfaced with the study by Dittrich, Streibart and Bathe (1974) on mite stimulation by DDT and carbaryl. Female *T. urticae* kept on residues of carbaryl and DDT showed significantly higher egg production than their untreated controls. The sex ratio of the F_1 generation was shifted toward females, and F_1 females produced more eggs than their untreated counterparts. As mites were shifted to new bean leaves every 4 days, the effect probably came as direct stimulation of the mite rather than influence through the plant. Their success in obtaining a mite response to treatment when other studies experienced inconsistent results was probably due to dosage selection based upon Luckey's calculations (Luckey, 1968). The biochemical target of such a stressor in the mite remains unresolved.

In studies which do not isolate the mechanism—hormoligosis or trophobiosis—one or both may be responsible for the final effect. Boykin and Campbell (1982) examined seven insecticide, fungicide and insecticide–fungicide combination treatments on field-treated peanut foliage. Peanut leaflets were excised and infested with *T. urticae* in the laboratory. Several treatments affected mite development time, survival, longevity and fecundity. When all factors were considered, the intrinsic rates of increase for mites exposed to the various treatments were not significantly different. However, the trends reported are similar to those in work done by Campbell (1978), and the authors indicated that, while increases in the intrinsic rate of increase were fairly small, such increases together with a decrease in other limiting factors might cause significant increases in mite populations (Boykin and Campbell, 1982). The excision of the peanut leaves may have also resulted in physiological changes that mask any trophobiotic effect.

Iftner and Hall (1984) found a reduction in *T. urticae* fecundity on fenvalerate-treated lima beans but saw an increase when mites reared on treated beans were transferred to untreated leaf discs. These treatments also shortened the development time by 1–2 days. As fenvalerate has been demonstrated to induce dispersal, it is quite possible that mites could be exposed to the effects of the pesticide on treated foliage and then disperse to nearby untreated foliage where the impact of the treatment results in spider mite outbreaks. Most studies have evaluated these pesticide effects under static or narrow environmental regimes. Undoubtedly, the application of these materials in an agricultural setting occurs under a wide range of environmental influences, host-plant responses and pesticide application techniques. McMurtry, Huffaker and van de Vrie (1970) warned that results in the laboratory may not provide a clear prediction of field results. Environmental conditions can significantly influence the organism's response to pesticide exposure. Avermectin B_1 results in accelerated egg development at lower temperatures and increased adult mortality at higher temperatures (El-Banhawy and Anderson, 1985). Similarly, Ibrahim and Knowles (1986) found that formamidines either inhibited or enhanced fecundity and egg hatch of *T. urticae*, depending upon the compound, concentration and post-treatment interval.

Spider mite population increases following the use of pesticides, especially synthetic pyrethroids, may be the result of complex interactions. The traditional explanations of natural enemy destruction and low toxicity against spider mites still hold true. However, these effects are more readily measured and quantified than are some of the direct and indirect effects on dispersal, host plants and population dynamics. These sublethal effects may be more pronounced when evaluated under actual field conditions (Moriarty, 1969) yet even more difficult to measure. The multi-interactive effects of pesticides, including groups other than insecticides, prompts a critical evaluation of each compound before incorporation into any cropping system. Studies need to evaluate the impact of pesticides under the variety of conditions frequently observed in each production scheme, including full regimes of other pesticides likely to be applied to the crop. Such studies, despite the difficulties involved, are essential to develop effective spider mite management strategies where pesticides are employed.

CROPPING PATTERNS, CULTURAL AND PRODUCTION PRACTICES

The pest status of spider mites is unquestionably related to various agricultural practices. Virtually all aspects of crop production can influence spider mite population dynamics (Huffaker, van de Vrie and McMurtry, 1970). In general, the improved care of crops has contributed to the frequency of spider mite outbreaks. Quality of host plants, sources of alternate hosts, soil–mineral–water relations, biochemical composition of leaves and, as previously reviewed, pesticide effects, all are agricultural parameters that influence spider mite outbreaks (van de Vrie, McMurtry and Huffaker, 1972). Jeppson, Keifer and Baker (1975) listed those agricultural practices and influences on spider mites as changes in plant vigour, new varieties, absence of predators, pesticides, presence of inert dusts and changes in soil organisms.

Previous studies and reviews provide sufficient documentation to caution us from proceeding with new agricultural production schemes without first evaluating their impact on spider mite populations. Agriculture is a dynamic system and will continue to change as the need for more efficient food production increases. Whereas older studies shed light on spider mite responses to cropping practices, more recent studies can provide insight into changes in spider mite pest status as new practices are employed.

New varieties and increased plant vigour are an integral part of modern agricultural production. While Jeppson, Keifer and Baker (1975) listed these as factors contributing to spider mite outbreaks, subsequent studies have more clearly defined this relationship.

Wrensch and Young (1978) found that host quality, in general, contributed to mite population increases. Good leaf quality during oviposition and development of offspring produced relatively faster developmental rates and higher survival of offspring. Host quality in response to fertilization also has a significant role in population dynamics.

The 'quality' of a host plant can be related to plant nutrition as discussed previously. However, it does not appear that such modification of fertility regimes to alter plant nutritional quality could be successfully used for mite management: to do so would require extreme plant nutritional regimes that would not be economically feasible.

As discussed earlier, the phenological stage of plant growth also influences suitability as a host for spider mite. This can be exploited in some cases through cultural practices such as delayed planting or use of late-maturing varieties to reduce the threat of spider mite attack (Schweissing, 1969, 1973;

Kattes, 1976); however, these practices often increase the risk of damage by early frost.

Irrigation has been evaluated as an alternative control measure, but the results have been variable. Kattes (1976) observed a significant reduction in *Oligonychus pratensis* in irrigated portions of maize on the Texas High Plains. Chandler *et al.* (1979) concluded that significant rainfall and/or irrigation can limit the development of spider mite populations (*O. pratensis* and *T. cinnabarinus*) on corn. *T. urticae* reared on greenhouse-grown peanut leaves exhibited significantly greater dispersal if leaves received seep irrigation and were not periodically washed with overhead irrigation (Boykin, Campbell and Nelson, 1983): this is probably the result of reduced leaf washing and an increase in accumulation of inert materials or dust on the leaves and, in the field, an accumulation of pesticide residue. Such is the case in almond production in California: improper irrigation resulted in excessive dust and accentuated feeding damage (Hoy, 1985b). Wilson *et al.* (1984) and Zalom *et al.* (1984a, b) reported that sprinkler irrigation may wash mites from lower portions of the trees.

Irrigation in almond orchards is usually withheld in midsummer to promote rapid hull drying and splitting as an aid to mechanical harvest. Such practices contribute greatly to spider mite outbreaks in almonds: not only is the washing effect eliminated, but the water-stressed foliage may also serve as a more nutrient-rich host (Hoy, 1985b). Similarly, in citrus, irrigation reduces the impact of mite feeding on trees. Overhead irrigation may significantly increase the relative humidity to levels conducive to epizootics of fungi pathogenic on tetranychid mites (Muma, 1970). Irrigation may also provide some dust control in the orchard.

While irrigation may create an environment less suitable for spider mites, it may also create appropriate conditions for mite predators. Wysoki (1985) reported that regular irrigation in strawberry fields in South Africa created favourable conditions for spiders to invade plants. Irrigation also produces indirect effects through differences in host quality, but these effects have been conflicting (e.g. Hollingsworth and Berry, 1982; Mellors and Propts, 1983). Oatman and Voth (1972) reported little effect of flooding following a thunderstorm on *T. urticae* or a predator *Phytoseiulus persimilis* Athias-Henriot on strawberries in southern California.

Numerous farm-management practices can drastically influence spider mite populations. Weed management in apple orchards in Michigan influences both pest (*T. urticae*) and predator (*A. fallacis*). Manipulation of the ground cover around the trees increased pest dispersal into the trees and reduced predator abundance (Croft and McGroarty, 1973). Huang (1978), as cited by McMurtry (1985), suggested that a cover crop of the weed *Ageratum conyzoides* L. would aid in the control of *Panonychus citri* on citrus by producing a pollen food source for the phytoseiid *Amblyseius neurami* and increasing the relative humidity, thereby favouring the predator. Similarly, the presence of grass around strawberry fields in South Africa encourages colonization by predacious spiders which aid in spider mite control (Wysoki, 1985). In European hop production, phytoseiid abundance is limited because of intensive weed control within the hop garden (Cranham, 1985). Disturbing weed borders by mowing in North Carolina increased the number of aerially dispersing spider mites (*T. urticae*) and increased movement into peanut fields (Boykin, Campbell and Nelson, 1984). Use of a herbicide on border areas also increased dispersal into peanut fields.

Many outbreaks in economic crops result from spider mite infestations in other non-crop sources. Infestations in cotton, for example, frequently originate from nearby field margins, dooryard ornamentals, home gardens, adjacent perennials or early season crops (Leigh, 1985). Brandenburg and Kennedy (1982a) found a relationship between infestations of *T. urticae* in field border vegetation and two crops (maize and peanuts) in North Carolina. *T. urticae* moved from field borders into corn early in the season and then used corn as a nursery crop to disperse aerially into adjacent peanut fields.

Crop rotation patterns and control of mites on hosts serving as sources of infestation have only limited current use in spider mite management. The broad host range of *T. urticae*, including both crop and non-crop plants (Jeppson, Keifer and Baker, 1975), creates many potential sources of infestation for crops. Manipulation of these sources of infestation has received some attention but is yet to be utilized as an additional management tool (Croft and McGroarty, 1973; Brandenburg and Kennedy, 1982a; Boykin, Campbell and Nelson, 1984). Strip cropping and adding habitat diversity (e.g. small corn patches) has reduced spider mite build-up and encouraged phytoseiid and stigmaeid predatory mite populations on cotton in Egypt (Elbadry, 1979). Irrigation, tillage and plant populations have also been observed to affect mite development.

Some cultural practices specific to a particular crop may also have a regulatory effect on *T. urticae* abundance. Peppermint production depends on the use of several unique practices: fall ploughing of fields is used in central Oregon to improve rhizome distribution and reduce winter frost injury; producers in western Oregon practice fall (autumn) flaming to reduce the spread of verticillium wilt (Hollingsworth and Berry, 1982). Ploughing killed mites and delayed population build-up but represented only half the cost of an acaricide and had little effect on subsequent peppermint growth; fall flaming also delayed population increases by reducing the overwintering mite population. In another crop, strawberry, pruning and removal of non-functioning leaves can reduce twospotted spider mite population by over 50% (Oatman *et al.*, 1967). Although these practices are used for agronomic reasons, in fact they provide an additional benefit by reducing the need for acaricide applications.

Crop rotations can actually enhance the potential for *T. urticae* outbreaks. Brandenburg and Kennedy (1982a) observed that *T. urticae* dispersed from peanut to overwinter on vegetation along field borders. Rotation schemes provide for maize, an excellent early season host, to be planted in that field the next spring.

Perhaps the most successful work involving the use of cultural practices has occurred in greenhouses. The environment of the greenhouse can be closely controlled and utilized in creating an environment that is less than optimal for spider mite outbreaks. Sanitation in the greenhouse is the first step (Hussey and Scopes, 1985a). In addition, because high humidity and moisture reduce spider mite outbreaks, programmes utilizing a misted water spray on greenhouse crops have shown potential (Tulisalo, 1974). Another area of potential involves the manipulation of diapause by controlling day-length in the greenhouse. Hussey (1972) found that a 2 h break in the middle of the scotophase, to reduce the scotophase to less than 8 h, reduced the incidence of diapause to 12%. By using artificial light in the greenhouse to maintain at least a 16 h photophase, the number of mites entering diapause could be reduced. Thus, when plants were removed at the end of their production, the remaining feeding mites would soon die, so the next crop would not be infested.

Crop resistance has long been viewed as an attractive arthropod management tool, because resistant varieties are easy to incorporate into current production schemes, are generally compatible with other management practices and are relatively cost effective. Even moderate levels of resistance can delay a pest outbreak and eliminate or reduce the need for pesticides.

Resistance to spider mites is present, to varying degrees, in many crops (dePonti, 1977). A thorough review concerning the current status of host-plant resistance to spider mites indicates that much potential exists for expansion of work in this area (dePonti, 1985). Breeding for resistance has been limited to a relatively few crops including cassava, cotton, strawberry, cucumber and peanuts. Excellent progress has been made in these few instances where intensive programmes are devoted to host-plant resistance. Although high levels of resistance have been found in some crops (strawberry and cucumber), other crop genotypes exhibit differences only when sprayed with an acaricide (cotton) (dePonti, 1985). Resistant germplasm has been identified in peanuts, which should be useful in future breeding programmes (Johnson, Campbell and Wynne, 1980, 1982; Johnson and Campbell, 1982). DePonti (1985) states 'because seemingly small changes in life history components can significantly reduce the population development of spider mites and other arthropods, resistant cultivars should become a major component of integrated control systems'. Perhaps some level of resistance could even be incorporated into crops that act as nursery crops and sources of infestation for other, more seriously damaged, crops (Brandenburg and Kennedy, 1982a; Kennedy and Margolies, 1985b).

Biological control, although an important and in many cases effective management tool, is not discussed in this review. Many authors have adequately reviewed this subject, with the most recent being Helle and Sabelis (1985b). In the light of agricultural considerations, all biological control programmes must consider the influence of many of the agricultural practices already mentioned in the review. Hamai and Huffaker (1978) found that when conditions enhance increases in the spider mite population, predators may be less successful in controlling outbreaks. The authors also cautioned that, while natural control may be successful from a biological perspective, the economic control may be much less reliable.

CROP DAMAGE, ECONOMIC THRESHOLDS AND SAMPLING

Crop damage

T. urticae feed preferentially on the underside of leaves, and the feeding results in typical light-coloured punctures that progress to white or greyish spots often called 'stippling'. The mites feed by inserting their stylets into plant cells and sucking out the cell contents. Feeding damage is mainly restricted to the spongy mesophyll but may also affect the lowest parenchymal layer (Hislop and Jeppson, 1976); this, in turn, can affect stomatal function and specific cell-component mechanisms. Storms (1971) suggested that saliva from *T. urticae* may play an important part by degrading plant cell contents, perhaps through enzymatic activity, but Sances, Wyman and Ting (1979a), while studying *T. urticae* on strawberry, indicated that damage was mechanical in nature. Rabbinge (1985) stated that mechanical damage was probably of limited importance, but DeAngelis *et al.* (1982) indicated that water loss through ruptured cells was significant.

Tomczyk and Kropczynska (1985) recently reviewed the literature concerning spider mite damage and damage mechanisms. Various changes in plant physiology are consistently observed as a result of mite feeding. Research has documented effects on photosynthesis and transpiration (Atanasov, 1971; Golik, 1975; Hall and Ferree, 1975). General effects include reduced photosynthesis and transpiration in severely damaged leaves but increased transpiration in moderately damaged leaves; in addition, chlorophyll content was usually reduced. More recently, studies have shown effects on leaf gas exchange and cellular ultrastructure to be important (Summers and Stocking, 1972; Tanigoshi and Davis, 1978; Sances, Wyman and Ting, 1979a, b; DeAngelis *et al.*, 1982). Sances, Wyman and Ting (1979b) indicated that an effect on stomata may result from disruption by *T. urticae* of the functional integrity of the stomatal guard cells, thus influencing transpiration and photosynthesis. In contrast, DeAngelis *et al.* (1982) cited water stress due to increased transpiration and water loss resulting directly from mite damage as a cause for the stomatal reaction. DeAngelis *et al.* (1982) studying *T. urticae* on *Mentha piperita* also found differences in stomatal and cuticular transpiration and suggested that high rates of night-time transpiration caused leaf water stress during the day. Inhibition of gas exchanges in injured leaves and the extraction of leaf chlorophyll through mechanical cell disruption (DeAngelis, Berry and Krantz, 1983a) also influences photosynthesis. Other workers, however, have considered the reduction in chlorophyll content not to be the major limiting factor in reducing photosynthesis (Hall and Ferree, 1975; Poskuta, Kolodziej and Kropczynska, 1975; Sances, Wyman and Ting, 1979b).

Although changes in the levels of basic elements (N, P, K), soluble sugars, starches and amino acids have been documented as occurring in association with mite feeding damage, the mechanisms involved are complex and poorly understood: for example, the water loss observed in mint from *T. urticae* feeding apparently results in changes in soluble plant carbohydrates (DeAngelis, Berry and Krantz, 1983b); an accumulation of soluble sugars was observed in the injured leaf tissue. Generally, such changes are observed as starch is degraded to simple sugars or proteins to amino acids, as when moisture stress occurs (Tomczyk and Kropczynska, 1985). However, neither DeAngelis, Berry and Krantz, (1983b), nor Tanigoshi and Davis (1978), found a change in leaf starch content after mite feeding, thereby indicating that starch degradation was not responsible for an increase in soluble sugars. Possible explanations include decreased transport of carbohydrates due to mechanical damage or the presence of soluble leaf carbohydrates in water-stressed leaves to help maintain cell turgor pressure through an osmotic adjustment (DeAngelis, Berry and Krantz, 1983b). Tomyczk and Kropczynska (1985) reviewed several studies concerning changes in CO_2 uptake and carbon metabolism: they suggested that increased content of soluble sugars in damaged tissue could also be the result of a decrease in starch synthesis. The authors concluded that it seems possible that different mechanisms might cause changes in different host species and at different mite densities.

Although this research provides valuable insight into the mechanisms of mite feeding, it does not provide data that are directly applicable to the development of economic or action thresholds. Its does help to explain differences in plant growth responses in a general way, but, in the absence of a quantitative understanding of the relationship between these physiological effects and crop growth and yield, the practical utility of this information is limited. Overall plant health is ultimately related to proper functioning of these processes, but agricultural producers are most concerned with actual 'end product' effects as measured under field conditions.

Changes in plant growth can frequently be observed in plants under spider mite attack. Van de Vrie, McMurtry and Huffaker (1972) and Jesiotr (1978b) found delayed growth of stems after spider mite feeding. *P. ulmi* reduced leaf area on fruit trees (Avery and Briggs, 1968a) and *Bryobia rubrioculus* had the same effect on almond (Summers and Stocking, 1972). Other studies indicate reduced leaf thickness, fewer branches and a reduced accumulation of dry matter following feeding (Tomczyk and Kropczynska, 1985). Fruit drop and ripening can also be affected by spider mites (van de Vrie, McMurtry and Huffaker, 1972; Hoyt and Tanigoshi, 1983). Jesiotr (1978a, b) found that spider mite feeding could reduce flower and stem size and result in fewer flowers on greenhouse-grown roses and carnations. Cotton yield can be affected through significant loss of fruit and squares (Leigh, 1985), and infestations in tea (Banerjee and Cranham, 1985) and cassava (Bellotti, 1985) can reduce shoot growth and even kill plants. Spider mite infestations on corn and sorghum cause premature death of leaves and major yield reductions (Owens, Ward and Teetes, 1976). In addition, feeding has been documented to affect the quality of crops such as hops (Cranham, 1985), grapes (Kinn *et al.*, 1974) and apples (Zwick, Fields and Mellenthin, 1976).

Although heavily infested plants express symptoms of obvious stress, the effects on crop yields are both complex and variable. In perennial crops, the real impact of mite feeding may not be observed until the following year.

Barnes and Andrews (1978) found no significant effect on untreated almond growth and productivity during the initial year of their study but found significantly reduced yields and terminal, shoot and trunk growth in mite-damaged trees the following year. Earlier studies on apple found a similar response on some varieties (Chapman, Lienk and Curtis, 1952; Lienk, Chapman and Curtis, 1956); Hoyt, Tanigoshi and Browne (1979) did not see this delayed effect, however, and Zwick, Fields and Mellenthin (1976) found no significant damage even with relatively large populations on two varieties of apple. Low populations of *T. urticae* enhanced the size and number of strawberries (Oatman *et al.*, 1982), the mean fruit size of peaches (Kovach and Gorsuch, 1985) and the oil content of mint leaves (DeAngelis *et al.*, 1983). In the case of peaches, mite feeding produces water stress, which reduces vegetative growth. It is possible that, when competition for water and photosynthates between vegetative and fruit tissues is reduced, fruit growth can be more rapid. Kovach and Gorsuch (1985) also noted a tendency toward increased flower bud initiation in peach at high mite densities.

Significant differences in plant response and susceptibility exist among crop species and varieties. Hoyt and Tanigoshi (1983) found pears to be very sensitive to mite feeding, and Banerjee and Cranham (1985) reported yield losses of 5–11% in tea, even when the crop was sprayed for mite control. In contrast is the example of strawberry yield enhancement during moderate twospotted spider mite infestations (Oatman *et al.*, 1982). Chapman, Lienk and Curtis (1952) and Lienk, Chapman and Curtis (1956) reported differences in varietal response of apple to mite feeding. Such differences bring to light the difficulties in evaluating the impact of spider mites on crops. Effects of feeding on yield components, as documented in many excellent studies, are often specific to particular environments, crops and varieties and thus are limited in their general predictive value. However, because of the relative importance of the economic or action threshold concept, the rest of this section examines the current status of such thresholds, their future in mite management and obstacles to their development.

Economic thresholds

Although economic injury levels (EILs) and economic thresholds (the pest density at which control measures should be applied to prevent an increasing pest population from reaching the economic injury level (Pfadt, 1985)), have been established for a number of insect pests on numerous crop hosts, their development in spider mite systems appears to have lagged. These two measurements, as well as action thresholds (the density reached by an insect pest or extent of its injury when, in the experience of entomologists, chemical control measures should be initiated to prevent serious damage of the crops (Pfadt, 1985)), form the basis of integrated pest management, and virtually every management decision is directed around the pests' abundance or potential abundance relative to these levels. Rather than review all the economic and action thresholds currently in use worldwide, our approach is to highlight several that appear to be quite sound and effective in practice.

Economic thresholds in glasshouse production of tomatoes and cucumbers are used quite successfully. *T. urticae* damage is evaluated by using leaf damage indices (1 = first signs of damage, 5 = entire leaf yellow and dry) (Hussey and Scopes, 1985a). Crop loss occurs in cucumbers when the damage index exceeds 1.9 (Hussey and Parr, 1963). No crop loss is detected until 5 weeks after the economic injury level is exceeded because, once flowers are set, fruit continues to grow and ripen. The significant crop loss occurs in a subsequent reduction in the number of flowers set (Stacey, Wyatt and Chambers, 1985). Using information on the rate at which damage increases in the absence of any control measures, a programme for acaricide use was developed (Hussey and Scopes, 1985a). The use of this programme reduces by one-half the average annual number of sprays (22) to control spider mites on cucumber. Of equal importance to the cost-saving benefits of the programme is the reduced selection pressure leading to acaricide resistance.

Although economic thresholds are most frequently associated with the application of pesticides, they can also be used to signify the need for biological control. The release of *P. persimilis* when the leaf damage index reached 0.4 allowed predator build-up that reduced spider mites before reaching the critical leaf damage index (1.9) (Hussey and Scopes, 1985a).

Tomatoes have a similar leaf damage index threshold of 2.0 (equivalent to 30% of the photosynthetic area damaged). Leaf damage indices can also be used to estimate spider mite abundance and to provide guidelines for predator introductions in tomatoes (Hussey and Scopes, 1985a). Stacey, Wyatt and Chambers (1985), however, report that a leaf damage index of 2.0 is equivalent to only a 10% loss of leaf area but would cause about a 9% loss of crop. In fact, regression lines of their data indicate that there is no damage threshold below which some crop loss does not occur.

Almond production in California utilizes a very different approach to decision making (action threshold): growers traditionally treated for mites as soon as stippling or webbing appeared, or on a preventative calendar basis. One current technique uses a simple presence–absence sampling for *Tetranychus* species and its phytoseiid predator *M. occidentalis*. Control is recommended when the proportion of infested leaves exceeds 0.436 with *M. occidentalis* present and 0.220 when *M. occidentalis* is absent (Zalom *et al.*, 1984a, b). This is based on data indicating a relationship between proportion of leaves infested and potential for leaf abscission.

A more refined decision-making technique uses the brush and count method and incorporates spider mite averages, spider mite days and predator–prey ratios into decisions (Hoy, 1985b). Damage to almond leaves becomes obvious after 120–130 spider mite days, and intensive defoliation occurs after about 400 spider mite days in well-irrigated orchards. Guidelines have been developed to manage spider mite populations at levels below those which cause visible damage. These guidelines incorporate the presence of native or multi-resistant predators. *M. occidentalis* may keep mites below levels that cause visible damage, but as almonds are typically water-stressed during July and August, the use of low rates of acaricides is likely. These low rates permit early season adjustments of the predator–prey ratio. These lower-than-label rates are not used if predators are scarce or absent (Hoy, 1985b).

Other crops utilize economic or action thresholds based on varying degrees of quantitative data. In north-east India, a treatment threshold on tea of 3–4 mites/cm^2 of mature leaf surface is used (Banerjee, 1979). Still other crops, such as ornamental plants, because of the importance of their appearance, have a threshold that is essentially zero (van de Vrie, 1985). The current decision for treatment in citrus is based upon experience rather than data on thresholds and predator–prey ratios (McMurtry, 1985). The guidelines provided by many college and university specialists for spider mite treatment for many crops rely upon varying amounts of data and experience. After operating under the IPM philosophy for many years, it seems unreasonable that so many commodities still lack sound economic injury levels and economic or action thresholds.

Unfortunately for *T. urticae*, the pest density–damage relationship is complex and difficult to evaluate quantitatively. Van de Vrie, McMurtry and Huffaker (1972) and Hoyt and Tanigoshi (1983) pointed out many of the factors that influence this relationship: these include weather, crop variety, crop population, management factors, crop use, crop load, natural enemies, competition, immigration, host-plant physiology, time of attack (stage of crop growth), pesticide interactions and disease control. Other factors pertaining specifically to orchard systems include tree vigour and pruning effects (Rabbinge, 1985). Different mite species can also have a varying impact on plants (Hoy, 1985a) and influence the amount of damage through interspecific competition or species exclusion (Croft and Hoying, 1977). Different sampling techniques and their effectiveness can also make quantitative evaluations difficult (Hoyt and Tanigoshi, 1983).

Rabbinge (1985) was critical of the qualitative nature of many studies attempting to develop economic injury levels. The author reviewed several works on EILs for *P. ulmi* on apple (Avery and Briggs, 1968a, b; Light and Ludlam, 1972; Zwick, Fields and Mellenthin, 1976). These studies found wide-ranging differences in the mite populations required to cause economic crop loss. The discrepancies reported appear so great as to defy explanation; however, several key parameters were not monitored. Information on age of the infestation, time of infestation, apple varieties involved, weather data and various agricultural practices were not well documented. As a reflection of the importance of these parameters, Rabbinge (1985) reported on a similar study in the Netherlands that showed no effect of even high populations of mites under drought conditions. In this situation, the effects of the drought obscured the impact of the mites. Despite these problems, other authors have developed economic threshold functions based on one year's data (e.g. Raworth, 1986).

Other specific influences include the timing of the infestation as reported on apples (Hoyt, Tanigoshi and Browne, 1979) and strawberries (Sances *et al.*, 1981, 1982a, b). Even cropping patterns, such as winter or summer planting of strawberries, can influence the host's susceptibility to damage

(Oatman *et al.*, 1981). Economic thresholds to minimize sprays and maximize yields are 20–25 mites/leaflet for winter-planted strawberries and 50 mites/leaflet for summer-planted strawberries (Oatman *et al.*, 1981, 1982).

To account for timing of infestation, Hoyt, Tanigoshi and Browne (1979) utilized the 'mite days' concept, which provides a cumulative measure of the duration and level of the infestation. They found that associated mite days and population peaks were linear in relationship. However, even with a more accurate appraisal of the current year infestations, some effects are not observed until the second year (Hoy, 1985a), and this further obscures the pest–damage relationship. Environmental stress in one year can influence the injury in both years. These second-year losses (Lienk, Chapman and Curtis, 1956) are probably the result of early mite infestations affecting bud formation for the following year; thus, timing of infestation plays a major part in plant response. In contrast, any second-year response may be overcome by crop thinning (Hoyt and Tanigoshi, 1983). Further, even a late-season attack that may not influence yield might provide substantial mite inoculum for next year and thus be of some economic consequence.

Hoyt, Tanigoshi and Browne (1979) outlined the basic problems associated with the development of quantitative injury data in replicated orchard work, including the following:

1. Naturally occurring populations are unreliable;
2. Manipulation of populations is difficult to replicate;
3. Within-tree variation is often significant (can be reduced by measuring the same individual fruit at each sampling and by selecting fruit for uniformity);
4. Among-tree variation is often significant (can be reduced by measuring fruit growth rather than size and by selecting uniform trees).

They found that fruit growth was correlated with mite days and fruit size at 35 days post-bloom. Even though measurements examine growth rate, within-tree variation in fruit size will still reduce the trial's ability to measure pest impact effectively.

Despite natural variation in size, yield and colour of apples, Hoyt and Tanigoshi (1983) made significant progress toward an accurate economic injury level by adjusting for the timing of infestation in weeks post-bloom. Such separation determines how a low-level population of long duration relates to a sudden large population later in the season. The authors discuss this relationship in terms of a fruit-growth model. A tree-performance model that simulates growth rate can be used to evaluate the impact of time of infestation (Tanigoshi and Browne, 1981). Further, it is possible to develop a series of economic injury level curves for the interaction of various factors: for instance, there is a direct relationship between growth rate and available nitrogen, soil moisture and size of fruit 35 days after bloom, and an inverse relationship between growth rate and crop load or mite density; the latter is dependent upon time of year (Hoyt and Tanigoshi, 1983). Such curves express a single factor or interaction of factors with pest density. As any number of other stresses or factors can interact, the use of such curves can be quite complex; perhaps the ability to control or eliminate some of these factors explains the success of developing economic thresholds in glasshouse production. With the complexities involved, the reliability of economic injury levels for many crops appears to be fairly weak; however, the concept and available data provide a reasonable basis for decision-making that is an improvement over experience alone, or a calendar-based spray schedule.

Rabbinge (1985) urged the development of simulation models that would study the consequences of spider mites on plant performance. Such models simulate the different combinations of population dynamics and crop growth, taking into account site and host condition, and are based upon the processes governing not only the internal physiology of the plant, but also the spider mite. Their use would provide insight into the behaviour of infested crops under various conditions. Poston, Pedigo and Welch (1983) stated that many of our current economic injury level research problems stem from a failure to consider the internal processes of the plant. If the plant is viewed as a set of interacting components, then the impact of spider mites on those components can be evaluated and a more realistic pest density–plant response relationship established. Spider mite models to simulate development and reproduction, dispersal, predation, mite–plant interactions and spatial distribution have been used successfully for several years in model-based mite management programmes (Welch, 1979).

The current lack of knowledge concerning many of the physiological processes limits our ability to develop such models. This deficiency is more obvious for perennial crops than annuals (Rabbinge, 1985). Development of such models on insufficient data will most certainly result in economic injury levels and thresholds that possess the same potential for error as more empirically derived values. In addition, models requiring large amounts of data input for accurate operation may not be practical if the cost (in time and money) of data collection at each site is great. The need for accurate guidelines for mite control decisions is essential, but for grower acceptance the benefits or advantages of new assessment techniques must outweigh their costs.

Sampling

Little can be added to the thorough update on spider mite sampling provided by Sabelis (1985e): that review covered all aspects of sampling and population assessment. From an agricultural and decision-making perspective, several points to consider were expressed. Although researchers may require absolute quantitative measurements of spider mite abundance, decisions for control in agriculture may need only presence–absence information. Other pest management programmes may require information on predators before decisions can be made (Zalom *et al.*, 1984a, b), and this information has been reviewed by Nachman (1985). In general, all IPM programmes involve sampling plans that require a minimum number of samples and processing time. Efficient, yet statistically adequate, plans are essential. Sequential sampling is frequently implemented to provide this information. Presence–absence sequential sampling decision schemes have been developed for *Tetranychus* spp. in at least three crops: snap bean (Bechinski and Stoltz, 1985), almond (Zalom *et al.*, 1984a, b) and cotton (Wilson *et al.*, 1983). For mites, presence–absence is preferred for field sampling because it eliminates the need to count mites (Wilson *et al.*, 1983). Other sampling techniques acceptable for field use may require counting only to a low level (e.g. more than 5 adult mites equal a plus; 5 or less equal a minus (Margolies, Lampert and Kennedy, 1984).

Sabelis (1985e) recommends that two areas should receive more attention in future studies on spider mite sampling: these include (1) evaluation of sampling plans based on realistically distributed populations of mites; and (2) selection of sampling methods that reduce the effects of sampler bias. The first area questions the validity of sampling schemes generated from equations based on normal distributions requiring large sample sizes. Practical application of sampling schemes requires assessment techniques of minimum effort (e.g. presence–absence). Sampling to develop and operate simulation models and for research on population dynamics may require higher levels of precision than are necessary for control decisions. The law of diminishing returns applies as sampling size is increased to gain precision, and costs may limit the attainable precision utilized in pest management programmes.

If mites were uniformly distributed over their host-plant habitat, sampling would be simplified. Because many ecological studies have revealed stratification of the habitat, sampling schemes can be developed to address that distribution. As distribution changes with time and plant development, so must sampling plans. Cultural practices such as irrigation can also influence mite distribution (Wilson *et al.*, 1984).

An area less understood but of equal significance is the impact of pesticides on spider mite distribution. Changes in distribution following pesticide application have been discussed. Trumble (1985) found that pesticide applications could produce changes in poulation dispersion. Such changes led to statistical errors that invalidated sampling programmes (both sequential and binomial) based on data from untreated fields. Researchers involved with the development of sampling plans for many agricultural systems must then consider modifications for such influences. Data should be collected both pre- and post-treatment.

PESTICIDE RESISTANCE

Pesticide resistance is an area that has received considerable attention over the past 30 years. The literature covering this subject is quite extensive and has been reviewed by Helle (1965), Helle and van de Vrie (1974), Dittrich (1975) and Jeppson, Keifer and Baker (1975). A review by Cranham and Helle (1985) includes more recent studies on this subject. Pesticide resistance in Phytoseiidae has also been recently reviewed by Fournier *et al.* (1985). To review here even the most recent literature would be redundant and beyond the scope of this article. Our purpose is to address several agricultural considerations in reducing the threat of resistance while managing *T. urticae*.

Utilization of integrated pest management approaches for pesticide use is essential in reducing the incidence of pesticide resistance (Georghiou and Taylor, 1986). However, we have seen an overall increase in the use of pesticides over the past 30 years (Pimentel, 1976). Pesticides provide cheap insurance against crop failure and their costs have decreased relative to other production inputs (Miranowski and Carlson, 1986).

T. urticae has a high reproductive rate, rendering it an excellent candidate for rapid development of resistance. However, its polyphagous nature may, in certain situations, help to slow the development of resistance. A smaller proportion of polyphagous mites is likely to be treated because some of the mites would be on nearby untreated hosts and would provide a reservoir for susceptible immigrants into the treated areas (Miller, Croft and Nelson, 1985; Georghiou and Taylor, 1986). Such a situation exists in a corn–peanut agroecosystem whereby *T. urticae* overwinters on feral hosts outside the cropping area and then utilizes two crop hosts, corn and peanuts, in a temporal–spatial sequence (Brandenburg and Kennedy, 1982a; Margolies and Kennedy, 1985).

Various attempts to delay the development of resistance have met with some success. Fitness, or the ability of the resistant genes to persist under reduced selection pressure, is quite strong (Cranham and Helle, 1985). However, if chemical selection is stopped at an early stage, natural selection will generally tend to favour non-resistant individuals. Limiting selection can be achieved by less frequent pesticide applications, using pesticides with short residual activity and by leaving untreated refuges for susceptible migrants. Incorporating resistant predators into an IPM system can lead to reduced sprays (Croft, 1982). This practice significantly encourages the use of compounds with short residual activity. In the absence of resistant predators, control with short residual sprays would require more frequent applications and thus would provide selection pressure equal to that from longer-persisting treatments. Additional selection pressure may result from host plants: Gould, Carroll and Futuyma (1982) found that *T. urticae* selected for survival on a resistant cucumber variety ('Marketmore 70') had a slightly higher resistance to several organophosphate pesticides. More studies on the effect of marginal hosts on the susceptibility of *T. urticae* are needed.

Rotation of compounds apparently delays the development of resistance only in terms of calendar time and not in terms of number of applications (Cranham and Helle, 1985). The impact of the immigration of susceptible individuals on pesticide resistance is not well understood. Kennedy and Smitley (1985) reviewed spider mite dispersal and indicate that wide-area and large-scale dispersal does occur. Immigration of susceptible individuals into an area under selective pressure can reduce the development of resistance in arthropods (Georghiou and Taylor, 1977a, b; Taylor and Georghiou, 1979; Tabashnik and Croft, 1982). Hoy, Groot and van de Baan (1985), however, showed that migration of native susceptible *M. occidentalis* had little impact on the persistence of the carbaryl-resistant population in almond orchards.

Cultural practices can be used to reduce selection pressure and to enhance the presence of susceptible mites in the cropping system. In central California strawberry production, susceptible *T. urticae* in diapause are reintroduced into new fields on nursery stock when the crop is disced under on 1- or 2-year rotations. Resistant mites colonize surrounding border vegetation. Frequently, the resistant mites migrate into the field and overwhelm the susceptible mites (due to the initial diapause state). When susceptible mites prevail, they persist only until the first acaricide application (Miller, Croft and Nelson, 1985). Cultural practices such as management of border vegetation without selecting for resistance, planting strawberries in isolated areas and crop rotation can enhance the presence of susceptible mites.

Perhaps the greatest prospect for the future management of pesticide resistance in spider mites is through simulation modelling. This would permit evaluation of various pesticide-use strategies and their outcome. Such factors as immigration could be considered in conjunction with low and high pesticide use of compounds of both long and short residual activity (Tabashnik and Croft, 1982). Tabashnik and Croft (1985) evaluated 12 different pesticide treatment rates and schedules on simulated pesticide resistance in *P. ulmi* and found scenarios whereby resistance could be delayed. With appropriate biological data, such models could be assembled for *T. urticae*.

Many factors influence selection for pesticide resistance in spider mites, and there are many social and economic considerations that influence our ability to manage the development of resistance (Miranowski and Carlson, 1986). While many programmes are making efforts to reduce selection pressure, more emphasis on integrating these efforts into current IPM programmes is necessary.

Conclusion

In summary, *T. urticae* is a pest that will continue to challenge our ability to manage damaging populations effectively as agroecosystems change through time. Numerous ecological factors enable the twospotted spider mite to exploit many cropping situations successfully. A wide host range, ability to produce large numbers of progeny under marginal conditions, adaptability to host plants, significant dispersal capability and a high reproductive rate are all factors that can be influenced by agricultural inputs. In addition, natural enemies play a significant part in regulating *T. urticae* populations and these, too, are affected by agricultural production practices.

When analysing spider mite outbreaks in a particular system, few if any extrinsic factors can be ignored. As new cropping systems are developed, the impact of these changes should be evaluated with regard to their effect on spider mite population dynamics. New systems may exacerbate an already existing problem or enhance a previously suppressed population. Because *T. urticae* is spatially dynamic, it is important that future research should take a 'wide-area' perspective on spider mite biology as many factors outside the system under study may influence its susceptibility to attack. It is important that management studies should not be inhibited by commodity orientation, but rather should be bound only by temporal and spatial limits of *T. urticae* populations.

Although the literature concerning *T. urticae* over the past few years has greatly increased our understanding of this pest, several areas require even more attention. The impact of the host plant on *T. urticae* population dynamics and pesticide susceptibility, pesticide resistance and economic injury levels are all areas that deserve additional work. Twospotted spider mite ecology is a complex of interactions closely associated with host plants, environment and natural enemies. Understanding of these interactions will enhance our ability to manipulate those factors most responsible for outbreaks. Success in these areas lies in future co-operative research among acarologists, entomologists, agronomists, plant physiologists and systems analysts.

Acknowledgement

We express our appreciation to F. Gould, J. DeAngelis and G. Rock for their helpful criticism of this manuscript.

References

ALINAIZEE, M. T. AND CRANHAM, J. E. (1980). Effect of four synthetic pyrethroids on a predatory mite, *Typhlodromus pyri*, and its prey, *Panonychus ulmi*, on apples in southeast England. *Environmental Entomology* **9**, 436–439.

ASQUITH, D., CROFT, B. A., HOYT, S. C., GLASS, E. H. AND RICE, R. E. (1980). The systems approach and general accomplishments toward better insect control in pome and stone fruits. In *New Technology of Pest Control* (C. B. Huffaker, Ed.), pp. 249–317. John Wiley & Sons, New York.

ATANASOV, N. (1971). Physiological functions of plants as affected by damage caused by *Tetranychus atlanticus* McGregor. In *Proceedings of the 3rd International Congress of Acarology, Prague, 1971* (M. Daniel and B. Rosicky, Eds), pp. 183–186. Publishing House of the Czechoslovakian Academy of Sciences, Prague.

ATTIAH, H. H. AND BOUDREAUX, H. B. (1964a). Population dynamics of spider mite influenced by DDT. *Journal of Economic Entomology* **57**, 53–57.

ATTIAH, H. H. AND BOUDREAUX, H. B. (1964b). Influence of DDT on egg-laying in spider mites. *Journal of Economic Entomology* **57**, 50–53.

AVERY, D. J. AND BRIGGS, J. B. (1968a). Damage to leaves caused by fruit tree red spider mite, *Panonychus ulmi* (Koch). *Journal of Horticultural Science* **43**, 463–473.

AVERY, D. J. AND BRIGGS, J. B. (1986b). The aetiology and development of damage in young fruit trees infested with fruit tree red spider mite, *Panonychus ulmi* (Koch). *Annals of Applied Biology* **61**, 277–288.

BANERJEE, B. (1979). Intratree variation in the distribution of the tea red spider mite *Oligonychus coffeae* (Nietne). *Acarologia* **21**, 216–220.

BANERJEE, B. AND CRANHAM, J. F. (1985). Tea. In *Spider Mites. Their Biology, Natural Enemies and Control* (W. Helle and M. W. Sabelis, Eds), volume 1B, pp. 371–374, Elsevier, New York.

BARNES, M. M. AND ANDREWS, K. L. (1978). Effects of spider mites on almond tree growth and productivity. *Journal of Economic Entomology* **71**, 555–558.

BARTLETT, B. R. (1968). Outbreaks of twospotted spider mites and cotton aphids following pesticide treatment. I. Pest stimulation vs. natural enemy destruction as the cause of outbreaks. *Journal of Economic Entomology* **61**, 297–303.

Bechinski, E. J. and Stoltz, R. L. (1985). Presence–absence sequential decision plans for *Tetranychus urticae* (Acari: Tetranychidae) in garden-seed beans, *Phaseolus vulgaris*. *Journal of Economic Entomology* **78**, 1475–1480.

Bellotti, A. C. (1985). Cassava. In *Spider Mites. Their Biology, Natural Enemies and Control* (W. Helle and M. W. Sabelis, Eds), volume 1B, pp. 333–338. Elsevier, New York.

Bernstein, C. (1985). A simulation model for an acarine predator–prey system: *Phytoseiulus persimilis, Tetranychus urticae*. *Journal of Animal Ecology* **54**, 375–390.

Boer, R. de (1982). Laboratory hybridization between semi-incompatible races of the arrhenotokous spider mite *Tetranychus urticae* Koch (Acari: Tetranychidae). *Evolution* **36**, 553–560.

Boer, R. de (1985). Reproductive barriers. In *Spider Mites. Their Biology, Natural Enemies and Control* (W. Helle and M. W. Sabelis, Eds), volume 1A, pp. 193–199. Elsevier, New York.

Boller, E. F., Janser, E. and Potter, C. (1985). Evaluation of side effects of vineyard herbicides on *Tetranychus urticae* and its predator *Typhlodromus pyri* under laboratory and semi-field conditions. *Zeitschrift für Pflanzenkrankheiten und Pflanzenschutz* **91**, 561–568.

Boudreaux, H. B. (1958). The effect of relative humidity on egg laying, hatching and survival in various spider mites. *Journal of Insect Physiology* **2**, 65–72.

Boudreaux, H. B. (1963). Biological aspects of some phytophagous mites. *Annual Review of Entomology* **8**, 137–154.

Boykin, L. S. (1983). *Ecology of the Twospotted Spider Mite*. Tetranychus urticae Koch, on Peanut in North Carolina. PhD dissertation, North Carolina State University, Raleigh. 97 pp.

Boykin, L. S. and Campbell, W. V. (1982). Rate of population increase of the twospotted spider mite (Acari: Tetranychidae) on peanut leaves treated with pesticides. *Journal of Economic Entomology* **75**, 966–971.

Boykin, L. S. and Campbell, W. V. (1984). Wind dispersal of the twospotted spider mite (Acari: Tetranychidae) in North Carolina peanut fields. *Environmental Entomology* **13**, 221–227.

Boykin, L. S., Campbell, W. V. and Beute, M. K. (1984). Effect of pesticides on *Neozygites floridana* (Entomophthorales: Entomophthoraceae) and arthropod predators attacking the twospotted spider mite (Acari: Tetranychidae) in North Carolina peanut fields. *Journal of Economic Entomology* **77**, 969–975.

Boykin, L. S., Campbell, W. V. and Nelson, L. A. (1983). Attempted dispersal of the twospotted spider mite, *Tetranychus urticae*, on greenhouse-grown peanut leaves in response to pesticides and irrigation. *Peanut Science* **10**, 1–3.

Boykin, L. S., Campbell, W. V. and Nelson, L. A. (1984). Effects of barren soil borders and weed border treatments on movement of the twospotted spider mite into peanut fields. *Peanut Science* **11**, 52–55.

Brandenburg, R. L. and Kennedy, G. G. (1981). Overwintering of the pathogen *Entomophthora floridana* and its host, the twospotted spider mite. *Journal of Economic Entomology* **74**, 428–431.

Brandenburg, R. L. and Kennedy, G. G. (1982a). Intercrop relationships and spider mite dispersal in a corn/peanut agroecosystem. *Entomologia experimentalis et applicata* **32**, 269–276.

Brandenburg, R. L. and Kennedy, G. G. (1982b). Relationship of *Neozygites floridana* (Entomophthorales: Entomophthoraceae) to twospotted spider mite (Acari: Tetranychidae) populations in field corn. *Journal of Economic Entomology* **75**, 691–694.

Brandenburg, R. L. and Kennedy, G. G. (1983). Interactive effects of selected pesticides on the twospotted spider mite and its fungal pathogen, *Neozygites floridana*. *Entomologia experimentalis et applicata* **34**, 240–244.

Cadapan, E. P. (1976). *The Effect of the Twospotted Spider Mite and Several Insects on the Yield of Soybeans*. PhD dissertation, University of California, Berkeley, 97 pp.

CAMPBELL, W. V. (1978). Effect of pesticide interactions on the twospotted spider mite on peanuts. *Peanut Science* **5**, 83–88.

CAMPBELL, W. V., BATTS, R. W., ROBERTSON, R. L. AND EMERY, D. A. (1974). Suppression of the twospotted spider mite on peanuts. *Peanut Science* **1**, 30–34.

CAREY, J. R. (1983). Practical application of the stable age distribution: analysis of a Tetranychid mite (Acari: Tetranychidae) population outbreak. *Environmental Entomology* **12**, 10–18.

CARNER, G. R. (1976). A description of the life cycle of *Entomophthora* sp. in the twospotted spider mite. *Journal of Invertebrate Pathology* **28**, 245–254.

CARNER, G. R. AND CANERDAY, T. D. (1968). Field and laboratory investigations with *Entomophthora fresneii*, a pathogen of *Tetranychus* spp. *Journal of Economic Entomology* **61**, 956–959.

CARNER, G. R. AND CANERDAY, T. D. (1970). *Entomophthora* sp. as a factor in the regulation of twospotted spider mite on cotton. *Journal of Economic Entomology* **63**, 638–640.

CARTER, C. D. AND SNYDER, J. C. (1985). Mite responses in relation to trichomes of *Lycopersicon esculentum* × *L. hirsutum* F_2 hybrids. *Euphytica* **34**, 177–185.

CARTER, C. D. AND SNYDER, J. C. (1986). Mite responses and trichome characters in a full-sib F_2 family of *Lycopersicon esculentum* × *L. hirsutum*. *Journal of the American Society for Horticultural Science* **111**, 130–133.

CASWELL, H. AND HASTINGS, A. (1980). Fecundity, developmental time and population growth rate: an analytical solution. *Theoretical Population Biology* **17**, 71–79.

CHABOUSSOU, F. (1966). Nouveau aspects de la phytiatrie et de la phytopharmacie. Le phénomène de la trophobiose. In *FAO Symposium on Integrated Pest Control*, pp. 1, 33–61. FAO, Rome.

CHANDLER, L. D., ARCHER, T. L., WARD, C. R. AND LYLE, W. M. (1979). Influences of irrigation practices on spider mite densities on field corn. *Environmental Entomology* **8**, 196–201.

CHAPMAN, P. J., LIENK, S. E. AND CURTIS, O. F. (1952). Responses of apple trees to mite infestations: I. *Journal of Economic Entomology* **45**, 815–821.

CHAPMAN, R. B. AND PENMAN, D. R. (1979). Toxicity of synthetic pyrethroid insecticides to phytophagous orchard mites. *Proceedings 32nd New Zealand Weed and Pest Control Conference*, pp. 240–244. New Zealand Weed and Pest Control Society, Inc., Ministry of Agriculture and Fisheries, Palmerton North, NZ.

CHARLES, J. G., COLLYER, E. AND WHITE, V. (1985). Integrated control of *Tetranychus urticae*, *Phytoseuilus persimilis* and *Stethorus bifidue* in commercial raspberry gardens. *New Zealand Journal of Experimental Agriculture* **13**, 385–393.

CLANCY, D. W. AND POLLARD, H. M. (1952). The effect of DDT on mite and predator populations in apple orchards. *Journal of Economic Entomology* **45**, 108–114.

COLE, L. (1954). The population consequences of life history phenomena. *Quarterly Review of Biology* **29**, 103–137.

CRAIG, R., MUMMA, R. O., GERHOLD, D. L., WINNER, B. L. AND SNETSINGER, R. (1986). Genetic control of a biochemical mechanism for mite resistance in geraniums. In *Natural Resistance of Plants to Pests. Roles of Allelochemicals: American Chemical Society Symposium Series 296*, (M. D. Green and P. A. Hedin, Eds), pp. 168–176. American Chemical Society, Washington, DC.

CRANHAM, J. E. (1985). Hop. In *Spider Mites. Their Biology, Natural Enemies and Control* (W. Helle and M. W. Sabelis, Eds), volume 1B, pp. 367–370. Elsevier, New York.

CRANHAM, J. E. AND HELLE, W. (1985). Pesticide resistance in Tetranychidae. In *Spider Mites. Their Biology, Natural Enemies and Control* (W. Helle and M. W. Sabelis, Eds), volume 1B, pp. 405–421. Elsevier, New York.

CROFT, B. A. (1977). Resistance in arthropod predators and parasites. In *Pesticide*

Management and Insecticide Resistance (D. L. W. Watson and A. W. A. Brown, Eds), pp. 337–393. Academic Press, New York.

CROFT, B. A. (1982). Arthropod resistance to insecticides: a key to pest control failures and successes in North American apple orchards. *Entomologia experimentalis et applicata* **31**, 88–110.

CROFT, B. A. AND BROWN, A. W. A. (1975). Responses of arthropod natural enemies to insecticides. *Annual Review of Entomology* **20**, 285–335.

CROFT, B. A. AND HOYING, S. A. (1977). Competitive displacement of *Panonychus ulmi* (Acarina: Tetranychidae) by *Aculus schlechtendali* (Acarina: Eriophyidae) in apple orchards. *Canadian Entomologist* **109**, 1025–1034.

CROFT, B.A. AND HOYT, S. C. (1978). Considerations for the use of pyrethroid insecticides for deciduous fruit pest control in the USA. *Environmental Entomology* **7**, 627–630.

CROFT, B. A. AND McGROARTY, D. L. (1973). A model study of acaricide resistance, spider mite outbreaks and biological control patterns in Michigan apple orchards. *Environmental Entomology* **2**, 633–638.

CROOKER, A. (1985). Embryonic and juvenile development. In *Spider Mites, Their Biology, Natural Enemies and Control* (W. Helle and M. W. Sabelis, Eds), volume 1A, pp. 149–163. Elsevier, New York.

DACOSTA, C. P. AND JONES, C. M. (1971). Cucumber beetle resistance and mite susceptibility controlled by the bitter gene in *Cucumis sativus* L. *Science* **172**, 1145–1146.

DABBOUR, A. I. (1977). Egg production of the red spider mite, *Tetranychus urticae* Koch, as affected by certain environmental factors. *Bulletin of the Entomological Society of Egypt* **61**, 31–35.

DAVIS, D. W. (1952a). Influence of population density on *Tetranychus multiselis*. *Journal of Economic Entomology* **45**, 652–654.

DAVIS, D. W. (1952b). Some effects of DDT on spider mite. *Journal of Economic Entomology* **45**, 1011–1019.

DEANGELIS, J. D., BERRY, R. E. AND KRANTZ, G. W. (1983a). Photosynthesis, leaf conductance and leaf chlorophyll content in spider mite (Acari: Tetranychidae)-injured peppermint leaves. *Environmental Entomology* **12**, 345–348.

DEANGELIS, J. D., BERRY, R. E. AND KRANTZ, G. W. (1983b). Evidence for spider mite (Acari: Tetranychidae) injury-induced leaf water deficits and osmotic adjustments in peppermints. *Environmental Entomology* **12**, 336–339.

DEANGELIS, J. D., LARSON, K. C., BERRY, R. E. AND KRANTZ, G. W. (1982). Effects of spider mite injury on transpiration and leaf water status in peppermint. *Environmental Entomology* **11**, 975–978.

DEANGELIS, J. E., MARIN, A. B., BERRY, R. E. AND KRANTZ, G. W. (1983). Effects of spider mite (Acari: Tetranychidae) injury on essential oil metabolism in peppermint. *Environmental Entomology* **12**, 522–527.

DEPONTI, O. M. B. (1977). Resistance in *Cucumis sativus* L. to *Tetranychus urticae* Koch. 1. The role of plant breeding in integrated control. *Euphytica* **26**, 633–640.

DEPONTI, O. M. B. (1985). Host plant resistance and its manipulation through plant breeding. In *Spider Mites. Their Biology, Natural Enemies and Control* (W. Helle and M. S. Sabelis, Eds), volume 1B, pp. 395–403. Elsevier, New York.

DITTRICH, U. (1975). Acaricide resistance in mites. *Zeitschrift für angewandte Entomologie* **78**, 28–45.

DITTRICH, U., STREIBERT, P. AND BATHE, P. A. (1974). An old case reopened: mite stimulation by insecticide residue. *Environmental Entomology* **3**, 534–540.

DONAHUE, D. J., McPHERSON, M. AND POE, S. L. (1985). Oviposition site selection responses of the twospotted spider mite to fenvalerate and permethrin residues on soybeans. 53rd Annual Meeting of the Virginia Academy of Science, Williamsburg, VA. *Virginia Journal of Science* **36**, 86.

DUNCOMBE, W. G. (1972). Red spider mite on cotton and its control. *Rhodesia Agricultural Journal* **69**, 7–10.

Duso, C. and Liquori, M. (1984). The mite fauna of grapevine in Veneto, Italy, and the influence of chemical treatments on populations of phytophagous and predatory mites. *Redia* **65**, 337–354.

Ehler, L. E. (1974). *A Review of the Spider Mite Problem on Grain Sorghum and Corn in West Texas. Texas Agricultural Experiment Station Bulletin 1149.* 15 pp.

Elbadry, E. A. (1979). Management of mite pests of cotton in Egypt. In *Recent Advances in Acarology* (J. G. Rodriguez, Ed.), pp. 49–57. Academic Press, New York.

El-Banhawy, E. M. and Anderson, T. E. (1985). Effects of avermectin B_1 and fenvalerate on the survival, reproduction and egg viability of the twospotted spider mite. *Tetranychus urticae* Koch (Acari: Tetranychidae). *International Journal of Acarology* **11**, 11–16.

Everson, P. R. and Addicott, J. F. (1982). Mate selection strategies by male mites in the absence of intersexual selection by females: a test of six hypotheses. *Canadian Journal of Zoology* **60**, 2729–2736.

Fournier, D., Pralvorio, M., Berge, J. B. and Cuany, A. (1985). Pesticide resistance in Phytoseiidae. In *Spider Mites. Their Biology, Natural Enemies and Control* (W. Helle and M. W. Sabelis, Eds), volume 1B, pp. 423–431. Elsevier, New York.

Franklin, E. J. and Knowles, C. O. (1984). Influence of formamidines on twospotted spider mite (Acari: Tetranychidae) dispersal behavior. *Journal of Economic Entomology* **77**, 318–323.

Fungo, N. K. and Curry, J. P. (1983). The effects of light fungicides on the glasshouse spider mite (*Tetranychus urticae* (Koch)). *Journal of Life Science Royal Dublin Society* **4**, 175–182.

Georghiou, G. P. and Taylor, C. E. (1977a). Genetic and biological influences in the evolution of insecticide resistance. *Environmental Entomology* **70**, 319–323.

Georghiou, G. P. and Taylor, C. E. (1977b). Operational influences in the evolution of insecticide resistance. *Environmental Entomology* **70**, 653–658.

Georghiou, G. P. and Taylor, C. E. (1986). Factors influencing the evolution of resistance. In *Pesticide Resistance, Strategies and Tactics for Management*, pp. 157–169. National Academy Press, Washington DC.

Gerson, U. (1985). Webbing. In *Spider Mites. Their Biology, Natural Enemies and Control* (W. Helle and M. W. Sabelis, Eds), volume 1A, pp. 223–232. Elsevier, New York.

Ghobrial, A. and Dittrich, V. (1980). Early and late season pest complexes on cotton, their control by aerial and ground application of insecticide and side effects on predator fauna. *Zeitschrift für angewandte Entomologie* **90**, 306–313.

Golik, Z. (1975). A study of the destructiveness of the fruit tree red spider mite, *Panonychus ulmi* (Koch) on apple. *Zesgyty Problemowe Postepou Nauk Rolniczych* **171**, 15–34.

Gould, F. (1978). Resistance of cucumber varieties to *Tetranychus urticae*: genetic and environmental determinants. *Journal of Economic Entomology* **71**, 680–683.

Gould, F., Carroll, C. R. and Futuyma, P. J. (1982). Cross-resistance to pesticides and plant defenses: a study of the twospotted spider mite. *Entomologia experimentalis et applicata* **31**, 175–180.

Grafton-Cardwell, E. E. and Hoy, M. A. (1983). Comparative toxicity of Avermectin B_1 to the predator, *Metaseiulus occidentalis* (Nesbitt) (Acari: Phytoseiidae) and the spider mites *Tetranychus urticae* Koch and *Panonychus ulmi* (Koch) (Acari: Tetranychidae). *Journal of Economic Entomology* **76**, 1216–1220.

Hall, F. R. (1979). Effects of synthetic pyrethroids on major insect and mite pests of apple. *Journal of Economic Entomology* **72**, 441–446.

Hall, F. R. and Ferree, D. C. (1975). Influence of twospotted spider mite populations on photosynthesis of apple leaves. *Journal of Economic Entomology* **68**, 517–520.

Hamai, J. and Huffaker, C. B. (1978). Potential of predation by *Metaseiulus occidentalis* in compensating for increased nutritionally induced power of increase of *Tetranychus urticae. Entomophaga* **23**, 225–237.

Harris, C. R. and Kinoshita, G. B. (1977). Influence of posttreatment temperature on the toxicity of pyrethroid insecticides. *Journal of Economic Entomology* **70**, 215–218.

Harrison, R. A. and Smith, A. G. (1961). The influence of temperature and relative humidity on the development of eggs and on the effectiveness of ovicides against *Tetranychus telarius* (L.) (Acarina: Tetranychidae). *New Zealand Journal of Science* **4**, 540–549.

Havelka, J. and Kindlmann, P. (1984). Optimal use of the 'pest in first' method for controlling *Tetranychus urticae* Koch (Acarina: Tetranychidae) on glasshouse cucumbers through *Phytoseiulus persimilis* A.-H. (Acarina: Phytoseiidae). *Zeitschrift für angewandte Entomologie* **98**, 254–263.

Hazan, A., Gerson, U. and Tahori, A. S. (1973). Life history and life tables of the carmine spider mite. *Acarologia* **15**, 414–440.

Headley, J. C. and Hoy, M. A. (1986). The economics of integrated mite management in almonds. *California Agriculture* **40**, 28–30.

Helle, W. (1965). Resistance in the Acarina Mites. In *Advances in Acarology* (J. A. Naegele, Ed.), volume 2, pp. 71–93. Comstock Publishers, Ithaca, New York.

Helle, W. (1967). Fertilization in the twospotted spider mite (*Tetranychus urticae*: Acari). *Entomologia experimentalis et applicata* **10**, 103–110.

Helle, W. and Pijnacker, L. P. (1985). Parthenogenesis, chromosomes and sex. In *Spider Mites, Their Biology, Natural Enemies and Control* (W. Helle and M. W. Sabelis, Eds), volume 1A, pp. 129–139. Elsevier, New York.

Helle, W. and Sabelis, M. W. (eds) (1985a). *Spider Mites. Their Biology, Natural Enemies and Control, Volume 1A.* Elsevier, New York. 405 pp.

Helle, W. and Sabelis, M. W. (Eds) (1985b). *Spider Mites. Their Biology, Natural Enemies and Control, Volume 1B.* Elsevier, New York. 458 pp.

Helle, W. and van de Vrie, M. (1974). Problems with spider mites. *Outlook on Agriculture* **8**, 119–125.

Herbert, H. J. (1981). Biology, life tables and innate capacity for increase of the twospotted spider mite. *Tetranychus urticae* (Acarina: Tetranychidae). *Canadian Entomologist* **113**, 371–378.

Hislop, R. G. and Jeppson, L. R. (1976). Morphology of the mouthparts of several species of phytophagous mites. *Annals of the Entomological Society of America* **69**, 1125–1135.

Hislop, R. G., Auditore, P. J., Weeks, B.L. and Prokopy, R.J. (1981). Repellency of pesticides to the predator *Amblyseius fallacis. Protection Ecology* **3**, 253–257.

Hollingsworth, C. S. and Berry, R. E. (1982). Twospotted spider mite (Acari: Tetranychidae) in peppermint: population dynamics and influence of cultural practices. *Environmental Entomology* **11**, 1280–1284.

Hoy, M. A. (Ed.) (1982a). *Recent Advances in Knowledge of the Phytoseiidae. University of California Publication 3284*, University of California, Berkeley, CA. 92 pp.

Hoy, M.A. (1982b). Aerial dispersal and field efficacy of a genetically improved strain of the spider mite predator *Metaseiulus occidentalis. Entomologica experimentalis et applicata* **32**, 205–212.

Hoy, M. A. (1985a). Recent advances in genetics and genetic improvement of the phytoseiidae. *Annual Review of Entomology* **30**, 345–370.

Hoy, M. A. (1985b). Almonds (California). In *Spider Mites. Their Biology, Natural Enemies and Control* (W. Helle and M. W. Sabelis, Eds), volume 1B, pp. 299–310. Elsevier, New York.

Hoy, M. A. and Cave, F. E. (1985). Laboratory evaluation of Avermectin as a selective acaricide for use with *Metaseiulus occidentalis* (Nesbitt) (Acarina: Phytoseiidae). *Experimental and Applied Acarology* **1**, 139–152.

Hoy, M. A., Groot, J. J. and van de Baan, H. E. (1985). Influence of aerial dispersal on persistance and spread of pesticide-resistant *Metaseiulus occidentalis* in California almond orchards. *Entomologia experimentalis et applicata* **37**, 17–31.

Hoy, M. A., Flaherty, D., Peacock, W. and Culver, D. (1979). Vineyard and laboratory evaluations of methomyl, dimethoate and permethrin for a grape pest management program in the San Joaquin Valley of California. *Journal of Economic Entomology* **72**, 250–255.

Hoyt, S. C. and Tanigoshi, L. K. (1983). Economic injury levels for apple insect and mite pests. In *Integrated Management of Insect Pests of Pome and Stone Fruit* (B. A. Croft and S. C. Hoyt, Eds), pp. 203–217. John Wiley and Sons, New York.

Hoyt, S. C., Tanigoshi, L. K. and Browne, R. W. (1979). Economic injury level studies in relation to mites on apple. In *Recent Advances in Acarology* (J. G. Rodriguez, Ed.), volume 1, pp. 3–12. Academic Press, New York.

Hoyt, S. C., Westigard, P. H. and Burts, E. C. (1978). Effects of two synthetic pyrethroids on the codling moth, pear psylla and various mite species in Northwest apple and pear orchards. *Journal of Economic Entomology* **71**, 431–434.

Huang, M. (1978). Studies on the integrated control of the citrus red mites with the predaceous mite as a principal controlling agent. *Acta entomologica sinica* **21**, 260–270 (in Chinese, with English summary).

Huffaker, C. B. and Spitzer, C. H., Jr (1950). Some factors affecting red mite populations on pears in California. *Journal of Economic Entomology* **43**, 819–831.

Huffaker, C. B., van de Vrie, M. and McMurtry, J. A. (1969). The ecology of tetranychid mites and their natural control. *Annual Review of Entomology* **14**, 125–174.

Huffaker, C. B., van de Vrie, M. and McMurtry, J. A. (1970). Ecology of tetranychid mites and their natural enemies: a review. II. Tetranychid populations and their possible control by predators: an evaluation. *Hilgardia* **40**, 391–458.

Hull, L. A. and Starner, V. R. (1983). Impact of four synthetic pyrethroids on major natural enemies and pests of apple in Pennsylvania. *Journal of Economic Entomology* **76**, 122–130.

Humber, R. A., Moraes, G. J. and dos Santos, J. M. (1981). Natural infection of *Tetranychus urticae* (Acarina: Tetranychidae) by a *Triplosporium* sp. (Zygomycetes: Entomophthorales) in northeastern Brazil. *Entomophaga* **26**, 421–425.

Hussey, N. W. (1972). Diapause in *Tetranychus urticae* Koch and its implications in glasshouse culture. *Acarologia* **13**, 344–350.

Hussey, N. W. and Parr, W. J. (1963). The effect of glasshouse red spider mite (*Tetranychus urticae*) on the yield of cucumbers. *Journal of Horticultural Science* **38**, 255–266.

Hussey, N. W. and Scopes, N. E. A. (1985a). Greenhouse vegetables (Britain). In *Spider Mites. Their Biology, Natural Enemies and Control* (W. Helle and M. W. Sabelis, Eds), volume 1B, pp. 285–297. Elsevier, New York.

Hussey, N. W. and Scopes, N. E. A. (1985b). *Biological Pest Control, the Glasshouse Experience*. Cornell University Press, Ithaca, New York. 240 pp.

Ibrahim, Y. B. and Knowles, C. O. (1986). Influence of formamidines on reproduction in twospotted spider mite (Acari: Tetranychidae). *Journal of Economic Entomology* **79**, 7–14.

Iftner, D. C. (1982). The effects of synthetic pyrethroids on the feeding behavior and dispersal of the twospotted spider mite. *Ohio Journal of Science* **82**, 3.

Iftner, D. C. and Hall, F. R. (1983). Effects of fenvalerate and permethrin on *Tetranychus urticae* Koch (Acari: Tetranychidae) dispersal behavior. *Environmental Entomology* **12**, 1782–1786.

Iftner, D. C. and Hall, F. R. (1984). The effects of fenvalerate and permethrin residues on *Tetranychus urticae* Koch fecundity and the rate of development. *Journal of Agricultural Entomology* **1**, 191–200.

Ignatowicz, S. and Helle, W. (1986). Genetics of diapause suppression in the twospotted spider mite, *Tetranychus urticae* Koch. *Experimental and Applied Acarology* **2**, 161–172.

Jeppson, L. R., Keifer, H. H. and Baker, E. W. (1975). *Mites Injurious to Economic Plants*. University of California Press, Berkeley. 614 pp.

Jesiotr, L. J. (1976). The influence of the host plant on the reproduction potential of the twospotted spider mite, *Tetranychus urticae* Koch (Acarina: Tetranychidae). *Ekologia Polska* **24**, 407–411.

Jesiotr, L. J. (1978a). Further study on the injurious effects of the twospotted spider mite (*Tetranychus urticae* Koch) on greenhouse carnations. *Ekologia Polska* **26**, 303–310.

Jesiotr, L. J. (1978b). The injurious effects of the twospotted spider mite (*Tetranychus urticae* Koch) on greenhouse roses. *Ekologia Polska* **26**, 311–318.

Jesiotr, L. J. (1979). The influence of the host plant on the reproduction potential of the twospotted spider mite, *Tetranychus urticae* Koch (Acarina: Tetranychidae). II. Responses of the field population feeding on roses and beans. *Ekologia Polska* **27**, 351–355.

Jesiotr, L. J. (1980). The influence of host plants on the reproduction potential of the twospotted spider mite, *Tetranychus urticae* Koch (Acarina: Tetranychidae) IV. Changes within different populations affected by a new species of host plants. *Ekologia Polska* **28**, 633–647.

Jesiotr, L. J., Suski, W. and Badowska-Czubik, T. (1979). Food quality influences on a spider mite population. *Recent Advances in Acarology* **1**, 189–196.

Johnson, D. R. and Campbell, W. V. (1982). Variation in the foliage nutrients of several peanut lines and their association with damage received by the twospotted spidermite, *Tetranychus urticae*. *Journal of Georgia Entomological Society* **17**, 69–72.

Johnson, D. R., Campbell, W. V. and Wynne, J. C. (1980). Fecundity and feeding preference of the twospotted spider mite on domestic and wild species of peanuts. *Journal of Economic Entomology* **73**, 575–576.

Johnson, D. R., Campbell, W. V. and Wynne, J. C. (1982). Resistance of peanuts to the twospotted spider mite (Acari: Tetranychidae). *Journal of Economic Entomology* **75**, 1045–1047.

Johnson, D. T. and Croft, B. A. (1976). Laboratory study of the dispersal behavior of *Amblyseius fallacis* (Acarina: Phytoseiidae). *Annals of the Entomological Society of America* **69**, 1019–1023.

Johnson, D. T. and Croft, B. A. (1981). Dispersal of *Amblyseius fallacis* (Acarina: Phytoseiidae) in an apple ecosystem. *Environmental Entomology* **10**, 313–319.

Karban, R. (1986). Induced immunelike resistance to spider mites in cotton. *California Agriculture* **50** (11 & 12), 13–15.

Karban, R. and Carey, J. R. (1984). Induced resistance of cotton seedlings to mites. *Science* **225**, 53–54.

Kattes, D. H. (1976). *Selected Factors Influencing the Abundance of Spider Mites (Acari: Tetranychidae) in Grain Sorghum on the Texas High Plains*. MS thesis, Texas A&M University, College Station. 66 pp.

Kennedy, G. G. and Margolies, D. C. (1985a). The twospotted spider mite: a case history. In *The Movement and Dispersal of Agriculturally Important Biotic Agents* (D. R. McKenzie, C. S. Barfield, G. G. Kennedy and R. D. Berger, Eds), pp. 443–452. Clartor's Publication Division, Baton Rouge, LA.

Kennedy, G. G. and Margolies, D. C. (1985b). Mobile arthropod pests: management in diversified agroecosystems. *Bulletin of the Entomological Society of America* **31**, 21–27.

Kennedy, G. G. and Smitley, D. R. (1985). Dispersal. In *Spider Mites. Their Biology, Natural Enemies and Control* (W. Helle and M. W. Sabelis, Eds), volume 1A, pp. 233–242. Elsevier, New York.

KINN, A. N., JOOS, J. L., DOUTT, R. L., SORENSEN, J. T. AND FOSKETT, M. J. (1974). Effects of *Tetranychus pacificus* and irrigation practices on yield and quality of grapes in northern coast vineyards of California. *Environmental Entomology* 3, 601–606.

KLOSTERMEYER, E. C. (1959). Insecticide-induced population changes in four mite species on alfalfa. *Journal of Economic Entomology* 52, 991–994.

KLOSTERMEYER, E. C. AND RASMUSSEN, W. B. (1953). The effect of soil insecticide treatments on mite population and damage. *Journal of Economic Entomology* 46, 910–912.

KOVACH, J. AND GORSUCH, C. S. (1985). Effect of *Tetranychus urticae* populations on peach production in South Carolina. *Journal of Agricultural Entomology* 2, 46–51.

LARSON, K. C. AND BERRY, R. E. (1984). Influence of peppermint phenolics and monoterpenes on twospotted spider mite (Acari: Tetranychidae). *Environmental Entomology* 13, 282–285.

LEIGH, T. F. (1985). Cotton. In *Spider Mites, Their Biology, Natural Enemies and Control* (W. Helle and M. W. Sabelis, Eds), volume 1B, pp. 349–358. Elsevier, New York.

LEIGH, T. F. AND WYNHOLDS, P. F. (1980). Insecticides enhance spider mite reproduction. *California Agriculture* 34, 14–15.

LESIEWICZ, D. S. (1981). *The Arthropod Fauna of the Field Corn Agroecosystem and Its Interactions with Soil-Applied Systemic Pesticides.* PhD dissertation, North Carolina State University. Raleigh. 90pp.

LEWONTIN, R. C. (1965). Selection for colonizing ability. In *The Genetics of Colonizing Species* (H. G. Baker and G. L. Stebbins, Eds), pp. 77–91. Academic Press, New York.

LIENK, S. E., CHAPMAN, P. J. AND CURTIS, O. F. (1956). Responses of apple trees to mite infestations: II. *Journal of Economic Entomology* 49, 350–353.

LIGHT, W. I. ST G. AND LUDLAM, A. B. (1972). The effects of fruit tree red spider mite (*Panonychus ulmi*) on yield of apple trees in Kent. *Plant Pathology* 21, 175–181.

LOGAN, J. A. (1982). Recent advances and new directions in phytoseiid population models. In *Recent Advances in Knowledge of the Phytoseiidae. University of California Publication 3284* (M. A. Hoy, Ed.), pp. 49–71. University of California, Berkeley, CA.

LUCKEY, T. D. (1968). Insecticide hormoligosis. *Journal of Economic Entomology* 61, 7–12.

McENROE, W. D. (1963). The role of the digestive system in the water balance of the twospotted spider mite. In *Advances in Acarology*, volume 1, pp. 225–231. Cornell University Press, Ithaca.

McENROE, W. D. AND DRONKA, K. (1971). Photobehavioural classes of the spider mite *Tetranychus urticae* (Acarina: Tetranychidae). *Entomologia experimentalis et applicata* 14, 420–424.

McMURTRY, J. A. (1982). The use of phytoseiids for biological control: progress and future prospects. In *Recent Advances in Knowledge of the Phytoseiidae. University of California Publication 3284* (M. A. Hoy, Ed.), pp. 23–48. University of California, Berkeley, CA.

McMURTRY, J. A. (1985). Citrus. In *Spider Mites. Their Biology, Natural Enemies and Control* (W. Helle and M. W. Sabelis, Eds), volume 1B, pp. 339–347. Elsevier, New York.

McMURTRY, J. A., HUFFAKER, C. B. AND VAN DE VRIE, M. (1970). Ecology of tetranychid mites and their natural enemies: a review I. Tetranychid enemies: their biological characters and the impact of spray practices. *Hilgardia* 40, 331–390.

MAGGI, U. L. AND LEIGH, T. F. (1983). Fecundity response of the twospotted spider mite to cotton treated with methyl parathion or phosphoric acid. *Journal of Economic Entomology* **76**, 20–25.

Margolies, D. C. and Kennedy, G. G. (1984). Population response of the twospotted spider mite, *Tetranychus urticae*, to host phenology in corn and peanut. *Entomologia experimentalis et applicata* **36**, 193–196.

Margolies, D. C. and Kennedy, G. G. (1985). Movement of the twospotted spider mite, *Tetranychus urticae*, among hosts in a corn-peanut agroecosystem. *Entomologia experimentalis et applicata* **37**, 55–61.

Margolies, D. C., Kennedy, G. G. and van Duyn, J. W. (1985). Effect of three soil-applied insecticides in field corn on spider mite (Acari: Tetranychidae) pest potential. *Journal of Economic Entomology* **78**, 117–120.

Margolies, D. C., Lampert, E. P. and Kennedy, G. G. (1984). Sampling programs for the twospotted spider mite (Acari: Tetranychidae) in peanut. *Journal of Economic Entomology* **77**, 1024–1028.

Mellors, W. K. and Propts, S. E. (1983). Effects of fertilizer level, fertility balance and soil moisture on the interaction of twospotted spider mites (Acarina: Tetranychidae) with radish plants. *Environmental Entomology* **12**, 1239–1244.

Metcalf, R. L. (1980). Changing roles of insecticides in crop protection. *Annual Review of Entomology* **25**, 219–256.

Miller, R. W., Croft, B. A. and Nelson, R. D. (1985). Effects of early season immigration on cyhexatin and formetanate resistance of *Tetranychus urticae* (Acari: Tetranychidae) on strawberry in central California. *Journal of Economic Entomology* **78**, 1379–1388.

Miranowski, J. A. and Carlson, G. A. (1986). Economic issues in public and private approaches to preserving pest susceptibility. In *Pesticide Resistance, Strategies and Tactics for Management*, pp. 436–448. National Academy Press, Washington, DC.

Mitchell, R. (1970). An analysis of dispersal in mites. *American Naturalist* **104**, 425–431.

Mitchell, R. (1972). The sex ratio of the spider mite *Tetranychus urticae*. *Entomologia experimentalis et applicata* **15**, 299–304.

Mitchell, R. (1973). Growth and population dynamics of a spider mite (*Tetranychus urticae* K., Acarina: Tetranychidae). *Ecology* **54**, 1349–1355.

Mori, H. (1961). Comparative studies on thermal reaction in four species of spider mites. *Journal of the Faculty of Agriculture Hokkaido University* **51**, 574–591.

Moriarty, F. (1969). The sublethal effects of insecticides in insects. *Biological Review* **44**, 321–357.

Mullin, C. A. and Croft, B. A. (1983). Host related alterations of detoxification enzymes in *Tetranychus urticae* (Acari: Tetranychidae). *Environmental Entomology* **12**, 1278–1282.

Muma, M. H. (1970). Preliminary studies on environmental manipulation to control injurious insects and mites in Florida citrus groves. *Proceedings Tall Timbers Conference on Ecological Animal Control by Habitat Management, 1970*, pp. 23–40. Tall Timbers Research Station, Tallahassee.

Nachman, G. (1985). Sampling Phytoseiidae. In *Spider Mites. Their Biology, Natural Enemies and Control* (W. Helle and M. W. Sabelis, Eds), volume 1B, pp. 175–182. Elsevier, New York.

National Academy of Sciences (1986). *Pesticide Resistance, Strategies and Tactics for Management*. National Academy Press, Washington, DC. 471 pp.

Neiswander, C. R., Rodriguez, G. and Neiswander, R. B. (1950). Natural and induced variations in twospotted spider mite populations. *Journal of Economic Entomology* **43**, 633–636.

Newsom, L. D. (1972). Predator insecticide relationships. *Entomophaga, Memoire Hors Serie* **7**, 13–23.

Nickel, J. L. (1960). Temperature and humidity relationships of *Tetranychus deserto-*

rum Banks with special reference to distribution. *Hilgardia* **30**, 41–100.

NOGGLE, G. R. AND FRITZ, G. J. (1976). *Introductory Plant Physiology*. Prentice-Hall, Englewood Cliffs, NJ. 688 pp.

OATMAN, E. R. AND VOTH, V. (1972). Effect of flooding on the twospotted spider mite and its predator on strawberry in southern California. *Environmental Entomology* **1**, 717–719.

OATMAN, E. R., McMURTRY, J. A., SHOREY, H. H. AND VOTH, V. (1967). Studies on integrating *Phytoseiulus persimilis* releases, chemical applications, cultural manipulations and natural predation for control of twospotted spider mite on strawberry in southern California. *Journal of Economic Entomology* **60**, 1344–1351.

OATMAN, E. R., WYMAN, J. A., BROWNING, H. W. AND VOTH, V. (1981). Effects of releases and varying infestation levels of the twospotted spider mite on strawberry yield in southern California. *Journal of Economic Entomology* **74**, 112–115.

OATMAN, E. R., SANCES, F. U., LaPRE, L. F., TOSCANO, N. C. AND VOTH, V. (1982). Effects of different infestation levels of the twospotted spider mite on strawberry yield in winter plantings in southern California. *Journal of Economic Entomology* **75**, 94–96.

OSBORNE, L. S. AND PETITT, F. L. (1985). Insecticidal soap and the predatory mite *Phytoseiulus persimilis* (Acari: Phytoseiidae) used in management of the twospotted spider mite (Acari: Tetranychidae) on greenhouse grown foliage plants. *Journal of Economic Entomology* **78**, 687–691.

OSMAN, A. A., ABO-KORAH, S. M. AND GHATTAS, A. (1985). Toxicity of some new pesticides to mites on cotton. *Indian Journal of Agricultural Science* **55**, 533–536.

OVERMEER, W. P. J. (1972). Notes on mating behaviour and sex ratio control of *Tetranychus urticae* Koch (Acarina: Tetranychidae). *Entomologische Berichten (Amsterdam)* **32**, 240–244.

OVERMEER, W. P. J. (1985). Alternative prey and other food resources. In *Spider Mites. Their Biology, Natural Enemies and Control* (W. Helle and M. W. Sabelis, Eds), volume 1B, pp. 131–139. Elsevier, New York.

OVERMEER, W. P. J., AND HARRISON, R. A. (1969). Notes on the control of the sex ratio in populations of the twospotted spider mite, *Tetranychus urticae* Koch (Acarina: Tetranychidae). *New Zealand Journal of Science* **12**, 920–928.

OWENS, J. C., WARD, C. R. AND TEETES, G. L. (1976). Current status of spider mites in corn and sorghum. *Proceedings 31st Annual Corn and Sorghum Research Conference*, pp. 38–64. American Seed Trade Association, Washington, DC.

PATTERSON, C. G., KNAVEL, D. E., KEMP, T. R. AND RODRIGUEZ, J. G. (1975). Chemical basis for resistance to *Tetranychus urticae* Koch in tomatoes. *Environmental Entomology* **4**, 670–674.

PENMAN, D. R. AND CHAPMAN, R. B. (1983). Fenvalerate-induced distributional imbalances of twospotted spider mite on bean plants. *Entomologia experimentalis et applicata* **33**, 71–78.

PENMAN, D. R., CHAPMAN, R. B. AND JENSON, K. E. (1981). Effects of fenvalerate and azinphosmethyl on twospotted spider mite and phytoseiid mites. *Entomologia experimentalis et applicata* **30**, 91–97.

PFADT, R. E. (1985). *Fundamentals of Applied Entomology*. Macmillan Publishing Company, New York. 742 pp.

PIELOU, P. P. (1960). The effect of DDT on oviposition and behaviour in the European red mite, *Panonychus ulmi* (Koch). *Canadian Journal of Zoology* **38**, 1147–1156.

PIMENTEL, D. (1976). World food crisis: energy and pests. *Bulletin of the Entomological Society of America* **22**, 20–26.

PIMENTEL, D., KRUMMEL, D. J., GALLAHAN, D., HOUGH, J., MERRILL, A., SCHREINER, I., VITTUM, P., KOZIOL, F., BACH, E., YEN, D. AND FIANCE, S. (1978). Benefits and costs of pesticide use in US food production. *BioScience* **28**, 772–784.

Poe, S. L. (1971). Influence of host plant phenology on populations of *Tetranychus urticae* infesting strawberry plants in peninsular Florida. *Florida Entomologist* **54**, 183–186.

Poe, S. L. (1979). *Tetranychus urticae* population development on vegetative and senescent host plants. *Proceedings of Fourth International Congress of Acarology* **4**, 215–220. Akadémiai Kiadó, Budapest.

Poskuta, J., Kolodziej, A. and Kropczynska, D. (1975). Photosynthesis, photorespiration and respiration of strawberry plants as influenced by infestation with *Tetranychus urticae* (Koch). *Fruit Science Report* **2**, 1–17.

Poston, F. L., Pedigo, L. P. and Welch, S. M. (1983). Economic injury levels: reality and practicality. *Bulletin of the Entomological Society of America* **29**, 49–53.

Potter, D. A. (1979). Reproductive behaviour and sexual selection in Tetranychine mites. *Recent Advances in Acarology* **1**, 137–145.

Putman, W. L. (1963). Lack of effect of DDT on fecundity and dispersion of the European red mite, *Panonychus ulmi* Koch (Acarina, Tetranychidae) in peach orchards. *Canadian Journal of Zoology* **41**, 603–610.

Putter, L., MacConnell, J. C., Preiser, F. A., Haidu, A. A., Ristich, S. S. and Dybas, R. A. (1981). Avermectins: novel insecticides, acaricides and nematicides from a soil micro-organism. *Experientia* **37**, 963–964.

Rabb, R. L. (1978). A sharp focus on insect populations and pest management from a wide-area view. *Bulletin of the Entomological Society of America* **24**, 55–61.

Rabbinge, R. (1985). Aspects of damage assessment. In *Spider Mites. Their Biology, Natural Enemies and Control* (W. Helle and M. W. Sabelis, Eds), volume 1B, pp. 261–272. Elsevier, New York.

Rabbinge, R. and Hoy, M. A. (1980). A population model for twospotted spider mite *Tetranychus urticae*, and its predator, *Metaseiulus occidentalis*. *Entomologia experimentalis et applicata* **28**, 64–81.

Raworth, D. A. (1986). An economic threshold function for the twospotted spider mite, *Tetranychus urticae* (Acari: Tetranychidae), on strawberries. *Canadian Entomologist* **118**, 9–16.

Riedl, H. and Hoying, S.A. (1980). Impact of fenvalerate and diflubenzuron on target and nontarget arthropod species on Bartlett pears in northern California. *Journal of Economic Entomology* **73**, 117–122.

Riedl, H. and Hoying, S. A. (1983). Toxicity and residual activity of fenvalerate to *Typhlodromus occidentalis* (Acari: Phytoseiidae) and its prey *Tetranychus urticae* (Acari: Tetranychidae) on pear. *Canadian Entomologist* **115**, 807–814.

Rishi, N. D. and Rather, A. Q. (1983). Life cycle of phytophagous mite *Tetranychus urticae* Koch and its predatory mite *Amblyseius finlandicus* Oudemans together with effect of certain acaricides/fungicides on them. *Journal of Entomological Research* **71**, 39–42.

Rodriguez, J. G. (1951). Mineral nutrition of the twospotted spider mite, *Tetranychus bimaculatus* Harvey. *Annals of the Entomological Society of America* **44**, 511–526.

Rodriguez, J. G. (1954). Radiophosphorus in metabolism studies in the twospotted spider mite. *Journal of Economic Entomology* **47**, 514–517.

Rodriguez, J. G., Chen, H. H. and Smith, W. T., Jr (1960). Effects of soil insecticides on apple trees and resulting effect on mite nutrition. *Journal of Economic Entomology* **53**, 487–490.

Rodriguez, J. G., Maynard, D. E. and Smith, W. T., Jr (1960). Effects of soil insecticides and absorbents on plant sugars and resulting effect on mite nutrition. *Journal of Economic Entomology* **53**, 491–495.

Rodriguez, J. G., Chaplin, C. E., Stoltz, L. P. and Lasheen, A. M. (1970). Studies of resistance of strawberries to mites. I. Effects of plant nitrogen. *Journal of Economic Entomology* **63**, 1856–1858.

Roush, R. T. and Hoy, M. A. (1978). Relative toxicity of permethrin to a predator,

Metaseiulus occidentalis and its prey *Tetranychus urticae*. *Environmental Entomology* **7**, 287–288.

ROUSH, R. T. AND HOY, M. A. (1981a). Genetic improvement of *Metaseiulus occidentalis*: selection with methomyl, dimethoate and carbaryl and genetic analysis of carbaryl resistance. *Journal of Economic Entomology* **74**, 138–141.

ROUSH, R. T. AND HOY, M. A. (1981b). Laboratory, greenhouse and field studies of artificially selected carbaryl resistance in *Metaseiulus occidentalis*. *Journal of Economic Entomology* **74**, 142–147.

SADA, F. (1961). Uber entwicklung und ruckgang der giftresistenz bei *Tetranychus urticae* Koch und deren abhangigkeit von der wirtspflanze. *Zeitschrift für angewandte Entomologie* **48**, 265–293.

SABELIS, M. W. (1985a). Reproductive strategies. In *Spider Mites. Their Biology, Natural Enemies and Control* (W. Helle and M. W. Sabelis, Eds), volume 1A, pp. 265– 278. Elsevier, New York.

SABELIS, M. W. (1985b). Reproduction. In *Spider Mites. Their Biology, Natural Enemies and Control* (W. Helle and M. W. Sabelis, Eds), volume 1B, pp. 73–83. Elsevier, New York.

SABELIS, M. W. (1985c). Predation on spider mites. In *Spider Mites. Their Biology, Natural Enemies and Control* (W. Helle and M. W. Sabelis, Eds), volume 1B, pp. 103–129. Elsevier, New York.

SABELIS, M. W. (1985d). Capacity for population increase. In *Spider Mites. Their Biology, Natural Enemies and Control* (W. Helle and M. W. Sabelis, Eds), volume 1B, pp. 35–41. Elsevier, New York.

SABELIS, M. W. (1985e). Sampling techniques. In *Spider Mites. Their Biology, Natural Enemies and Control* (W. Helle and M. W. Sabelis, Eds), volume 1A, pp. 337–350. Elsevier, New York.

SABELIS, M. W. AND DICKE, M. (1985). Long range dispersal and searching behaviour. In *Spider Mites. Their Biology, Natural Enemies and Control* (W. Helle and M. W. Sabelis, Eds), volume 1B, pp. 141–160. Elsevier, New York.

SAINI, R. S. AND CUTKOMP, L. K. (1966). The effects of DDT and sublethal doses of dicofol on reproduction of the twospotted spider mite. *Journal of Economic Entomology* **59**, 249–253.

SANCES, F. V., WYMAN, J. A., TING, I. P. (1979a). Morphological responses of strawberry leaves to infestations of twospotted spider mite. *Journal of Economic Entomology* **72**, 710–713.

SANCES, F. V., WYMAN, J. A. AND TING, I. P. (1979B). Physiological responses to spider mite infestations on strawberries. *Environmental Entomology* **8**, 711–714.

SANCES, F. V., WYMAN, J. A., TING, I. P., VAN STEENWYK, R. A. AND OATMAN, E. R. (1981). Spider mite interactions with photosynthesis, transpiration and productivity of strawberry. *Environmental Entomology* **10**, 442–448.

SANCES, F. V., TOSCANO, N. C., LAPRE, L. F., OATMAN, E. R. AND JOHNSON, M. W. (1982a). Spider mites can reduce strawberry yields. *California Agriculture* **36**, 14–16.

SANCES, F. V., TOSCANO, N. C., OATMAN, E. R., LAPRE, L. F., JOHNSON, M. W. AND VOTH, V. (1982b). Reductions in plant processes by *Tetranychus urticae* (Acari: Tetranychidae) feeding on strawberry. *Environmental Entomology* **11**, 733–737.

SCHWEISSING, F. C. (1969). *Cultural Practice for Possible Use in Reducing Spider Mite Damage in Corn. Progress Report PR69–72, Colorado State University*. 2 pp.

SCHWEISSING, F. C. (1973). *Approaches to the Management of Banks Grass Mite on Corn. Progress Report PR73–41, Colorado State University*. 3 pp.

SHAW, P. B. (1984). Simulation model of a predator-prey system comprised of *Phytoseiulus persimilis* Anthias–Henriot (Acari: Phytoseiidae) and *Tetranychus urticae* Koch (Acari: Tetranychidae). I. Structure and validation of the model. *Researches on Population Ecology* **26**, 235–259.

SHAW, P. B. (1985). Simulation model of a predator-prey system comprised of *Phytoseuilus persimilis* and *Tetranychus urticae*. II. Model sensitivity to variations in the life history parameters of both species and to variations in the functional response and components of the numerical response. *Researches on Population Ecology* **27**, 1–24.

SILBERMAN, L. L. (1983). *Aerial Dispersal of the Twospotted Spider Mite*, Tetranychus urticae (*Koch*), *from its Host Plants*. MS thesis, North Carolina State University, Raleigh. 68 pp.

SITES, R. W. AND CONE, W. W. (1985). Vertical dispersion of twospotted spider mites, *Tetranychus urticae* on hops *Humulus lupulus* throughout the growing season. *Journal of the Entomological Society of British Columbia* **82**, 22–25.

SMITH, J. W. AND FURR, R.E. (1975). Spider mites and some natural control agents found on cotton in the delta area of Mississippi. *Environmental Entomology* **4**, 559–560.

SMITH MEYER, M. K. P. (1981). Mite pests of crops in southern Africa. *Plant Protection Research Institute, Pretoria, Science Bulletin* **397**, 59–86.

SMITLEY, D. R. AND KENNEDY, G. G. (1985). Photo-oriented aerial-dispersal behavior of *Tetranychus urticae* (Acari: Tetranychidae) enhances escape from the leaf surface. *Annals of the Entomological Society of America* **78**, 609–614.

SMITLEY, D. R. AND KENNEDY, G. G. (1987). Identification of factors leading to the aerial dispersal of the twospotted spider mite (*Tetranychus urticae* Koch) from field corn. *Experimental and Applied Acarology*, in press.

SMITLEY, D. R., BROOKS, W. M. AND KENNEDY, G. G. (1986). Environmental effects on production of primary and secondary conidia, infection and pathogenesis of *Neozygites floridana*, a pathogen of the twospotted spider mite, *Tetranychus urticae*. *Journal of Invertebrate Pathology* **47**, 325–332.

SMITLEY, D. R., KENNEDY, G. G. AND BROOKS, W. M. (1986). Role of entomogenous fungus, *Neozygites floridana*, in population declines of the twospotted spider mite, *Tetranychus urticae*, on field corn. *Entomologia experimentalis et applicata* **41**, 255–264.

SNYDER, J. C. AND CARTER, C. D. (1984). Leaf trichomes and resistance of *Lycopersicon hirsutum* and *L. esculentum* to spider mites. *Journal of the American Society for Horticultural Science* **109**, 837–843.

SPADAFORA, R. R. AND LINDQUIST, R. K. (1972). Ovicidal action of benomyl on eggs of the twospotted spider mite. *Journal of Economic Entomology* **65**, 1718–1720.

STACEY, D. L., WYATT, I. J. AND CHAMBERS, R. J. (1985). The effect of glasshouse red spider mite damage on the yield of tomatoes. *Journal of Horticultural Science* **60**, 517–523.

STORMS, J. J. H. (1971). Some physiological effects of spider mite infestation on bean plants. *Netherland Journal of Plant Pathology* **77**, 154–167.

SUMMERS, F. M. AND STOCKING, C. R. (1972). Some immediate effects on almond leaves of feeding by *Bryobia rubrioculus* (Scheuten). *Acarologia* **14**, 170–178.

SUSKI, Z. W. AND BADOWSKA, T. (1975). Effect of the host plant nutrition on the population of the twospotted spider mite. *Tetranychus urticae* Koch (Acarina: Tetranychidae). *Ekologia Polska* **23**, 185–209.

SUSKI, Z. W. AND NAEGELE, J. A. (1963). Light response in the twospotted spider mite. II. Behavior of the sedentary and dispersal phases. In *Advances in Acarology* (J. A. Naegele, Ed.), volume 1, pp. 193–198. Cornell Press, Ithaca, New York.

TABASHNIK, B. E. AND CROFT, B. A. (1982). Managing pesticide resistance in crop–arthropod complexes: interactions between biological and operational factors. *Environmental Entomology* **11**, 1137–1144.

TABASHNIK, B. E. AND CROFT, B. A. (1985). Evolution of pesticide resistance in apple pests and their natural enemies. *Entomophaga* **30**, 37–49.

TAKAFUJI, A. AND KAMBAYASHI, M. (1984). Life cycle of a nondiapausing population of the twospotted spider mite. *Tetranychus urticae* Koch in a pear orchard. *Researches on Population Ecology* **26**, 113–123.

TANIGOSHI, L. K. AND BROWNE, R. W. (1981). Coupling the cytological aspects of spider mite feeding to economic injury levels on apple. *Protection Ecology* **2**, 29–40.

TANIGOSHI, L. K. AND DAVIS, R. W. (1978). An ultrastructural study of *Tetranychus medanieli* feeding injury to the leaves of 'red delicious' apple (Acari: Tetranychidae). *International Journal of Acarology* **4**, 47–56.

TAYLOR, C. E. AND GEORGHIOU, G. P. (1979). Suppression of insecticide resistance by alteration of dominance and migration. *Journal of Economic Entomology* **72**, 105–109.

TOMCZYK, A. AND KROPCZYNSKA, D. (1985). Effects on the host plant. In *Spider Mites. Their Biology, Natural Enemies and Control* (W. Helle and M. W. Sabelis, Eds), volume 1A, pp. 312–330. Elsevier, New York.

TOWNSEND, J. F. AND LUCKEY, T. D. (1960). Hormoligosis in pharmacology. *Journal of the American Medical Association* **173**, 44–48.

TRICHILO, P. J. AND LEIGH, T. F. (1985). The use of life tables to assess varietal resistance of cotton to spider mites. *Entomologia experimentalis et applicata* **39**, 27–33.

TRUMBLE, J. T. (1985). Implications of changes in arthropod distribution following chemical application. *Researches on Population Ecology* **27**, 277–285.

TULISALO, U. (1974). Control of the twospotted spider mite *Tetranychus urticae* Koch by high air humidity or direct contact with water. *Annales entomologici Fennici* **40**, 158–162.

UNWIN, B. (1971). Biology and control of the twospotted spider mite. *Tetranychus urticae* Koch. *Journal of the Australian Institute of Agricultural Science* **37**, 192–211.

VAN DE VRIE, M. (1985). Greenhouse ornamentals. In *Spider Mites. Their Biology, Natural Enemies and Control* (W. Helle and M. W. Sabelis, Eds), volume 1B, pp. 273–278. Elsevier, New York.

VAN DE VRIE, M., MCMURTRY, J. A. AND HUFFAKER, C. B. (1972). Ecology of tetranychid mites and their natural enemies: a review III. Biology, ecology and pest status and plant relations of tetranychids. *Hilgardia* **41**, 343–432.

VEERMAN, A. (1985). Diapause. In *Spider Mites. Their Biology, Natural Enemies and Control* (W. Helle and M. W. Sabelis, Eds), volume 1A, pp. 279–316. Elsevier, New York.

WELCH, S. M. (1979). The application of simulation models to mite pest management. In *Recent Advances in Acarology* (J. G. Rodriguez, Ed.), pp. 31–40. Academic Press, New York.

WERMELINGER, B., OERTLI, J. J. AND DELUCCHI, V. (1985). Effect of host plant nitrogen fertilization on the biology of the twospotted spider mite *Tetranychus urticae*. *Entomologia experimentalis et applicata* **38**, 23–28.

WILSON, L. T., GONZALEZ, D., LEIGH, T. F., MAGGI, U., FORISTIERE, C. AND GOODELL, P. (1983). Within-plant distribution of spider mites (Acari: Tetranychidae) on cotton: a developing implementable monitoring program. *Environmental Entomology* **12**, 128–134.

WILSON, L. T., HOY, M. A., ZALOM, F. G. AND SMILANICK, J. M. (1984). Sampling mites in almonds: I. The within-tree distribution and clumping pattern of mites with comments on predator–prey interactions. *Hilgardia* **52**, 1–13.

WONG, S.W. AND CHAPMAN, R. B. (1979). Toxicity of synthetic pyrethroid insecticides to predaceous phytoseiid mites and their prey. *Australian Journal of Agricultural Research* **30**, 497–501.

WRENSCH, D. L. (1979). Components of reproductive success in spider mites. In *Recent Advances in Acarology* (J. G. Rodriguez, Ed.), volume 2, pp. 155–165. Elsevier, New York.

Wrensch, D. L. (1985). Reproductive parameters. In *Spider Mites. Their Biology, Natural Enemies and Control* (W. Helle and M. W. Sabelis, Eds), volume 1A, pp. 165–170. Elsevier, New York.

Wrensch, D. L. and Young, S. S. Y. (1975). Effects of quality of resource and fertilization on some fitness traits in the twospotted spider mite, *Tetranychus urticae* Koch, *Oecologia* (Berlin) 18, 259–269.

Wrensch, D. L. and Young, S. S. Y. (1978). Effects of density and host quality on rate of development, survivorship and sex ratio in the carmine spider mite. *Environmental Entomology* 7, 499–501.

Wrensch, D. L. and Young, S. S. Y. (1983). Relationship between primary and tertiary sex ratio in the twospotted spider mite (Acarina: Tetranychidae). *Annals of the Entomological Society of America* 7, 786–789.

Wysoki, M. (1985). Other outdoor crops. In *Spider Mites. Their Biology, Natural Enemies and Control* (W. Helle and M. W. Sabelis, Eds), volume 1B. pp. 375–384. Elsevier, New York.

Young, S. S. Y. and Wrensch, D. L. (1981). Relative influence of fitness components on total fitness of the twospotted spider mite in different environments. *Environmental Entomology* 10, 1–5.

Young, S. S. Y., Wrensch, D. L. and Kongchuensin, M. (1985). Geographic variations and combining abilities in the twospotted spider mite *Tetranychus urticae*. *Entomologia experimentalis et applicata* 39, 109–113.

Zalom, F. G., Hoy, M. A., Wilson, L. T. and Barnett, W. W. (1984a). Sampling *Tetranychus* spider mites in almonds. *California Agriculture* 38, 17–19.

Zalom, F. G., Hoy, M. A., Wilson, L. T. and Barnett, W. W. (1984b). Sampling mites in almonds. II. Presence–absence sequential sampling for *Tetranychus* mite species. *Hilgardia* 52, 14–24.

Zwick, R. W., Fields, G. J. and Mellenthin, W. M. (1976). Effects of mite population density on 'Newton' and 'Golden Delicious' apple tree performance. *Journal of the American Society for Horticultural Science* 101, 123–125.

10
Population Dynamics and Management of *Heterodera glycines*

D. P. SCHMITT* AND R. D. RIGGS†

Department of Plant Pathology, Box 7631, North Carolina State University, Raleigh, NC 27695, USA and †Department of Plant Pathology, University of Arkansas, Fayetteville, AR 72701, USA

Embryogenesis and development
Factors influencing the life stages of *Heterodera glycines*
Seasonal population changes
Managing populations of *Heterodera glycines*
References

Heterodera glycines Ichinohe is probably indigenous to the orient. The nematode is currently known to occur in Japan, the People's Republic of China, Korea, USA, (Schmitt and Noel, 1984), the Soviet Union, Colombia, Indonesia and Canada (Anderson and Welacky, 1988). It was found in 26 states of the USA between 1954 and 1987. Incidence of isolated occurrence and apparent rapid spread have been used as evidence that this nematode may be indigenous to some of these areas. DNA analyses of several populations, however, lead to the interpretation that it was introduced to the USA (Radice, 1987).

The disease, 'yellow dwarf', caused by *H. glycines* was reported in Japan in 1916 (Hori, 1916) but had been noticed as early as 1881 (Riggs, 1977). The causal nematode was referred to as *Heterodera schachtii* Schmidt (Filipjev and Schuurmans-Stekhoven, 1941) and as *H. gottingiana* Liebscher (Goffart, 1951). It was named *H. glycines* in 1952 (Ichinohe, 1952). It became a severe pest of soybean in the USA when production of this crop was intensified and expanded over a wide geographical area.

Early research on *H. glycines* was directed largely at control. Some biological, ecological and epidemiological aspects were characterized recently (Schmitt and Noel, 1984; Francl, 1986). This species is highly variable genetically. This variability impacts *H. glycines*-induced yield losses which are modified by environmental factors, especially soil moisture and texture

342   D. P. Schmitt and R. D. Riggs

(Schmitt, Ferris, and Barker, 1987). The genetic diversity also complicates control of *H. glycines*. Ability to survive long periods without a host further frustrates control efforts. For example, diapause is a major means of survival for the nematode, especially during autumn, winter and spring.

Options for management of *H. glycines* on soybean, a low cash value crop, are limited. Therefore, knowledge of the nematode's biology and population dynamics is critical in developing management programmes. The programme should capitalize on the knowledge of the relationship between numbers of the nematode and associated yield losses. Available control tactics can be integrated with modifications in cultural practices to give practical and economical management of *H. glycines*.

Embryogenesis and development

Heterodera glycines progresses through a series of stages in its life cycle to develop from oocytes to mature males and females. Oogenesis is similar to that in most other animals, with some exceptions (Triantaphyllou, 1971). Oogonia multiply in the germinal zone of the ovary. As oocytes migrate through the ovary, they increase in size and move into a large sperm-filled spermatheca to be fertilized. Most or all of the eggs of *Heterodera* species are maintained in elongated uteri where they undergo embryogenesis (Triantaphyllou, 1971).

The first two juvenile stages of *H. glycines* are completed within the egg. The first-stage juvenile (J1) moults and becomes a second-stage juvenile (J2). Further development of the J2 and subsequent stages occurs after hatching and is dependent upon the establishment of a feeding site in a compatible host (Lauritis, Rebois and Graney, 1983). Penetration by the J2 most frequently occurs at or behind the zone of maturation of the root, where tissue is young but differentiated (*Figure* 1). The J2 moves into the root and establishes a feeding site near the pericycle. It begins to swell and then moults to become a J3. The posterior portion of the J3 begins to break through the surface of the root. The J3 stage lasts less than 24 hours before moulting to become a J4, at which time sex differentiation is evident. J4 males moult, 1–3 days earlier than J4 females. The young female produces a gelatinous matrix from the vulval region. Males exit from the fourth-stage cuticle and entwine around the gelatinous matrix of the female. Copulation occurs and sperm migrate to the spermatheca where eggs are fertilized. Eggs fill the body of the white female, and some are deposited in the gelatinous matrix. Females change in colour from white to yellow and when they die they become brown cysts which protect the eggs contained within them.

Factors influencing the life stages of *Heterodera glycines*

Heterodera glycines, a poikilothermic organism, is closely regulated by environmental conditions. Temperature and moisture are the factors most dynamically involved in the life processes.

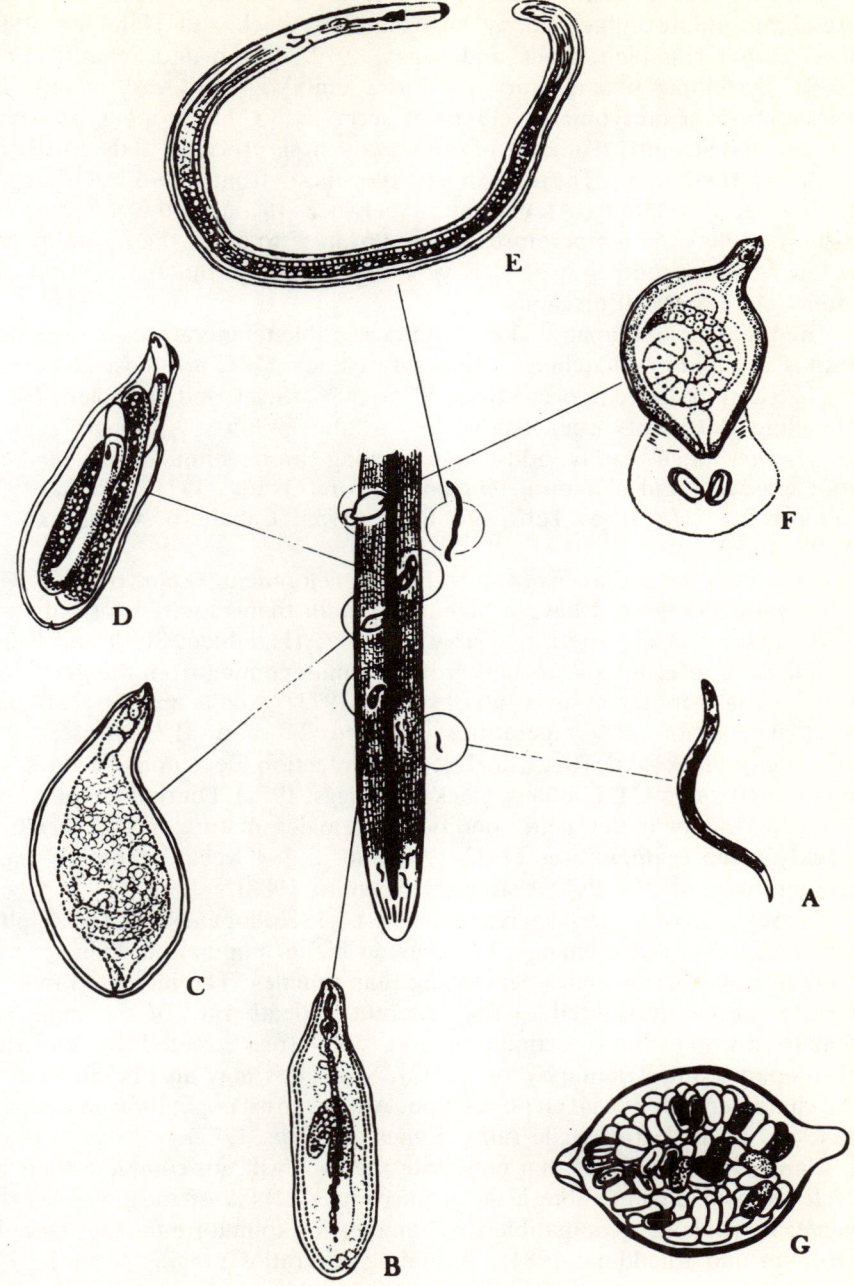

Figure 1. Life cycle of *Heterodera glycines*. (A) Second-stage juvenile; (B) third-stage juvenile; (C) fourth-stage female juvenile; (D) fourth-stage male juvenile; (E) male; (F) female; (G) egg-filled cyst.

Embryogenesis, hatching and post-infection development of the nematode are differentially influenced by environment (Slack and Hamblen, 1961; Ross, 1964; Hamblen, Slack and Riggs, 1972; Alston and Schmitt, 1987, 1988). Development and mortality during embryogenesis vary at different temperatures. Embryonic development occurs at 5°C but at a very slow rate (Alston and Schmitt, 1988). At 10°C embryogenesis proceeds to the formation of the J1 then stops. The rate of embryogenesis from a two-celled egg to J2 is directly related to temperature between 15° and 30°C (Alston and Schmitt, 1988). At temperatures ranging from 5° to 30°C, the mortality rate is low. Embryogenesis at 36°C is very rapid initially, but the embryo dies after only a few cell divisions.

The breaking of diapause along with favourable temperatures in the spring results in rapid egg hatching. Temperatures near 24°C are most favourable for hatching, but it can occur from 20° to 30°C (Slack and Hamblen, 1961). Hatching does not occur below 16° or above 36°C. Although eggs of *H. glycines* hatch readily, additional hatching can be stimulated by soybean root exudates and a natural hatching factor (Okada, 1971; Okada, 1972; Fukuzawa *et al.*, 1985; Tefft and Bone, 1985; Caballero, Osmanski and Tefft, 1986).

Root penetration and post-infection development occur over a wide temperature range but have a higher optimum than required for hatching. The highest rate of penetration occurs at 28°C (Hamblen, Slack and Riggs, 1972). Root infection occurs earlier by juveniles coming from the gelatinous matrix than from the cysts (Ishibashi *et al.*, 1973). Adults appear 12–14 days after inoculation at temperatures between 24° and 31°C (Ross, 1964; Hamblen, Slack and Riggs, 1972). Post-infection development does not occur at 10° or 35°C (Hamblen, Slack and Riggs, 1972). Diurnal temperatures fluctuate widely under field conditions. Females mature in 2–3 weeks at weekly mean temperatures of 22–29°C and in 3–4 weeks at weekly mean temperatures of 17–22°C (Alston and Schmitt, 1988).

The sex ratio of *H. glycines* is genetically 1 : 1 (Koliopanos and Triantaphyllou, 1972). Sex ratios change, however, at higher population densities with a greater number of males developing than females. The increased male to female ratio is attributed to the differential death rate of the male and female juveniles under conditions of food stress created by crowding (Koliopanos and Triantaphyllou, 1972). Sex ratios may also be affected by the characteristics of a given population. An Arkansas population consistently gave a 3 : 1 male to female ratio (Evans and Fox, 1977).

Many of the juveniles that penetrate the root will not complete their life cycle even in a compatible host. Approximately 14% of the juveniles that penetrate roots of a compatible soybean develop to mature females (Acedo, Dropkin and Luedders, 1984). Assuming sex ratios ranging from 1 : 1 to 3 : 1 (male : female), an additional 14% to 42% of the juveniles become males. Thus, only 28–56% survive to maturity.

Data on the effects of soil moisture on penetration and development of *H. glycines* are confusing. More juveniles penetrate roots in wet soils than in drier soils (Yoshihara and Kegasawa, 1974). After penetration, on the other hand, development is favoured by drier conditions. The greatest development occurs in the upper 15 cm of soil at -0.4 bars moisture tension (Heatherly *et al.*, 1982). In contrast, at 15–30 cm deep, soils wetter or drier than -0.3 to -0.4 bars are more favourable for cyst development (Heatherly *et al.*, 1982). *H. glycines* penetrates and develops readily in hemp sesbania growing in flooded rice fields in Arkansas (Riggs, unpublished).

Seasonal population changes

Increase and decline in numbers of various life stages of *H. glycines* are related primarily to host-plant vigour and selected environmental conditions. J2 densities fluctuate more than those of eggs or cysts (*Figure 2*). The ephemeral nature of the J2 in soil probably accounts for the rapid changes in its numbers over time (*Figure 2*). The soil J2 population decreases rapidly as they penetrate roots or die. Soil J2 population fluctuations vary in amplitude throughout the growing season (Bonner and Schmitt, 1985). Although generations are not synchronous, changes in numbers of J2 are somewhat co-ordinated with generation time.

Cysts and eggs gradually increase during the growing season of a host and slowly decrease in number during periods when a host is absent. Carryover from autumn to spring is primarily in the egg stage with most

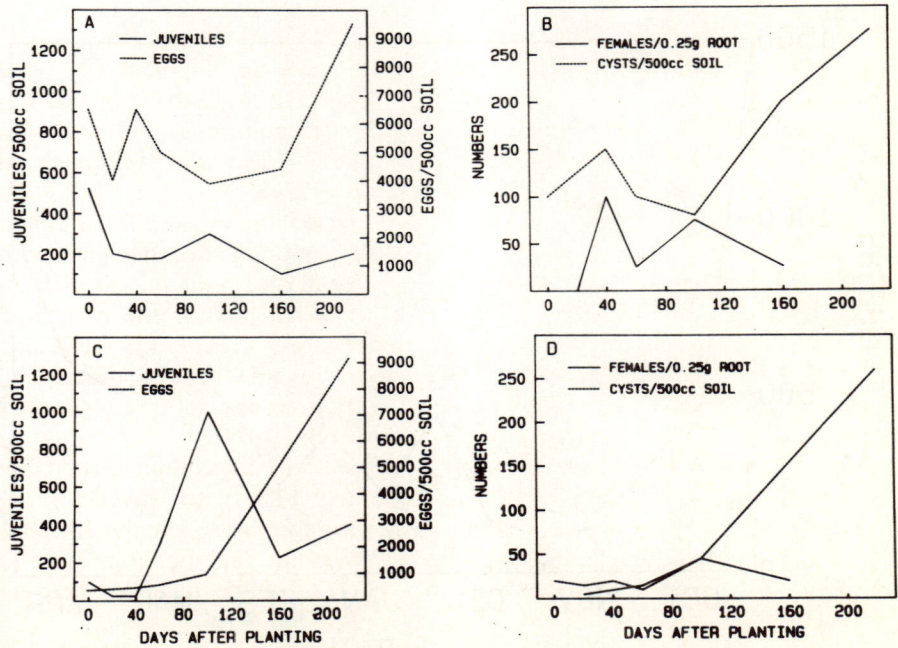

Figure 2. Population fluctuations of cysts, eggs and juveniles of *Heterodera glycines*. (A) and (B) High initial population density field; (C) and (D) low initial population density field (source Bonner and Schmitt, 1985).

eggs being encysted. Numbers of eggs decrease early in the growing season because environmental conditions are favourable for hatching (Slack and Hamblen, 1961; Ross, 1963; Okada, 1972). Egg numbers begin to increase when the first generation is completed but do not change appreciably until late in the growing season (Schmitt, Corbin and Nelson, 1983; Bonner and Schmitt, 1985; Bostian, Schmitt and Barker, 1986b). The hot, dry soils (temperatures greater than 30°C) that commonly occur in midsummer are unfavourable for *H. glycines* development (Alston and Schmitt, 1987). Soil temperatures in early autumn tend to be optimal for hatching and development (Slack and Hamblen, 1961; Ross, 1963). The rate of reproduction increases on plants producing fruit (Hill, 1985). Egg numbers peak late in the season (*Figure 2*). The J2 population density decreases rapidly when the crop is maturing, soil temperatures are cooling and the photoperiod is shortening (Ross, 1963; Bonner and Schmitt, 1985; Riggs, unpublished). Diapause is initiated when temperatures are declining (*Figure 3*) (Ross, 1963; Koenning and Schmitt, 1985), although photoperiod and host senescence also appear to be involved (*Figure 3*). This diapause aids winter survival of *H. glycines*. Hatching will increase slightly during the winter and spring, with significant hatching beginning in the spring (Ross, 1963; Koenning and Schmitt, 1985). In Illinois, USA, eggs overwinter in diapause with little mortality (Noel, personal communications). In North Carolina, winter survival is variable with high survival some winters and low survival

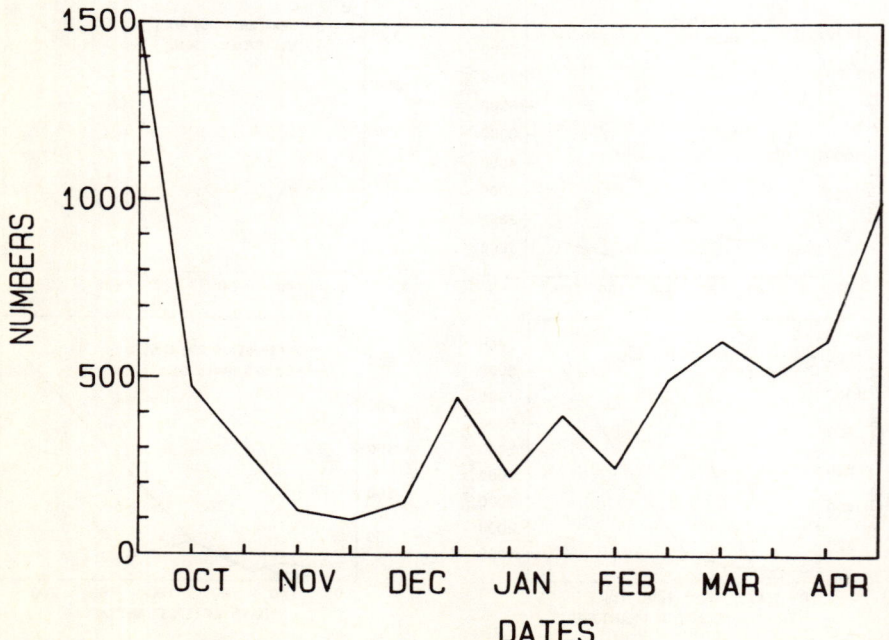

Figure 3. Juvenile emergence from cysts of *Heterodera glycines* from October to April (source Ross, 1963).

in others (Schmitt, unpublished). Eggs with fully developed J2 appear to be the primary stage for survival of *H. glycines* when a host is absent (Schmitt, unpublished). In the spring after diapause is broken, some eggs hatch spontaneously (Ross, 1963; Bonner and Schmitt, 1985), but soybean root diffusates stimulate additional hatching (Okada, 1972; Tefft and Bone, 1985).

The most basic component in the population dynamics of *H. glycines* is the availability of a suitable host. The nematodes population levels decline on non-hosts (Ross, 1962b), poor hosts such as pea (*Pisum sativum*) and resistant cultivars of soybeans. Prolonged use of cultivars resistant to *H. glycines* results in a shift in the nematode genotype (Epps and Chambers, 1965; Epps and Chambers, 1966; Riggs, Hamblen and Rakes, 1977; Sortland and MacDonald, 1987). For example, reproductive potential of *H. glycines* populations changed from 12% to 71% in 3 years of continuous cropping with resistant cultivars (Elliott, Phipps and Terrill, 1986). Selection also occurs under laboratory conditions (Triantaphyllou, 1975).

Population dynamics of *H. glycines* varies among geographical locations and years because soil factors, soil moisture and temperature (Chelbi, 1984) and crop production systems differ. The soybean planting date also affects the population dynamics of the nematode (Schmitt, unpublished). Initial population density is important because of its relationship to root damage and subsequent root growth. High initial levels cause extensive root damage, thereby limiting food supply and subsequent nematode reproduction. A low initial population causes little root damage; thus, an abundant food supply is available for subsequent generations.

Date of planting and cultivar maturity date affect population growth (Hill, 1985) (*Table 1*). A late-maturing soybean cultivar planted early provides a food substrate for *H. glycines* for up to 6 months. Five to eight nematode generations could develop during this period. In contrast, an early maturing cultivar planted late, a situation that occurs frequently with soybeans double-cropped with wheat, could be grown for as little as 3 months. There are situations in which end-of-season populations may not be affected by planting date (Hussey and Boerma, 1983). Soil moisture and temperature are probably

Table 1. End of season numbers of eggs of *Heterodera glycines* on a Group V soybean planted in late June, and a Group VII soybean planted in mid-May in North Carolina, USA

	Craven County	Perquimans County	Washington County
Group V	16 683 b	11 800 b	33 809 b
Group VII	51 233 a	41 467 a	42 650 a

Group V is an early maturity (Oct. 1–15) and Group VII is a late maturity (Nov. 1–15) cultivar for North Carolina, USA.
Numbers in a column followed by the same letter are not significantly different at $P = 0.05$.

major factors in accounting for differences in population change as related to planting date because of the impact they have on hatching, penetration and development.

Pesticides are common inputs for soybean production and these often influence both target and non-target organisms. Nematicides are designed to alter and modify population levels of target organisms such as *H. glycines*. In addition, some herbicides affect the population dynamics of *H. glycines* (Schmitt, Corbin and Nelson, 1983). The general configuration of the population curve may not be changed, but the magnitude is altered (*Figure 4*) (Schmitt, Corbin and Nelson, 1983). When two or more pesticides are applied to the same site for different purposes, the effect on the target can change. Fenamiphos applied alone prevents an increase in egg numbers of *H. glycines* at mid-season, whereas the population treated with fenamiphos and the herbicide alachlor behaves similarly to untreated controls and also results in a large end-of-season population. Development in the root is completely inhibited by fenamiphos applied alone but addition of alachlor allows development to proceed, although at a slower rate (Sipes, 1987). Some herbicides alone affect hatching and penetration of *H. glycines*. Alachlor (Bostian, Schmitt and Barker, 1986a) and trifluralin (Kraus, Noel and Edwards, 1982; Riggs and Oliver, 1982) enhance hatching of *H. glycines*. Alachlor increases root penetration by J2 (Bostian, 1982).

The mechanism by which pesticides affect *H. glycines* population dynamics is not the subject of this review. Nevertheless, research is needed in this

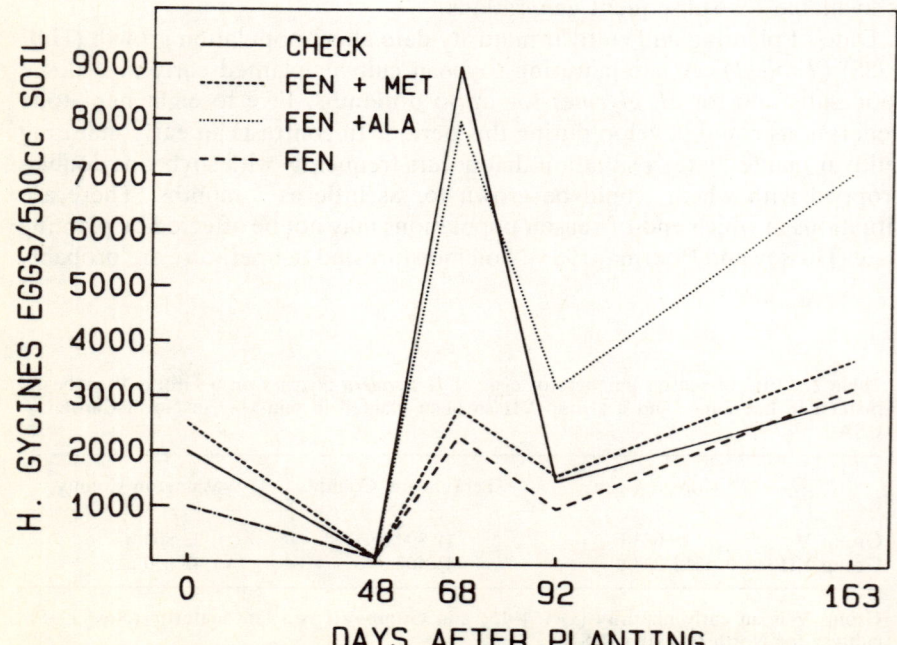

Figure 4. Changes in population density of *Heterodera glycines* eggs influenced by pesticides. FEN = fenamiphos, MET = metribuzin, ALA = alachlor (source Schmitt, Corbin and Nelson, 1983).

area from a practical aspect. An understanding of the positive and negative facets of pesticides will allow manipulations to provide better nematode management.

Cultural practices used in soybean production vary widely among growers. Each practice impacts micro-environments and the overall agroecosystem. Intensity of tillage, fertilizer application and cultivars have major ecological consequences for the population dynamics of *H. glycines*. The effects of tillage may be evident only after an extended period, but some difference in population fluctuations may be noted after only one season (Baird and Bernard, 1984; Schmitt and Nelson, 1987). Significant changes in the population may be exhibited only after three or more years (Schmitt, unpublished). With time, root development is affected by changes in soil structure due to tillage. Soil may become compacted in no-till, thus inhibiting root growth (Paul Denton, personal communication).

Soil fertility affects *H. glycines* population dynamics in at least two ways. The quantity of nutrients affects plant vigour which, in turn, affects *H. glycines* (Bernstein and Ogata, 1966). Some fertilizers at high rates may be toxic to the nematode. Reproduction of *H. glycines* on non-nodulating soybean isolines is much less with excessive rates of $NaNO_3$ per hectare (1882 kg) compared to low rates (171 and 470 kg ha^{-1}) (Ross, 1959). With more typical field rates, a general linear increase in population occurs with increase in nitrogen rates (0–95 kg ha^{-1}) (Ross, 1959). The increase in nematode population is directly related to enhanced plant growth. Adequate fertility is necessary for good root growth; thus, enhanced vigour of a plant should provide more feeding sites for *H. glycines*, enabling it to produce more progeny (Luedders, Shannon and Baldwin, 1979).

Heterodera glycines survives for relatively long periods without a host. This nematode was found in a field after 4 years of fallow (Inoue, 1959) and in soil maintained in plastic bags, both air-dried and at natural field moisture conditions, for up to 9 years (Inagaki and Tsutsumi, 1971). In Arkansas, USA, it survived 7 years of fallow and then increased to damaging levels in 1 year of a host (Slack, Riggs and Hamblen, 1981). Populations decrease rapidly in the first 2 years of a non-host (Ross, 1962b; Slack, Riggs and Hamblen, 1981). The cyst wall of *H. glycines* contains a hatch inhibitor (Okada, 1972), which is more effective from cysts collected in winter than in the spring (Kaul, 1962).

Management of the soybean cyst nematode can be greatly enhanced when the biological and ecological aspects of the nematode are considered, even if our understanding of them is incomplete. As we learn more about *H. glycines* and its behaviour, we will be able to better manipulate its population dynamics to assure economically profitable production of soybean, the primary agricultural host of this important pest.

Managing populations of *Heterodera glycines*

Yield losses due to *H. glycines* occur wherever the nematode is found (Riggs, 1977; Schmitt and Noel, 1984; Riggs and Schmitt, 1987). Damage is related to the population at planting (Noel, 1982; Schmitt, Ferris and Barker, 1987) and usually is severe only in irregular patches, but can involve entire fields. Management decisions are most economical when they are based on the loss predicated for an entire field.

Management of *H. glycines* involves nematicide application and cultural practices which influence the population dynamics of *H. glycines* and also influence the ability of the soybean plant to withstand attack by this nematode. Cultural practices include the use of resistant cultivars, crop rotation, fertilizer application, tillage, planting date and irrigation. Resistance is a highly effective tactic, but continuous planting of resistant cultivars results in the selection of races which can parasitize them. Other cultural practices help manipulate the nematode population and limit damage. Nematicides used to reduce the population level are costly and some are not effective.

Rotation of host plants with non-hosts of *H. glycines* is very effective in reducing the nematode population (Ozaki and Asai, 1962; Ross, 1962b; Slack, Riggs and Hamblen, 1981). Since *H. glycines* survives for long periods in the absence of a host, 2 years of non-host are often necessary to reduce nematode populations below damaging levels (Slack, Riggs and Hamblen, 1981). This time interval reduces the population density by about 90% (Slack, Riggs and Hamblen, 1981). A resistant soybean cultivar can be used in place of a non-host for one of the 2 years. Weed control is important because a number of weeds (e.g. hemp sesbania, joint vetch and common lespedeza) found in soybean fields are hosts (Riggs and Hamblen, 1962, 1966). The population level of *H. glycines* should be monitored annually to determine if a host can be planted. In addition, the cost of production and value of the rotation crop need to be considered in the total crop production system. Value and cost must be calculated over the total period of the rotation and not just for individual crop years. Non-host crops that may be economically grown in rotation with soybean to control *H. glycines* include grain sorghum, corn, cotton, rice, alfalfa, sunflower, small grain, tobacco and certain vegetables. Planting a tolerant cultivar in the third year of the rotation would help insure against damage by the residual population (Boerma and Hussey, 1984).

The date of soybean planting was discussed earlier (p. 259). This cultural practice can be used to avoid the nematode. Soybeans are normally planted when the soil temperature is around 18–22°C during the day and the nematode is coming out of diapause. If they are planted earlier while the soil is relatively cool and the nematode is still in diapause, soybean roots may grow sufficiently to get the crop well established, thereby escaping serious damage. In contrast, a late planting date may result in as much as 50% mortality of J2 that hatch early and 1–2 generations of the nematode may be prevented (Schmitt, unpublished). When planting late, soils are also generally drier and hotter, consequently less favourable for nematode movement, penetration, development and reproduction.

Other practices that may aid in reducing damage by *H. glycines* are: (1) summer fallow; (2) subsoiling to break hard pans that prevent roots from penetrating into deeper soil layers where there are fewer nematodes and additional moisture; and (3) irrigation and fertilization to provide plants with sufficient nutrition in spite of nematode-damaged roots.

Regardless of the cultural practice used, cultivar selection must fit the production system for the geographic area. Manipulation of cultivar, aside from resistance, can aid in the management of *H. glycines* populations.

Resistance to *H. glycines* is race specific. Soybean cultivars are commercially available with resistance to races, 1, 3 and 4. Control with resistance is complicated by variation in the nematode (Ross, 1962a). This variation has now become so complex that 16 races of *H. glycines* are described (Riggs and Schmitt, 1988). Resistance in soybean to *H. glycines* is regulated by as many as 10 genes; 1–4 genes may be needed for resistance to any one race (Thomas, 1974). All commercial resistance is obtained from two of the host differentials (Peking and PI 88788). However, different sets of genes for resistance are available and should be introduced into breeding programmes (Anand, personal communication).

Resistance is typically expressed as a necrotic response in infected roots (Ross, 1964; Endo, 1965) which is temperature independent (Melton, Jacobsen and Noel, 1986). When *H. glycines* juveniles invade root tissue, syncytia are initiated, but this tissue quickly becomes necrotic and most juveniles die.

Tolerance enables soybean plants to produce relatively good yields in spite of being good hosts of *H. glycines* (Boerma and Hussey, 1984). The mechanism of tolerance needs elucidation. One possible mechanism might involve the root tissue in which the syncytia develop. They may develop in the cortex rather than in the stele, and might be less disruptive of plant function (Kim, Kim and Riggs, 1986).

Further development, improvement and increased availability of tolerant cultivars would be advantageous for producers because they could avoid the race-shift problem found with resistant cultivars. The resistant cultivar will control most individual nematodes in a field, but a few juveniles have the genetic capacity to mature and reproduce. Continuous planting for 3–5 years will enable the population to shift and damage the previously resistant cultivar.

Nematicides are chemicals that are toxic or nematistatic to the nematode. *H. glycines* is generally controlled better with fumigants than with non-fumigants. The fumigants apparently kill the juveniles and probably the eggs. Non-fumigants inhibit penetration by the nematode into the roots, or inhibit development of the nematode after penetration (Wright, 1981; Sipes, 1987). Few nematicides, especially fumigants, remain on the market. High cost and lack of consistent performance have resulted in a reluctance to use chemicals to control *H. glycines*. Aldicarb will generally produce a growth response in soybean (Barker *et al.*, 1988), but it does not always produce a yield increase.

Organophosphates and carbamates generally are assumed to inhibit acetylcholinesterase at the cholinergic synapses of the nematode nervous system (Wright, 1981). These chemicals could act to narcotize the host-finding system in the nematode, preventing the nematode from parasitizing the plant. However, penetration of soybean roots by J2 in the field treated with aldicarb, fenamiphos or ethoprop was not affected (Bostian, 1982). After penetration, uptake of fenamiphos narcotizes the J2 and they do not develop (Sipes, 1987).

Nematicides give insufficient control of *H. glycines* to be economical. The return may pay for the cost of the chemical but full potential production may not be achieved (Schmitt, Ferris and Barker, 1987; Riggs and Schmitt, unpublished).

As indicated earlier, herbicides and nematicides may interact and (1) reduce the effectiveness of nematicides; (2) increase nematicidal activity; or (3) have no effect. Application of the herbicides metribuzin, alachlor, and linuron to aldicarb-treated soil results in lower final populations of *H. glycines* than with aldicarb alone. The interaction is particularly strong between aldicarb and metribuzin (Schmitt and Corbin, 1981; Schmitt, Corbin and Nelson, 1983). Control of *H. glycines* by fenamiphos is reduced by alachlor even when alachlor is applied 2 days after the nematicide. In some cases, herbicides increase the hatching of *H. glycines* eggs, exposing more juveniles to the action of nematicides (Kraus, Noel and Edwards, 1982; Riggs and Oliver, 1982).

Available nematicides, in addition to giving inadequate control, are expensive to use on soybean. Rotations are effective in controlling *H. glycines*, but are not popular with many growers. Resistant cultivars are also effective, but continuous planting has resulted in selection of more virulent races of *H. glycines*. Therefore, new approaches are needed for managing *H. glycines* populations.

A new approach to chemical control would include chemicals that affect the endocrine system, ecdysteroids to alter nematode development, juvenile hormones to inhibit growth and development, vertebrate steroid hormones to shift male : female ratios, and chemicals that inhibit sterol metabolism (Chitwood, 1987). Even if these different types of chemicals are effective in reducing *H. glycines* populations, they must also be economical. Another approach would be to exploit the attraction–repellency stimuli (Jansson, 1987). This could take the form of host alteration to change its attraction or an application of chemical to the soil to negate the sensory perception of the nematode. Certain bacteria repel nematodes and could be applied as seed treatments. When these bacteria colonize the roots, they would impede *H. glycines* penetration into soybean roots (Dusenbery, 1987). Avermectins, lactones from *Streptomyces avermitilis*, are effective against nematodes but are very expensive (Stretton, Campbell and Babu, 1987). Intervention in the communication system by the use of synthetic pheromones may eventually be feasible (Bone, 1987).

A control measure currently being considered is the application of a hatching factor to induce the hatching of *H. glycines* eggs (Perry, 1987). Glycinoeclepin A induces hatching of *H. glycines* eggs (Masamune *et al.*, 1982; Fukuzawa *et al.*, 1985), and could be applied to the soil prior to planting a non-host. Most *H. glycines* eggs would hatch and juveniles would die because no host would be available. However, economical production and application of glycinoeclepin A will be a major factor in the use of the material.

Biotechnology offers some far-reaching possibilities, at least conceptually, for control of *H. glycines*. A gene could be inserted into a soybean chromosome which would make the soybean plant immune to *H. glycines*.

Biocontrol through the use and manipulation of natural enemies could be a practical approach for future control of *H. glycines*. A number of predaceous and parasitic fungi have been isolated from *H. glycines* (Morgan-Jones, Gintis and Rodriguez-Kabana, 1981; Gintis, Morgan-Jones and Rodriguez-Kabana, 1983; Kim and Riggs, 1987). Insects that feed on *H. glycines* females and cysts, such as earwigs and springtails, should be investigated.

New approaches to the management of *H. glycines* populations may be available in the near future. Still, a concerted effort by researchers in various disciplines will be necessary to bring them to fruition.

References

ACEDO, J. R., DROPKIN, V. H. AND LUEDDERS, V. D. (1984). Nematode population attrition and histopathology of *Heterodera glycines*–soybean associations. *Journal of Nematology* **16**, 48–57.

ALSTON, D. G. AND SCHMITT, D. P. (1987). Population density and spatial pattern of *Heterodera glycines* in relation to soybean phenology. *Journal of Nematology* **19**, 336–345.

ALSTON, D. G. AND SCHMITT, D. P. (1988). Development of *Heterodera glycines* life stages as influenced by temperature. *Journal of Nematology* **20**, 366–372.

ANDERSON, T. R. AND WELACKY, T. W. (1988). First report of *Heterodera glycines* on soybeans in Ontario, Canada. *Plant Disease* **72**, 453.

BAIRD, S. M. AND BERNARD, E. C. (1984). Nematode population and community dynamics in soybean–wheat cropping and tillage regimes. *Journal of Nematology* **16**, 379–386.

BARKER, K. R., KOENNING, S. R., BOSTIAN, A. L. AND AYERS, A. R. (1988). Growth and yield responses of soybean to aldicarb. *Journal of Nematology* **20**, 421–431.

BERNSTEIN, L. AND OGATA, G. (1966). Effects of salinity on nodulation, nitrogen fixation, and growth of soybeans and alfalfa. *Agronomy Journal* **58**, 201–203.

BOERMA, H. R. AND HUSSEY, R. S. (1984). Tolerance to *Heterodera glycines* in soybean. *Journal of Nematology* **16**, 289–296.

BONE, L. W. (1987). Pheromone communication in nematodes. In *Vistas on Nematology* (J. A. Veech and D. W. Dickson, Eds), pp. 147–152. Society of Nematologists, Jay, FL.

BONNER, M. J. AND SCHMITT, D. P. (1985). Population dynamics of *Heterodera glycines* life stages on soybean. *Journal of Nematology* **17**, 153–158.

BOSTIAN, A. L. (1982). *Population dynamics of* Heterodera glycines *as affected by the interactions of phenamiphos or ethoprop with alachlor*. Ph.D. Thesis, North Carolina State University, Raleigh, NC.

BOSTIAN, A. L., SCHMITT, D. P. AND BARKER, K. R. (1986a). *In vitro* hatch and survival of *Heterodera glycines* as affected by alachlor and phenamiphos. *Journal of Nematology* **18**, 22–26.

BOSTIAN, A. L., SCHMITT, D.P. AND BARKER, K. R. (1986b). Population changes of *Heterodera glycines* and soybean yields resulting from soil treatment with alachlor, fenamiphos and ethoprop. *Journal of Nematology* **18**, 458–463.

CABALLERO, L. G., OSMANSKI, E. A. AND TEFFT, P. M. (1986). Factors influencing plant-induced egg hatching in *Heterodera glycines*. *Journal of Nematology* **18**, 636 (abstract).

Chelbi, Mustapha (1984). *Influence of soil temperature and moisture on penetration, development, and reproduction of* Heterodera glycines *in soybean*. M.S. Thesis, University of Georgia, Athen, GA.

Chitwood, D. J. (1987). Inhibition of steroid or hormone metabolism or action in nematodes. In *Vistas on Nematology* (J. A. Veech and D. W. Dickson, Eds), pp. 122–130. Society of Nematologists, Jay, FL.

Dusenbery, D. B. (1987). Prospects for exploiting sensory stimuli in nematode control. In *Vistas on Nematology* (J. A. Veech and D. W. Dickson, Eds), pp. 131–135. Society of Nematologists, Jay, FL.

Elliott, A. P., Phipps, P. M. and Terrill, R. (1986). Effects of continuous cropping of resistant cultivars on reproduction potentials of *Heterodera glycines* and *Globodera tabacum solanacearum*. *Journal of Nematology* **18**, 375–379.

Endo, B. Y. (1965). Histological responses of resistant and susceptible varieties, and backcross progeny to entry and development of *Heterodera glycines*. *Phytopathology* **55**, 375–381.

Epps, J. M. and Chambers, A. Y. (1965). Population dynamics of *Heterodera glycines* under various cropping sequences in field bins. *Phytopathology* **55**, 100–103.

Epps, J. M. and Chambers, A. Y. (1966). Comparative rates of reproduction of *Heterodera glycines* on twelve host plants. *Plant Disease Reporter* **50**, 608–610.

Evans, D. M. and Fox, J. A. (1977). The sex ratio of *Heterodera glycines* at low population densities. *Journal of Nematology* **9**, 207–210.

Filipjev, I. N. and Schuurmans-Stekhoven, J. H. (1941). *A Manual of Agricultural Helminthology*. Brill, Leiden.

Francl, L. J. (1986). *Heterodera glycines* population dynamics and relation of initial population to soybean yield. *Plant Disease* **70**, 791–795.

Fukuzawa, A., Furusaki, A., Ikura, M. and Masamune, T. (1985). Glycinoeclepin-A, a natural hatching stimulus for the soybean cyst nematode. *Journal of the Chemical Society, Chemical Communications 1985*, pp. 222–224.

Gintis, B. O., Morgan-Jones, G. and Rodriguez-Kabana, R. (1983). Fungi associated with several developmental stages of *Heterodera glycines* from an Alabama soybean field soil. *Nematropica* **13**, 181–200.

Goffart, H. (1951). *Nematoden der Kulturpflanzen Europas*. Parley-Verlag, Berlin.

Hamblen, M. L., Slack, D. A. and Riggs, R. D. (1972). Temperature effects on penetration and reproduction of soybean-cyst nematode. *Phytopathology* **62**, 762 (abstract).

Heatherly, L. G., Young, L. D., Epps, J. M. and Hartwig, E. E. (1982). Effect of upper-profile soil water potential on numbers of cysts of *Heterodera glycines* on soybeans. *Crop Science* **22**, 833–835.

Hill, N. S. (1985). The influence of temperature and soybean phenology on inducing dormancy of *Heterodera glycines* eggs. *Journal of Nematology* **17**, 498 (abstract).

Hori, S. (1916). Phytopathological notes. 5. Sick soil of soybean caused by nematodes. *Journal of Plant Protection* **2**, 927–930.

Hussey, R. S. and Boerma, H. R. (1983). Influence of planting date on damage to soybean caused by *Heterodera glycines*. *Journal of Nematology* **15**, 253–258.

Ichinohe, M. (1952). On the soybean nematode *Heterodera glycines* n. sp. from Japan. *Magazine of Applied Zoology* **17**, 1–4.

Inagaki, H. and Tsutsumi, M. (1971). Survival of the soybean cyst nematode, *Heterodera glycines* Ichinohe (Tylenchida:Heteroderidae) under certain storing conditions. *Applied Entomology and Zoology* **6**, 156–162.

Inoue, H. (1959). Surveys of the soybean cyst nematode. VIII. Cyst population and soybean growth in fallowed fields. *Annual Report of Plant Protection in North Japan* **10**, 111–112.

Ishibashi, N., Kondo, E., Muraoka, M. and Yokoo, T. (1973). Ecological significance of dormancy in plant-parasitic nematodes. I. Ecological difference between eggs in gelatinous matrix and cyst of *Heterodera glycines* Ichinohe (Tylenchida:Heteroderidae). *Applied Entomology and Zoology* **8**, 53–63.

JANSSON, H. (1987). Receptors and recognition in nematodes. In *Vistas on Nematology* (J. A. Veech and D. W. Dickson, Eds), pp. 153–158. Society of Nematologists: Jay, FL.

KAUL, R. (1962). Untersuchungen über einen aus Zysten des Kartoffelnematoden (*Heterodera rostochiensis* Woll.) isolierten phenolischen Komplex. *Nematologica* **8**, 288–292.

KIM, D. G. AND RIGGS, R. D. (1987). Control of *Heterodera glycines* by a fungus parasite. *Journal of Nematology* **19**, 535 (abstract).

KIM, Y. H., KIM, K. S. AND RIGGS, R. D. (1986). Morphological characteristics of syncytia in susceptible hosts infected by the soybean cyst nematode. *Phytopathology* **76**, 913–917.

KOENNING, S. R. AND SCHMITT, D. P. (1985). Hatching and diapause of field populations of *Heterodera glycines*. *Journal of Nematology* **17**, 502 (abstract).

KOLIOPANOS, C. N. AND TRIANTAPHYLLOU, A. C. (1972). Effect of infection density on sex ratio of *Heterodera glycines*. *Nematologica* **18**, 131–137.

KRAUS, R., NOEL, G. R. AND EDWARDS, D. I. (1982). Effects of preemergence herbicides and aldicarb on *Heterodera glycines* population dynamics and yield of soybean. *Journal of Nematology* **14**, 452 (abstract).

LAURITIS, J. A., REBOIS, R. V. AND GRANEY, L. S. (1983). Development of *Heterodera glycines* Ichinohe on soybean, *Glycine max* (L.) Merr., under gnotobiotic conditions. *Journal of Nematology* **15**, 272–281.

LUEDDERS, V. D., SHANNON, J. G. AND BALDWIN, C. H., JR. (1979). Influence of rate and source of potassium on soybean cyst nematode reproduction on soybean seedlings. *Plant Disease Reporter* **63**, 558–560.

MASAMUNE, T., ANETAI, M., TAKASUGI, M. AND KATSUI, N. (1982). Isolation of a natural hatching stimulus, glycinoeclepin A, for the soybean cyst nematode. *Nature* **297**, 495–496.

MELTON, T. A., JACOBSEN, B. J. AND NOEL, G. R. (1986). Effects of temperature on development of *Heterodera glycines* on *Glycine max* and *Phaseolus vulgaris*. *Journal of Nematology* **18**, 468–474.

MORGAN-JONES, G., GINTIS, B. O. AND RODRIGUEZ-KABANA, R. (1981). Fungal colonization of *Heterodera glycines* cysts in Arkansas, Florida, Mississippi and Missouri soils. *Nematropica* **11**, 155–164.

NOEL, G. (1982). Relating numbers of soybean cyst nematode to crop damage. *Proceedings of the Fifth Cyst Nematode Workshop*, pp. 17–19. North Carolina State University, Raleigh.

OKADA, T. (1971). The effect of root diffusate and pre-soaking periods on the hatching of *Heterodera glycines* Ichinohe. *Japanese Journal of Applied Entomology and Zoology* **15**, 215–221.

OKADA, T. (1972). Hatching stimulant in the egg of the soybean cyst nematode, *Heterodera glycines* Ichinohe (Tylenchida:Heteroderidae). *Applied Entomology and Zoology* **7**, 234–237.

OZAKI, K. AND ASAI, K. (1962). Studies on the rotation systems. II. The relationships between crop sequence and the soybean cyst nematode population in the soil. *Research Bulletin of the Hokkaido National Agricultural Experiment Station* **81**, 11–21.

PERRY, R. N. (1987). Host-induced hatching of phytoparasitic nematode eggs. In *Vistas on Nematology* (J. A. Veech and D. W. Dickson, Eds), pp. 159–164. Society of Nematologists, Jay, FL.

RADICE, A. D. (1987). *Intraspecific diversity of* Heterodera glycines *(Nematoda: Heteroderidae) detected by restriction fragment analysis of mitochondrial DNA and isozyme phenotypes*. Dissertation, University of Arkansas, Fayetteville, AR.

RIGGS, R. D. (1977). Worldwide distribution of soybean-cyst nematode and its economic importance. *Journal of Nematology* **9**, 34–39.

Riggs, R. D. and Hamblen, M. L. (1962). Soybean-cyst nematode host studies in the family Leguminosae. *Arkansas Agricultural Experiment Station Report Series 110*.

Riggs, R. D. and Hamblen, M. L. (1966). Further studies on the host range of the soybean-cyst nematode. *Arkansas Agricultural Experiment Station Bulletin 718*.

Riggs, R. D. and Oliver, L. R. (1982). Effect of trifluralin (Treflan) on soybean cyst nematode. *Journal of Nematology* **14**, 466 (abstract).

Riggs, R. D. and Schmitt, D. P. (1987). Nematodes. In *Soybeans: Improvement, Production and Uses* (R. Wilcox, Ed.), pp. 757–778. Agronomy Publishers: Madison, WI.

Riggs, R. D. and Schmitt, D. P. (1988). Complete characterization of the race scheme for *Heterodera glycines*. *Journal of Nematology* **20**, 392–395.

Riggs, R. D., Hamblen, M. L. and Rakes, L. (1977). Development of *Heterodera glycines* pathotypes as affected by soybean cultivars. *Journal of Nematology* **9**, 312–318.

Ross, J. P. (1959). Nitrogen fertilization on the response of soybeans infected with *Heterodera glycines*. *Plant Disease Reporter* **43**, 1284–1286.

Ross, J. P. (1962a). Physiological strains of *Heterodera glycines*. *Plant Disease Reporter* **46**, 766–769.

Ross, J. P. (1962b). Crop rotation effects on the soybean cyst nematode population and soybean yields. *Phytopathology* **52**, 815–818.

Ross, J. P. (1963). Seasonal variation of larval emergence from cysts of the soybean cyst nematode, *Heterodera glycines*. *Phytopathology* **53**, 608–609.

Ross, J. P. (1964). Effect of soil temperature on development of *Heterodera glycines* in soybean roots. *Phytopathology* **54**, 1228–1231.

Schmitt, D. P. and Corbin, F. T. (1981). Interaction of fensulfothion and phorate with preemergence herbicides on soybean parasitic nematodes. *Journal of Nematology* **13**, 37–41.

Schmitt, D. P. and Nelson, L. A. (1987). Chemical control of selected plant-parasitic nematodes in soybeans double-cropped with wheat in no-till and conventional tillage systems. *Plant Disease* **71**, 323–326.

Schmitt, D. P. and Noel, G. R. (1984). Nematode parasites of soybeans. In *Plant and Insect Nematodes* (W. R. Nickle, Ed.), pp. 17–59. Marcel Dekker Inc., New York.

Schmitt, D. P., Corbin, F. T. and Nelson, L. A. (1983). Population dynamics of *Heterodera glycines* and soybean response in soils treated with selected nematicides and herbicides. *Journal of Nematology* **15**, 432–437.

Schmitt, D. P., Ferris, H. and Barker, K. R. (1987). Response of soybean to *Heterodera glycines* races 1 and 2 in different soil types. *Journal of Nematology* **19**, 240–250.

Sipes, B. S. (1987). *Post-infection development of* Heterodera glycines *as affected by alachlor and fenamiphos*. M.S. Thesis, North Carolina State University, Raleigh.

Skotland, C. B. (1957). Biological studies of the soybean cyst nematode. *Phytopathology* **47**, 623–625.

Slack, D. A. and Hamblen, M. L. (1961). The effect of various factors on larval emergence from cysts of *Heterodera glycines*. *Phytopathology* **51**, 350–355.

Slack, D. A., Riggs, R. D. and Hamblen, M. L. (1981). Nematode control in soybeans: Rotation and population dynamics of soybean cyst and other nematodes. *Arkansas Agricultural Experiment Station Report Series 263*.

Sortland, M. E. and MacDonald, D. H. (1987). Effect of crop and weed species on development of a Minnesota population of *Heterodera glycines* race 5 after one to three growing periods. *Plant Disease* **71**, 23–27.

Stretton, A. O. W., Campbell, W. C. and Babu, J. R. (1987). Biological activity and mode of action of avermectins. In *Vistas on Nematology* (J. A. Veech and D. W. Dickson, Eds), pp. 136–146. Society of Nematologists, Jay, FL.

TEFFT, P. M. AND BONE, L.W . (1985). Plant-induced hatching of eggs of the soybean cyst nematode *Heterodera glycines. Journal of Nematology* **17**, 275–279.

THOMAS, J. D. (1974). *Genetics of resistance to races of the soybean-cyst nematode.* M.S. Thesis, University of Arkansas, Fayetteville, AR.

TRIANTAPHYLLOU, A. C. (1971). Genetics and cytology. In *Plant Parasitic Nematodes,* Volume II (B.M. Zuckerman, W. F. Mai and R. A. Rohde, Eds), pp. 1–34. Academic Press, New York.

TRIANTAPHYLLOU, A. C. (1975). Genetic structure of races of *Heterodera glycines* and inheritance of ability to reproduce on resistant soybeans. *Journal of Nematology* **7**, 356–364.

WRIGHT, D. J. (1981). Nematicides: Mode of action and new approaches to chemical control. In *Plant Parasitic Nematodes,* Volume III, (B. M. Zuckerman and R. A. Rohde, Eds), pp. 421–449. Academic Press, New York.

YOSHIHARA, T. AND KEGASAWA, K. (1974). Studies on the influence of soil moisture conditions on soybean cyst nematodes. Part 2. Nematode development and soybean growth in the surface soil and sub-soil of well-drained volcanic ash soil of the Tokachi region under two moisture levels. *Research Bulletin of the Hokkaido National Agricultural Experiment Station No. 1099,* pp. 19–26.

Index